23

Springer Series in Chemical Physics

Edited by Fritz Peter Schäfer

Springer Series in Chemical Physics

Editors: V. I. Goldanskii R. Gomer F. P. Schäfer J. P. Toennies

Volume 1 **Atomic Spectra and Radiative Transitions** By I. I. Sobelman

Volume 2 **Surface Crystallography by LEED** Theory, Computation and Structural Results By M. A. Van Hove, S. Y. Tong

Volume 3 **Advances in Laser Chemistry** Editor: A. H. Zewail

Volume 4 **Picosecond Phenomena** Editors: C. V. Shank, E. P. Ippen, S. L. Shapiro

Volume 5 **Laser Spectroscopy** Basic Concepts and Instrumentation By W. Demtröder

Volume 6 **Laser-Induced Processes in Molecules** Physics and Chemistry Editors: K. L. Kompa, S. D. Smith

Volume 7 **Excitation of Atoms and Broadening of Spectral Lines** By I. I. Sobelman, L. A. Vainshtein, E. A. Yukov

Volume 8 **Spin Exchange** Principles and Applications in Chemistry and Biology By Yu. N. Molin, K. M. Salikhov, K. I. Zamaraev

Volume 9 **Secondary Ion Mass Spectrometry SIMS II** Editors: A. Benninghoven, C. A. Evans, Jr., R. A. Powell, R. Shimizu, H. A. Storms

Volume 10 **Lasers and Chemical Change** By A. Ben-Shaul, Y. Haas, K. L. Kompa, R. D. Levine

Volume 11 **Liquid Crystals of One- and Two-Dimensional Order** Editors: W. Helfrich, G. Heppke

Volume 12 **Gasdynamic Laser** By S. A. Losev

Volume 13 **Atomic Many-Body Theory** By I. Lindgren, J. Morrison

Volume 14 **Picosecond Phenomena II** Editors: R. M. Hochstrasser, W. Kaiser, C. V. Shank

Volume 15 **Vibrational Spectroscopy of Adsorbates** Editor: R. F. Willis

Volume 16 **Spectroscopy of Molecular Excitons** By V. L. Broude, E. I. Rashba, E. F. Sheka

Volume 17 **Inelastic Particle-Surface Collisions** Editors: E. Taglauer, W. Heiland

Volume 18 **Modelling of Chemical Reaction Systems** Editors: K. H. Ebert, P. Deuflhard, W. Jäger

Volume 19 **Secondary Ion Mass Spectrometry SIMS III** Editors: A. Benninghoven, J. Giber, J. László, M. Riedel, H. W. Werner

Volume 20 **Chemistry and Physics of Solid Surfaces IV** Editors: R. Vanselow, R. Howe

Volume 21 **Dynamics of Gas-Surface Interaction** Editors: G. Benedek, U. Valbusa

Volume 22 **Nonlinear Laser Chemistry** Multiple - Photon Excitation By V.S. Letokhov

Volume 23 **Picosecond Phenomena III** Editors: K. B. Eisenthal, R. M. Hochstrasser, W. Kaiser, A. Laubereau

Picosecond Phenomena III

Proceedings of the Third International Conference on Picosecond Phenomena
Garmisch-Partenkirchen, Fed. Rep. of Germany
June 16–18, 1982

Editors
K.B. Eisenthal R.M. Hochstrasser W. Kaiser
A. Laubereau

With 288 Figures

Springer-Verlag Berlin Heidelberg New York 1982

Series Editors

Professor Vitalii I. Goldanskii

Institute of Chemical Physics
Academy of Sciences
Vorobyevskoye Chaussee 2-b
Moscow V-334, USSR

Professor Robert Gomer

The James Franck Institute
The University of Chicago
5640 Ellis Avenue
Chicago, IL 60637, USA

Professor Dr. Fritz Peter Schäfer

Max-Planck-Institut für
Biophysikalische Chemie
D-3400 Göttingen-Nikolausberg
Fed. Rep. of Germany

Professor Dr. J. Peter Toennies

Max-Planck-Institut für Strömungsforschung
Böttingerstraße 6–8
D-3400 Göttingen
Fed. Rep. of Germany

Conference Chairmen and Editors
Professor **R.M. Hochstrasser,** University of Pennsylvania, Philadelphia, PA 19104, USA
Professor Dr. **W. Kaiser,** Technische Universität München,
D-800 München, Fed. Rep. of Germany

Program Co-Chairmen and Editors
Professor **K.B. Eisenthal,** University of New York, New York, NY 10027, USA
Professor Dr. **A. Laubereau,** Universität Bayreuth
D-8580 Bayreuth, Fed. Rep. of Germany

Sponsored by
Deutsche Forschungsgemeinschaft
Bayerisches Staatsministerium für Unterricht und Kultus

Supported by Grants from
U.S. Army Research Office (Durham)

Industrial Support

Sohio	Quantronix	Xerox Corporation
E.I. DuPont de Nemours	Spectra Physics	Quantel
Hamamatsu	Newport Research	Coherent Radiation

ISBN 978-3-642-87866-4 ISBN 978-3-642-87864-0 (eBook)
DOI 10.1007/978-3-642-87864-0

Library of Congress Cataloging in Publication Data. International Conference on Picosecond Phenomena (3rd : 1982 : Garmisch-Partenkirchen, Germany) Picosecond phenomena III. (Springer series in chemical physics ; v. 23) Bibliography: p. Includes index. 1. Picosecond pulses–Congresses. I. Eisenthal, K. B. II. Title. III. Title: Picosecond phenomena 3. IV. Title: Picosecond phenomena three. V. Series. QC689.5.L37I57 1982 535.5'8 82-16889

2153/3130-543210

We dedicate this volume to two of our most distinguished colleagues who met untimely deaths this year:

H. MAHR   S.L. SHAPIRO

Both will be sadly missed by all in the picosecond field and our deepest sympathy goes out to their families.

Preface

The third international conference devoted to picosecond phenomena was held June 16-18, 1982 in Garmisch-Partenkirchen, West Germany. Scientists from widely varying disciplines, physicists, chemists, biologists, and engineers came together to share their common interest in picosecond and subpicosecond processes. The meeting attracted approximately 250 scientists from numerous countries around the globe.

More than 100 papers were concerned with the latest advances in the experimental and theoretical understanding of ultrafast phenomena. New discoveries in femtosecond and picosecond pulse generation and new results in chemical dynamics, solid-state physics, and nonlinear optics were presented. The quality of the scientific reports, the enthusiasm of the participating scientists, as well as the magnificent surroundings of the Bavarian alps guaranteed a successful and pleasant conference.

Numerous people have helped to make the conference a success. Special thanks are due to Carin von Oberkamp for doing a superb job in implementing the meeting arrangements and to the program committee for the selection and organisation of the scientific presentations.

The financial support of the Deutsche Forschungsgemeinschaft and of the Bayerische Staatsministerium für Unterricht und Kultur is gratefully acknowledged.

New York, NY
Philadelphia, PA
Munich, Fed. Rep. of Germany
Bayreuth, Fed. Rep. of Germany
August, 1982

K.B. Eisenthal
R.M. Hochstrasser
W. Kaiser
A. Laubereau

Contents

Part I Advances in the Generation of Ultrashort Light Pulses

Part II Ultrashort Measuring Techniques

Part VII Applications in Solid-State Physics

Part I

Advances in the Generation of Ultrashort Light Pulses

Moving from the Picosecond to the Femtosecond Time Regime

C. V. Shank, R. L. Fork, and R. T. Yen

Bell Laboratories, Holmdel, NJ 07733, USA

Considerable progress has taken place in the last decade and a half in the generation of ultrashort optical pulses. A steady progression of developments has led to ever shorter optical pulses and subsequent improvements in our ability to resolve fast spectroscopic events. In this discussion I will describe progress that has taken place in our laboratory which has led to the generation of optical pulses of less than 0.1 picoseconds in duration, pushing well into the femtosecond time regime. Three significant advances have taken place in the last two years which have made possible experimental investigations on a femtosecond time scale. The first key advance was the improvement of the passively modelocked dye laser [1] using the colliding pulse modelocked dye laser configuration [2]. With this laser the first pulses with a duration of less than 0.1 picoseconds were generated. The second important result is an improvement in amplification technique which has permitted the generation of femtosecond optical pulses of gigawatt intensities. The final point of discussion will be the use of pulse compression techniques to generate a 30 femtosecond optical pulse, the shortest optical pulse duration yet reported.

1. Colliding Pulse Modelocking (CPM)

The diagram of a CPM ring dye laser is shown in Fig. 1. This laser utilizes the interaction or "collision" of two counterpropagating optical pulse trains in a thin saturable absorber stream. The interaction of the counterpropagating pulses creates a transient grating of the population of the absorber molecules, which synchronizes, stabilizes, and shortens the pulses. The gain medium is a conventional flowing stream of Rhodamine 6G dissolved in ethylene glycol. The saturable absorber is 3,3'-Diethyloxadicarbocyanine iodide dissolved in ethylene glycol. The dye laser was pumped with a cw argon laser using 3-7 watts at 514.5 nm. When the first results of the operation of this

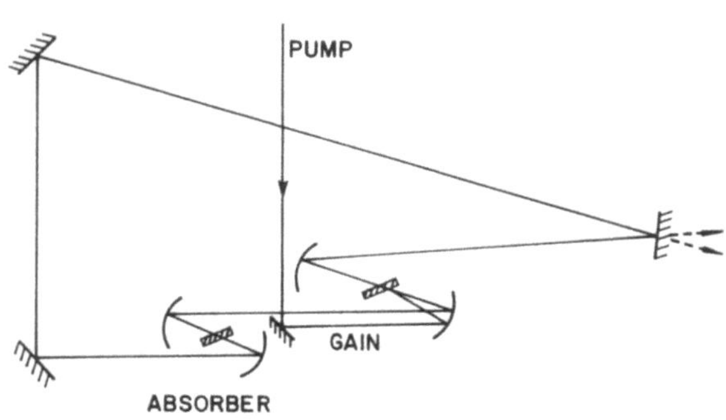

Fig. 1 CPM laser cavity configuration

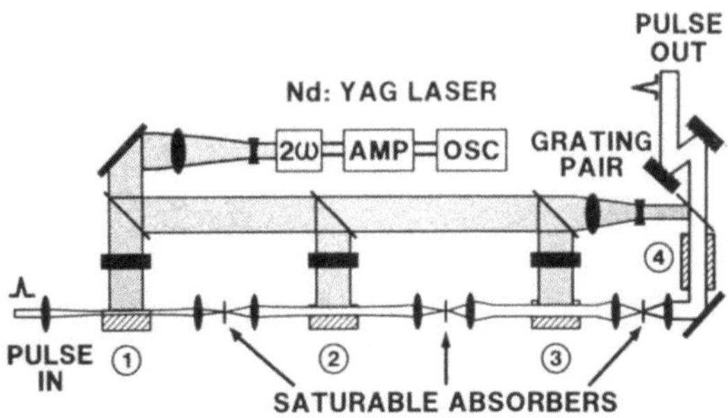

Fig. 2 Femtosecond pulse amplifier

laser were reported a pulsewidth of 90 femtoseconds was measured. It has subsequently been found that by empirical substitution of laser mirrors the pulsewidth could be reduced to 65 femtoseconds. Apparently the operation of these short pulse lasers is very sensitive to mirror coatings. Most laser mirrors are fabricated with only the reflectivity specified. The frequency dependence of the phase change upon reflection is not accurately controlled. This can lead to dispersion in the optical cavity and subsequent pulse broadening. Some future improvement in femtosecond laser operation is almost certainly to come by designing laser mirrors to minimize cavity loss and dispersion.

2. Femtosecond Pulse Amplification

For many experimental situations it is highly desirable to have intense optical pulses in the gigawatt range. Earlier we reported techniques for amplifying subpicosecond optical pulses (0.5 psec in duration) while maintaining the pulse duration [3]. The short duration and high electric field intensities encountered in amplifying femtosecond optical pulses introduce new problems in amplifier design. Group velocity dispersion in the dye solvent and amplifier optics causes significant temporal broadening.

The diagram of a femtosecond pulse amplifier is shown in Fig. 2. As in earlier versions of this amplifier, the four stages are pumped with a frequency doubled Nd:YAG laser. Each stage is isolated with saturable absorbers. The temporal broadening due to group velocity dispersion alone can be compensated with a grating pair compressor [4]. In addition to linear dispersion arising from propagating through the amplifier chain, nonlinear amplification effects can introduce distortions of the incident pulse more complex than a linear frequency chirp. Our approach is to seek the necessary amplification while minimizing these distortions which interfere with recompression [5].

The measured autocorrelation functions of our amplified pulses are shown in Fig. 3. During amplification an incident 90 femtosecond optical pulse is broadened to 0.4 picoseconds in the course of being amplified to an energy of one millijoule. Using the grating compressor the pulsewidth can be restored to 70-90 femtoseconds with a peak intensity at gigawatt levels. This pulse is quite useful for generating a picosecond continuum as described by Fork, et al. in this volume [6].

3. Compression of Femtosecond Optical Pulses

Using optical compression techniques [7,8,9,10] we have succeeded in generating and measuring an optical pulse of only 30 femtoseconds in duration. This is

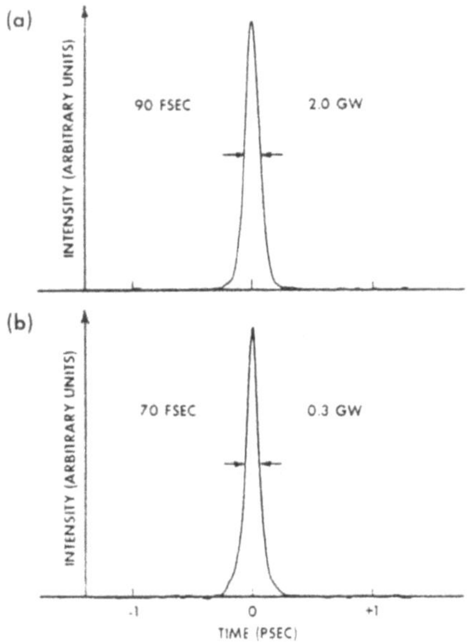

Fig. 3 Amplified pulse autocorrelation functions

the shortest optical pulse ever generated and corresponds to only 14 optical cycles.

The experimental arrangement for pulse compression is shown in Fig. 4. Optical pulses of 90 fsec duration are obtained from a CPM dye laser and amplified as described above. The amplified pulse is attenuated and focused into a 15 cm long single mode optical fiber [11]. For a few nJ energy coupled into the optical fiber the optical spectrum is observed to broaden significantly by the process of self phase modulation. The light from the fiber is recollimated with a lens and passed through a grating compressor. Nakatsuka, et al. have shown that optical fibers are a nearly ideal medium for applying a well-controlled chirp of frequency sweep on the optical pulses [12]. The optical elements used to recollimate the beam exiting from the fiber and to direct the pulses into the autocorrelator all contribute to the group velocity dispersion experienced by the pulse. Fortunately, the grating compressor used in this experiment compresses the chirped pulse and provides a means of compensating other dispersive elements in the beam as well. The results of an autocorrelation measurement of a 30 femtosecond optical pulse are shown in Fig. 5.

Conclusion

We expect that pulses on the femtosecond time scale will open the way to new investigations in physics, chemistry and biology. Pulses of the duration

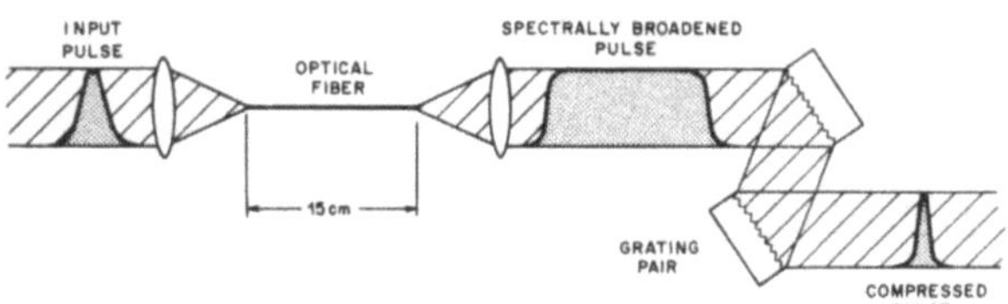

Fig. 4 Experimental arrangement for pulse compression

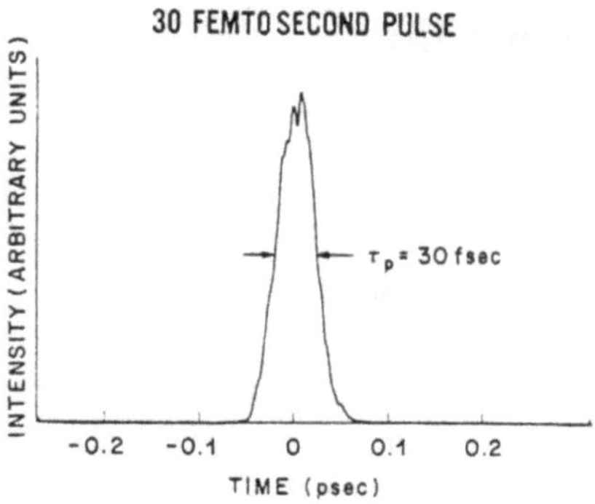

Fig. 5 Autocorrelation function of a compressed 30 femtosecond optical pulse

described here are comparable to the vibrational period of many phonon excitations. Already the work of Tang and that of Greene in this conference have shown the value of femtosecond spectroscopy in unraveling the nonlinear optical properties of simple molecules. Even shorter pulses appear possible in the future by straightforward extensions of the techniques described above.

References

1. E. P. Ippen, C. V. Shank, and A. Dienes, Appl. Phys. Lett. 21, 348 (1972).
2. R. L. Fork, B. I. Greene, and C. V. Shank, Appl. Phys. Lett. 38, 671 (1981).
3. C. V. Shank, R. L. Fork, R. F. Leheny, and J. Shah, Phys. Rev. Lett. 42, 112 (1979).
4. E. B. Treacy, IEEE J. Quan. Elec. QE-5, 454 (1969).
5. R. L. Fork, C. V. Shank, and R. T. Yen, Appl. Phys. Lett. 41, 273 (1982).
6. R. L. Fork, C. V. Shank, R. T. Yen, C. Hirlimann, and W. J. Tomlinson, (This volume).
7. F. Gires and P. Tournois, Compt. Rent. 258, 6112 (1964).
8. J. A. Giordmaine, M. A. Duguay, and J. W. Hansen, Quan. Elec. 4, 252 (1968).
9. R. A. Fisher, P. L. Kelly, and T. K. Gustafson, Appl. Phys. Lett. 14, 140 (1969).
10. A. Laubereau, Phys. Lett. 29A, 539 (1969).
11. C. V. Shank, R. L. Fork, R. T. Yen, R. H. Stolen, and W. J. Tomlinson, Appl. Phys. Lett. 40, 761 (1982).
12. H. Nakatsuka, D. Grischkowsky, and A. C. Balant, Phys. Rev. Lett. 47, 1910 (1981).

Femtosecond Optical Pulses: Towards Tunability at the Gigawatt Level

A. Migus, J.L. Martin, R. Astier, A. Antonetti, and A. Orszag

Laboratoire d'Optique Appliquée, Ecole Polytechnique - ENSTA, F-91120 Palaiseau, France

Since 1975 different kinds of dye lasers have been developed to generate subpicosecond pulses : CW pumped passively mode locked oscillators [1], synchronously pumped lasers or more recently CW mode locked ring lasers [2] which generate pulses of duration less than 100 femtoseconds. Many more applications become accessible when these pulses are amplified to powers in the gigawatt range [3,4] and particularly if one can utilize nonlinear optical phenomena to generate pulses at different wavelengths. In this paper, we describe new techniques for producing high power wavelength tunable subpicosecond pulses, starting from the amplified outputs of CW mode locked linear or ring dye lasers, generating a white light continuum pulse and amplifying any selected spectral part of it to GW peak power.

As initial laser we have first used a linear passively mode locked CW dye laser [1]. After passive compression through a grating pair the resulting pulses of energy 1 nJ and duration typically 0.5 psec were close to transform limited. The amplifier configuration [4] used with this oscillator employs three amplifier stages separated by saturable absorbers which help to prevent self-oscillation or overamplification of the leading edge. The pumping is realized with a Q-switched Nd-Yag laser which produces pulses of 1.7 J at 1.06 µm (14 ns duration) and 500 mJ at 0.53 µm with one single KDP. This set up which generates subpicosecond pulses of up to three GW peak power in the 610-620 nm spectral range has been used in time resolved spectroscopy experiments because such intense pulses have the capability of generating white light pulses of same duration. Up to now this continuum light has always been used as a weak probe (or interrogation) beam while the intense pump (or excitation) was at the fundamental wavelength or with much less energy at the second harmonic or at Stokes shifted frequencies. We have developed a technique which allows to produce high peak power subpicosecond pulses at a selected wavelength all over the visible spectrum by using one fraction of the 615 nm amplified beam to generate a continuum and by amplifying a selected part of it. In our experiment an interferential filter selected a few microjoules in a 7 nm spectral band centered at 575 nm. A second KDP crystal allows a pumping source of 200 mJ at 0.53 µm from the remaining 1.2 J at 1.06 µm. This green light pumped transversally one single 3 cm long amplifier stage containing a circulating solution of Rhodamine 6G. The transverse amplified pulse area was 0.15 cm². We can notice that synchronization is automatically realized once the initial amplifier is synchronized.

A gain saturation experiment has been performed by measuring the output pulse energy E (in µJ) as a function of the input energy E_0 (in µJ) with calibrated filters. The experimental points are in very good agreement with the theoretical saturation curve

$$E \sim 200\, \ell n \left[1 + 3.3\, E_0 \right] \qquad \text{where } E \text{ and } E_0 \text{ are expressed in } \mu J$$

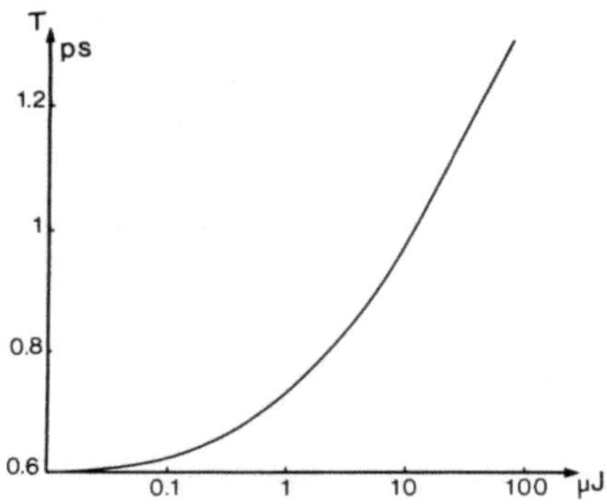

Fig. 1 Computed pulse broadening due to gain saturation. This curve does not take into account broadening due to group velocity dispersion.

obtained by following the computations of ref [5]. Output energies as high as 1 mJ have been obtained.

Pulsewidths were measured using background free SHG autocorrelation techniques. The propagation of such a large spectrum is associated with pulse temporal broadening due to group velocity dispersion in the dye solvent and the optics between the continuum generation and the KDP crystal. By extending the case of gaussian pulses, the pulsewidth τ, after propagating in different media of length L_m and group velocity dispersion D_m, can be approximated by

$$\tau = \left[\tau_p^2 + \left(\Delta\lambda \sum_m L_m D_m \right)^2 \right]^{1/2}$$

where τ_p = 0.6 ps is the initial pulsewidth (fig. 2) and $\Delta\lambda$ the amplified spectrum width. In our experimental conditions $\sum_m L_m D_m \sim 50$ fsec/nm.

While this effect can be compensated by the grating pair technique, this is not the case for gain saturation which is another strong factor of pulse broadening. Assuming a sech² pulse shape we have numerically [5] computed the output pulse width as a function of input energy (fig. 1). The theoretical prediction which takes into account all the broadening factors is then in very good agreement with the observed pulsewidths (fig. 2).

This technique can directly be extended to the infrared and the blue part of the spectrum by using the 355 nm obtained by tripling the remaining 1.06 µm.

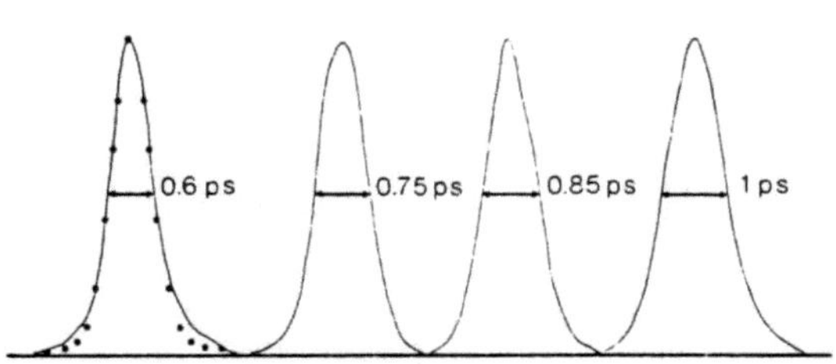

Fig. 2 Starting left : autocorrelation curve of the initial 1 mJ 615 nm pulse. The dotted curve corresponds to the autocorrelation of a sech² of duration 0.6 ps. The following autocorrelation curves demonstrate the effect of gain saturation (and of dispersion) on the pulse duration. They respectively correspond to the following energies : 100 µJ, 200 µJ and 400 µJ in a spectral band of 7 nm centered at 575 nm

This same technique can also be applied to much shorter pulses. Following the recent discovery of techniques for generating pulses in the femtosecond regime [2] and their amplification [6] we have built a ring laser in (fig.3) [7] and amplified the output pulses of duration 100 fsec at 618 nm to the mJ regime while keeping the same order of pulsewidth. Our oscillator is different from the one reported already [2] in that we did not include an intracavity pellicle etalon to avoid losses in the cavity. Instead we found the use of an external interference filter very effective to remove the red part of the spectrum. The output pulse is then transform limited with very clean rise and decay and a duration very independent of the laser fluctuations.

Our amplifier (fig. 3) is similar to the one developed by R.L. FORK and coworkers [6] in that it includes four stages of amplification separated by saturable absorber jets. The introduction of one more stage is of prime importance because of the low energetic input pulse (less than 50 pJ). As already noticed group velocity dispersion in the amplifier and in the optics before the KDP crystal implies temporal broadening. In our set up the estimated dispersion of 110 fsec/nm is compensated by a dispersive delay line composed of two 1200 g/mm diffraction gratings blazed at 630 nm, used at 50° and separated by a distance of 37 mm. Different solutions can be adopted : this compressor can be introduced after the last stage [6] but it implies then a loss factor which can be quite important (above five). It can be introduced at the oscillator output, but the resulting low energetic pulse is then difficult to separate from the fluorescence after the first stage of amplification. We have found that the optimum solution was to introduce the dispersive delay line after the second stage. In that case the satura-

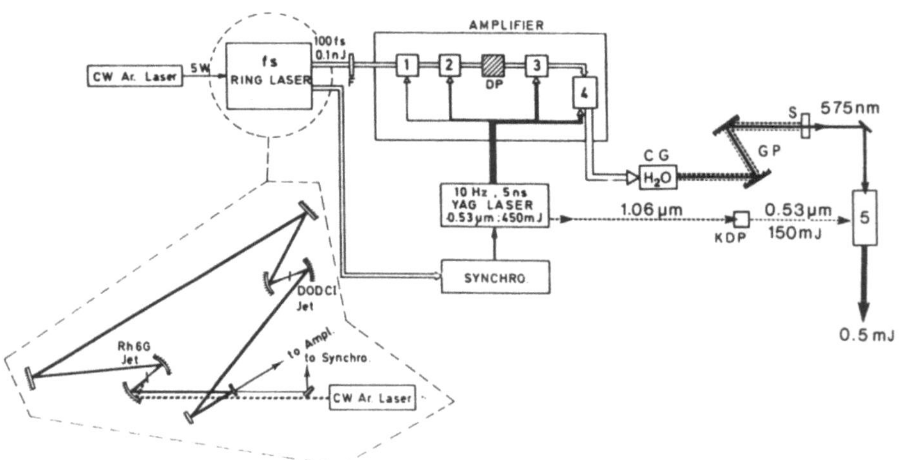

Fig. 3 Ring laser and amplification set up. F is an interferential filter. 1, 2, 3 are three transversally pumped amplifier stages followed by saturable absorber jets. 4 is a longitudinally pumped amplifier stage. DP is the dispersion compensation with a grating pair. The right part of the figure represents the set up for tunable continuum amplification. CG is the continuum generation cell, GP is a grating pair which compensates the further group velocity dispersion and S is the spectral selection. The stage 5 is pumped transversally by 150 mJ at 0.53 μm obtained from a second KDP crystal.

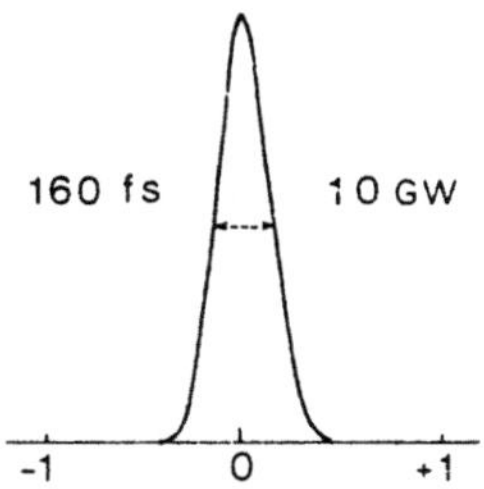

Fig. 4 Autocorrelation of the 10 Hz amplified pulse at 618 nm. Peak powers as high as 10 GW have been obtained

tion of the gain in the next two stages allows to recover a large part of the losses at the expense of a small temporal broadening. The output pulse with an energy well above 1 mJ has then a duration of 160 fsec (fig. 4).

This new set up has been applied in time resolved spectroscopy experiments using as excitation source the frequency doubled output pulses. In that case we found even shorter 309 nm UV pulses (fig. 4 [7]) with a duration of 100 fsec and an energy up to 20 μJ (0.2 GW peak power). The generation of less than 200 fsec tunable pulses by continuum amplification has been undertaken. To that effect we have modified our original set up (fig. 3) by introducing a grating pair (1000 g/mm and 25 mm separation) which takes care of the 45 fs/nm dispersion at 575 nm. On the opposite, we have then to shorten to 22 mm the separation of the gratings in the four stage amplifier.

In conclusion, we have shown that the main problem of broadening associated with group velocity dispersion can be overcome, and demonstrated that the technique of continuum amplification is a powerful tool to generate less than 200 fs tunable pulses in the GW range.

This work has been supported by the Direction des recherches, Etudes et Techniques.

References

1 C.V. Shank, E.P. Ippen, Appl. Phys. Lett. 24, 373-375, (1977).
2 R.L. Fork, B.I. Greene, C.V. Shank, Appl. Phys. Lett. 38, 671, (1981).
3 E.P. Ippen, C.V. Shank, "Subpicosecond Spectroscopy" in Picosecond Phenomena, ed. C.V. Shank, E.P. Ippen, S.L. Shapiro, Springer Verlag, 103-107, (1978).
4 A. Migus, C.V. Shank, E.P. Ippen, R.L. Fork, J. of Quan.Elec. 18, 101 (1982).
5 A. Migus, J.L. Martin, R. Astier, A. Orszag, in Picosecond Phenomena II, Springer Verlag, 59-63 (1980).
6 R.L. Fork, C.V. Shank, R.T. Yen, to be published in Appl. Phys. Lett.
7 J.L. Martin, C. Poyart, A. Migus, Y. Lecarpentier, R. Astier, J.P. Chambaret. This issue.

Femtosecond Continuum Generation

R.L. Fork, C.V. Shank, R.T. Yen, and C. Hirlimann*

Bell Laboratoires, Holmdel, NJ 07733, USA

W.J. Tomlinson

Bell Laboratories, Allentown, NJ 18103, USA

We obtain white light continuum pulses with durations as short as 80 fsec. The broad spectral range (0.19μ-1.6μ) minimal chirp, stable repetitive character, and availability of powers extending to the gigawatt range suggest these pulses will provide a powerful new tool for femtosecond spectroscopy. We obtain these pulses by using short (65 fsec) pulses from our colliding pulse mode locked laser [1] which have been amplified to gigawatt powers in an amplifier designed specifically for short pulse amplification [2]. The short duration and high intensity of these amplified pulses permit us to generate continuum pulses in a short length of nonlinear medium with the consequence that pulse distortion by group velocity dispersion is minimal. We thus generate pulses which are temporally short and which also have almost negligible chirp over broad spectral regions. Such a feature is essential for femtosecond time resolution in most practical applications. This approach is also of special interest in that the temporal distribution of the continuum is determined primarily by the generation mechanism rather than by group velocity dispersion in the generating or analyzing media. We can thus present convincing evidence, e.g., that self phase modulation plays a prominent role in the continuum generation process.

1. Experimental Apparatus

We generate and measure our continuum pulses as illustrated in Fig. 1. A pump pulse at 627 nm of 80 fsec duration and 1.2 GW power is focused into a thin (500μ) jet of ethylene glycol. The focusing mirror has a radius of 25 cm producing intensities at the jet of 10^{13}-10^{14} W/cm^2. Because the continuum has an angular divergence approximately twice that of the pump beam we use a 10 cm mirror for recollimating the continuum. Aluminum coatings are used on the mirrors which reflect the continuum and dielectric or aluminum coatings on the other mirrors. A reference pulse is split off from the pump pulse, delayed by a stepper motor driven stage and focused by a 25 cm mirror on a thin (100μ) KDP crystal. Another 25 cm mirror focuses and overlaps the continuum pulse with the reference pulse at the KDP crystal. Crosscorrelation functions are obtained by varying the stepper motor driven delay and observing the upconverted signal on an OMA2 optical multichannel analyzer. The finite bandwidth of the KDP crystal and its angular orientation serve to select a given spectral region of the continuum. The spectral bandwidth of the upconverted light is sufficiently broad to permit adequate

---

* Permanent Address: Université Pierre et Marie Curie, Laboratoire de Physique des Solides, C.N.R.S. (L.A. 154) 4 Place Jussieu T13-2, 75230 Paris Cedex 05, France

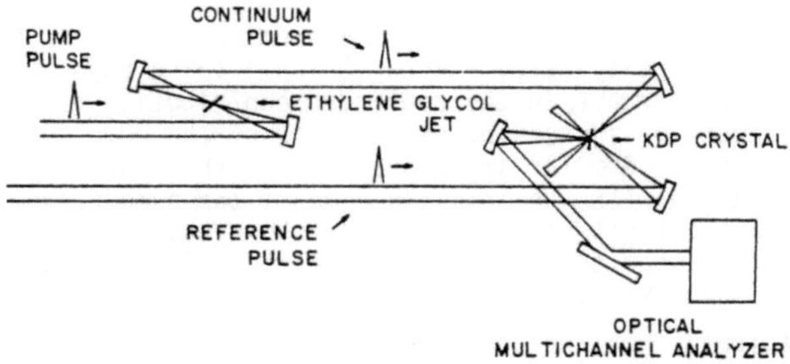

Fig. 1 Experimental arrangement for continuum generation and measurement

temporal resolution. For our KDP crystal which was cut to double 620 nm at normal incidence we can upconvert continuum light over the range .44μ to 1.1μ. (A slightly greater range can be obtained by using the crystal at highly oblique angles; however, we avoided these conditions because of the relatively long path for the continuum light in the crystal.) The crosscorrelation of the continuum with the reference pulse was obtained by selecting a given spectral region and relative pulse delay, integrating 500 pulses in the OMA2, changing the delay and repeating the measurement. The crystal angle was then changed and the process repeated at another wavelength until the time dependence of the continuum was mapped out.

2. Continuum Properties

We illustrate the temporal properties of the continuum by the crosscorrelation traces shown in Fig. 2. The data points for a given trace have been connected by a smooth curve. Here we have crosscorrelated a segment of the continuum centered at 4694Å with the reference pump pulse. We also show the crosscorrelation of a segment of the continuum centered at 10093Å with the reference pulse. The zero of time is indicated by the autocorrelation function of the pump pulse taken at an intensity below threshold for continuum generation.

We see both the infrared and blue portions of the continuum have durations closely approximating that of the 80 fsec pump pulse. We also see the infrared portion of the continuum coincides with the leading edge of the pump pulse and the blue portion of the continuum coincides with the trailing edge

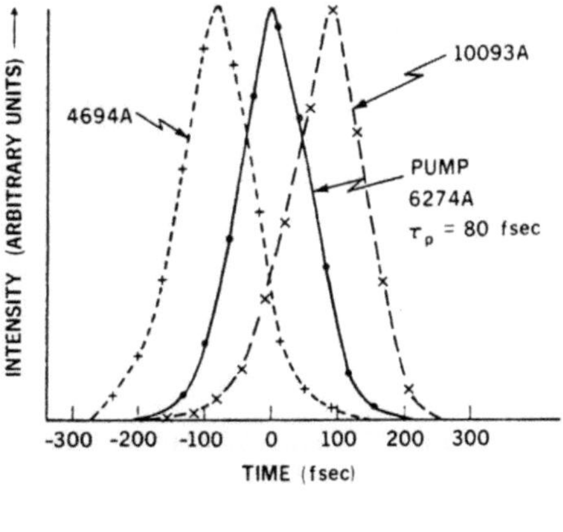

Fig. 2 Crosscorrelation traces for representative blue (+) and infrared (x) portions of the continuum. An autocorrelation trace of the pump pulse (●) at low intensity is also shown.

of the pump pulse. This temporal behavior is consistent with a model of continuum generation in which self phase modulation plays a prominent role. That is, the rapid increase in intensity associated with the leading edge of the pump pulse causes a rapid increase in the nonlinear contribution to the refractive index. This causes an increasing retardation of the pump light and hence a shift to the red. Similar arguments predict the blue portion of the continuum will coincide with the trailing edge of the pump pulse. The crosscorrelation trace for the light at 4694Å clearly shows such a result. (The positions of the crosscorrelation traces have been shifted slightly to correct for dispersion in the KDP crystal, see Fig. 3).

We plot the temporal distribution of the continuum in time in Fig. 3. The data points were obtained from a series of crosscorrelation traces taken as described above. The uncorrected data points are shown by (x). The data points corrected for the apparent time shift introduced by group velocity dispersion in the KDP are shown by (●). These points give the actual distribution of the continuum in time and are connected by a smooth curve drawn through the origin. We also include an indication of the position of the continuum if a correction is introduced for the temporal shift caused by group velocity dispersion in the generating jet (▼). The principal point to be made is that the dominant factor determining the temporal distribution of the continuum is the generating mechanism rather than group velocity dispersion in the generating or measuring apparatus. We also see that the chirp is small amounting to <30 fsec/1000Å in the blue and <10 fsec/1000Å in the red.

The continuum is generated with high efficiency (>50%), is sufficiently reproducible to permit detection of spectral changes as small as one part in ten thousand (500 shots averaged), and falls off relatively slowly with increasing frequency shift from the pump. The intensity decreases by approximately an order of magnitude for an increase in optical frequency of 2×10^{14} Hz and somewhat more rapidly for decreasing frequency. Given the gigawatt power level in the continuum this yields experimentally useful light from 0.19μ to 1.6μ. We have recently used continuum from .65μ to 1.55μ to measure a relaxation time of 160 fsec in polyacetylene, e.g. [3]

3. Pump-Probe Experiment

We have also performed pump-probe experiments in which the KDP crystal is replaced by a second liquid sample. In that case a portion of the continuum pulse selected by a narrow band pass filter probes the region of the sample excited by the reference pump pulse. We find, e.g., that the probe is upshifted if it coincides with the leading edge of the pump pulse and downshifted if it coincides with the trailing edge of the pump pulse. We also find evidence for a frequency dependence of the nonlinear index.

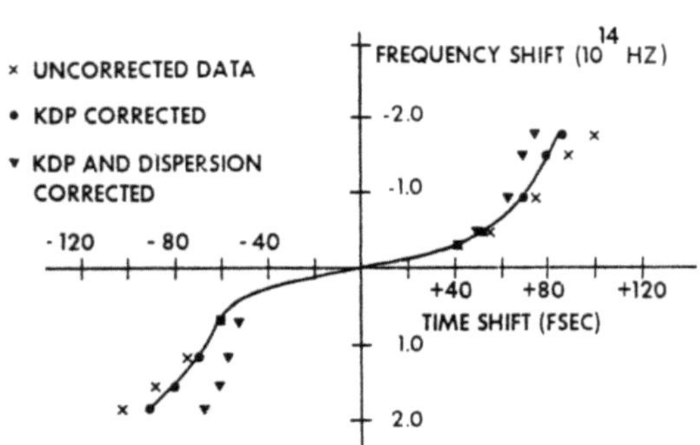

Fig. 3 Plot of the continuum frequency shift versus time

4. Calculation

Our calculation of continuum generation using the nonlinear Schrödinger equation [4] predicts approximately the observed spectrum. Given the very large frequency shifts, variations in the focused pump intensity, and the possibility of some self focusing, it is difficult to make precise comparisons with theory. We suggest that while self phase modulation clearly plays a major role other nonlinear mechanisms, such as self steepening, free electron generation, and parametric mixing must also be considered as having significant influence.

Special thanks are due F. A. Beisser for essential technical support.

References

1. R. L. Fork, B. I Greene, and C. V. Shank, Appl. Phys. Lett. 38, 671 (1981).

2. R. L. Fork, C. V. Shank, and R. T. Yen, Appl. Phys. Lett. 41, 273 (1982).

3. C. V. Shank, R. T. Yen, R. L. Fork, J. Orenstein, and G. L. Baker, to be published.

4. L. F. Mollenauer, R. H. Stolen, and J. P. Gordon, Phys. Rev. Lett. 45, 1095 (1980).

New Picosecond Sources and Techniques

A.E. Siegman

Stanford University, Stanford, CA 94305, USA

H. Vanherzeele

Vrije Universiteit Brussel, B-1050 Brussel, Belgium

1. Introduction

In this paper we report on two new techniques for generating picosecond optical pulses, that is, colliding pulse mode-locking in a special "antiresonant ring" cavity, and mode-locking of a self-pumped phase conjugate resonator. We also give a preliminary report on a possible new technique for picosecond spectroscopy, namely the photoacoustic detection of picosecond pulse effects.

2. Colliding Pulse Mode-Locking in an Antiresonant Ring Cavity

By far the most important recent advance in mode-locking has been the development of colliding pulse mode-locking as described earlier by SHANK, FORK and YEN. We report here a useful variant of this technique in which the colliding pulse function is obtained using an "antiresonant ring" laser cavity (Fig.1).

In this interferometer (also known as a cyclic or SAGNAC interferometer), a signal at any frequency entering through the beam splitter from either external arm will divide and travel around the ring in opposite directions, thus providing the colliding pulse function for a Brewster angle saturable absorber cell located at the midpoint of the ring. For a fifty-fifty beam splitter and a ring with symmetrical focusing elements, these signals then always recombine in such a fashion that all the energy returns back into the same external arm, independent of the frequency or pulse envelope of the radiation (hence the "antiresonant" character of the ring). In the example shown, the ring thus acts as a lossless end mirror for the laser medium, with no energy loss out the other external arm indicated by the dashed lines.

This concept has now been implemented on flash-pumped, passively mode-locked Nd:YAG lasers in experiments first by VANHERZEELE and VAN ECK at the

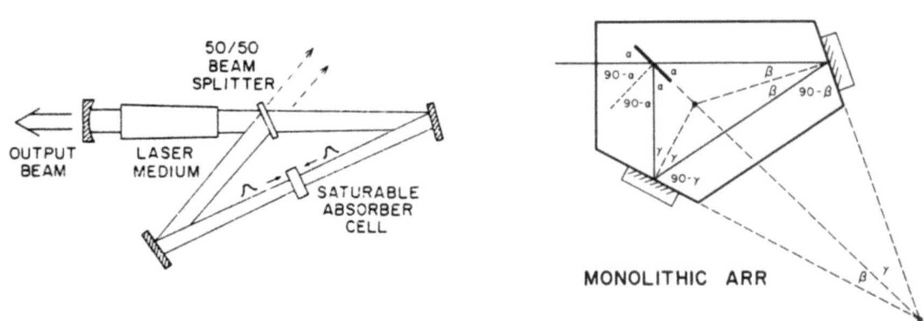

Fig.1 Antiresonant Ring Laser Cavity

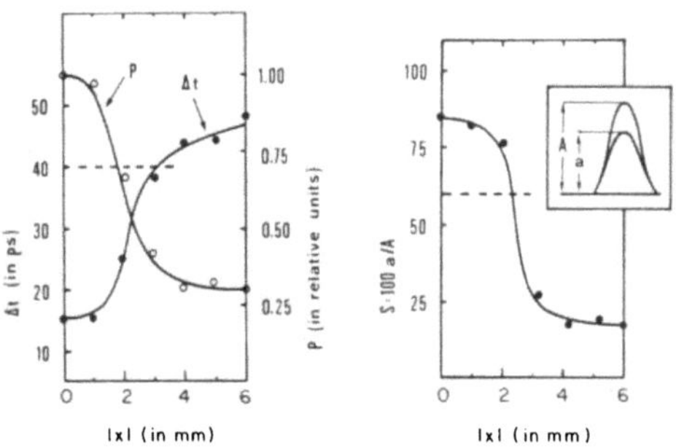

Fig.2 Mode-locked pulse width (Δt), peak power (P) and stability (S)

Vrije Universiteit Brussel [1,2], and more recently at Stanford University. In the experiments at Brussels, TPF pulsewidth measurements made for various axial positions of the passive mode-locking cell around the exact central point in the antiresonant ring gave results for pulsewidth, peak pulse intensity, and shot-to-shot stability as plotted in Fig. 2. The dashed line indicates the typical 40 ps mode-locked pulsewidth produced by the same laser using a conventional contacted saturable absorber cell. The $\leq$ 15 ps pulsewidth measured in the colliding pulse case with the cell centered in the ring is an average over all the pulses in approximately 30 laser shots with a non-flowing saturable absorber cell. A two-photon fluorescence trace for a single laser shot with fresh dye indicates a pulsewidth closer to 10 to 12 ps, approaching the bandwidth limit for a Nd:YAG laser. The shot-to-shot pulse amplitude was also substantially more stable in the colliding pulse configuration than in the standing wave cavity as indicated by the second plot in Fig.2.

Advantages of the antiresonant ring configuration for colliding pulse mode-locking include the fact that all the energy can be taken out of the cavity in a single pulse train, and the pulses can double-pass the gain medium for each single pass through the resonator. We have found the antiresonant ring configuration easy to align and use. Fig.1(b) shows a simple monolithic design which, once aligned, can function in effect as one end mirror for a laser cavity. An effective form of single-pulse cavity dumping for the antiresonant ring is also available [3].

One disadvantage of this approach for femtosecond dye laser pulses will be the unavoidable dispersion of even a very thin beam splitter substrate. We believe that this design should be the cavity of choice, however, at least for all mode-locked solid-state lasers in the future.

3. Mode-Locking of a Self-Pumped Phase Conjugate Resonator

A second new mechanism for generating ultrashort optical pulses makes use of four-wave mixing in a "self-pumped" phase conjugate resonator as illustrated in Fig.3.

It is well understood that three optical pulses colliding in a four-wave mixing cell can produce a reflected and phase conjugated pulse that is significantly shorter than any of the other pulses. Suppose such a shortened and phase-conjugated pulse is redirected back through a laser gain medium to become the pump pulse for a subsequent collision, as shown by the shaded path in Fig.3, while the previous pump pulse becomes the new probe pulse. Once appropriate initial pulse conditions can be established, the phase conjugated pulse #4 produced by the four-wave cell on one pulse collision can provide the

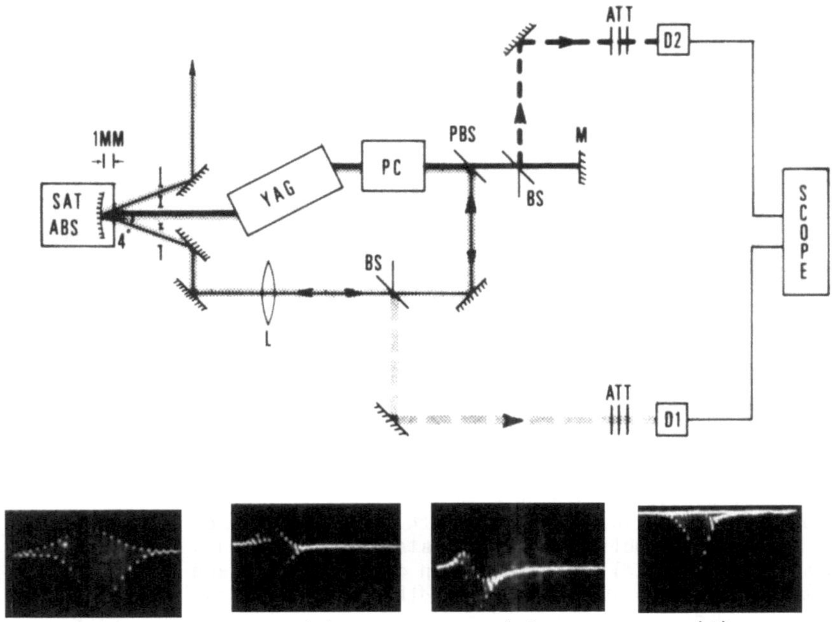

Fig.3 Combined Nd:YAG laser and phase conjugate resonator

necessary feedback for oscillation of cavity. Self-sustained oscillation, Q-switching and mode-locking of this resonator, with spatial phase conjugation as well, will then ensue.

This concept has now been demonstrated experimentally using a Nd:YAG laser [5], with results as shown in Fig.3. Here the beam path indicated by the heavy line represents a conventional flash-pumped and mode-locked Nd:YAG laser with a saturable absorber cell, which also functions as a phase conjugate cell, in close contact with the laser mirror at one end. A Pockels cell (PC) and polarizing beam splitter (PBS) were then added to the laser cavity to split off a fraction of the circulating energy and create the phase conjugate probe beam path shown by the lower shaded line, thus forming a folded and self-pumped phase conjugate cavity.

In our experiments the Nd:YAG laser was fired and the Pockels cell then switched at various times before or during the laser burst so as to couple a varying fraction of the circulating energy (up to 100%) from the original laser cavity (dark line) into the self-pumped phase conjugate cavity (shaded line). In part (a) for example the Pockels cell couples a fixed 50% of the conventional laser cavity energy into the phase conjugate probe beam. The upper trace then shows the conventional mode-locked laser burst, while the lower trace shows the phase conjugate reflection of the shaded probe beam from the saturable absorber cell. Part (b) shows the behavior with the Pockels cell triggered so as to block the conventional laser cavity at the instant when the conventional laser intensity reached its maximum value. The pulses in the second half of this trace thus show essentially self-sustained oscillations in the shaded-path phase-conjugate cavity. Parts (c) and (d) show similar results with a second Nd:YAG laser head added in the lower shaded part of the path. Almost no initial signal is needed in the primary laser cavity to initiate these "self-pumped" phase conjugate oscillations.

The pulse widths obtained in the self-pumped phase conjugate cavity ranged from 30 to more than 300 ps long depending upon various adjustments The behavior of this system is complex, and deserves further study. Delay processes in the phase-conjugating medium itself and strong cumulative thermal gratings in the saturable absorber cell may play significant roles. In any case this represents a new and interesting type of optical mode-locking and, we believe, the first demonstration of mode-locking in a phase conjugate laser cavity.

4. Picosecond Spectroscopy Using Photoacoustic Detection

Picosecond lifetimes are often measured by observing some comparatively slow secondary effect produced by a pair of picosecond optical pulses - for example, the total fluorescent emission from some excited state - as one varies the arrival time delay between two ultrashort pulses. We are attempting to make such picosecond spectroscopic measurements using photoacoustic detection - that is, measurement of the total acoustic emission produced by the two optical pulses - to provide the necessary signal detection.

Photoacoustic spectroscopy (PAS) in its conventional form, using chopped cw or long-pulse optical excitation, offers widely recognized advantages for measuring very weak absorptions in liquids, gases and solids. If extended to picosecond pulse excitation, photoacoustic detection may offer advantages for picosecond spectroscopy in nonfluorescing or very weakly absorbing samples. In general, photoacoustic detection, because it focuses on the nonradiative part of the energy delivered to the sample, may yield different information than purely optical or fluorescent techniques.

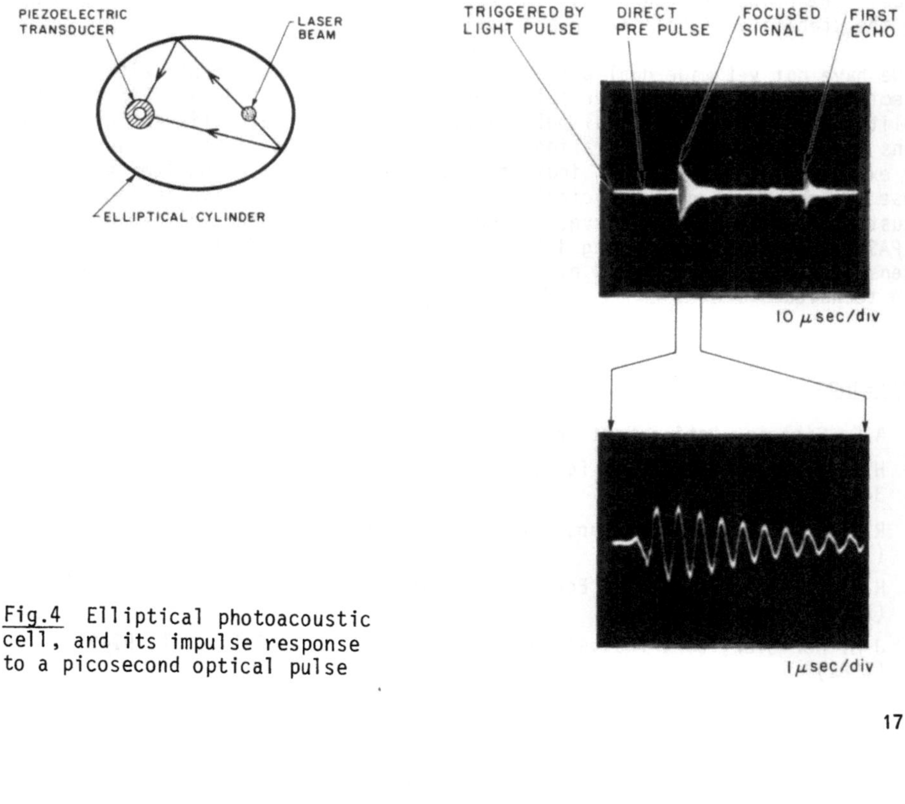

Fig.4 Elliptical photoacoustic cell, and its impulse response to a picosecond optical pulse

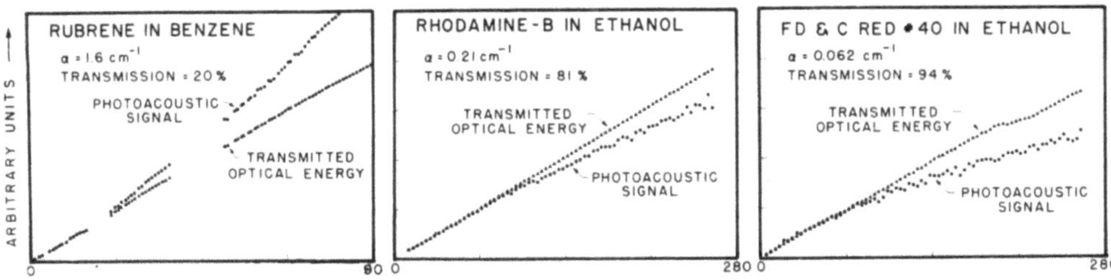

Fig.5 Photoacoustic Cell Response vs. Incident Optical Flux

A useful tool for this type of measurement would seem to be a photoacoustic cell with a clean and sensitive response to short pulse excitation. Fig.4 shows a novel photoacoustic cell with an elliptical cross section, in which the cylindrical pressure wave produced by a laser pulse passing along one focal line is focused upon a cylindrical piezoelectric transducer located along the other focal line. The upper trace in Fig. 4 shows the impulse response of this cell to either a Q-switched or a single mode-locked and doubled Nd:YAG laser pulse passing through a weak solution of Rhodamine 6G in carbon tetrachloride. The main response appears at ~23 μsec after the laser pulse, followed ~46 μsec later by an acoustic echo that has reflected from the transducer and made one complete round trip in the cavity. One can also observe a faint precursor signal before the main impulse, resulting from acoustic energy travelling directly from the optic axis to the acoustic axis without wall reflection. The arrival of each pulse causes the transducer itself to ring at its resonance frequency of ~1 MHz, as shown in the lower trace. The response shown is a time exposure over ~100 laser shots, indicating the stability of both laser and cell.

We have not yet made dual-pulse or variable-delay measurements with this detector. Fig.5 shows however output optical pulse amplitude and PAS signal amplitude versus input optical pulse amplitude for several different dye solutions in this cell. The supralinear response of the PAS signal in rubrene, for example, we interpret as indicating the presence of an excited state whose absorption is more efficient than ground state absorption in generating acoustic response. These curves indicate at least the potential capability of PAS detection for measuring different excitations and different saturation intensities than are measured by conventional fluorescence or optical saturation techniques.

References

1. A.E. Siegman, Optics Lett. 6, 334-335 (1981).
2. H. Vanherzeele, J.L. Van Eck and A.E. Siegman, Appl. Optics 20, 3483-3486 (1981).
3. R. Trutna and A.E. Siegman, IEEE J. Quant. Electr. QE-13, 955-962 (1977).
4. H. Vanherzeele, J.L. Van Eck and A.E. Siegman, Optics Lett. 6, 467-469 (1981).
5. J-M. Heritier, J.E. Fouquet and A.E. Siegman, Appl. Optics 21, 90-93 (1982).

Generation of Coherent Tunable Picosecond Pulses in the XUV

T. Srinivasan, K. Boyer, H. Egger, T.S. Luk, D.F. Muller, H. Pummer, and C.K. Rhodes

University of Illinois at Chicago, P.O. Box 4348, Chicago, IL 60680, USA

Coherent extreme ultraviolet (XUV) radiation is useful in a wide range of applications to physical measurements including areas such as high resolution spectroscopy [1] and solid state surface studies [2]. Nonlinear optical processes, in which ultraviolet radiation serves as the fundamental, have been proven [3-5] to be an efficient means of upconverting ultraviolet laser radiation to the XUV spectral range. Furthermore, excimer lasers, which serve as attractive sources of radiation at 193 nm and 248 nm, have demonstrated [6-8] performance essentially at the fundamental limits governing spectral brightness. Naturally, the peak power of these high brightness systems can be very greatly increased, if the energy can be extracted in a short pulse. In this paper we report the development of such a source and the generation of coherent, tunable, narrow linewidth radiation in the XUV using the radiation from this laser system.

The schematic of a high brightness source at 193 nm with pulse duration ∿ 10 ps is shown in Fig. 1. The output of a synchronously pumped mode-locked dye laser (Coherent Radiation 599-04, λ ∿ 580 nm, pulse duration < 10 ps) is pulse amplified in a three stage, XeF* excimer laser pumped dye amplifier chain. In order to suppress amplified spontaneous emission, cells (250 µm thick) containing DQOCI as a saturable absorber are installed between two consecutive amplifier stages. The ∿ 1-mJ 580-nm output pulse (pulse duration ∿ 6 ps) is focussed into a strontium heat pipe, where ∿ 2 nJ of the third harmonic at 193 nm is produced. This radiation is amplified in a double pass ArF* amplifier, spatially filtered and passed through a grating

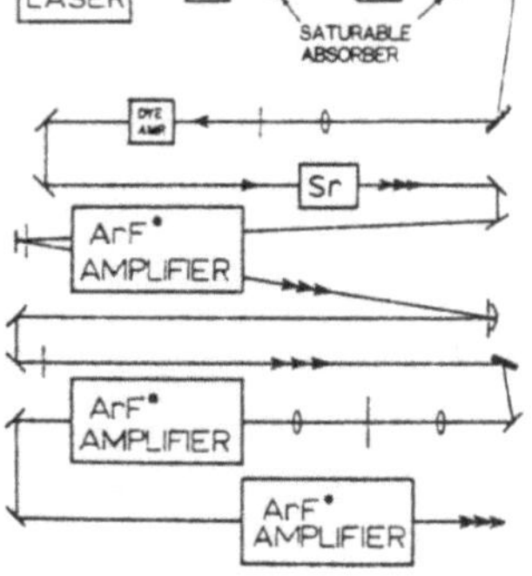

Fig. 1 Schematic of tunable picosecond ArF* laser

pinhole combination with a 25-cm^{-1} bandpass. After further amplification in two single pass amplifiers, the output energy in the short pulse is typically in the range of 30 ± 10 mJ, a value which is a substantial fraction of the maximum energy available for extraction (∿ 50 mJ) in a pulse shorter than the excimer lifetime. This fact is further supported by the observation of very strong saturation in the amplified spontaneous emission occurring immediately after the passage of the 10 ps pulse.

The XUV radiation is produced via harmonic generation and mixing in various gases. Such experiments involve two basic technical difficulties, namely, the lack of suitable window materials and self-absorption at the XUV frequency in the nonlinear medium. The window problem is readily solved by using the well known technique of differential pumping shown in Fig. 2, in which the nonlinear gas flows through a small pinhole into the differential pumping chamber. This allows the fundamental radiation to be focussed into the region near the pinhole where the XUV radiation is generated. Absorption of the generated XUV light is minimized by using pinholes in thin foils (thickness < 100 μm) so that the gas tends to expand into a relatively large solid angle, instead of rather thick plates, which cause the gas stream into the differential pumping stage to be more directed. Even for optimized geometries, however, absorption in this expanding gas plume can represent a substantial loss for the generated XUV radiation. In cases for which a buffer gas with low absorption cross section is available, the pinhole in Fig. 2 can be replaced by the simple transverse-flow geometry shown in Fig. 3. Here, the nonlinear medium is confined to the conversion zone by easily established flow patterns. A more detailed description of this arrangement has been given earlier [3].

Using the picosecond ArF* laser as a pump source, the generation of third and fifth harmonic radiation has been observed in various gases. In the transverse flow geometry, with neon as the buffer gas, peak powers up to 20 kW at 64 nm have been obtained with hydrogen or several other gases as the nonlinear medium. Due to the lack of a suitable buffer gas for 38.7 nm radiation, fifth harmonic generation has been studied in the conventional pinhole geometry (refer to Fig. 2). In this configuration, peak powers up to 200 W at 38.7 nm have been obtained with argon as the nonlinear medium. The list of gases in which fifth harmonic generation was observed, along

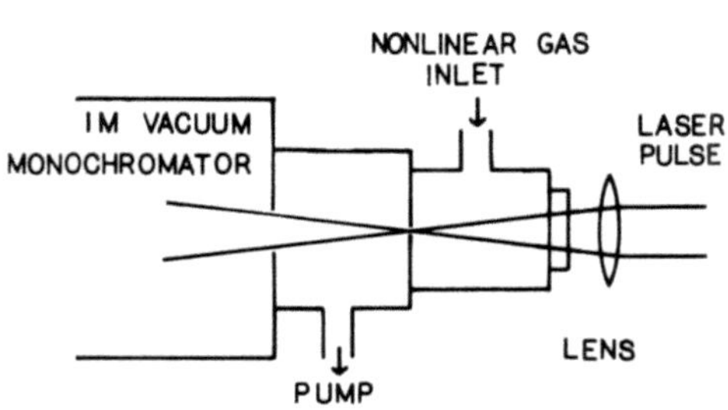

Fig. 2 Simple differential pumping system for XUV generation via nonlinear processes

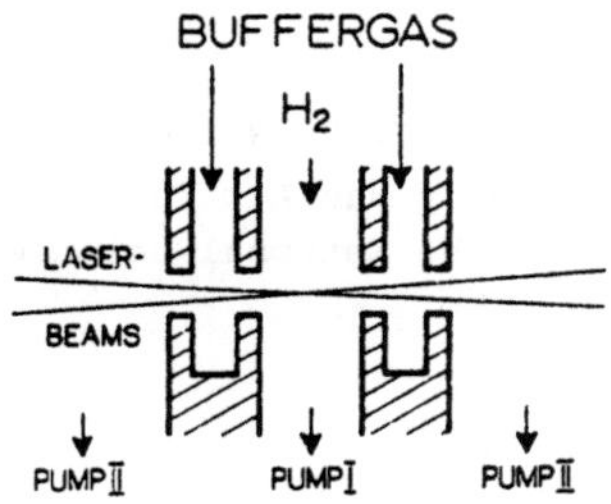

Fig. 3 Transverse flow cell. The nonlinear medium is confined to the core flow and the absorption of the generated XUV radiation is minimized

with the approximate harmonic power, is given in Table 1. We note that for the first time, under 193-nm irradiation, He and Ne have produced third and fifth harmonic outputs. In earlier studies with 200-mJ, 10 ns ArF* pulses, no harmonic radiation could be detected with these two gases.

The results summarized in Table 1 have been obtained at different pressures, depending on the capacity of the differential pumping system for the specific gas. With all gases, an increase in harmonic outputs has been observed with increasing gas pressure. In addition, no attempt has been made to phase match the harmonic and fundamental beams. Estimates indicate that the present conversion efficiency can be substantially enhanced in this manner.

The XUV radiation can be tuned over a frequency range corresponding to the tuning range of the ArF* laser ($\sim$ 200 cm^{-1}). Tunability over a considerably broader interval can be achieved with frequency mixing processes. Previous experience with mixing of ArF* radiation and dye radiation in 10 ns pulses indicates that peak powers analogous to those of the third and fifth harmonic can be produced.

Table 1 Harmonic generation with 193 nm 10 ps pulses

GAS	RELATIVE POWER	
	3rd harmonic (64 nm)	5th harmonic (38 nm)
H_2	10^5	weak
He	10^2	10
Ne	10^2	10^2
Ar	10^5	10^3
N_2	10^3	10^2
CO	10^2	0

5th Harmonic power in Ar is 200 W.

Acknowledgements

The authors wish to acknowledge the technical assistance of J. R. Wright, S. W. Vendetta and M. J. Scaggs. This work was supported by the Air Force Office of Scientific Research, the Department of Energy, the National Science Foundation, and the Office of Naval Research.

References

1. C. K. Rhodes and P. W. Hoff in "Excimer Lasers" edited by C. K. Rhodes (Springer-Verlag, Berlin 1979) p. 175
2. "Synchrotron Radiation Research", edited by H. Winick and S. Doniach (Plenum Press, N.Y. 1980); C. K. Rhodes, "Atomic, Molecular and Condensed Matter Studies Using Ultraviolet Excimer Lasers" in Proceedings of the 29th Midwest Solid State Conference - Novel Materials and Techniques in Condensed Matter (Elsevier North Holland, New York, 1982) p. 151
3. H. Pummer, T. Srinivasan, H. Egger, K. Boyer, T. S. Luk and C. K. Rhodes, Opt. Lett 7, 93 (1982)
4. K. Boyer, H. Egger, T. S. Luk, D. F. Muller, H. Pummer, T. Srinivasan, and C. K. Rhodes, "High Spectral Brightness Extreme Ultraviolet Generation With Excimer Lasers", in Proceedings of the International Conference on Lasers '81
5. J. Reintjes, Appl. Opt. 19, 3389 (1980)
6. H. Egger, T. Srinivasan, K. Hohla, H. Scheingraber, C. R. Vidal, H. Pummer, and C. K. Rhodes, Appl. Phys. Lett. 39, 37 (1981)
7. R. T. Hawkins, H. Egger, J. Bokor and C. K. Rhodes, Appl. Phys. Lett. 36, 391 (1980)
8. R. G. Caro, and M. C. Gower, Opt. Lett. 11, 557 (1981)

New Infrared Dyes for Synchronously Pumped Picosecond Lasers

A. Seilmeier, B. Kopainsky, W. Kranitzky, and W. Kaiser
Physik Department der Technischen Universität München,
D-8000 München, Fed. Rep. of Germany

K.H. Drexhage
Physikalische Chemie der Universität Siegen
D-5900 Siegen, Fed. Rep. of Germany

Introduction

In recent years substantial progress has been made in the synthesis of new infrared dyes of good photochemical stability which exhibit absorption and fluorescence bands beyond 1 µm. Infrared dyes with very short ground-state recovery times have been used for switching of Nd-lasers for many years /1,2/. Of special interest is the application of dyes with long fluorescence lifetimes in infrared dye lasers.

New Infrared Dyes

We have investigated a series of new polymethine cyanine dyes /3/. Detailed data on the absorption and fluorescence properties, the fluorescence lifetimes and the photochemical stability were collected.

In Fig. 1 examples of two groups of molecules are shown which are of particular interest. Dye No. 24 is a representative of group I dyes. These dyes are well suited for switching purposes due to their short fluorescence lifetimes. Dyes of group IV exhibit longer lifetimes. They are candidates for infrared dye lasers. In the lower part of Fig.1 dye No. 26 is shown as an example. This molecule was used in a laser system.

Group I

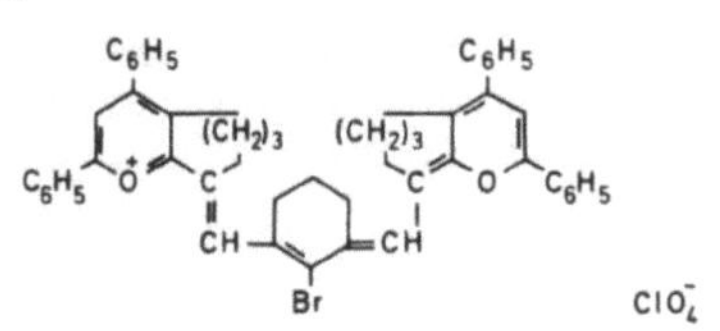

No. 24

Group IV

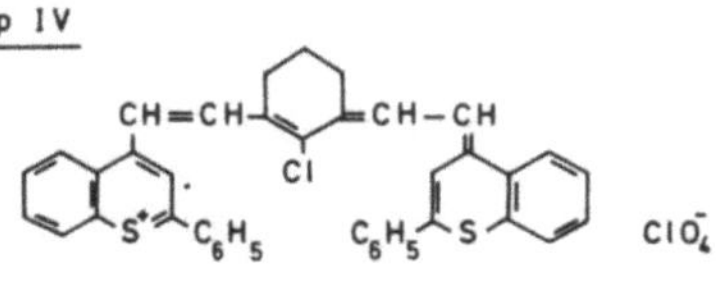

No. 26

Fig.1 Molecular structure of a group I and a group IV dye

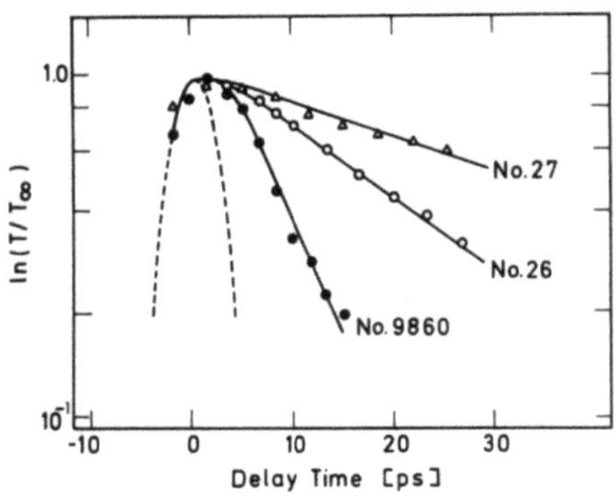

Fig.2 Bleaching recovery in dyes 9860, No. 26, and No. 27

Group I dyes are pyrylium dyes with four freely rotating phenyl groups. It is known that molecules with freely rotating parts have short fluorescence lifetimes due to rapid internal conversion. In group IV a benzene ring is rigidly attached to the pyrylium group and there are only two phenyl groups. In consequence these molecules dissipate less energy via internal conversion.

We measured the fluorescence lifetimes of the new infrared dyes in dichloroethane solution. In our experiment a picosecond pulse at 1.06 μm reversibly bleaches the dye. A second pulse at the same frequency monitors the absorption recovery as a function of time /1/. In Fig.2 the change of transmission as a function of delay time is shown for the commercial dye 9860 and for two dyes from group IV. For dyes No. 26 and No. 27 we find relaxation times of 22 ps and 50 ps, respectively. Dye No. 27 has the same structure as dye No. 26 but the sulfur atoms are replaced by oxygen. The broken line is the time resolution of our experiment.

Table 1 summarizes the properties of four of the most interesting dyes. All dyes are strongly absorbing around 1 μm. For group I dyes we find lifetimes below 10 ps. The lifetimes in group IV are longer than 20 ps. All new dyes have a higher photochemical stability compared to 9860. Dyes of group IV represent the most stable infrared dyes reported on so far.

Infrared Dye Lasers

Considering the values in Table 1 dye No. 26 appears to be a good dye for a picosecond laser-system. The photostability is excellent and the fluorescence lifetime is sufficiently long.

Table 1

Group	Dye No.	Absorption λ_{max}[nm]	Lifetime τ[ps]	Relative Stability
I	9860	1060	7 ± 1	1
I	24	1120	8.5 ± 1	100
IV	26	1080	22 ± 1.5	$>10^4$
IV	27	985	50 ± 5	$>10^4$

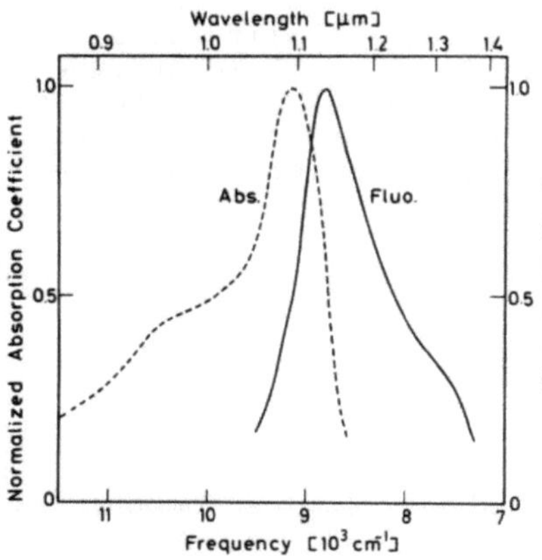

Fig.3 Absorption and fluorescence spectrum of dye No. 26

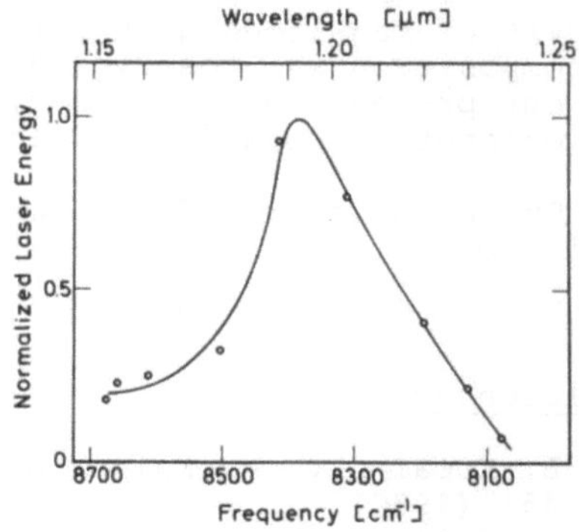

Fig.4 Tuning range of the IR dye-laser using dye No. 26

The absorption and fluorescence spectrum /2/ of dye No. 26 dissolved in dichloroethane is depicted in Fig.3. It shows a strong absorption at the wavelength of the Nd-laser. The fluorescence spectrum extends from 1.1 μm to 1.4 μm.

In our dye laser experiment /4/ dye No. 26 was synchronously pumped by a pulsed Nd:YAG laser. The cavity consists of two mirrors with high reflectivity and a prism as dispersive element. The emission spectrum and the pulse duration were observed.

Fig.4 shows the tuning curve of dye No. 26. The normalized output energy of the dye laser is plotted as a function of the frequency. We found laser action between 1.15 μm and 1.24 μm. To our knowledge picosecond pulses at 1.24 μm generated by a dye laser have never been reported previously.

The properties of our IR dye-laser are summarized in Table 2. We pumped dye No. 26 at 1.06 μm and found a tuning range from 1.15 μm to 1.24 μm. The pulse duration of the dye laser was 5 ps using pump pulses of 21 ps duration. An energy conversion of approximately 1% and a divergence of 2 mrad were observed.

Table 2

Tuning range	1.15 μm to 1.24 μm
Pulse duration	5.5 ps (pump 21 ps)
Power conversion of No. 26	~4 %
Energy conversion of No. 26	~1 % at 1.2 μm
DNTPC	~2 % at 1.1 μm
Divergence	2 mrad
Stability, relative to DNTPC	$>10^5$

Most important for the application is the stability of a laser dye. We found the photostability of No. 26 to be at least five orders of magnitude higher than of the commercial dye DNTPC. No decomposition was observed after weeks of operation. According to our present results the synthesis of IR dyes for even longer wave-lengths appears possible.

The authors gratefully acknowledge the contributions of B. Sens and P.Qiu.

References

1 B. Kopainsky, W. Kaiser,and K.H. Drexhage, Optics Commun. 32, 451 (1980)
2 A. Seilmeier, B. Kopainsky, and W. Kaiser, Appl. Phys. 22, 355 (1980); B. Kopainsky, A. Seilmeier, and W. Kaiser, in "Picosecond Phenomena II, eds. R. Hochstrasser, W. Kaiser, C.V. Shank, Springer Verlag, Heidelberg 1980, p. 7
3 B. Kopainsky, P. Qiu, W. Kaiser, B. Sens, and K.H. Drexhage, to be published in 1982
4 W. Kranitzky, B. Kopainsky, W. Kaiser, K.H. Drexhage, and G.A. Reynolds, Optics Commun. 36, 149 (1981)

Acousto-Optic Stabilization of Mode-Locked Pulsed Nd: YAG Laser

H.P. Kortz

Quantel International, Inc., 395 Reed Street,
Santa Clara, CA 95050, USA

1. Introduction

Increasing complexity and accuracy of experiments involving picosecond lasers have led to the need for more reliable and more stable picosecond sources than available thus far. Improvements of the stability of a passively mode-locked Nd:YAG laser by adding both Q-Switch and active Q-Modulation has been reported already in 1974 [1]. A Pockels cell was used for the modulation requiring high voltage in the 100MHz range. Later, the availability of appropriate acousto-optic modulators made experimental set-ups with less complexity possible. Improvements of the pulse train amplitude stability of Nd:Glass lasers down to ±5% was reported [2,3]. Nd:YAG lasers are only briefly mentioned in reference 3. An actively/passively mode-locked Nd:YAG laser operated at 1Hz repetition rate is described in reference [4], but no characteristics of the laser are given.

Here an extensive description of the characteristics of an actively/passively mode-locked Nd:YAG laser operated at a repetition rate of 10Hz is given. An output stabilization of ±1% was achieved, without any sacrifice in pulse duration or energy.

2. Theoretical Considerations

When the system is optimized the transmission of the modulator is periodically varied with the laser round trip frequency ν_{Res}. The single pass transmission of the modulator can be written as:

$$T_{mod}(t) = (1 - a) + a \cdot \sin (2\pi\nu_{Res}t) \qquad (1)$$

having the modulation depth 2a. The combination of the output mirror with reflectivity R and the active modulator can be considered as one element at the position of the output mirror with periodically variable reflectivity.

$$R_{mod}(t) = \{ (1 - a)^2 + a (1 - a) 2 \sin (2\pi\nu_{Res}t) \cdot k + (k^2 - 1) a^2 + a^2 \sin^2 (2\pi\nu_{Res}t) \} \cdot R \qquad (2)$$

The factor $k = \cos(\pi\, d/L_{Res})$ contains the distance d between active modulator and mirror. If $d \neq 0$ both the modulation depth is lower and an additional constant loss is generated.

$$LOSS_{const} = -(k^2 - 1)a^2 \qquad (3)$$

In the case of high gain lasers like pulsed Nd lasers the additional constant loss is negligible compared to the regular diffraction loss (10-40% per round trip) and the output loss (50-90% per round trip), even if the distance between output mirror and modulator is relatively large (e.g. 1/10 of the resonator length). This is of course different in low gain lasers where the losses per round trip have to be much smaller.

3. Experiment

A standard acousto-optic modulator (70MHz RF, 140MHz optical) was integrated into a passively mode-locked Nd:YAG laser (Quantel YG400). The modulator was placed inside the oscillator close to the output mirror. The output mirror was mounted on a precision translation stage allowing the exact matching of resonator round trip frequency and modulator frequency. The output signal was given to a photodiode in order to monitor pulse train and stability with an oscilloscope. The pulse length was measured with a well known SHG method [5,6]. In this method the output beam is split in two parts, one variably delayed against the other. The two parts are recombined and sent through a doubling crystal in which doubling occurs only if field components of both beams are overlapping.

4. Results

4.1 Threshold

The most spectacular effect of the acousto-optic modulator is a dramatic change in the threshold behavior of the laser. At dye concentrations for which the thresholds for mode-locking and for double pulse-train mode-locking are broad and overlapping without additional active modulation, these thresholds become very sharp and clearly separated when the modulator is turned on. The sharpness of the threshold is shown in fig.1. The sharpness is defined as $\{(E_{5\%} + E_{95\%})/2\}/\{E_{95\%} - E_{5\%}\}$, with $E_{5\%}$ and $E_{95\%}$ being the pump energies at which the probability for mode-locking is 5% or 95% respectively. The sharpness increases with increasing dye concentration. The threshold pump energy goes through a minimum when the RF drive power of the modulator is increased. There is also a sharp minimum for the threshold for proper matching of the resonator length with respect to the modulation frequency.

4.2 Stability

Since the build-up process of mode-locked pulse trains is of a statistical nature, the energy output of a mode-locked oscillator without active modulator can have variations of up to ±30%, depending on the specific parameters of operation. With the acousto-optic modulator on, the output stability can be considerably improved down to about ±1%. The improvement becomes even more evident when the frequency doubled output is monitored. The resonator length has to be matched to the modulator frequency with an accuracy of better than ±100μm in order to achieve optimum stabilization (see fig.2). The stabilization reaches an optimum for a specific dye concentration and is better when the laser is pumped well above threshold.

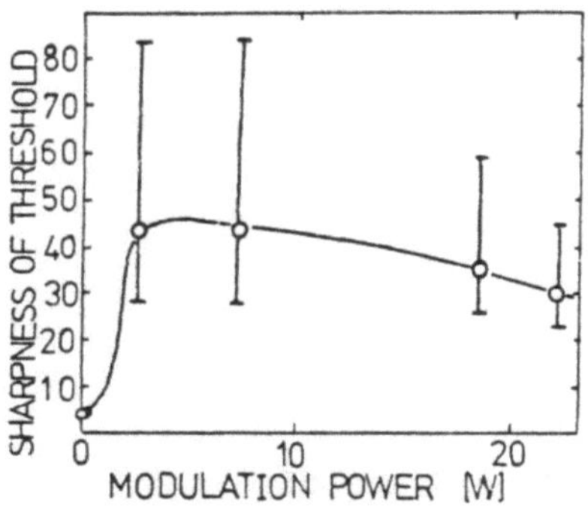

Fig.1 Sharpness of threshold vs modulation power

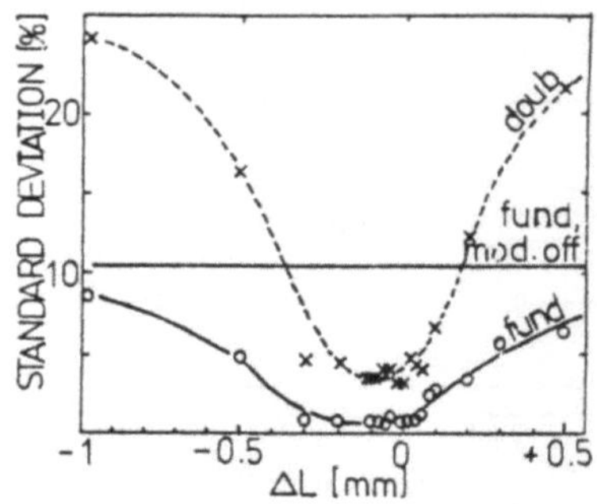

Fig.2 stability vs resonator-length mismatch

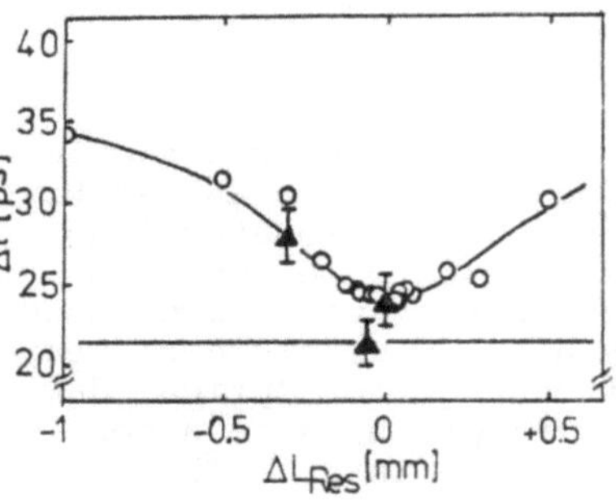

Fig.3 Pulsewidth vs resonator-length mismatch

4.3 Pulse-Length

The influence of the active modulation on the build-up process of the mode-locked pulse train can cause a lengthening of the pulses. This effect is minimized when the resonator length is properly adjusted (see fig.3). The data points with error bars were directly obtained with the delayed pulse method as described above, while the circles are indirect data points obtained by comparison of non-saturation second harmonic conversion efficiencies.

4.4 Distance Between Modulator and Mirror

In order to see the influence of the distance between output mirror and modulator, this distance was varied and the parameters (stability, pulse-length and output energy) were measured (see fig.4). The output fluctuations slightly increase with increasing distance, the pulse-length remains constant, and the output energy remains constant up to a distance of 100mm, then decreases due to the higher resonator loss. These results confirm the theoretical predictions which say that for this type of laser the position of the active modulator is not critical as long as it is reasonably close to one of the mirrors.

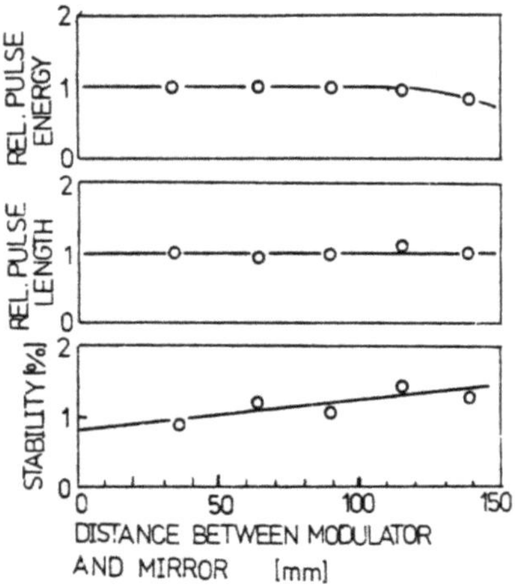

Fig.4 Dependence of output energy, pulse-length and stability on the position of the active modulator inside the laser resonator

5. Conclusion

It has been shown that acousto-optic Q-modulation is a simple and very effective way of improving the performance and stability of mode-locked pulsed Nd:YAG lasers. The resonator length or the modulator frequency respectively have to be kept constant to within $\pm 10^{-4}$. The threshold behavior was improved so that stable mode-locking operation is already possible at very low dye concentrations meaning that the danger of optical damage is now almost completely eliminated. This results in a consierable improvement of the reliability of the laser. The increase of the pulse-length with the acousto-optic modulator was found to be only 10% with pulses of ~25 ps still possible. At a repetition rate of 10Hz the pulse to pulse stability of the full train output energy was ±1%.

Acknowledgement

The author wants to thank J.M. Heritier of E.L. Ginzton Lab, Stanford University for his active cooperation in the early stages of these experiments and for helpful discussions.

References

1 B.C. Johnson, W.D. Fountain; Int. Electron. Device Meeting (Dec 1974) Tech. Digest 322.

2 S. Kishida, T. Yamane; Opt. Comm. 18, 19, 1976.

3 W. Seka, J. Bunkenburg; J. Appl. Phys. 49, 4, 1978.

4 B.B. Craig, W.L. Faust, L.S. Goldberg, P.E. Schoen, R.G. Weiss; Proceedings of the Second International Conference on Picosecond Phenomena; 1980, Cape Cod, MA.,(R. Hochstrasser, W. Kaiser, C.V. Shank ed), pp. 253-258, Springer-Verl. Berlin, Heidelberg, New York, 1980.

5 H.P. Weber; J. Appl. Phys. 38, 2231, 1967.

6 J.A. Armstrong; Appl. Phys. Lett. 10, 16, 1967.

Active Mode Stabilization of Synchronously Pumped Dye Lasers

A.I. Ferguson and R.A. Taylor

Clarendon Laboratory, Parks Road,
Oxford OX1 3PU, United Kingdom

1. Introduction

Synchronously pumped mode-locked dye lasers have proved to be a reliable way of producing ultrashort light pulses throughout the visible region of the spectrum. Dramatic improvements in pulse duration have resulted from the use of rf generators of high spectral purity and very fine control of the dye laser cavity length with respect to the pump laser. However, synchronously pumped dye lasers have not yet reached the extremely short pulse duration typical of passive systems and appear to suffer from more timing jitter than their passive counterparts. We have explored methods of improving the properties of synchronously pumped dye lasers which are common in single mode dye lasers [1]. We expect such stabilization schemes to give rise to shorter, cleaner and more coherent pulse trains with less timing jitter. Many other stabilization methods have been proposed and demonstrated [2, 3, 4, 5]. These schemes attempt either to stabilize the ion pump laser or to stabilize the dye laser. It is clearly crucial that the pump laser must first be stabilized before the dye laser can be operated at its best and indeed most of the efforts have been directed at improving the pump laser. In this paper we propose and demonstrate a stabilization scheme which is applied to the dye laser only. The scheme can be applied to any mode-locked laser and has particular utility in experiments where the phase coherence of the pulse train is important. We have demonstrated the use of a stabilized dye laser in performing coherent multiple pulse Doppler-free spectroscopy. This source should also be ideal for coherent transient experiments using multiple pulse interactions [6].

2. Principle of the Method

The time averaged spectrum of a train of picosecond light pulses such as is produced by a synchronously pumped dye laser consists of a comb of modes each separated by the laser repetition rate and spanning the band width of an individual pulse in the train. The mode-locking process ensures that adjacent modes are rigidly locked together. However the absolute optical frequency of the modes is free to drift and will be determined by the dye laser cavity length. Thus a perturbation in the dye laser cavity length does not appreciably perturb the mode spacing but rather moves the whole mode profile. If it is possible to stabilize one of the laser modes then all other modes will also be stabilized through the mode-locking process. A laser stabilized in this way will certainly have improved temporal coherence but we also believe that such stabilization will give rise to cleaner and shorter pulses with less timing jitter.

Two methods have been proposed to accomplish frequency stabilization of mode-locked lasers [2]. One of these methods is to use a series of interferometers to filter out a single mode and then actively stabilize this mode to a reference interferometer. The second method uses a heterodyne signal between one of the laser modes and a stabilized single mode laser. The beat frequency, which will be at less than half the laser mode spacing, can then be used to stabilize the mode-locked train. The method which we demonstrate here is closely related to the first of the above proposals except that we have been able to lock the laser using only one reference interferometer.

The main problem in the use of a single interferometer as a stabilizer is that it should be capable of resolving the laser modes, separated by about 100 MHz and which span a region of about 500 GHz. An interferometer used in the conventional way would have to have a finesse of at least 5000 to be of any use in such a scheme. However, since the laser is mode-locked, adjacent modes are equally spaced. Provided the interferometer can be adjusted such that its free spectral range is an integer multiple of the mode spacing the interferometer will be comensurate with the mode spacing and a considerable simplification in the transmission of the interferometer will be observed. This requirement on the interferometer is not very restrictive as the maximum change in free spectral range that could be called for is half a mode spacing.

We have chosen to demonstrate the stabilization method using a confocal interferometer. The critical adjustment in such an interferometer is that the separation between the mirrors should be accurately matched to their common radius of curvature. The maximum value, $|\delta\ell|$, of the departure from the exact confocal condition which can be tolerated without allowing the TEM_{mn} mode of the interferometer to resonate at an observably different mirror spacing from that at which the TEM_{oo} mode is resonant is given by [7]

$$|\delta\ell| = \pi R/2\ (1 + m + n)F, \tag{1}$$

where R is the radius of curvature of the mirrors and F is the finesse of the interferometer. The change in length $\delta\ell$ which will bring about a change in free spectral range of at most f/2 is given by $\delta\ell = 2\ell^2 f/c$, where ℓ is the separation between the mirrors of the interferometer and f is the laser repetition rate. Setting this change in length equal to the maximum allowed by (1) we see that provided ℓ is chosen to satisfy the inequality

$$\ell < \pi c/4fF\ (1 + m + n) \tag{2}$$

it will be possible to adjust the free spectral range to the interferometer to make it comensurate with the laser mode spacing without destroying the confocal condition. The interferometer will then resolve the laser modes despite the band width of an individual pulse in the train being very much more than the interferometer free spectral range.

3. Experiment

We have demonstrated the stabilization scheme using the apparatus represented schematically in fig. 1. The pump laser was an argon ion laser operated at 228, 759, 468 Hz which corresponded to these being three pulses within the ion laser cavity at any instant [8]. The dye laser was operated with rhodamine 6 G and was a modified version of a commercially available model (CR-599) in which the fold mirror was replaced by a small mirror mounted on a high frequency piezoceramic for fast cavity length corrections. Slow corrections were accomplished using a plate mounted at Brewster's angle on a scanning galvanometer. The laser pulse duration was 1.5 ps which corresponded to a line width of 300 GHz.

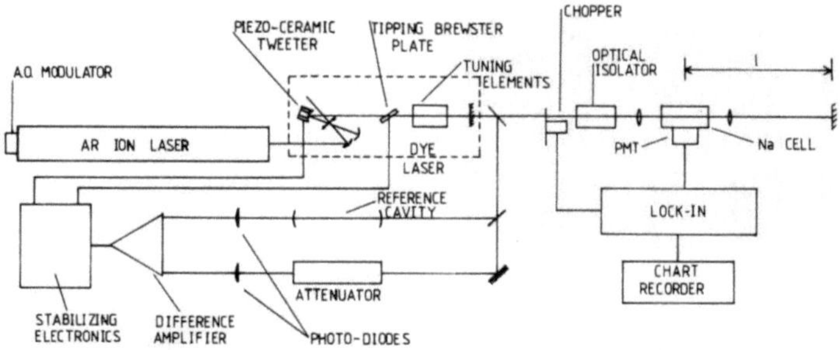

Fig. 1 Schematic diagram of apparatus used for active mode stabilization and coherent multiple pulse spectroscopy of sodium

The output from the dye laser was directed into a confocal interferometer. (Tec Optics SA 7.5) which had a finesse of 100 and a free spectral range of 7.5 GHz. Inserting these numbers into (2) it can be seen that this interferometer is capable of resolving the laser modes and of being adjusted to be comensurate with the mode spacing of 228 MHz provided that high order transverse modes of the cavity are not excited. This ties in with our observation that the laser modes are easily observed but that the alignment is considerably more critical than when a single mode laser is used. It has also been possible to mode match into the confocal interferometer.

A frequency error signal was obtained by comparing the transmission through the interferometer with an attenuated portion of the laser beam on two similar photodiodes. The error signal was applied to a commercially available electronic servo system of the kind usually used with single mode lasers (CR 699 - 21). Figure 2(a) shows an oscilloscope trace of the difference signal as the reference interferometer was scanned through a small portion of its free spectral range. Each peak corresponds to a superposition of laser modes and is separated from its neighbour by 228, 749, 468 Hz. In this particular example the interferometer was mode matched. The error signal obtained when the interferometer was no longer scanned and the servo system was locked is shown in fig.2(b). This is 50 times more sensitive than the trace in fig.2(a). From this figure it is possible to estimate the laser frequency jitter relative to the reference cavity. We estimate an rms jitter of less than $\pm$ 350 kHz in a dc to 10 kHz band width. This corresponds to approximately 25 MHz when the laser was not locked.

The improvement in stability of the dye laser was demonstrated by using it in a high resolution Doppler-free experiment on the 3S to 4D transition in sodium. The experimental technique has been described previously [9]. The arrangement is shown in figure 1. The pulse train was focused into a sodium cell and retroflected with a delay adjusted such that each pulse met its predecessor at the focal region in the cell. A Doppler-free signal was observed in this standing wave region by monitoring the fluorescence from the 4D level.

The levels involved in the 3S to 4D transition are shown in fig.3(a). The four allowed transitions are also shown together with their expected relative strengths. Figure 3(b) shows the 4D fluoresence as the laser modes were electronically scanned. In multiple pulse spectroscopy each transition gives rise to a comb of resonances and so the complete spectrum is obtained by scanning through half a mode spacing. The resonances are labelled according to the letters used in fig.3(a). From this data we have estimated the hyperfine splitting in the ground state to be 1771.0 $\pm$ 1.1 MHz and the fine structure splitting in the 4D state to be 1028.1 $\pm$ 1.0 MHz.

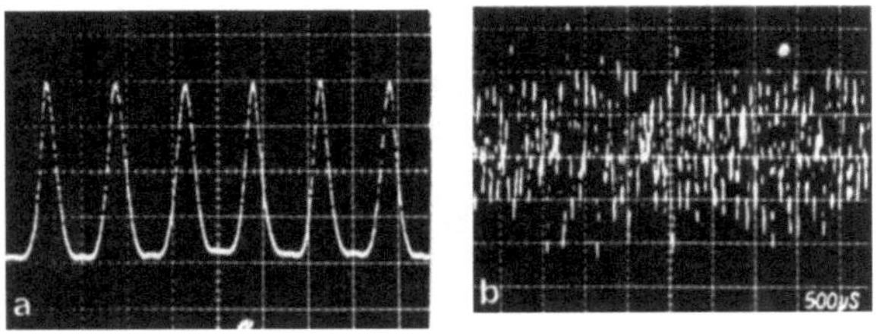

Fig.2(a) Difference signals as the reference cavity was scanned, the interferometer being mode-matched (228 MHz between peaks), 1V/div
(b) The difference signal when the laser was locked to the resonator

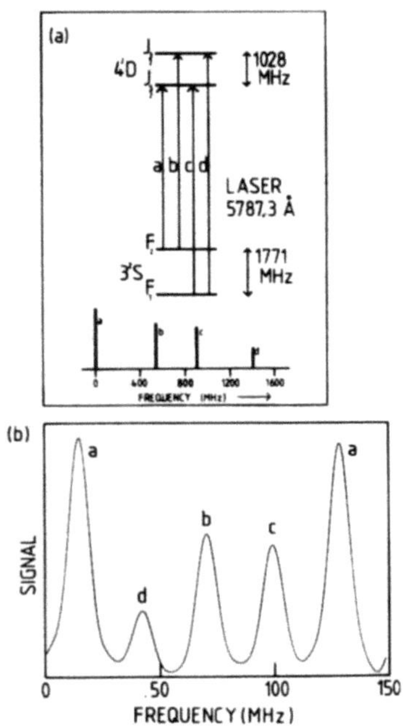

Fig.3(a) Energy levels involved in the 3S to 4D transition in sodium, together with expected single-mode laser spectrum
(b) Multiple pulse spectrum of the above transition using 6 ps pulses versus laser frequency change. Integration time was 1s and the linewidths are 10 MHz FWHM

4. Conclusions

We have demonstrated that the mode spectrum of a synchronously pumped dye laser can be readily locked using variations on well established technology. The reduction in frequency jitter corresponds to an improvement in the temporal coherence of the laser by almost two orders of magnitude. The method can be easily applied to other types of laser system. One of the main reasons for our investigation of this technique has been to establish the influence of the laser frequency jitter on the pulse duration, cleanliness and timing jitter of synchronously pumped lasers. However, these properties are at present limited by the quality of pulses from the pump laser. We believe that if we can reduce the noise of our pump laser then the active frequency stabilization methods described above will lead to shorter and clearer pulses from synchronously pumped dye lasers.

Acknowledgements

We gratefully acknowledge the financial support of SERC in the form of an Advanced Fellowship (AIF) and Studentship (RAT).

References

1. A.I. Ferguson and R.A. Taylor, Opt. Comm. 41, 271 (1982).
2. A.I. Ferguson, J.N. Eckstein and T.W. Hänsch, J. Appl. Phys. 49, 5389 (1978).
3. B. Couilland, A. Ducasse, L. Sarger and D. Bosher, Appl. Phys. Lett. 36, 1 (1980).
4. S.R. Rotman, C.B. Roxlo, D. Bebelaar, T.K. Yee and M.M. Salour, *Picosecond Phenomena* II, eds. R.M. Hochstrasser, W. Kaiser and C.V. Shank, Chemical Physics Vol.14 (Springer-Verlag, Berlin, Heidelber, New York, (1980.)
5. H. Klann, J. Kuhl and D. von der Linde, Opt. Comm. 38, 390 (1981).
6. W.H. Hesselink and D.A. Wiersma, Phys. Rev. Lett. 43, 1991 (1979).
7. M. Hercher, Appl, Opt. 7, 951 (1968).
8. A.I. Ferguson, Opt. Comm. 38, 387 (1981).
9. J.N. Eckstein, A.I. Ferguson and T.W. Hänsch, Phys. Rev. Lett. 40, 847 (1978).

Spectral Hole Burning in the Saturation Region of Mode-Locked Nd-Glass Lasers

A. Penzkofer and N. Weinhardt

Naturwissenschaftliche Fakultät II-Physik, Universität Regensburg, D-8400 Regensburg, Fed. Rep. of Germany

1. Introduction

Mode-locked Nd-glass lasers generate bandwidth limited picosecond light pulses only in the early part of the pulse train [1-3]. Towards the maximum of the pulse train the spectra are broadened and irregular spectra are observed beyond it [1-5]. Self-phase modulation causes the spectral broadening and structuring. The spectral broadening might be enhanced by self-focusing [2] and nonlinear dispersion [6].

Here we studied the spectral development of picosecond light pulses along pulse trains of passively mode-locked Nd-glass lasers. The experimental spectra beyond the pulse train maximum may be explained by taking into account spectral hole burning in the inhomogeneous gain profile of the active medium in addition to self-phase modulation.

2. Experiments

The investigated laser oscillator is depicted in Fig. 1. Different types of Nd-glass rods were used. A silicate (Schott LG 630), a phosphate (Schott LG 703) and a fluorophosphate (Schott LG 802) Nd-glass rod were used as active media AM (doping 3 weight-% Nd_2O_3). The saturable absorber cell SA is contacted to a flat 100 % mirror M (dye Eastman 9860, thickness 0.1 mm). The output mirror M2 has 30 % reflectivity and a curvature of 3 m.

Single picosecond light pulses are selected from the pulse train at varying positions with an electro-optic shutter. The spectra are analysed with a spectrograph and an optical spectrum analyser. The pulse durations are determined by the two-photon fluorescence technique. Values of $\Delta t = 6 \pm 2$ ps (FWHM) are obtained. They practically do not change with switching position. The intensity of the light pulses is determined by a

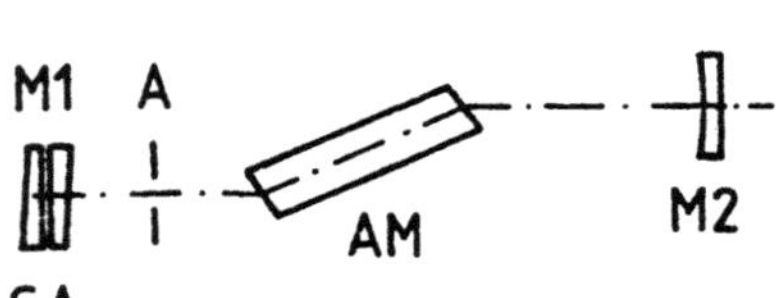

Fig.1 Laser oscillator

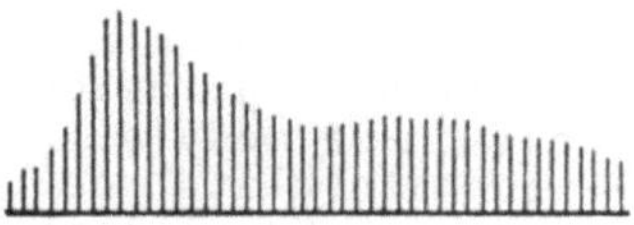

Fig.2 Typical pulse train

saturable absorber technique. The maximum peak intensity at the Nd-glass rod inside the oscillator is $I_{OP} \simeq 7\times10^9$ W/cm^2 (single pass saturable dye transmission T = 0.85).

3. Results

The spectra of the selected pulses change along the pulse train. In the early part of the train (pulse heights less 0.2 maximum pulse height) nearly bandwidth limited pulses are generated. In the following region up to the pulse train maximum the half-width of the spectra increases, wings become detectable and modulate the spectrum (self-phase modulation). A small spectral shift to higher frequencies is observable (temporal asymmetry of pulse shape). Beyond the maximum a new small spectral peak builds up preferably at the high frequency side about 20 cm^{-1} away from the central frequency (spectral hole burning in active medium). Sometimes two intense peaks at both sides of the central frequency are observed upon a broad pedestal. Frequently the pulse trains form a second temporal maximum. The spectra around this second maximum often show a small spectral peak upon a broad pedestal at the lower frequency side.

Fig. 2 depicts a typical pulse train. In Fig. 3 pulse spectra at the first pulse maximum (a), the trailing part (b) and

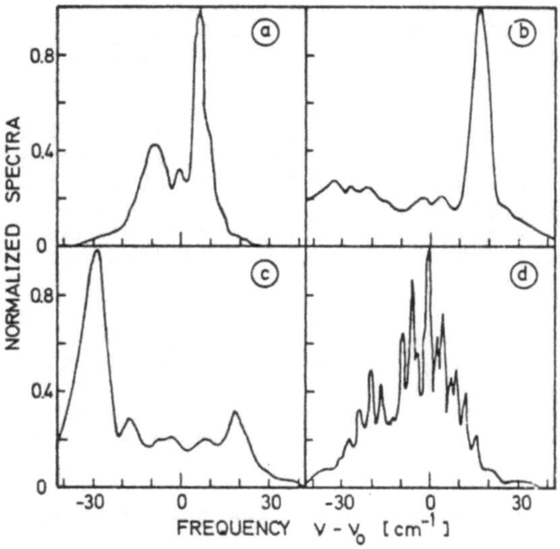

Fig.3 Single picosecond pulse spectra with Schott LG 703 laser glass
a) at pulse train maximum, b) at trailing part, c) behind second maximum, d) at pulse train maximum with 2 cm CS_2 in oscillator.

the second maximum (c) are depicted. Spectrum (d) was obtained at the first train maximum with a 2 cm long cell of CS_2 inside the oscillator to increase the effects of self-phase modulation. The depicted spectra belong to a Nd-phosphate glass rod (Schott LG 703). Similar spectra were obtained with Nd-silicate and Nd-fluorophosphate glasses.

4. Discussion

The spectral development along the pulse train may be explained by taking into account spectral hole burning in the inhomogeneously broadened gain profile of the active medium [7-9].

In the linear build-up region the laser spectrum is narrowed by spectral mode selection. Light amplification occurs only in a narrow spectral range at the maximum of the gain profile since the laser is operated only slightly above threshold. The statistically fluctuating spontaneous emission is amplified in the linear phase.

The nonlinear region begins when the strongest fluctuation spike is intense enough to bleach the saturable absorber. This spike is preferably amplified due to its reduced losses. During the opening time of the absorber the laser loss is reduced for all frequency components of the emission and effective amplification takes place over a wide frequency range. Spontaneous emission at the spectral wings is amplified. As long as saturation effects do not occur the gain is highest at the center frequency. Towards the pulse train maximum the central frequency components of the generated emission deplete the inversion of the inhomogeneously broadened laser medium (inhomogeneous width $\approx 200\ cm^{-1}$, homogeneous width $\approx 15\ cm^{-1}$ [8], cross-relaxation time $\approx 70\ \mu s$ [9]) and a hole is burned in the center of the gain profile. The wings of the emission are more strongly amplified than the central part.

Towards the maximum of the pulse train the intensity of the pulses becomes large and self-phase modulation broadens, modulates and slightly shifts the spectra [3-5,10]. The observed small spectral shift of the peak frequency to higher values indicates an asymmetric pulse shape with steeper rising than decaying parts [10]. The peak pulse intensity is limited by two-photon absorption of the Nd^{3+} ions [11].

For the trailing part of the pulse train optimum gain acts in the wings of the emission during the opening period of the absorber. The gain at the central frequency is below laser threshold (net gain <1) beyond the first train maximum. The spectral wings are built-up and emission maxima occur at shorter and longer wavelengths.

Computer simulations verify the described spectral development.

References

1. D. von der Linde, IEEE J. Quant. Electron. QE-8, 328 (1972).
2. R.C. Eckard, C.H. Lee, and J.N. Bradford, Opto-Electron. 6, 67 (1974).
3. W. Zinth, A. Laubereau, and W. Kaiser, Opt. Commun. 22, 161 (1977).
4. M.A. Duguay, J.W. Hansen, and S.L. Shapiro, IEEE J. Quant. Electron. QE-6, 725 (1970).
5. V.A. Korobkin, A.A. Malutin, and A.M. Prikhorov, Sov. Phys. JETP Lett. 12, 150 (1970).
6. R.R. Cubeddu and O. Svelto, IEEE J. Quant. Electron. QE-5, 495 (1969).
7. W.H. Keene and J.A. Weiss, Appl. Opt. 3, 545 (1964).
8. V.I. Nikitin, M.S. Soskin, and A.I. Khizhnyak, Sov. Techn. Phys. Lett. 3, 5 (1977).
9. A.A. Mak, D.S. Prilezhaev, V.A. Serebryakov and A.D. Starikov, Opt. Spectrosc. 33, 689 (1972).
10. S.A. Akhmanov, R.V. Khokhlov, and A.P. Sukhorukov, in Laser Handbook, ed. by F.T. Arrecchi and E.O. Schultz-Dubois (North-Holland, Amsterdam, 1972).
11. A. Penzkofer and W. Kaiser, Appl. Phys. Lett. 21, 427 (1972).

Single and Double Mode-Locked Ring Dye Lasers; Theory and Experiment

K.K. Li*, G. Arjavalingam*, Andrew Dienes**, and J.R. Whinnery*

* Electronic Research Laboratory, U.C. Berkeley, Berkeley, CA 94720, USA

**Department of Electrical and Computer Engineering, U.C. Davis, Davis 95616, USA

1. Introduction

A passively modelocked ring dye laser has been shown[1] to produce optical pulses in the femtosecond regime. We have undertaken experimental and theoretical investigations of this configuration with both single and double modelocked[2] operation. In the set up shown in Fig.1, Rh6G was used as the gain dye, with DODCI or cresyl violet as the saturable absorber in the single and double modelocked lasers respectively. The Rh6G was pumped by a cw argon laser. At certain positions of the dye jets and critical tuning (615nm), narrow modelocked pulses were observed. At other tunings with pulse widths of the order of a fraction of the roundtrip time are formed. These pulses are very stable with respect to pump power. The exact behavior changes with the relative positions of the gain and absorber dyes. A computer program was written to analyze the system and provide a better understanding of the pulse formation.

2. Theory of single and double modelocked lasers

The dyes are each modeled as two-level systems whose dynamics are governed by the following rate equations[3]:

$$\left(\frac{\partial}{\partial x} + \frac{1}{c}\frac{\partial}{\partial t}\right) I = (\sigma_e n_1 - \sigma_a n_0) I \qquad (1)$$

$$\left(\frac{\partial}{\partial x} + \frac{1}{c}\frac{\partial}{\partial t}\right) I_p = - \sigma_p n_0 I_p \qquad (2)$$

$$\left(\frac{\partial}{\partial t} + \frac{1}{T_1}\right) n_1 = \sigma_p n_0 I_p - (\sigma_e n_1 - \sigma_a n_o) I \qquad (3)$$

$$n_0 + n_1 = n_T \qquad (4)$$

where I is the laser photon flux and I_p is the pump photon flux in photon/cm^{2}sec; n_0 is the ground-state population density, n_1 is the excited state population density, n_T is the total density of molecules in molecules/cm^3. σ_e, σ_a, and σ_p are the emission, absorption, and pump absorption cross-sections respectively. T_1 is the relaxation time of the dye.

These four equations are applied to the gain and absorber separately with appropriate conditions for single and double

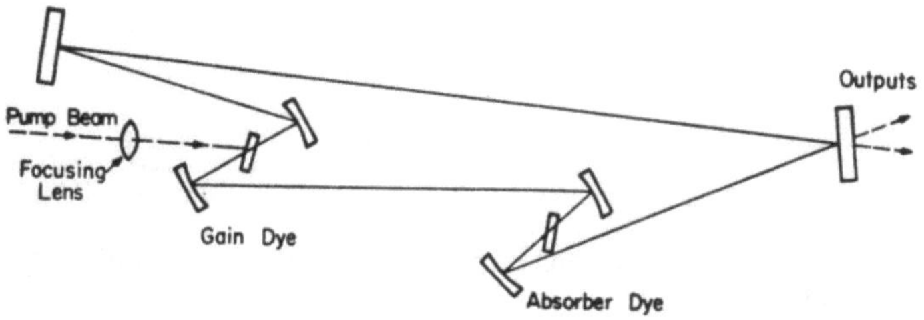

Fig.1 The ring dye laser

modelocked cases. For example, for cresyl violet in the double modelocked case, I is the cresyl violet flux and I_p is the Rh6G flux. Dispersion is introduced into the system in the form of a digital filter. For initial conditions, we used both random noise ensembles and low level cw radiations with the same end result. The computation converges to short optical pulses for some parameters and to long pulses (as long as half a cavity roundtrip time) for others.

3. Numerical results of single modelocked laser

For a single modelocked laser, there is a cw pump for the Rh6G; lasing intensity I is omitted for the absorber dye DODCI.

3.1 Dependence on initial conditions

Fig.2 shows the evolution of laser pulses inside the cavity with both cw and noise spike initial conditions for a particular set of parameters which result in long square pulses. We see in Fig.2a that the pulse slowly evolves from a continuous wave to square pulses. In Fig.2b, starting from a noise input, the evolution is quite different. The noise spikes persist throughout the first fifty roundtrips, but the envelope of the noise spikes begins to indicate the general form of the steady-state solution as early as the 30th roundtrip. The

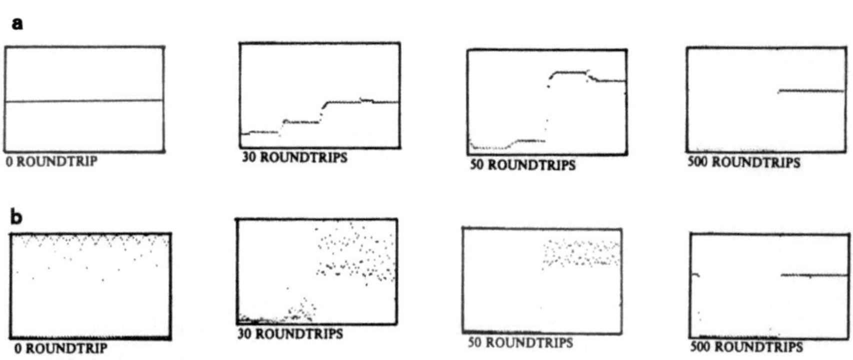

Fig.2 Evolution in the long pulse region with low level cw and noise initial conditions. In (b), the dots indicates the upper and lower boundaries of the noise spikes.

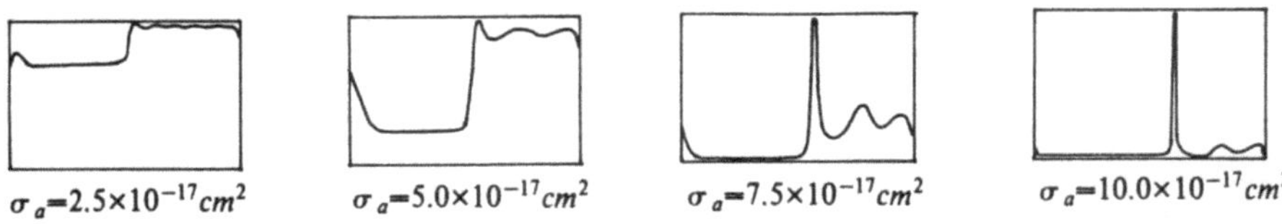

Fig.3 Output of laser at 40th roundtrip with different values of absorber cross-section

final steady-state solution at the 500th roundtrip is the same as in Fig.2a. So the solution converges to the same value for these two quite different initial conditions. This fact gives us some confidence in the validity of our model.

3.2 Dependence on the absorber cross-section and focusing

Fig.3 shows the pulse shapes of the laser at the 40th roundtrip for different values of absorber cross-section. The steady-state solution has not yet been reached but the final solutions are already indicated. For (a), the pulses will evolve into square shapes as in the case shown in Fig.2. For (d), the pulses narrow rapidly and have already reached the resolution of our computations. With finer resolutions a steady-state picosecond pulse will result. Quite apart from the question of resolution, it is not clear that the physical model used here is complete enough to explain the shortest pulses obtained with the ring laser. In particular, the transient grating effect proposed by Fork et al.[1] is not included. However, our model does show clearly the crucial mechanisms for short pulse formation in the early pulse evolution stage. We see that a small absorber cross-section favors long pulse formation and a large cross-section (or stronger focusing) favors short pulse formation due to changes in the effective saturation energy.

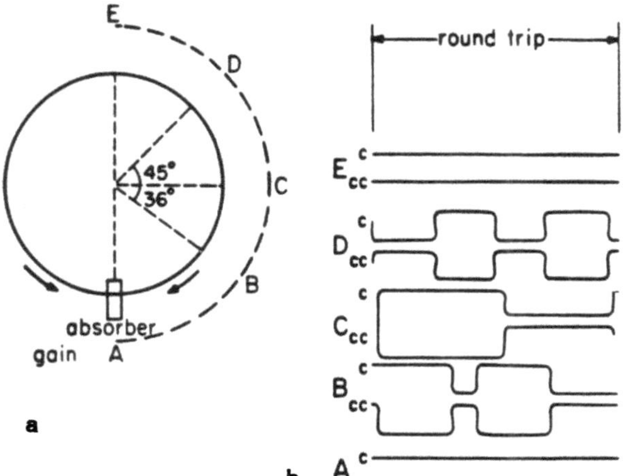

Fig.4

a. Position of the absorber with respect to the gain dye

b. Numerically calculated output of the laser. c: clockwise cc: counter-clockwise

3.3 Dependence on relative positions of gain and absorber dyes

In the long pulse situations, the calculations show a strong dependence on the positions of the two dyes. Fig.4a shows the relative positions and Fig.4b shows the calculated output of the laser for both the clockwise and counter-clockwise directions. The origin in Fig.4b is always set at the gain dye's edge. The anti-symmetry of the two trains of pulses suggests gain sharing in the gain medium with the laser operating near the linear region. This represents a saturation too weak to favor short pulse evolution.

4. Experiments with single modelocked laser

The experimental set up is shown in Fig.1. Fig.5 shows the outputs as measured with a photodiode-oscilloscope combination with a resolution of about 1 ns. In (a), the laser is tuned to about 580 nm, and the pulse widths are half roundtrip time of the cavity. In (b), with tuning at 615 nm, the pulses are resolution limited. This is consistent with the calculation because the cross-section of DODCI is larger at 615 nm than at 580 nm[3] due to the formation of photoisomers.

Different relative positions of the dyes were also tried. When the dyes were mixed or placed directly opposite, no pulses were oberved. When the absorber was one quarter roundtrip away from the gain, pulse widths equal to half the roundtrip time were observed as shown in Fig.5a. These results are again in agreement with the numerical calculation.

5. Double modelocked lasers - theory and experiment

Calculations were also made to model the double modelocked laser[3], in which the absorber also lases. Both long- and short-pulse behaviors are again predicted. In the short-pulse case, the pulses from the cresyl violet dye seem to form faster than the pulse from the Rh6G. For experimental investigations, cresyl violet was substituted for DODCI. Trains of yellow and red pulses were observed.

Preliminary correlation measurements show Rh6G pulses about 1 ps duration with improved stability over the linear cavity configuration[2].

This research was supported in part by NSF grants ECS81-14526 and ECS79-21177.

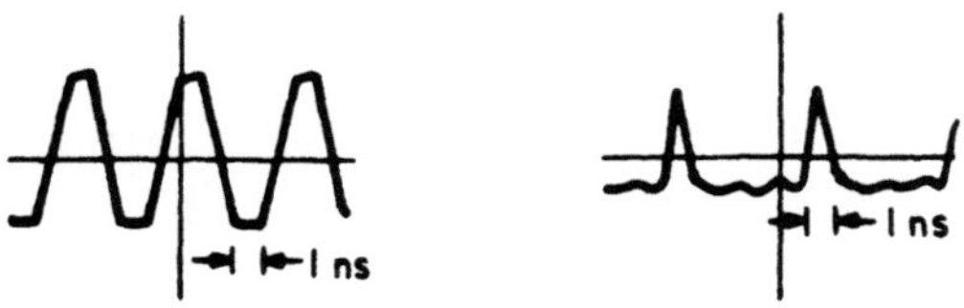

a b

Fig. 5

a. square pulses from the ring dye laser with the absorber placed one quarter round trip from the gain dye

b. Narrow pulses from ring laser tuned to about 615nm

References:

1. R.L. Fork, B.I. Greene, and C.V. Shank, Appl. Phys. Lett. 38(9), 671, 1981
2. Z.A. Yasa, A. Dienes, and J.R. Whinnery, Appl. Phys. Lett. 30,24,1977
3. D.J. Bradley, "Ultra-short light pulses", edited by S.L. Shapiro, Topics in Applied Physics. (Springer New York 1977) vol.18

Theoretical and Experimental Investigations of Colliding Pulse Mode-Locking (CPM)

W. Dietel, D. Kühlke, W. Rudolph, and B. Wilhelmi
Sektion Physik der Friedrich-Schiller-Universität Jena
DDR-6900 Jena, German Democratic Republic

1. Introduction

Due to its large bandwidth, the cw dye laser is very suitable for generating subpicosecond optical pulses. The shortest pulses obtained to date have been produced by the method of "colliding pulse mode-locking" (CPM) recently reported by FORK et al. [1]. This method is based on the coherent interaction of two counter-propagating pulses in the adsorber dye used for the passive mode-locking of a cw dye laser. The counter-propagating pulses form a transient grating in the population of the adsorber that scatters a portion of the pulse travelling in one direction into the other pulse. Because the scattered and original waves are in phase, constructive interferences occur. A regime of pulses counter-running in the mode-locking absorber can be achieved either by an absorber optically contacted with an end mirror, by a Fabry-Perot laser where m pulses oscillate and the absorber is placed at a submultiple m of the cavity length, or by a ring laser with two counter-propagating pulses [1]. The essential condition for CPM is a sufficiently thin saturable absorber; i.e., the optical path in the absorber must be of the order of (or less than) the pulse width.

2. Theoretical Results

Our theoretical treatment of the action of a thin saturable absorber as mode-locker of a cw ring laser within the framework of the approximations made in [2,3] gives the following results: When the spacing between the absorbing and the amplifying media is a quarter of the perimeter of the ring resonator, two counter-running pulses of a $sech^2$ shape with equal durations and energies occur which meet inside the absorber (CPM regime). In comparison to a cw dye laser mode-locked by an absorber which is long compared to the pulse width, the stability region of the CPM regime is more favourable, and assuming equal pulse energies, the pulse duration is expected to be shorter by a factor of about 3 (Figs.1,2). From Fig.1 it is evident that

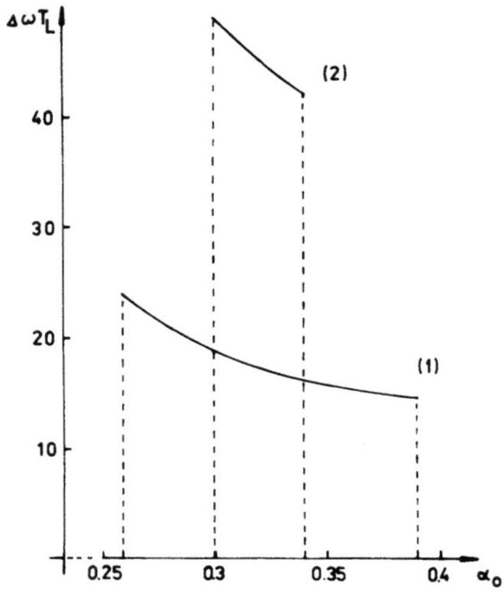

Fig.1. Pulse duration T_L in dependence on the small signal gain α_0 for the CPM regime (curve 1) and for mode-locking with a long absorber (curve 2). The small signal absorption ($\varkappa_0$) has been chosen to be 0.2 for the two cases. (The pulse duration is normalized to $\Delta\omega^{-1}$ where $\Delta\omega$ is the spectral width of any bandwidth-limiting element in the cavity)

for a bandwidth of about 70 nm, typical for dye lasers, a pulse duration ranging from 40 to 60 fs should be expected. (The bandwidth of the cavity was assumed to be determined by the bandwidth of the gain line.)

If the spacing between the absorber and the amplifier deviates from one fourth of the perimeter, the $sech^2$ shape remains only approximately valid. The pulse for which the amplifier is more recovered gets the greater stationary energy, and its duration is slightly prolonged. The stability range of the CPM regime decreases with increasing deviations (Fig.2). Above the upper boundary of the CPM regime, the ring laser is expected to work in a unidirectional regime or in a regime where the counter-travelling pulses do not meet in the absorber. The parameters of those pulses correspond to the case of mode-locking by a long absorber.

3. Experimental Results

The experimental investigations involved a cw-pumped rhodamine 6G laser mode-locked by use of DODCI. In order to determine the shortening of the pulses by the transient absorber grating (CPM), a passively mode-locked dye laser with Fabry-Perot resonator was optimized with respect to shortest pulse duration. To avoid CPM, the distance between the DODCI jet and the end mirror was chosen to be 3 cm. The shortest pulses were $sech^2$ shaped with 0.42 ps FWHM.

For the CPM we used a ring resonator with a DODCI jet of 30-50 μm thickness at a distance of approximately one quarter of the ring perimeter from

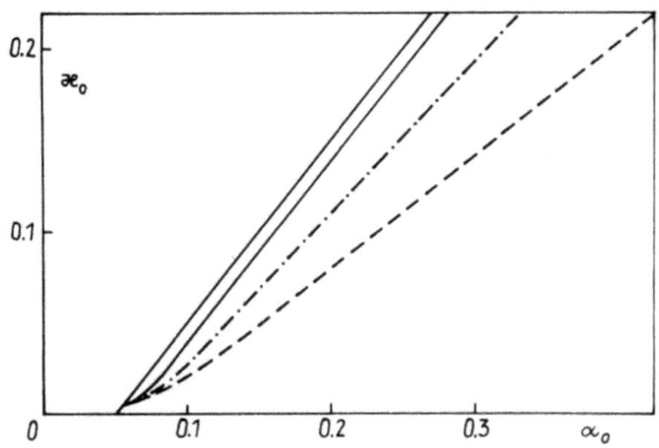

Fig.2. Stability range of the CPM regime in the $(\alpha_0, \varkappa_0)$ plane (α_0 is the small signal gain, $\varkappa_0$ the small signal absorption). The straight line indicates the laser threshold. The solid line denotes the lower boundary of the stability range, the dashed line the upper boundary for the case of counter-running pulses amplified in the same manner (the spacing between the two media is one quarter of the perimeter u), and the dot-dashed line is the upper limit for the case when the spacing is 0.21 u

the rhodamine jet. The wavelength and bandwidth of laser oscillation can be controlled by a prism. Sech^2shaped pulses of 0.1 ps were generated.

In both cases similar conditions were chosen: pump power 0.9-1.2 W at 514 nm; 1.3% output mirror; 3-5 mW output; identical spherical mirrors; round trip time 10 ns and 13 ns, respectively; distance between the two jets 0.7 m and 1 m, respectively.

Obviously the action of the transient absorber grating shortens the generated pulses by a factor of about 4. This is somewhat better than expected from the theoretical results. The absolute theoretical values for the minimum pulse duration are in good agreement with new experimental results reported in [4]. However, it should be mentioned that the calculated pulse durations sensitively depend on some parameters which are not known. Besides the pump and loss parameters this concerns in particular the ratio of the saturation energies of the two media. By adjustment of the prism the generation maximum can be shifted within narrow limits. The pulses have been found to have a down chirp, which was observed via the shortening of the generated pulse passing a medium with normal dispersion. The influence of non-linear optical effects on the pulse length can be ruled out. In Fig.3 the FWHM τ_{out} of the autocorrelation function of the pulses having passed a 17-cm glass block is plotted versus the autocorrelation width τ_{in} before entering. Pulses of τ_{in} = 0.4 ps, for example, are shortened to 0.26 ps; for pulses of 0.24 ps the dispersion of the glass block compensates the

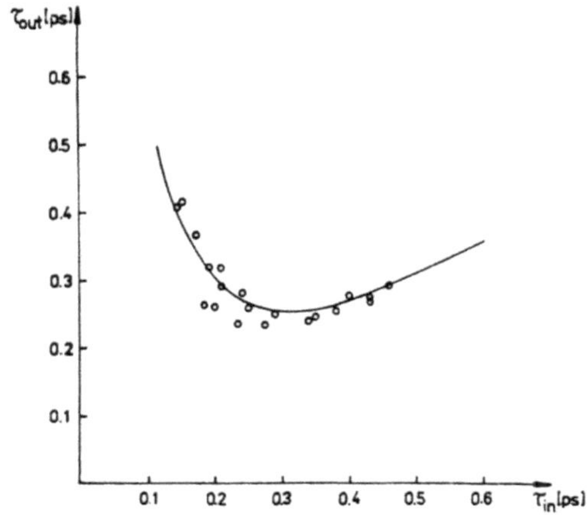

Fig.3. The FWHM of autocorrelation of the pulses generated in a passively mode-locked dye laser after passing a BK 5 glass block of 17 cm length (τ_{out}) vs the autocorrelation width before entering (τ_{in}). The solid line shows the theoretical dependence of the FWHM of the autocorrelation of down-chirped Gaussian pulses fitted to the experimental data

down chirp and then broadens the pulses to the initial width; and pulses with $\tau_{in} < 0.24$ ps are broadened.

Fitting the theoretical dependence of the FWHM of the autocorrelation of down-chirped Gaussian pulses to the experimental data (Fig.3) gives a wavelength sweep over the pulse duration of about 3 nm for 0.1-ps pulses. This should also be a good estimate for the sech^2-shaped pulses measured. These experimental results show that (1) the chirp should be taken into account in the theory, (2) for the generation of very short pulses the chirp should be avoided or the pulses should be compressed, and (3) even not very long pathways in glass may affect the pulse-width measurement.

References

1 R.L. Fork, B.I. Greene, C.V. Shank: Appl. Phys. Lett. *38*, 671 (1981)
2 J. Herrman, F. Weidner, B. Wilhelmi: Appl. Phys. B*26*, 197 (1981)
3 H.A. Haus: IEEE J. Quantum Electron. *QE-11*, 736 (1975)
4 C.V. Shank, R.L. Fork, R.T. Yen: This volume

Picosecond Carrier Dynamics and Laser Action in Optically Pumped Buried Heterostructure Lasers

T.L. Koch, L.C. Chiu, Ch. Harder, and A. Yariv

California Institute of Technology, Pasadena, CA 91125, USA

Over the past several years there has been a continued interest in the generation of picosecond pulses from semiconductor diode lasers. We report here some preliminary results from an experiment to investigate the picosecond carrier and lasing dynamics of high quality buried heterostructure (BH) diode lasers using techniques similar to those outlined in earlier work with stripe lasers[1]. The GaAs-AlGaAs BH lasers, shown in Fig. 1, were grown in our lab with a transparent contact over the active waveguiding region to allow optical pumping as well as the usual electrical operation, and had typical threshold currents in the 20-30 ma range. Electrical operation was required for alignment into the optical measurement system to be described below, but was used only for this purpose.

Gigawatt level subpicosecond pulses were generated from a CW synchronously modelocked dye laser and Nd:YAG pumped amplifier chain combination at a 10 hz repitition rate[2]. The tunable system was operated at 598 nm and split into two paths. One was Raman shifted with deuterated acetone to 686 nm and was then focussed onto the BH diode, passing through the transparent contact and $Al_{0.4}Ga_{0.6}As$ upper cladding region to be absorbed directly in the active region of the device. The second pulse was sent through a variable delay arm and then combined collinearly with the IR diode laser output in a $LiIO_3$ crystal phase matched for sum frequency generation. The sum frequency was fed into a signal averaging system, effectively synthesizing a slow sweep picosecond resolution sampling oscilloscope by varying the delay of the short amplified dye laser pulse.

We find that the picosecond dynamics of optically pumped semiconductor lasers are not as strongly dependent upon the diode cavity length as has been suggested

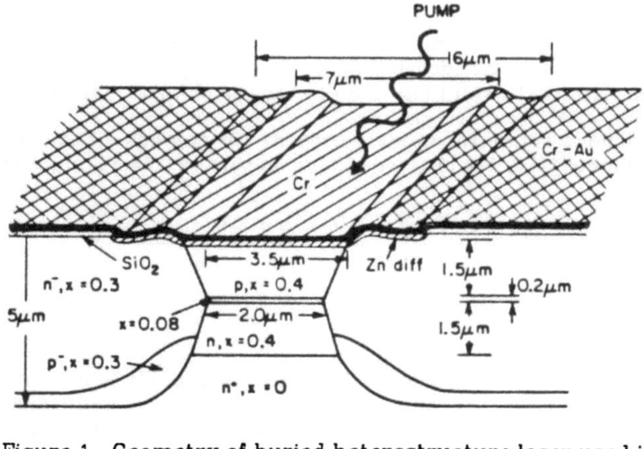

Figure 1. Geometry of buried heterostructure laser used in experiment

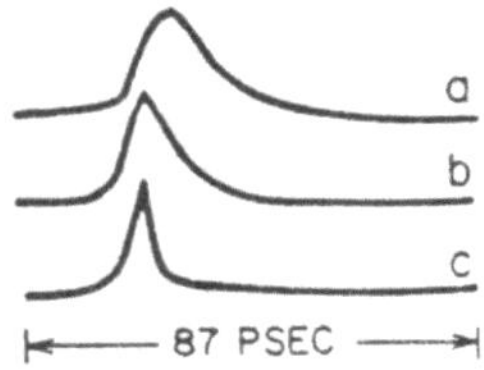

Figure 2. BH diode laser output; **a**, **b**, and **c** are the long, medium, and short wavelength normalized outputs with pulsewidths of 15.6, 9.0, and 4.5 psec respectively

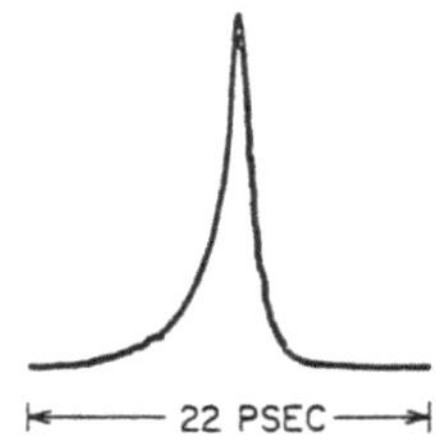

Figure 3. Short wavelength BH laser output under high pumping conditions. Pulse FWHM is 1.2 psec, and note unusual shape

by other investigators working with thin film GaAs and InGaAsP [3,4]. The important quantities for short pulse operation are fast and significant gain/loss transients to initiate laser action and a short photon lifetime for termination. Fig. 2 shows the results from a 200 μm device which, even at this long length, already has a short photon lifetime of ~ 1.3 psec. The three curves are the normalized simultaneous output at three wavelengths obtained by angle tuning the $LiIO_3$ crystal. The pulse FWHM are 4.5, 9.0, and 15.6 psec at λ_0, $\lambda_0 + 11$ nm, and $\lambda_0 + 23$ nm respectively, where $\lambda_0 \approx 743 \pm 15$ nm. This confirms the trend towards shorter pulses at higher photon energies[3] which appears to be an essential feature when high carrier densities are injected well above the bandgap. We estimate our carrier density to be $\sim 5\times10^{18}$ cm^3, although this figure is only approximate due to the uncertainties of the transparent contact injection scheme.

When the pumping level is varied, the general behavior at a fixed wavelength is a forward shift in time and a shortening of the pulse as expected from a usual transient gain/loss analysis. We estimate the upper limit on the carrier density to be $\lesssim 10^{19}$ cm^3 due to a dynamic Burstein-Moss bleaching effect at the pump wavelength in the $Al_{0.08}Ga_{0.92}As$ active region. As the diodes are pumped up to this limit, several interesting features begin to appear. As expected from band filling, shorter wavelength pulses are generated, but the pulses also develop unusual temporal profiles. Fig. 3 shows the ~700 nm output of an 80 μm long BH diode on an expanded scale with a FWHM of 1.2 psec. Since this approaches the pump pulsewidth ($\lesssim$ 1 psec), the deconvolved diode pulse is also probably subpicosecond. Note that the rise time is more gradual than the fall time, in sharp contrast to the usual rate equation predictions of gain/loss switched laser action. At long wavelengths, the pulses are longer to start with, but they begin to acquire a distinct tail as the excitation density increases, even to the point of developing a secondary peak at the longest wavelengths of emission. This is shown in Fig. 4, and note the change in scale.

We find that a straightforward computer model accounts for the basic behavior as depicted in Fig. 2. The model includes the broadband nature of the stimulated

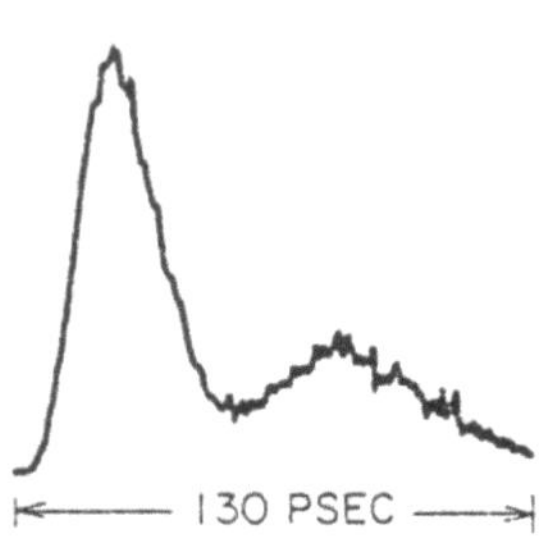

Fig 4. Long wavelength BH laser output under high pumping conditions. Primary peak of pulse has a 21.6 psec FWHM, but note the secondary peak

emission from the optically pumped semiconductor laser output, but makes the simplest assumptions possible while still retaining the essential physics of the device; i.e., we employ all the standard results of parabolic band, direct gap semiconductor theory, and assume that the electron-hole plasma maintains an equilibrium Fermi-Dirac distribution at all times. The plasma temperature T, number density N, and the spatially averaged signal photon density $\varphi(\nu)$ are the dynamic variables, and cooling is given by the usual optical phonon emission cooling formula[5] which we modified to include Fermi-Dirac statistics and screening by the large free carrier density.

$$\frac{dN}{dt} = -\frac{I_p(t)}{h\nu_p}\alpha(\nu_p,N,T) - \int_0^\infty [\frac{c}{n}\alpha(\nu,N,T)\,\varphi(\nu) + r(\nu)\,]\,d\nu \tag{1}$$

$$\frac{d\varphi(\nu)}{dt} = [\,\Gamma\,\frac{c}{n}\alpha(\nu,N,T)\,\varphi(\nu) + \beta r(\nu)\,] - \frac{\varphi(\nu)}{\tau_{ph}} \tag{2}$$

$$\frac{dT}{dt} = \frac{1}{\frac{\partial U}{\partial T}}\Big[(\frac{E_g}{h\nu_p} - 1)I_p(t)\alpha(\nu_p,N,T) - \int_0^\infty [\frac{c}{n}\alpha(\nu,N,T)\,\varphi(\nu) + r(\nu)\,]\,h\nu\,d\nu - \frac{\partial U}{\partial N}\frac{dN}{dt} + \frac{dU}{dt}_{\text{phonon cooling}}\Big] \tag{3}$$

Here $I_p(t)$ is the pump intensity, $U = U(N,T)$ is the total energy density of the electron-hole plasma, τ_{ph} is the photon or cavity lifetime, β is the fraction of spontaneous emission into the lasing modes,and Γ is the mode confinement factor, while the gain/loss $\alpha(\nu,N,T)$, recombination rate $r(\nu)$, $\partial U/\partial T$, and $\partial U/\partial N$ are all well known expressions involving Fermi-Dirac distributions and integrals involving these distributions.

This simple model predicts most of the observed features of the short pulse laser performance. In particular, the strong λ dependence of the pulsewidth is correctly predicted, even in detail down to the unusual shapes as shown in Fig. 3. The important physical effects seem to arise just from the cooling of the plasma and the downward motion of the Fermi level. One unusual and perhaps non-intuitive computed result is that after a preliminary cooling period of a few picoseconds before most of the laser action has initiated, the lasing begins and this actually re-heats the plasma thereby reducing the gain. Lasing action slows or subsides and the plasma then cools again. However, this mechanism will lead to double pulsing on the time scale shown in Fig. 4 only if the carrier cooling is considerably slower than the theoretical value. We believe that the double pulsing as observed in Fig. 4 is due to many-valley effects in the AlGaAs band structure. In this model the delayed peak evolves after the return of electrons from L valleys which are partially filled by the hot electron distribution during the pump pulse. We have included these effects in a more sophisticated set of integro-differential rate equations which appear to support this interpretation[6].

Since any ultrashort pulse driving scheme for semiconductor diode lasers without optical feedback involves large and rapid gain and carrier population transients, these hot electron effects possibly have more generality than the particular injection scheme considered here and could affect the ultimate short pulse performance of diode lasers.

In conclusion we have used picosecond spectroscopic techniques to observe some novel picosecond laser dynamics in high quality BH diode structures. We also have shown that a relatively simple model seems to account for most of the observed features of our experiment.

This work was supported by a grant from the National Science Foundation.

REFERENCES

1. T. L. Koch, D. P. Wilt, and A. Yariv, " Q-Switching of Semiconductor Lasers with Picosecond Light Pulses " in *Picosecond Phenomena II*, Springer Series in Chemical Physics, Vol.**14**, Ed. by R. Hochstrasser, W. Kaiser, and C. V. Shank, Springer-Verlag, Berlin, Heidelberg, New York, p.34, 1980.
2. T. L. Koch, L. C. Chiu, and A. Yariv, Opt. Commun. **40**, 364 (1982).
3. T. C. Damen, M. A. Duguay, J. Shah, J. Stone, J. M. Wiesenfeld, and R. A. Logan, App. Phys. Lett. **39**, 142 (1981).
4. J. Stone, J. M. Wiesenfeld, A. G. Dentai, T. C. Damen, M. A. Duguay, T. Y. Chang and E. A. Caridi, Opt. Lett. **6**, 534 (1981).
5. E. M. Conwell, *High Field Transport in Semiconductors*, Supp. 9 in Solid State Physics Series, Ed. by F. Seitz and D. Turnbull, and H. Ehrenreich, Acedemic Press, New York, 1967.
6. T. L. Koch, L. C. Chiu, Ch. Harder and A. Yariv, " Picosecond Carrier Dynamics and Laser Action in Optically Pumped Buried Heterostructure Semiconductor Lasers," to appear in July 1 App. Phys. Lett. (1982).

Optically Pumped Semiconductor Platelet Lasers in External Cavities

M.M. Salour[1]

Department of Electrical Engineering and Computer Science

and

Research Laboratory of Electronics, Massachusetts Institute of Technology, Cambridge, MA 02139, USA

The new type of laser under discussion is based upon optical pumping in semiconductors. Various individual features of such a device have already been realized [1] and the attractiveness of these individual features is that they potentially combine pleasingly in the newly developed laser [2,3].

The most singular feature of semiconductor lasers [4-7] is that one is not dealing with gain centers (atoms, ions, molecules, complexes) sparsely distributed in a passive medium or empty space, but rather with the problem of inverting the atoms in an entire block of solid, unlike any other kind of laser. Since the absorption and gain lengths in a semiconductor laser are very small indeed compared to other lasers, this central fact bears on the choice of pumping scheme: active medium, heatsinking, cavity design, and sample geometry as semiconductors are crystals, and their ordering implies spatial anisotropy (selection rules), and polarization, since polarization effects all depend on crystal orientation in general.

A single bulk semiconductor is the simplest amplifier medium: it is cheap, readily available, and requires far less processing than heterostructure. To date, most optically pumped semiconductor lasers have used either crystal faces [1] or closely attached mirrors [8] as the cavity reflectors; this prevents the insertion of tuning elements into the cavity and lowers its optical quality. We recently reported the first cw optically pumped semiconductor laser in an external cavity [2]. The lasing medium was a cadmium sulfide platelet pumped longitudinally by an Ar^+ laser. The laser had an output power of 9 mW and a linewidth of 0.1 nm. The power conversion efficiency was 10%. The output is both prism- and temperature-tunable over a range of 9 nm.

The crystals used were very thin (<5 μm) cadmium sulfide platelets chosen for flatness and parallelism by observing under a microscope the interference patterns created by a sodium lamp. The best crystals varied in thickness by less than 3 wavelengths over an area of 5 mm by 5 mm. The crystals were mounted onto a piece of sapphire using a thin film of silicone oil (refer to Fig.1). The same side of the sapphire was dielectrically coated with a maximum reflectivity mirror. The pump beam was focused onto the crystal to a spot size of ∿5 μm by a 10X microscope objective, which also served to collimate the crystal fluorescence. This beam was separated from the pump beam by a polarizing beamsplitter which transmitted 98% of the CdS emission. In order to make this possible, the c-axis of the CdS crystal was vertically oriented. Then its fluorescence, which primarily has $E \perp c$, was polarized perpendicularly to the vertically polarized Ar^+ laser beam.

---

[1] Alfred P. Sloan Fellow.

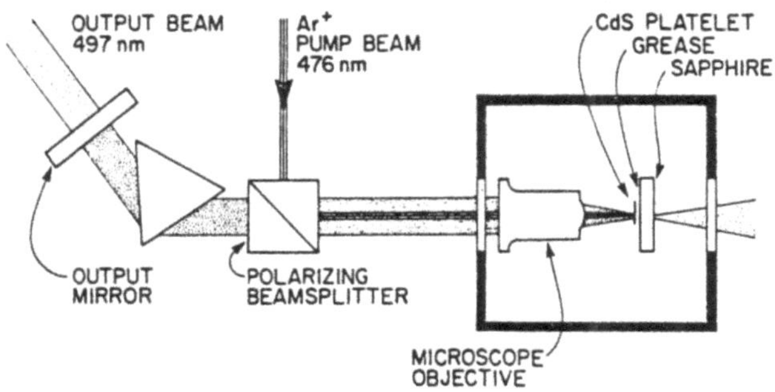

Fig.1 Cavity design for experiments with prism tuning

The semiconductor laser beam passed through a prism and could be tuned by rotating the output mirror about the vertical axis [2]. The sapphire mirror was held by copper pieces which were attached to a liquid nitrogen chamber using copper braid. It could also be tilted in two directions to align this end of the cavity. This arrangement allowed sample temperatures as low as 94 K. The microscope objective was also held in the vacuum, which was sealed by two AR-coated windows.

This work demonstrated the first prism-tunable optically pumped semiconductor laser. We have extended this technique to CdSe and CdSSe crystals, using a 514-nm Ar^+ pump. Then several crystals were mounted adjacent to each other on the same mirror, yielding a laser which is easily tunable from 500 to 700 nm. Heating problems in the crystal might be further reduced by the use of epitaxial layers of CdSe grown on sapphire [10].

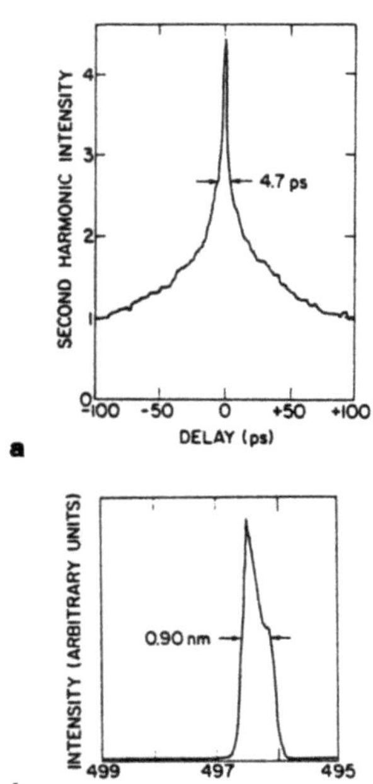

Fig.2 (a) Autocorrelation trace taken from laser with 3.2 mW output power. A pulse width of 4.7 ps is calculated if a single-sided exponential shape is assumed but the large signals in the tails of the pulse indicate an 8 ps pulse width. (b) Simultaneous spectrum, taken with 0.08 nm resolution

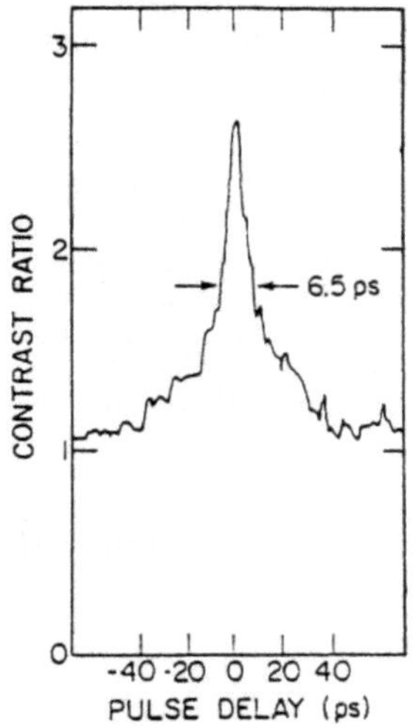

Fig.3 Autocorrelation trace obtained from InGaAsP laser operating at 1.19 μm

For synchronous pumping the Ar^+ laser was actively mode-locked to produce 100-psec pulses on the 476-nm line, and the cavity was lengthened to 1.8 m. The pulse width was measured in an autocorrelator using a lithium formate crystal. Typical spots gave stable 4-10 psec (assuming a single-sided exponential shape; see Fig.2) pulses with 3-5 mW output power and 50 mW of pump power. Up to 26 mW of output power could be maintained for a short while.

Tunable, bandwidth-limited pulses of 6-ps duration (see Fig.3) were also obtained at 1.1 to 1.3 μm from synchronously pumped InGaAsP lasers in external cavity with ACW output powers of 2 mW. Using 2-5 μm thick LPE quaternary layers, longitudinally pumped by a Kr^+ laser, peak output power of over 15 W directly tunable over a 2-nm range was achieved [11].

The cavity length can be changed by 500 μm without adversely affecting the pulse shape (Fig.4). This is in sharp contrast to synchronously pumped dye lasers, where changes of only microns can drastically alter the pulse shape. This indicated that passive pulse shaping may be taking place. The output spectrum is much broader than that encountered in cw operation, and often

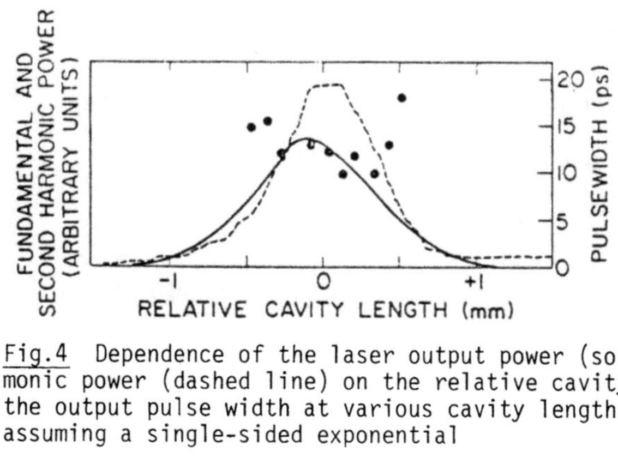

Fig.4 Dependence of the laser output power (solid line) and peak second harmonic power (dashed line) on the relative cavity length. The dots refer to the output pulse width at various cavity lengths as measured by autocorrelation assuming a single-sided exponential

lasing in more than one mode of the crystal Fabry-Perot occurs. We have demonstrated that antireflection coating of the crystal face eliminates these modes, increases the cavity bandwidth, and results in shorter pulses.

These lasers differ from dye lasers in that no jet fluctuations are present, eliminating a very strong source of noise. They can be operated completely in a vacuum, eliminating atmospheric pressure fluctuations present in dye laser cavities. In addition, the spontaneous emission spectrum is narrower than those of dyes, allowing a stabilized single-frequency laser to operate with fewer wavelength-selecting elements while tuning can be done by varying the temperature. We believe that lasers of this type have the capability for single-frequency operation tunable throughout most of the visible and near IR.

The author thanks C. B. Roxlo, D. Bebelaar, R. Putnam, D. A. Johnson, D. C. Reynolds, A. Mooradian, H. A. Haus, and E. P. Ippen. This work was supported by the U.S. Air Force Office of Scientific Research.

References

1. M. R. Johnson, N. Holonyak: J. Appl. Phys. 39, 3977 (1968); S. R. Chinn, J. A. Rossi, C. M. Wolfe, A. Mooradian: IEEE J. Quant. Electron. QE-9, 294 (1973); N. Menyuk, A. S. Pine, A. Mooradian: IEEE J. Quant. Electron. QE-11, 477 (1975)

2. C. B. Roxlo, D. Bebelaar, M. M. Salour: Appl. Phys. Lett. 38, 307 (1981)

3. C. B. Roxlo, M. M. Salour: Appl. Phys. Lett. 38, 738 (1981)

4. J. A. Rossi, S. R. Chinn, A. Mooradian: Appl. Phys. Lett. 20, 84 (1974)

5. C. E. Hurwitz: Appl. Phys. Lett. 8, 121 (1966)

6. A. Passner, H. M. Gibbs, A. C. Gossard, S. L. McCall, T. N. C. Venkatesan, W. Wiegman: IEEE J. Quant. Electron. QE-16, 1283 (1980); H. M. Gibbs, S. S. Targ, J. L. Jewell, D. A. Weinberger, K. Tai, A. C. Gossard, S. L. McCall, A. Passner, W. Wiegmann (to be published, August 1 issue Appl. Phys. Lett. 1982); T. C. Damen, M. A. Duguay, J. M. Wiesenfeld, J. Stone, C. A. Burris: Picosecond Phenomena II, R. M. Hochstrasser et. al. (eds.) (Springer Verlag, Berlin, 1980), p. 38

7. E. P. Ippen, D. J. Eilenberger, R. W. Dixon: Picosecond Phenomena II, R. M. Hochstrasser et al. (eds.) (Springer Verlag, Berlin, 1980), p. 21

8. J. Stone, C. A. Burris, J. C. Campbell: J. Appl. Phys. 51, 3038 (1980)

9. R. F. Leheny, J. Shah: Phys. Rev. Lett. 37, 871 (1976)

10. V. N. Martynor, S. A. Modvedev, L. L. Aksenova, Tu. D. Avchukhov: Sov. Phys. Crystallogr. 24, 743 (1979)

11. R. S. Putnam, C. B. Roxlo, M. M. Salour, S. H. Groves, M. Plonko (to be published)

Two Photon Pumped Bulk Semiconductor Laser for the Generation of Picosecond Pulses

Wei-Lou Cao[1], Fei-Ming Tong[2], De-Sen Shao[3], Scott A. Strobel, V.K. Mathur, and Chi H. Lee

Electrical Engineering Department, University of Maryland, College Park, MD 20742, USA

Recently considerable attention has been devoted to optically pumped bulk semiconductor lasers. The primary motivation has been to obtain high peak power, which has been a main drawback of diode lasers due to their small active volume confined to the junction region. Single photon longitudinal pumping of the bulk semiconductor platelets have yielded tunable semiconductor lasers with pulse widths of ~4-6 picoseconds [1,2], however, peak powers did not exceed a few tens of watts. This performance is difficult to improve through single photon longitudinal pumping due to poor heat dissipation and low thermal conductivity, and any further increase in the power of pump laser leads to the destruction of the medium. We underscore that the single photon longitudinal pumping, which is adapted from a dye laser system can not be easily adopted for optically pumped semiconductor lasers for the following two reasons: 1) heat dissipation does not present any problem in a dye system because it is flowing, 2)the absorption coefficient, a factor which is fixed in a semiconductor can be manipulated by adjusting the concentration of the dye. Thin platelets, of the order of 10μm, have been used for semiconductor lasers, which is quite thick compared to an absorption depth of 1μm. Consequently the gain medium has a nonuniform population inversion.

We suggest, therefore, that a two photon pumping in a transverse configuration is more suitable for the generation of high peak power optically pumped semiconductor lasers. Single photon pumping in a transverse configuration has the disadvantage of large diffraction losses because the thickness of the active volume is of the same magnitude as the wavelength of emission. Moreover such a system is difficult to align in a ring cavity and surface recombination will deplete the inverse population. The distinct advantages of the two photon pumping scheme are the deeper penetration of the pump beam and the flexibility of the two photon absorption coefficient $K^{(2)}$ ($=\beta I$) which is intensity dependent. $K^{(2)}$ can vary from 1 cm^{-1} to 10^7 cm^{-1} in contrast to $K^{(1)}$, single photon absorption coefficient, which ranges from 10^4-10^6 cm^{-1}. Peak powers as high as one Megawatt, with pulse durations of 7-10 ps have been demonstrated by us [3] under two photon pumping in a transverse configuration. Both synchronous mode-locking and amplified spontaneous emission (ASE) have been observed in GaAs, with an output power density in excess of 100 MW/cm^2.

This peak power is several orders higher than the peak power reported [2] with single photon longitudinal pumping. We report here our recent results

---

[1] On leave from Shanghai Institute of Optics and Fine Mechanics, Chinese Academy of Sciences, Shanghai, PRC.

[2] On leave from Shanghai Institute of Technical Physics, Chinese Academy of Sciences, Shanghai, PRC.

[3] On leave from Peking Research Institute of Materials and Technology, Peking, PRC.

obtained by the extension of this technique to $CdS_{0.5}Se_{0.5}$. The mixed crystal of cadmium sulphide selenide is attractive because of its potential of lasing over a broad spectral range from green to red depending on the ratio of sulphur to selenium.

In the present experiment no external cavity was formed. Instead the opposite emitting surfaces of the crystal were anti-reflection coated and were deliberately cut so that they are not parallel to each other. We believe that the stimulated emission observed is due to ASE. The device would lase even when only a portion of the crystal was pumped. A study of the output power versus pumping length confirms the characteristics of amplified spontaneous emission.

A 10x5x4mm crystal of high resistivity ($\approx 10^8$ Ωcm) $CdS_{0.5}Se_{0.5}$ was mounted in a cryostat which was cooled by liquid nitrogen. The end faces (5x4mm) of this crystal were A.R. coated (R≈.01) for its intrinsic emission. The crystal was pumped from the long side (10x5mm) in a transverse configuration. The 1.06μm laser spot was focussed along the length of the crystal by using a dove prism and a cylindrical lens. The complete mode-locked pulse train from a Nd: Glass laser was used to pump the crystal. The emission from the crystal was collected and image relayed to the monochromator, where it is detected by a photomultiplier. The schematic of the experimental arrangement has been described earlier [3]. The spatial profile of the ASE was obtained by replacing the monochromator by a camera and photographing the image of the end face of the crystal under a magnification of about 3.5.

Fig. 1 shows oscilloscopic traces of the ASE and pump pulse trains (upper and lower traces respectively) by two high speed ITT photodiodes. The peak intensity of the pump pulse train is estimated to be several hundred MW/cm^2. The vertical scale of the two traces is not the same since the intensity of $CdS_{0.5}Se_{0.5}$ emission is more than three orders of magnitude less than the pump intensity. The total energy output in ASE was measured to be $1x10^{-5}$ Joule for the whole pulse train. The energy of pump pulse was measured to be 0.1 J of which 85% is absorbed in the crystal over its thickness of 0.4 cm. The output pulse train from $CdS_{0.5}Se_{0.5}$ is well resolved and resembles the pump pulse train. This is interesting in view of the long

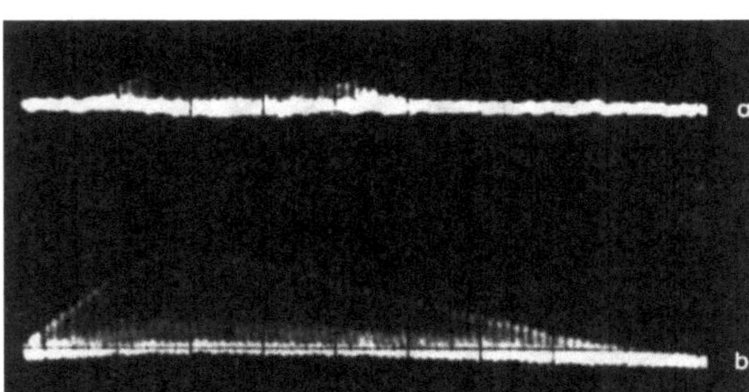

Fig. 1 Oscilloscopic trace of ASE pulse train and pump pulse train. Sweep rate 50 ns/div.
(a) ASE (upper trace)
(b) pump (lower trace)
(Vertical scale for two traces is different)

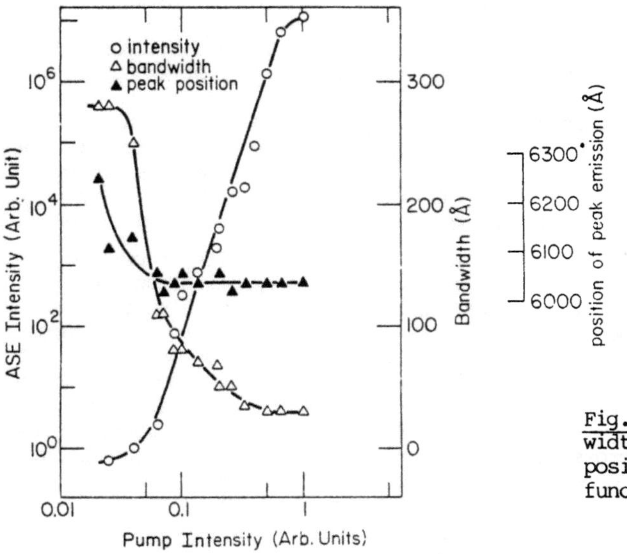

Fig. 2 ASE intensity, bandwidth (FWHM) and the peak position of emission as a function of pump intensity

lifetime of excited carriers reported earlier [4]. Moreover, the ASE pulse train emerges later and disappears earlier than pumping laser train. We believe that this behaviour is related to the threshold of ASE. At the leading and trailing part of the pump pulse train, the pump intensity is less than the threshold, therefore no stimulated emission is observed.

The gain characteristic of $CdS_{0.5}Se_{0.5}$ was further investigated by studying the band-width (FWHM) of emission and the output intensity as a function of pump intensity. The plots of bandwidth and output intensity versus pump intensity are shown in Fig. 2. The pump intensity was varied by inserting neutral density filters in the path of the exciting laser. At very low pump intensity the bandwidth is very large (≈250Å), a characteristic of spontaneous emission. As the pump intensity is increased the band-width decreases rapidly in the beginning and then slowly to a value of 30 Å. At the same time, the output intensity increases abruptly and begins to saturate at high pump intensity. Our results are similar to the results obtained by SHAKLEE and LEHENY [5]. Both bandwidth and output intensity measurements exhibit a threshold behavior. The threshold is, however, not sharp. Moreover, the peak of the emission moves towards shorter wavelengths at higher excitation levels. The spontaneous emission was observed to peak around 6240 Å at lower pumping intensity, while at higher pumping intensity the peak of emission occurred at 6040 Å as shown in Fig. 2. This shift can be attributed to the band filling.

We have measured two photon absorption coefficient of 1.06μm radiation in $CdS_{0.5}Se_{0.5}$ with a view to extend the technique of the generator-amplifier configuration to obtain high peak powers. The results are tabulated in Table I.

Table 1 Measurement of β for $CdS_{0.5}Se_{0.5}$ Crystal of thickness 0.4 cm

I_0 MW/cm^2	Transmission	β cm/MW
312	15%	0.045
50	52%	0.046
20	73%	0.045

The measured value of β reported here is consistent with the value reported earlier [6]. The above results were used to determine the pump intensity for the amplifier which should be kept below the threshold of ASE. Our preliminary results indicate that an amplification factor of 2-5 can easily be achieved.

To demonstrate the deep penetration of the two photon pump beam the spatial profile of the ASE was monitored at different pump intensity. The results are described in Fig. 3. Fig. 3a, is a shadow of the emitting face obtained by shining light from the opposite ends of the crystal. The upper and lower edges of the shadow are not visible due to the structure of the cold finger.

The crystal is pumped from the right side. At low intensity of pumping, the whole of the emitting crystal surface shows uniform illumination indicating deep and uniform generation of carriers in the bulk which recombine radiatively to give spontaneous emission. As the intensity of the beam pump is gradually increased, an intensity contrast begins to show up on the emitting surface. The part of the surface towards the incident pump radiation appears brightest, finally ASE appears as a bright patch on the right side. It seems that at very high pump intensity there is a considerable carrier gradient from right to left across the crystal and, the active volume for ASE is confined to a depth of the order of one millimeter on the right, i.e., the side from which pump radiation is incident on the crystal.

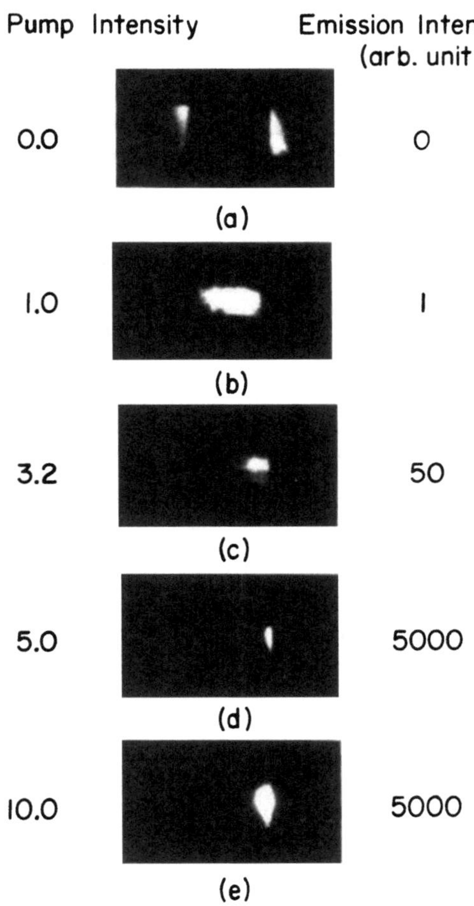

Fig. 3 Spatial profile of ASE. pump intensity was varied by inserting Neutral Density filters in its path. In the same way, appropriate a Neutral Density filters were placed before the camera. (a) is a shadow of the crystal face (0.4 cm wide) from which emission was monitored. Successive pictures show how the relative intensity of ASE changes with respect to spontaneous emission as the pump intensity increases. In (b) the emission is predominantly spontaneous, while in (e) it is ASE.

It may be remarked that in conjunction with a monochromator and using a pinhole camera method, the above technique can be extended to monitor simultaneously the spectral and spatial profile in a single shot. This experiment is now underway in our laboratory.

Simultaneous measurements of photoconductivity and photoluminescence excited by a single picosecond 1.06μm laser pulse indicates that the radiative life-time of the carriers is of the order of a few nanoseconds because the luminescence ceases within several nanoseconds. However the photoconductivity signal lasts for several milliseconds which signifies trapping and detrapping of carriers before final recombination. It has been observed that ASE pulses are not well resolved under low pump intensity but at high pump intensity individual pulses are well defined as shown in Fig. 1. This appears to be due to the decrease in lifetime with the increase in the density of photoexcited carriers, a result which conforms with the recent work of OLSON and TANG [7].

To summarize, we have demonstrated the potentiality of a two photon pumping scheme in a transverse configuration for the generation of high peak power picosecond pulses. This scheme can be extended to a generator-amplifier configuration to further increase the peak powers.

Authors thank Aileen Vaucher for her valuable assistance. This work was supported in part by National Science Foundation under Grant NO. ENG-78-06862 and by Minta Martin Aeronautical Research Fund from the College of Engineering, the University of Maryland.

References

1. J. Stone, J.M. Wiesenfeld, A.G. Dentai, T.C. Damen, M.A. Duguay, T.Y. Chang and E.A. Caridi, Opt. Lett. 6, 534 (1981).

2. C.B. Roxlo, R.S. Putnam and M.M. Salour, IEEE J. Quantum Electron. QE-18, 338 (1982).

3. Wei-Lou Cao, A.M. Vaucher and Chi H. Lee, Appl. Phys. Lett. 38, 653 (1981).

4. V.K. Mathur, P.S. Mak and Chi H. Lee, J. Appl. Phys., 51, 4889 (1980).

5. K.L. Shaklee and R.F. Leheny, Appl. Phys. Lett. 18, 475 (1971).

6. Chi H. Lee and S. Jayaraman, Opto-electronics, 6, 115 (1974).

7. N.A. Olsson and C.L. Tang, IEEE. J. Quantum. Electron. QE-18, 971 (1982).

The Pulse Duration of a Distributed Feedback Dye Laser Under Single Pulse Conditions

Zs. Bor, B. Rácz, G. Szabó

JATE University, Dept. of Experimental Physics, Dóm tér 9,
H-6720 Szeged, Hungary

A. Müller

Max-Planck-Institut für biophysikalische Chemie, Abteilung Laserphysik,
Am Fassberg, D-3400 Göttingen, Fed. Rep. of Germany

1. Introduction

As we have shown recently [1-3], distributed feedback dye lasers (DFDLs) are a new type of source for the generation of picosecond light pulses. DFDLs are relatively simple in construction and produce stable single pulses without need for a special pulse selecting device. Their range of operation extends through the visible to the near uv part of the spectrum [4]. Smooth tuning can be achieved by several techniques [2]. It is easy to set up an oscillator-amplifier system for DFDL pulses which is pumped by a single N_2-laser [5]. Synchronization is greatly facilitated this way.

A disadvantage of DFDLs as compared to mode-locked lasers has been, so far, the long duration of the output pulses of typically 80 to 100 ps. As has been indicated in [1] the rate equation theory of the DFDL predicts the shortening of the laser pulses with decreasing pump pulse duration. It is the goal of the present study to verify this prediction experimentally and to produce the shortest pulses possible with the pump sources presently available to us.

2. Rate Equation Model of the DFDL

The influence of various parameters, viz. pump intensity, fluorescence lifetime, geometrical dimensions etc., on the temporal and energetic properties of DFDL output has been studied in detail [3]. The good agreement observed between theory and experiment encouraged us to apply the rate equation model of the DFDL also to the present investigation.

The rate equations (cf. equations (1-6) of [3]) were solved numerically. The parameters of Rhodamine 6G as given in [3] were used. Pumping pulses of various durations in the range from 0.5 to 5 ns (FWHM) were assumed to be Gaussian.

3. Experimental

The experimental layout of the DFDL is shown in Fig. 1. Five different pumping sources were employed:

a) A low pressure N_2-laser (LAMBDA PHYSIK M-1000) with pulse duration of 3.5 ns.

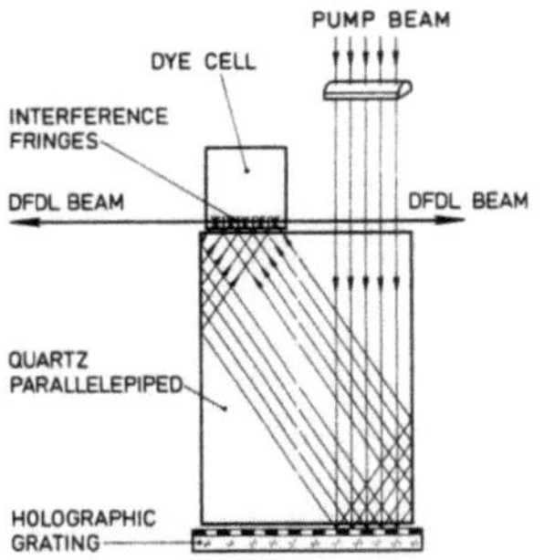

Fig.1 Experimental arrangement of the DFDL

b) An oscillator-amplifier system comprising a TEA-N_2-laser, built in our laboratory, as oscillator and the low pressure N_2-laser as amplifier. The TEA-N_2-laser produced pulses of 1.1 ns which were broadened in the amplifier to about 1.8 ns. A telescope was inserted between oscillator and amplifier.

c) The same setup as in (b) with an additional cuvette containing a solution of saturable absorber (bis-MSB) in the focal plane of the telescope. Pulse duration was 1.2 ns.

d) A similar system as in (c), but incorporating a different TEA-N_2-laser producing pulses of 0.7 ns duration.

e) A frequency-tripled Nd-laser system (J. K. LASERS, System 2000) consisting of a passively mode-locked Nd:YAG oscillator and two Nd:glass amplifiers. A single pulse was selected. Its duration was 16 ps for the third harmonic of the Nd:YAG frequency.

The concentration c of Rhodamine 6G and the pumped length L of the DFDL had the following values during different experiments: (a,b,c): $c = 3.5 \times 10^{-3}$ M/ℓ, L = 3.5 mm; (d): $c = 5 \times 10^{-3}$ M/ℓ, L = 2 mm; (e): $c = 4 \times 10^{-3}$ M/ℓ, L = 3 mm. Solvent mixtures were used in order to adjust the output wavelength to 590 nm. The height of the pumped volume was 0.025 mm in all cases.

Pulse durations were measured with a streak camera-image intensifier system (HADLAND IMACON 600 + EMI T2001) coupled to an optical multichannel analyzer (PAR OMA) and a DEC PDP-11/34 computer. Instrumental time resolution of the system was 11 ps. In the case of picosecond pulse pumping (e) DFDL output was analyzed by a second order correlation technique using diffraction gratings in Littrow mounting as delay elements for the subsequent SHG [6]. This arrangement represents a modification of the technique described in [7] and allows to measure single shots with a resolution of the order of 0.1 ps.

4. Results and Discussion

The output characteristic of a DFDL is governed by the pump power level [1,3]. The number of pulses generated increases with pump power. We observed very good agreement between the results of the rate equation model and experiments. This situation is not altered when the pump pulse duration is varied. Stable single pulse output can be obtained when the pump power is adjusted for the threshold of the second pulse [3]. Figure 2 shows some typical examples under this condition.

With even shorter pump pulses the duration of DFDL output approaches the time resolution limit of our streak camera

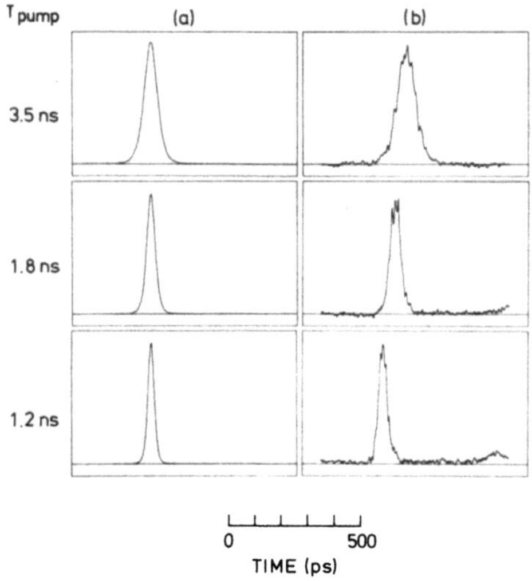

Fig.2 Single pulses of the DFDL. a) Results of the rate equation model. b) Streak camera recordings. Pump pulse duration decreases from top to bottom of the figure corresponding to pump sources (a), (b) and (c), respectively

recording system (Fig.3). In the typical example shown FWHM is 8.8 OMA channels. Considering a streak speed of 1.4 channels/ps and an instrumental resolution of 11 ps we obtain a deconvoluted pulse duration of 6 ps. The energy of the single pulses was about 40 nJ corresponding to 7 kW peak power.

Experiments with picosecond pulse pumping (e) are presently in progress. Preliminary results indicate a pulse duration of 1 ps and a time-bandwidth product of $\Delta\nu \cdot \Delta t \approx 0.6$ [6]. Further shortening of the pulses appears to be feasible by pumping the DFDL in a travelling wave excitation arrangement. In this case no limitation of the shortest pulse duration is expected by transit time effects, which presently might impose a lower limit of the order of a picosecond.

Figure 4 summarizes our results and demonstrates the correspondence with the model computations.

Pulse shortening of the DFDL pulses relative to pump pulse duration is about the same as that for cw synchronously pumped mode-locked dye lasers. The use of DFDLs as simple and rela-

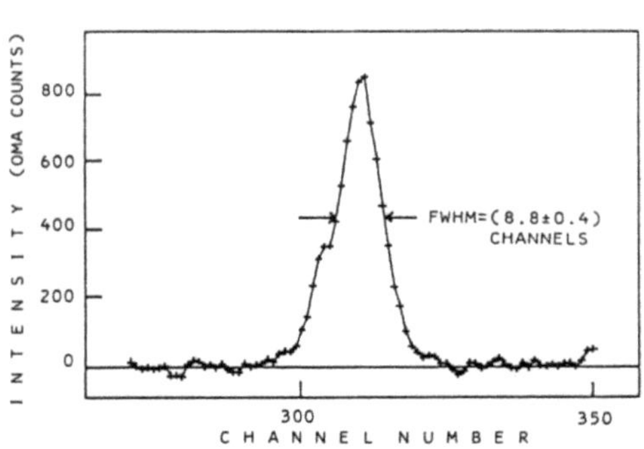

Fig.3 Single DFDL pulse obtained with pump pulse of 0.7 ns duration

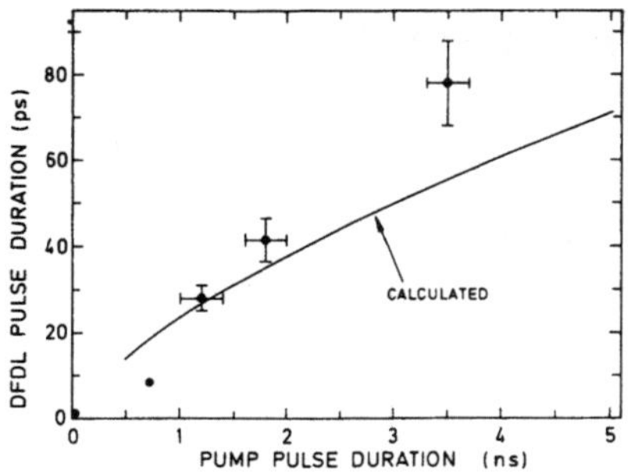

Fig.4 Dependence of DFDL pulse duration on pump pulse duration. The experimental points marked with error bars represent averages of 20 shots each

tively inexpensive sources of ultrashort laser pulses could considerably facilitate the progress in picosecond laser spectroscopy.

5. Acknowledgments

This work has been supported by a joint project of the Deutsche Forschungsgemeinschaft and the Hungarian Academy of Sciences. We thank Prof. F. P. Schäfer for his interest and Dr. W. Zapka for his aid at an early stage of this work.

6. References

1. Zs. Bor: IEEE J. Quant. Electron. QE-16, 517 (1980)
2. Zs. Bor: Opt. Commun. 29, 103 (1979)
3. Zs. Bor, A. Müller, B. Rácz, F. P. Schäfer:
   Part I: Appl. Phys. B27, 9 (1982)
   Part II: Appl. Phys. B27, 77 (1982)
4. Zs. Bor, A. Müller, B. Rácz: Opt. Commun. 40, 294 (1982)
5. Zs. Bor, B. Rácz, F. P. Schäfer: Sov. J. Quant. Electron. (submitted)
6. G. Szabó, A. Müller, Zs. Bor: (to be published)
7. R. Wyatt, E. E. Marinero: Appl. Phys. 25, 297 (1981)

Picosecond Distributed Feedback Dye Laser Tunable in a Broad Spectral Range

A.N. Rubinov, I. Chesnulyavichus, and T.Sh. Efendiev
Institute of Physics, Academy of Sciences of the Byelorussian SSR, Minsk, 220602, USSR

The rapid development of picosecond spectroscopy requires efficient and convenient sources of ultrashort pulses that are easily tunable over a wide spectral range. The generation of tunable picosecond pulses in dye lasers has been demonstrated in a number of experiments, in particular with cw dye lasers. Unfortunately, laser systems allowing single ultrashort high-power pulses that are tunable in a broad spectral region are still too complicated and expensive, which makes them inconvenient for practical use and unavailable for many laboratories.

The operation of a simple and reliable pump-induced, distributed-feedback (PIDF) dye laser which produces single ultrashort pulses tunable in a wide spectral range is reported here.

The applicability of the PIDF dye laser for the generation of train or single tunable ultrashort pulses was first demonstrated in [1,2]. The mode-locked ruby [1] and Nd^{3+}:glass [2] lasers operating in a single short regime were utilized as a pumping source in those early experiments. We now report on the operation of a picosecond PIDF dye laser pumped with a Nd^{3+}:YAG laser with a 12.5-pps repetition rate.

The possibilities of shortening the PIDF dye laser pulse by decreasing the length of the DFB structure in a solution, and also by introducing time delay between interfering pumping beams, are demonstrated experimentally. The generation of the narrowest picosecond pulse so far obtained in a PIDF-type dye laser is also reported.

The second harmonic of a passively mode-locked Nd^{3+}:YAG laser with a 12.5-pps repetition rate was used to pump a PIDF dye laser. A single pulse of 30 ps duration, 2 MW peak power and 1 Å spectral width was picked out of a train of Nd^{3+}:YAG laser pulses. The pulse duration of both pumping and PIDF lasers was measured by the conventional technique of noncollinear second harmonic generation in an ADP crystal. The wavelength of dye laser oscillation was registered by the diffraction grating spectrograph with 4 Å/mm linear dispersion.

Stable generation of ultrashort pulses with the repetition rate of the pumping source was observed in a PIDF laser using rhodamine 6G(R-6G), R-B and R-4C dyes in ethanol as an active medium. The absorption coefficient of the laser solution at the wavelength of pumping was in the range of 60-120 cm^{-1}.

Spectral tuning of PIDF laser emission from 548 nm to 590 nm for R-6G, 582-610 nm for R-B and 583-620 nm for R-4C was easily obtainable with a linewidth varying from 1.5 Å to 4 Å, depending on the excitation conditions.

The pumping radiation was converted into stimulated emission of the dye with an efficiency of 10%, and thus tunable picosecond pulses of up to 0.2 MW peak power were obtained.

The pulse duration of the PIDF laser was found to be in almost linear dependence on the length of the periodical structure in a dye solution. Thus, with periodical structure lengths of 1 mm, 2 mm, and 3 mm, the lasing pulse durations were 15 ps, 35 ps, and 45 ps, respectively. A decrease in the pulse duration from 45 ps to 15 ps was followed by broadening of the spectrum from 1.5 Å to 4 Å.

A possibility of further narrowing the dye laser pulse by optically delaying one of the interfering beams with respect to another was studied. With a delay of 12 ps, the lasing pulse width was reduced to 9 ps, while the duration of the pumping pulse was 30 ps. Shortening of the PIDF laser pulse was followed by some decrease in the output energy.

Still shorter pulses were obtained in a PIDF laser when the dye was excited by the second harmonic of a mode-locked Nd^{3+}:phosphate-glass laser. With a pumping pulse duration of 6 ps, a dye laser pulse duration of 3 ps was observed. At present this is the shortest tunable pulse obtained in a PIDF-type dye laser.

In conclusion it can be said that the application of pump-induced distributed-feedback dye lasers in conjunction with mode-locked Nd^{3+}:YAG or Nd^{3+}:glass lasers is very efficient and is apparently the simplest method of obtaining powerful picosecond pulses that are tunable in a broade spectral range.

References

1 V.A. Zaporozhchenko, A.N. Rubinov, T.Sh. Efendiev: Pis'ma Zh. Tekh. Fiz. *5*, 114 (1977)

2 B.A. Bushuk, V.A. Zaporozhchenko, A.L. Kiselevskii, A.N. Rubinov, A.P. Stupak, T.Sh. Efendiev: Pis'ma Zh. Tekh. Fiz. *5*, 880 (1979)

Modelocking of a Wavelength Tunable High-Pressure CO_2-Laser by Synchronous Modulation of a Broadband Intracavity Saturable Absorber

J.K. Ajo, Y. Hefetz, and A.V. Nurmikko

Division of Engineering, Brown University,
Providence, R.I. 02912, USA

In recent years, both passive and injection modelocking has been applied to the multiatmospheric CO_2-laser for generation of subnanosecond pulses of high power radiation at the 10.6 µm wavelength, thus taking advantage of the strongly pressure broadened bandwidth (1,2). On the other hand, continuous wavelength tuning over significant range from 9 to 11 µm of the freerunning laser can be readily obtained by operating the laser in the 10 atm pressure range. In spite of the many potential and attractive applications, for example to time-resolved spectroscopy of molecules, there appears to be no reported success of short pulse generation with such wavelength tunability. We have combined in a simple scheme elements of both forced and passive modelocking to provide a versatile source with good tuning characteristics. The difficulties with reliability in modelocking which are generally connected with the relatively high gain and its short lifetime in high pressure lasers were here reduced by additional external optical modulation of absorption losses in intracavity p-Ge.

In our approach a simple linear resonator with a high resolution grating provides the wavelength selective structure for the high-pressure laser. The modelocking is achieved by using the intensity dependent absorption associated with intervalence band transitions in p-Ge. Earlier experimental work by Keilmann has shown that the saturation intensity in such an inhomogeneous absorber is approximately 4-10 MW/cm^2 at 10.6 µm and room temperature, while the homogeneous width obtained from holeburning experiments with discretely tunable CO_2 lasers is large in comparison with the adjacent line separation (3). The doping of germanium is here assumed to be sufficiently low so that acoustic and optical phonon emission dominate the hot hole relaxation rate. In our experiments, a 3 mm thick Brewster angle plate of germanium ($p = 3 \times 10^{15} cm^{-3}$) was placed near the grating end of the 10 atm laser. In addition, external radiation from an acousto-optically modelocked, linetunable TEA CO_2-laser was also directed at the absorber. The discharge circuits of the lasers were synchronized in such a way that the modelocked pulses ($t_p \gtrsim 2$nsec) from the TEA laser could reach the p-Ge before any significant oscillations had taken place in the resonator of the high pressure laser. The presence of this modelocked emission affected the nonlinear absorption periodically according to the resonator length of the TEA laser. This method of loss modulation is similar to that used earlier by Keilmann and Kuhl to control emission from an HF laser outside the resonator [4].

Under conditions of resonator length matching of the two lasers, we have obtained distinct modelocking from the high pressure laser. This emission

*Research supported by NSF/ECS 80-17519. J. K. Ajo is partly supported by Wihuri-foundation of Finland.

is characterized by high intensity subnanosecond pulses and behavior which indicates that both externally applied and self-induced nonlinear absorption in the p-Ge is responsible. The duration and details of the modelocked trains have been found to depend strongly on the intensity of the controlling TEA at the fixed wavelength of the injection source. In contrast, the synchronous loss modulation readily yielded significantly shorter pulses whose wavelength could be tuned approximately 5 cm^{-1} about the particular line chosen for the TEA laser in its 9 or 10 micron band. While this tuning range is substantially less than the previously reported holeburning width measured for p-Ge (3), it nevertheless demonstrated how continuous wavelength tuning of the modelocked high-pressure laser is possible by accompanying discrete wavelength adjustment of the TEA laser.

Our study of the temporal characteristics of this modelocked emission shows how pulse narrowing occurs during the formation of the train of subnanosecond pulses. While the details of this are sensitive, for example, to the wavelength tuning, there appears to be a gradual transition from an forcibly modelocked to a passively modelocked regime during the pulse train. An example of the resulting pulse compression is illustrated in Figure 1 laser and its timing relative to the onset of laser oscillations in the high pressure laser. Typically, considerable improvement in repeatability, when compared with purely passive modelocking, would be observed when the external loss modulation was added by the presence of the nanosecond pulses from the TEA laser at the p-Ge. However, care had to be exercised to prevent conventional injection locking from occurring as a result of accidental scattering, from the surface of the absorber, of the TEA laser radiation directly to the gain medium. In this instance, relatively long pulses were obtained (t_p=1nsec) which compares a single pulse from the TEA laser to that from the middle of the train of the high-pressure laser. The latter is not time resolved by the 1 GHz bandwidth of the detection electronics. By using additional pulsewidth diagnostics (spectral width measurements and second harmonic correlation in Te) we estimate that the lower limit to our pulsewidths is in the range of 100 psec, consistent with earlier results of passive modelocking at a fixed wavelength. Precise temporal measurements by averaging methods were here made somewhat difficult by the finite fluctuations of the pulsewidths from shot to shot; this behavior is ascribed to problems inherent to the operation of a truly stable, reproducible discharge in the high-pressure environment. The intensities associated with successful modelocking were high, estimated to exceed 100 MW/cm^2 within the laser resonator itself, thus presenting a ready challenge to the damage resistance of present infrared component and coating technology.

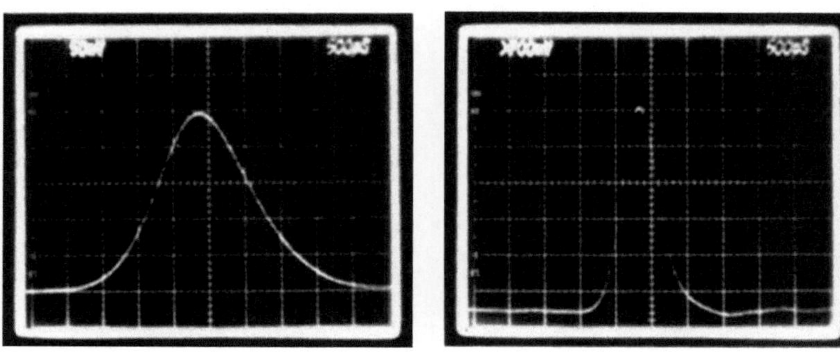

Fig. 1: Single pulses from the TEA laser (left)and the high-pressure laser (right; not time resolved). Horizontal scale 500 psec/div.

References:

1. A. J. Alcock and A. C. Walker, Appl. Phys. Lett. 25, 299 (1974)

2. P. B. Corkum and A. J. Alcock, D. F. Rollin and H. D. Morrison, Appl. Phys. Lett. 32, 27 (1978

3. F. Keilmann, IEEE J. Quant. Electr. QE-12, 592 (1976)

4. F. Keilmann and J. Kuhl, IEEE J. Quant. Electr. QE-14, 203 (1978)

The Non-Mode-Locked Picosecond Laser

F. Armani, F. De Martini, and P. Mataloni

Quantum Optics Laboratory, Istituto di Fisica, "G. Marconi",
I-00185 Roma, Italy

In the present paper we report the first application of the regenerative compression technique to the self-injected Nd-YAG laser to generate band-limited picosecond pulses [1,2]. With this technique we are able to reduce the typical 30 nanosecond pulse duration of a normal Pockels Cell Q-switched laser down to less than 6ps, realizing a pulse compression by about 3 orders of magnitude. This is achieved by first stepping down the Q-switched pulse to a seed pulse about 1ns long using the self-injection (i.e. cavity flipping) technique. Then this seed pulse is amplified by the active medium and, at the same time, undergoes a further compression due to the nonlinear transmission characteristics of a saturable dye flowing in a cell inserted in the laser cavity. Since the shortening process is quasi-adiabatic, very high pulse peak power can be obtained, limited basically by damage to the optical components and by effects of nonlinear loss and self actions arising in the active medium [2]. In addition to that, the system has very good stability characteristics.

The basic laser is the self-injected Nd-YAG system described in a previous paper [1]. A 3x50mm Nd-YAG rod with AR coated end faces is pumped in a double elliptical cavity by a couple of simmered flashlamps. For our present application, the optical cavity was designed in such a way as to provide the adequate balance between the laser intensities seen by the active medium and by the saturable dye flowing in a cell. This is required by the different values of the gain and absorption cross sections for the two media respectively and by the need of reaching simultaneously a regime of saturation for both gain and absorption in order to obtain good compression performance [3,4]. Two equally totally reflecting spherical mirrors with radii R_1=99.9cm and R_2=67.5 cm determined the geometrical size of the cavity (see fig.1). With the geometrical parameters given in fig. 1, the ratio of beam areas in the dye cell and in the rod was ~ 7,6. The dye we used was the Eastman-Kodak no. 9740 Q-switching solution in dichloroethane. The dye concentration was adjusted to an adequate value, different from the one which caused self Q-switching and mode-locking in the laser. The Pockels Cell (PC) was a Lasermetrics 1057 FV driven by a krytron circuit capable of delivering

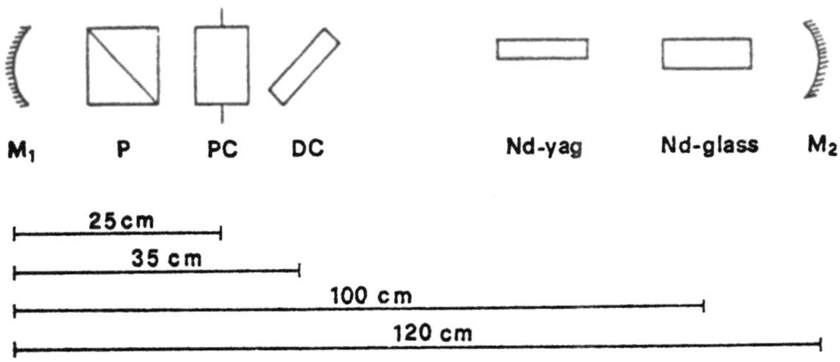

Fig.1. Laser cavity scheme used for regenerative pulse compression in the self-injected laser

the waveform which is necessary for a combined Q-switching, cavity flipping and cavity dumping operations[1]. The laser operated typically at 3 pps and the output was monitored by an ITL 1850 photodiode and a Tektronix 519 travelling wave oscilloscope. The overall risetime of the detection system was 320ps. Ultrashort pulse duration measurements were performed by a standard triangular two photon fluorescence technique using a cell filled with a 10^{-3}M/l solution a Rhodamine 6G in methanol.

The laser was initially adjusted for operation under self-injection conditions by flowing pure dichloroethane through the dye cell. In this way the light pulse evolved in the laser cavity as a train of 2.5 ns pulses as shown in figs.2a and 2b, with the saturable dye flowing into the cell, we acted upon the flashlamp pumping voltage to find the condition of best pulse compression. This one has been found to correspond to driving the laser just above threshold. In this condition the pulse train of figs.2c and 2d was obtained. The two photon fluorescence pattern is shown in fig.3 and corresponds to the highest pulse shown in fig.2c. Drawing the corresponding densitometric trace we have measured a pulse duration of 15 ps using Nd-YAG as active medium. In these conditions the peak power was found to be ~1 GW. A further pulse shortening (of a factor ~ 2,5) has been obtained by inserting in the laser cavity an independently flashpumped 6% doped Nd-Phosphate (4mmx60mm) glass rod. In this configuration the Nd-YAG ac-

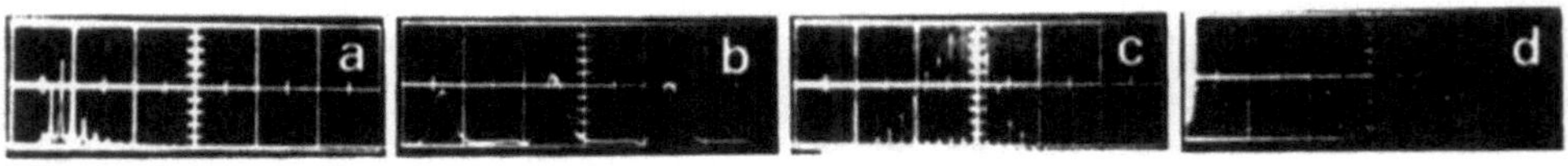

Fig.2(a) Pulse train in self-injection operation without dye in the cavity (50ns/div).(b) Same conditions as in (a) but showing the self-injection operation with dye solution in the cavity (50 ns/div).(d) Detail of the pulses near the peak of (c) (5ns/div). Pulse shape is determined by the detection system bandwidth.

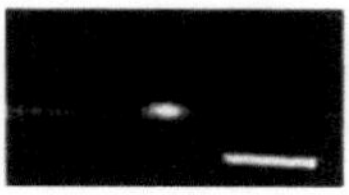

Fig.3. Two photon fluorescence measurement pattern. The length sign at the bottom of the figure corresponds to 50 ps.

tive medium was responsible for the low threshold generation of the seed pulse while the Nd phosphate rod provided large band intracavity amplification for pulse shortening. Details on the behaviour of this new, two media, picosecond laser will be given in a forthcoming paper. When only the Nd-YAG was operating, the envelope of the pulse train shown in fig.2c had a shape which is quite typical of the physical processes which are at the base of the behaviour of our device [3]. This shape can also give a fair indication of the good pulse shortening performance of the laser. The slow rise of the leading edge of the envelope at low pulse power indicates that the overall, low level, laser gain is small. This is a condition for reaching the maximum intensity of the train with a large number (>100) of shortening passages.

In summary, we have presented a new technique for production of high power picosecond pulses. In the present application we observed a pulse compression by a factor $\sim$100 with respect to the normal self-injection operation, while the peak power gain in the compression process was 50 . This implied an energy reduction in this process by a factor of only 2. The obtained short pulses were very stable and showed low jitter ($\pm$10ns) relative to the Q-switching HV pulse applied to the Pockels Cell.

We stress here the basic difference existing between our laser and the usual mode-locked lasers. In our system the pulse doesn't develop from quantum noise but rather is the result of a nonlinear frequency-time processing in the cavity of a coherent seed pulse. This one keeps its coherence properties since the starting of the shortening process. Apart from obvious technical advantages (no need for Brewster cut rods and components), our device shows remarkable characteristics of pulse stability due to the above considerations on coherence. We believe that our device opens new perspectives in the field of laser physics and technology.

Work supported by Consiglio Nazionale delle Ricerche, Italy.

References

1 C.H. Brito Cruz, E. Palange and F. De Martini: Optics Comm. 39, 331 (1981).
2 C.H. Brito Cruz, F. De Martini, H.L. Fragnito and E. Palange: Optics Comm. 40, 298 (1982).
3 J.E. Murray and D.J. Kuizenga: Appl. Phys. Lett. 37, 27 (1980).
4 J.E. Murray: Lawrence Livermore Lab. UCRL-77, 210-219 (1977).

A Novel Method for Generating Sub-Transform Limited Picosecond Nd: YAG Laser Pulses

S.C. Hsu and H.S. Kwok

Department of Electrical and Computer Engineering,
State University of New York at Buffalo,
Amherst, NY 14226, USA

1. Introduction

Picosecond Nd:YAG pulses have been traditionally generated by mode-locking. For high power pulses, passive mode-locking with a saturable dye solution is usually employed. However, being a statistical process, this method is plagued with fluctuations in both the pulse train reproducibility and the delay time in the pulse formation. We propose a method to generate these picosecond pulses using a scheme which has been proven to be successful for the CO_2 laser.

In this demonstrated method, a fast transient is introduced into the nanosecond pulses by a triggerable plasma shutter [1]. This fast transient is then filtered out with a suitable high pass spectral filter. For the case of the CO_2 laser, another CO_2 gas cell is used as a resonantly absorbing filter and pulses as short as 30 ps can be produced routinely [2]. Effectively, a picosecond slice is taken out of the nanosecond pulse. The ultimate duration of the slice is determined by the speed of the plasma shutter.

The advantages of applying the same scheme to the case of the Nd:YAG system are: (1) there is a possibility of generating pulses shorter than 30 ps, provided that the plasma shutter is fast enough; (2) no pulse selector is necessary to switch out a single pulse out of the pulse train; and (3) most importantly, the switching of this picosecond pulse can be triggered externally and therefore it can be synchronized with another picosecond laser system (e.g., the OFID CO_2 system) to perform double resonance experiments. It is estimated that picosecond pulses with peak powers of 20 MW can be produced by this method which is comparable to mode-locked pulses.

2. The Plasma Shutter

The main uncertainty in the proposed system is the speed of the plasma shutter. It is well known that a gas breakdown plasma can truncate the laser pulse ever since the early days of the giant pulse Ruby laser. However, the rapidity of the truncation had not been studied carefully since most attentions were focussed on the plasma itself.

In the case of the CO_2 laser plasma, the truncation time was measured to be 10 ps, corresponding to a plasma front propagation speed of 8.5×10^7 cm/sec. However, this result cannot be applied to the present situation without modification. Most importantly, the CO_2 laser plasma is overdense, i.e., $\omega_p > \omega_{laser}$ where ω_p is the plasma frequency. For the Nd:YAG laser, the plasma is underdense even assuming that all electrons are stripped off the gas molecules. The truncation of the laser pulse in this case is due primarily

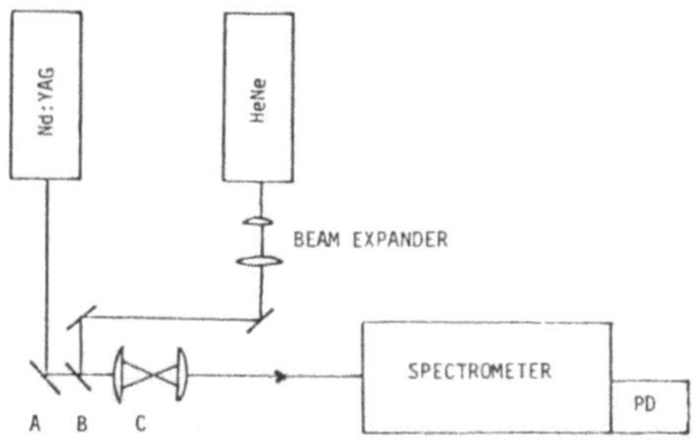

Fig. 1 Experimental set-up to demonstrate the truncation of a weak laser beam by a strong laser plasma

to inverse Bremsstrahlung absorption. (At intensities typical in our experiments, stimulated Brillouin and Raman scattering should be negligible.)

We have measured the transmitted pulse with a 100 ps resolution photodiode together with a Tektronix 519 oscilloscope. The observed falltimes for both the 1.06μm fundamental and the 0.532μm second harmonic pulses were ∿0.8 ns. The laser plasma was generated by focussing a 20 ns 0.3J pulse with a 50 mm focal length lens in clean air. The estimated peak power was 10^{12} W/cm^2. Since this falltime is quite close to the risetime of the oscilloscope, we believe that the actual falltime may be shorter. Interestingly, it was found that upon focussing the same laser pulse into a liquid cell, there was no truncation observed. Different liquids such as H_2O, methanol and acetone were tried. The rapid quenching of the plasma in a liquid must be responsible for the lack of truncation.

We have also performed one interesting measurement of the propagation of the plasma beyond the focal point. This experiment showed that it is possible to truncate a weak laser pulse with the plasma produced by another powerful laser. The experimental arrangement is depicted in Fig. 1.

A cw 10 mW HeNe laser was combined with the 1.06μm laser pulse by a dichroic mirror. The overlap of the two foci could be adjusted by a preçision angular orientation mount. The spectrometer was tuned to pass the 6328Å HeNe laser frequency. However, a small amount of the 1.06μm leaked through the spectrometer and could be detected. A typical trace of the detector output is shown in Fig. 2. A positive and a negative pulse can clearly be identified.

The positive pulse is the truncated 1.06μm while the negative pulse is the truncated 6328Å. The time delay between the two pulses could be varied by adjusting the distance between the two foci. There was a range of separations where no apparent delay could be observed as the two pulses were truncated simultaneously. Fig. 3 shows the time delay between the two truncation events versus the separation between the two beams. The risetime of the system was limited to 2 ns because a slower oscilloscope (Tektronix 485) was used. From the slope of this curve, the lateral propagation speed of

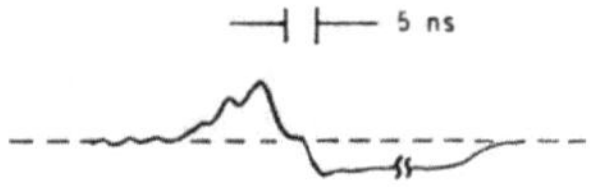

Fig. 2 Photodiode output. The positive pulse is the 1.06μm driver and the negative pulse is the truncated 6328Å

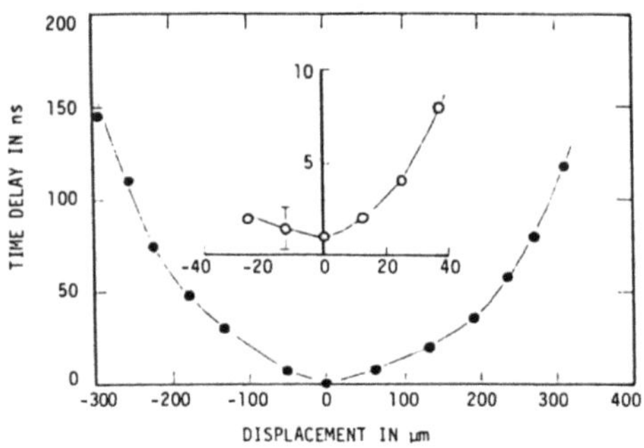

Fig. 3 Delay time between the truncation events vs. separation between the two foci

the truncating plasma can be estimated. It varies from 2.4 x 10^6 cm/sec at a distance of 20μm to 10^5 cm/sec at a distance of 300μm away. Propagation speed at distances closer than 20μm are inaccurate due to the uncertainty in the delay time. The size of the focus was estimated to be ∿10μm. Presumably, when the two foci are exactly aligned, the truncation time of the HeNe will be the same as the 1.06μm pulse which is <0.8 ns. Therefore, the ultimate plasma speed must be >1.4 x 10^6 cm/sec. Incidentally, these speeds are all within the range of the hydrodynamic or shock wave propagation mechanism [3].

The result of this experiment does not give us the speed of the plasma shutter. However, it directly demonstrates that the laser plasma can be used to truncate a much weaker cw laser beam. This result will have important applications in fast transient studies of atoms and molecules. For example, a single mode, highly coherent cw dye laser can be truncated and used to observe the free induction decay of atoms. This introduces a new option for fast transient spectroscopy.

3. Formation of Ultrashort Pulses

Once the rapidly truncated pulse is obtained, the spectral filtering and ultrashort pulse formation can be carried out in a straight-forward manner. Three different methods had been employed in the past: (1) A Michelson Interferometer can be used to produce a variable duration pulse provided the path difference is given by $2(L_1-L_2) = (n + 1/2)\lambda$ where n is an integer [4]. The corresponding square pulse duration is $2(L_1-L_2)/c$ where c is the speed of light; (2) A Fabry-Perot used in reflection can be used to produce a triangular pulse [5]. For reasonable reflectivities, the pulse duration is given by the round trip transit time inside the cavity $2\ell/c$; (3) An antiresonant ring also produces a square pulse with duration $(L_1-L_2)/c$ [6]. However, the plasma shutter is inside the ring and makes it more difficult to align. The above interference schemes can either be external or internal to the laser cavity.

In all of the above methods, the ultimate pulse duration that can be produced is given by the falltime of the plasma shutter. Moreover, they all depend on the constructive and destructive interference of two laser beams. Since the coherence length of a Q-switched Nd:YAG laser without any longitudinal mode control is about 1 cm, this puts an upper limit of 30 ps on the maximum pulse duration. Obviously, the spatial and longitudinal mode of the laser has to be improved before such schemes can be realized. Presumably, a single longitudinal mode laser should be used as the driver. This may diminish the attractiveness of the proposed system.

To demonstrate the feasibility of the proposed method, further work has to be done in (1) optimizing the speed of the plasma shutter by changing the gas pressure and/or type of gas molecules; and (2) limit the lasing to a single longitudinal mode. Experimental work is being carried out at this time.

Support by NSF Grant No. CPE 8103623 is gratefully acknowledged.

References

1. H.S. Kwok and E. Yablonovitch, Appl. Phys. Lett. 27, 583 (1975).
2. H.S. Kwok and E. Yablonovitch, Appl. Phys. Lett. 30, 158 (1977).
3. Yu P. Raiser, Laser-Induced Discharge Phenomena, Plenum Press, New York, 1977.
4. A. Szoke, J. Goldbar, H.P. Grieneisen and N.A. Kurnit, Opt. Comm. 6, 131 (1972).
5. R.A. Fisher and B.J. Feldman, Opt. Lett. 1, 161 (1977).
6. R. Trutna and A.E. Siegman, IEEE J. Quant. Elect. AE-13, 955 (1977).

Optical Dephasing in Inorganic Glasses

R.M. Shelby and R.M. Macfarlane

IBM Research Laboratory, 5600 Cottle Road,
San Jose, CA 95193, USA

In recent years it has become clear that the low temperature dynamical properites of glasses are quite different from those of crystals and that this has a profound effect on the relaxation behavior of ions and molecules present as dopants in glassy media [1,2]. In particular, the presence of 'two level systems' (TLS) — double well potentials associated with defects in glasses with a wide range of tunneling splittings [3] — has been proposed to explain the linewidths seen in optical experiments at low temperatures. Fluorescence line narrowing [1,2] and hole burning [4,5] widths are observed corresponding to dephasing times (T_2) in the nanosecond and sub-nanosecond range, as much as four orders of magnitude faster than for the same ion or molecule in a crystal.

These measurements have had experimental timescales ranging from the excited state lifetime (i.e. for fluorescence line narrowing) to many minutes (for long-lived hole burning experiments), but no time-domain measurements of T_2 have been reported. We have measured optical homogeneous linewidths at liquid helium temperatures for the rare earth ions Nd^{3+} and Eu^{3+} in silicate glass using picosecond coherent transients and long-lived optical hole burning techniques. The observed optical dephasing times are assigned for Nd^{3+} to population decay to other electronic states by multi-phonon emission and for Eu^{3+} to interactions with the TLS modes.

For the Nd-glass, measurements were made for Nd^{3+} concentrations of 0.5% and 2.0% at temperatures between 1.3 and 3.5K on the ${}^4I_{9/2} \leftrightarrow {}^4G_{5/2}$ band at 5850A. Dephasing measurements were made with the picosecond accumulated photon echo technique [6]. With this technique photon echoes are stimulated from a population distribution which varies sinusoidally with optical frequency. This distribution is produced by the action of high repetition rate picosecond pulse pairs which burn the pattern into the inhomogeneous line by storing population in a third, metastable level. The population storage level for Nd^{3+} was the ${}^4F_{3/2}$ level with a 200 μ sec lifetime. The photon echo decay curves were wavelength dependent, being longest (T_2 = 100psec) at the long wavelength edge of the band, and becoming faster and non-exponential at shorter wavelengths.

The photon echo data was in good agreement with linewidths measured in optical hole burning experiments where permanent (> 1 hour) holes were burned into the line by ~ 30min exposure with ~ 150W/cm^2 of single frequency cw laser light. These holes are attributed to a photo-induced rearrangement of the local environment of the excited ions. The change in environment shifts the optical absorption frequency within the broad inhomogeneous line. This type of hole burning is facilitated in glasses by the relative ease of conversion among a wide range of available environments. The potential barrier associated with the relaxation of the local environment is sufficiently large that at liquid helium temperatures the hole lifetimes are very long. This hole burning mechanism has been observed by others in organic glasses [4,5], but our measurements in Nd^{3+} and Pr^{3+} samples are the first for inorganic glasses.

The agreement of the hole burning and accumulated photon echo results shows that no additional relaxation of local environments in the glass contributes to widths observed on the (much longer) hole-burning timescale. This may not be surprising for Nd^{3+} since the T_2 values are likely dominated by population relaxation among the various 'crystal field' states of $^4G_{5/2}$ or to the nearby ^2H_J or ^2F_J manifolds. This interpretation is consistent with the observed concentration and temperature independence of our results over the range studied, and with estimates of the $^4G_{5/2}$ lifetime based on an energy gap law or on the fluorescence quantum yield. The wavelength dependence can be attributed to varying contributions from more than one crystal field level whose inhomogeneously broadened spectra overlap. The fast non-exponential behavior represents fast relaxation to the lowest crystal field level of this manifold.

The situation is quite different in Eu^{3+} doped silicate glass where we have observed hole burning in the $^7F_0 \leftrightarrow {}^5D_0$ transition with a recovery time of 20 seconds due to optical pumping of the Eu^{3+} nuclear quadrupole levels. At 2K the hole width is 50MHz, corresponding to T_2 = 13nsec. This value of the homogeneous linewidth (Γ = 24MHz) falls on a line extrapolated from higher temperature fluorescence line narrowing data [2] using a T^2 dependence, showing that the same T^2 law, i.e. $\Gamma = 5T^2$ MHz is obeyed from below 2K to over 200K. This dependence is often ascribed to dephasing by impurity ion - TLS interactions. Once again, the agreement between the fluorescence line narrowing and hole burning results is significant in view of the difference in experimental timescales of 10^4. The observation of a smooth T^2 temperature dependence over the entire temperature range is in disagreement with recent theoretical work [7] which predicts a crossover from quadratic to linear temperature dependence at these low temperatures. Clearly more work in both theory and experiment will be required to understand this dephasing mechanism in terms of the TLS model.

The authors wish to thank M.J. Weber for providing some of the samples used in these experiments.

REFERENCES

1. J. Szeftel and H. Alloul, Phys.Rev.Lett. *34*, 657 (1975); J. Hegarty and W.M. Yen, Phys.Rev.Lett. *43*, 1126 (1979).
2. P.M. Selzer, D.L. Huber, D.S. Hamilton, W.M. Yen,and M.J. Weber, Phys.Rev.Lett. *36*, 813 (1976).
3. P.W. Anderson, B.I. Halperin and C.M. Varma, Phil.Mag. *25*, 1(1972).
4. B.M. Karlamov, R.I. Personov, and L.A. Bykovskaya, Opt.Commun. *12*, 191 (1974).
5. J.M. Hayes, R.P. Stout, and G.J. Small, J.Chem.Phys. *74*, 4266 (1981).
6. W.H. Hesselink and D.A. Wiersma, Phys.Rev.Lett. *43*, 1991 (1979).
7. P. Reineker and H. Morawitz, Chem.Phys.Lett. *86*,359(1982).

Part II

Ultrashort Measuring Techniques

Picosecond Holographic Grating Experiments in Molecular Condensed Phases

M.D. Fayer

Chemistry Department, Stanford University, Stanford, CA 94305, USA

In this article some recent results from a nonlinear approach to the application of subnanosecond laser pulses to the investigation of molecular interactions and excited state dynamics are briefly described. The method involves the optical generation of a transient holographic diffraction grating in a sample, and the observation of various time and frequency dependent phenomena via subsequent Bragg diffraction from the induced grating. The basic experiment works in the following manner. Two time coincident picosecond laser pulses of the same wavelength are crossed inside of the sample to set up an optical interference pattern. The fringe spacing, d, of the interference pattern is determined by the angle between the beams, θ, and the wavelength, λ, of the excitation pulses, i.e.,

$$d = \lambda/2\sin(\theta/2). \qquad (1)$$

The interaction of the radiation field with the sample can produce a number of different changes in the sample, depending on the nature of the sample and the wavelength, λ. Electronic excited states can be produced (1), internal molecular vibrations can be excited (2), or acoustic waves, i.e., phonons, the collective vibrations of the medium, can be generated (2,3). In some experimental situations more than one of the above types of excitations are simultaneously produced (3d).

In all cases, the excitations generated in the sample have a spatial periodicity which mimics the periodicity of the optical interference pattern used to excite the sample. Excitation results in a spatially periodic change in the physical properties of the system. This in turn produces a periodic variation in the sample, s complex index of refraction, $\tilde{n}$ (3d),

$$\tilde{n} = n + iK. \qquad (2)$$

The periodic variation in $\tilde{n}$ acts as a Bragg diffraction grating for a picosecond probe pulse (3d). The probe pulse is brought into the sample to meet the Bragg diffraction condition for the holographic transient grating produced by the excitation beams. A part of the probe pulse is diffracted and leaves the sample in a unique direction as a collimated beam. The intensity of the diffracted beam is the observable in the transient grating experiment. The probe pulse can be delayed in time various amounts, and the intensity of the diffracted beam as a function of probe pulse delay can be related to the system's dynamics (1). In addition, the probe pulse can be brought in at a fixed delay time, and the wavelength of either the probe (3d), or excitation beams (2), can be varied. In this manner various types of spectroscopic measurements can be made.

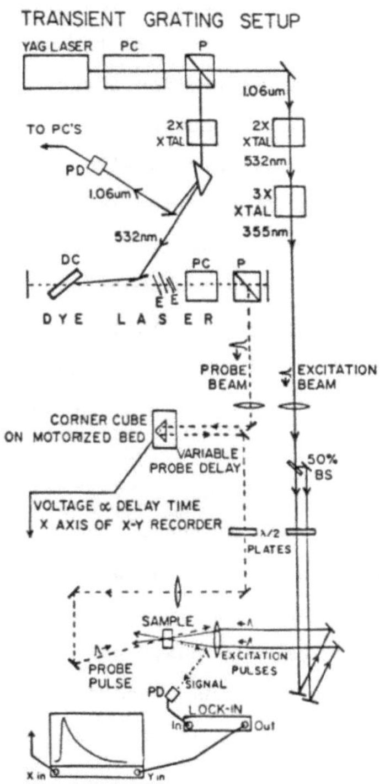

The transient grating experimental setup is shown in Fig. 1. The system operates at 500 Hz. A single 1.06 μm pulse is selected from the YAG mode-locked pulse train. Generally, one of its harmonics is employed (shown here with 3×) although in some experiments the fundamental or a tunable mode-locked dye laser pulse may be used. The single pulse is then split into two excitation pulses. These excitation pulses are recombined at the sample, creating the transient grating. The remainder of the pulse train is frequency doubled to synchronously pump a tunable dye laser whose output probes the grating after a variable delay. In some experiments a YAG harmonic is used as a probe. The Bragg-diffracted part of the probe pulse is the transient grating signal. (In the figure, PC ≡ Pockels cell; P ≡ polarizer; PD ≡ photodiode; DC ≡ dye cell; E ≡ etalon; BS ≡ beam splitter.)

The contributions to the transient grating diffraction efficiency arising from excited state amplitude grating effects (changes in the imaginary part of the index of refraction, K) and from phase grating effects (changes in the real part of the index of refraction, n) are demonstrated experimentally in Fig. 2 (3d). The sample is a mixed molecular crystal (solid solution) of pentacene in the host p-terphenyl. The inset in the figure shows the absorption spectrum. The transient grating excitation employed doubled Nd:YAG pulses at 532 nm. The diffraction efficiency $\eta_{ex}(\omega)$ due to the resulting excited state grating was measured at fixed time delay (500 psec) as a function of probe wavelength, ω, using tunable dye laser pulses as the probe. In Fig. 2a, the circles with the solid line are the experimentally measured excited state grating diffraction intensity as a function of probe wavelength near the S_0 to S_1 transition of pentacene. The dashed curve is theoretically calculated from the absorption spectrum (inset) (3d). The dash-dot curve is the calculated amplitude grating contribution to the diffraction efficiency. This demonstrates the significant contribution of phase grating effects. In Fig. 2b, the points with the solid line are the phase grating contribution to the diffraction intensity obtained by subtracting the curves in 2a. The predicted m shape curve associated with excited state phase grating diffraction is clearly observed. The dashed curve is theoretically calculated from the absorption spectrum. On the red side where the transition is isolated the agreement is good. On the blue side, interference from the next spectral peak (see inset), which was not included in the calculation, influences the dispersion effect. Although excited state phase gratings have been discussed by a number of authors (4), experimental observations have been sketchy. The results presented here provide the clearest characterization of the wavelength dependence of excited state phase grating diffraction. In many experimental situations, failure to properly account for both phase and amplitude grating effects can lead to erroneous interpretations of data. This can be true in more general four-wave mixing experiments as well as in transient grating experiments.

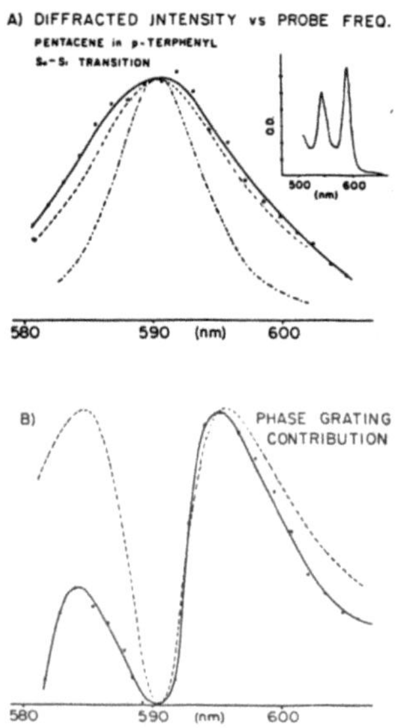

Picosecond optical gratings can provide a convenient method for optical generation of ultrasonic waves in transparent or light-absorbing liquids and solids. The acoustic frequency can be continuously and easily varied from about 3 MHz to 30 GHz with our experimental apparatus, and a considerably wider range should be possible. In anisotropic media any propagation direction can be selected.

The technique, called Laser Induced Phonons (LIPS), is based on the transient grating experiment. Energy deposited into the system via optical absorption or stimulated Brillouin scattering results in the launching of counterpropagating ultrasonic waves (phonons) whose wavelength and orientation match the interference pattern geometry. The acoustic wavelength is given by Eq. 1. The acoustic wave propagation causes time-dependent, spatially periodic variations in the material density, and since the sample's optical properties (real and imaginary parts of the index of refraction) are density-dependent, the irradiated region of the sample acts as a Bragg diffraction grating. The propagation of the ultrasonic waves can be optically monitored by time-dependent Bragg diffraction of a variably delayed probe laser pulse (3d).

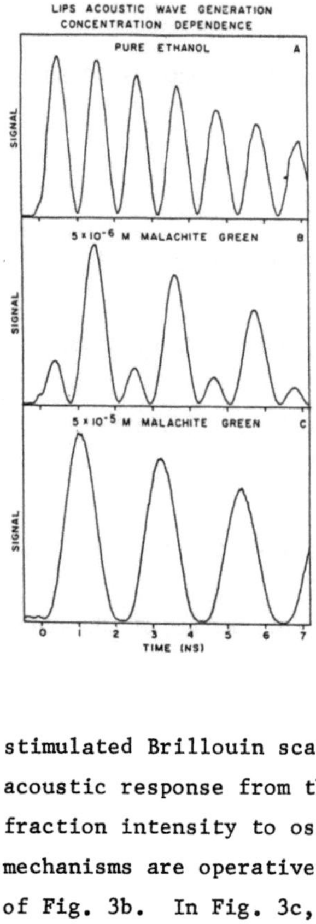

Figure 3 shows LIPS transient grating data from pure ethanol and solutions of malachite green (MG) in ethanol. The excitation wavelength was 532 nm and the probe wavelength was 566 nm. The fringe spacing (2.47 μm) and the ethanol velocity of sound produce an acoustic cycle time, τ_{ac} = 2.13 nsec. Experimental conditions for the data sets a - c were identical except for the MG concentration. In pure ethanol (Fig. 3a) there is no optical absorption. Stimulated Brillouin scattering is responsible for the generation of a standing acoustic wave with wavelength equal to the grating fringe spacing. The standing wave causes the diffraction intensity to oscillate twice each acoustic cycle, τ_{ac}. In Fig. 3b optical absorption by the MG and the subsequent rapid (2 ps) deposition of heat, as well as stimulated Brillouin scattering, contribute to the acoustic response. The acoustic response from the optical absorption mechanism causes the diffraction intensity to oscillate only once each acoustic cycle. When both mechanisms are operative to comparable degrees, the data has the appearance of Fig. 3b. In Fig. 3c, because of increased MG concentration, the optical absorption mechanism completely dominates and there is one oscillation per τ_{ac}.

The LIPS technique is an extremely versatile tool for controlled optical generation of ultrasonic waves in condensed media. LIPS experiments have been performed on transparent and absorbing solutions, organic and inorganic crystals, glasses, and plastics. The effects have been observed from liquid helium temperatures to room temperature. The optically generated acoustic waves can be optically amplified, cancelled, or phase shifted (3c). LIPS has been used to measure anisotropic elastic constants, acoustic attenuation parameters, photoelastic constants and spectra of weakly absorbing materials. We are currently using LIPS to investigate excited state-phonon interactions, thermal diffusivity in molecular crystals at liquid helium temperature and structural properties of phospholipid bilayers.

Acknowledgment

I would like to thank the National Science Foundation (DMR 79-20380) for support of this research.

References

1. Lutz, D. R., Nelson, K. A., Gochanour, C. R., Fayer, M. D. 1981, Chem. Phys. 58:325.
2. Miller, R. J. D., Casalegno, R., Nelson, K. A., Fayer, M. D. 1982, Chem. Phys., in press.
3. a) Nelson, K. A., Fayer, M. D. 1980, J. Chem. Phys. 72:5202;
   b) Nelson, K. A., Lutz, D. R., Fayer, M. D. 1981, Phys. Rev. B 24:3261;
   c) Nelson, K. A., Miller, R. J. D., Lutz, D. R., Fayer, M. D. 1982, J. Appl. Phys. 53:1144;
   d) Nelson, K. A., Casalegno, R., Miller, R. J. D., Fayer, M. D. 1982, J. Chem. Phys., in press.
4. Hammer, J. M. 1968, Appl. Phys. Lett. 13:318.

Self-Diffraction from Laser-Induced Orientational Gratings in Semiconductors

A.L. Smirl, T.F. Boggess, B.S. Wherrett*, G.P. Perryman, and A. Miller†

Center for Applied Quantum Electronics, North Texas State University, Denton, Texas 76203, USA

We have resolved new ultrafast structure in the picosecond excitation-probe response of thin semiconductor wafers. This structure, located near zero delay and observed only at the very highest excitation intensities (~10 GW/cm^2), can be understood only in terms of self-diffraction from a transient orientational grating produced by anisotropic state-filling. This anisotropic state-filling, as opposed to band-filling, arises from a δ-function-like spike in energy and a directional dependence in momentum of the carrier distribution function, caused by the nearly monochromatic polarized nature of the exciting radiation. This is the first observation of anisotropic state-filling of which we are aware.

A single pulse at 1.06 µm with a duration of 8 psec (FWHM), produced by a mode-locked Nd:glass laser, was divided by a beam splitter and a variable delay was introduced into one path. The delayed pulse (probe) was attenuated by a factor greater than 1,000. The two pulses, the excitation and the probe, were recombined at a small angle θ after focusing on the surface of a 5.7-µm-thick wafer of crystalline germanium. The experimental procedure was to measure the probe transmission as a function of time delay after the excitation pulse with excitation and probe electric field polarizations arranged parallel. The peak excitation pulse fluence was measured to be 60 mJ/cm^2. Notice that this is just short of the damage threshold and is higher than excitation levels previously reported.

Inspection of Fig.1a reveals three distinct features. The most prominent of these is a rapid rise and fall in probe transmission (~2 psec, FWHM) centered about zero delay. This spike has been observed previously [1] and has been interpreted [2,3] as a parametric coupling (or self-diffraction) of the excitation beam into the probe beam caused by a carrier concentration grating produced by the interference of the two pulses when they are temporally and spatially coincident near zero delay. This narrow spike is followed by a gradual rise and fall of the probe transmission lasting hundreds of picoseconds. This slower structure has been studied previously and is not the subject of our investigations here.

A more careful examination of the structure near zero delay in Fig.1a shows that the narrow feature is superposed on a broader rise and fall in the probe transmission approximately 10 psec (FWHM) wide. This structure

---

*Permanent Address: Department of Physics, Heriot-Watt University, Edinburgh, Scotland.

†Present Address: Royal Signals and Radar Establishment, Malvern, Worcs., England.

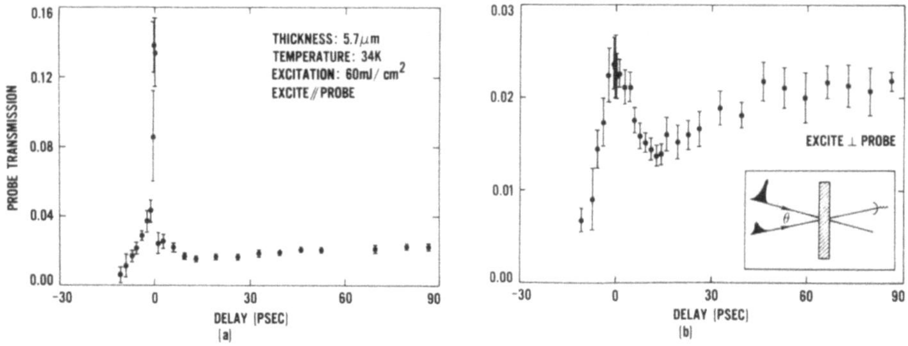

Fig.1 Probe transmission as a function of time delay between the excitation and probe pulses

has not been previously observed. That this structure is distinct from the narrower correlation spike can be demonstrated by repeating the measurements in Fig.1a with the probe polarization rotated perpendicular to that of the excitation pulse. The results are shown in Fig.1b. Clearly, the narrower spike disappears when the polarizations are crossed while a broader structure remains.

Similar measurements (Fig.2) in the direction of the background-free self-diffracted pulse ($-\theta$) confirm that both features are produced by self-diffraction from laser-induced transient gratings and substantiate the polarization dependence. The self-diffracted signal at $-\theta$ is shown in Fig.2b for perpendicular relative polarizations for the excitation and probe fields. Again notice that the narrow spike has disappeared but the broader spike remains for crossed polarizations.

The origin of the broader spike, and hence of the entire signal in the configuration of Fig.2b, we attribute to the presence of an orientational grating. We hold that the excited carriers in the valence band are distri-

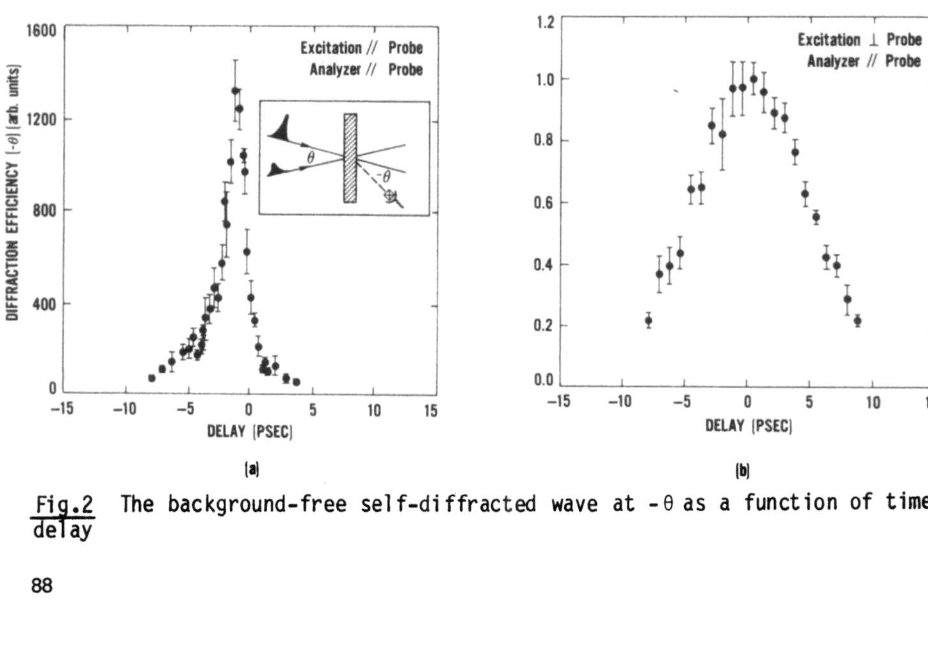

Fig.2 The background-free self-diffracted wave at $-\theta$ as a function of time delay

buted in k-space with a preferred orientation and that this orientation modulates across the irradiated region. To show how this anisotropy can come about we concentrate on the heavy-hole states and model the interband transitions by those of a set of independent two-level systems. Saturation is described by performing iterative calculations to third-order in the electromagnetic field, within the slowly-varying-wave approximation. We account for (i) de-excitation of the saturated two-level systems by recombination or by scattering away from the optically coupled states; (ii) spatial diffusion and (iii) reorientational diffusion. The third-order polarization, at time t, takes the form:

$$\underset{\sim}{P}^{(3)}(t) \propto |p_{cv}|^4 \sum_{ijk\ell} \exp[i(\underset{\sim}{k}_j-\underset{\sim}{k}_k+\underset{\sim}{k}_\ell)\cdot\underset{\sim}{r}]\hat{\varepsilon}_i E_j(t) \int_{-\infty}^{t} E_k^*(t')E_\ell(t')A(t-t')dt' \;,$$

$$A(t) = \left\{Y_{ij}^{(1)}Y_{k\ell}^{(1)}[1 - \exp(-t/T_\theta)] + Y_{ijk\ell}^{(3)}\exp(-t/T_\theta)\right\} \exp[-t(T_v^{-1} + T_{k\ell}^{-1})] \,.$$

Here p_{cv} is the interband momentum matrix element, and $\underset{\sim}{k}_j$ is the propagation direction of the component of the field of polarization $\hat{\varepsilon}_j$. The orientational diffusion time is T_θ, T_v is the de-excitation lifetime and $T_{k\ell}$ is the spatial diffusion time. The latter is present only if the fields E_k and E_ℓ propagate in different directions. The parameters $Y^{(1)}$ and $Y^{(3)}$ have the symmetries of $\chi^{(1)}$ and $\chi^{(3)}$ optical susceptibilities, respectively,

$$Y_{ij}^{(1)} = \langle\hat{p}_{vc}\cdot\hat{\varepsilon}_i^*\;\hat{p}_{cv}\cdot\hat{\varepsilon}_j\rangle;\; Y_{ijk\ell}^{(3)} = \langle\hat{p}_{vc}\cdot\hat{\varepsilon}_i^*\;\hat{p}_{cv}\cdot\hat{\varepsilon}_j\;\hat{p}_{vc}\cdot\hat{\varepsilon}_k^*\;\hat{p}_{cv}\cdot\hat{\varepsilon}_\ell\rangle \,.$$

The average < > is over all $\underset{\sim}{k}$ orientations. Note that the effect of reorientation is to change the tensor nature of $P^{(3)}$. The values of the Y-parameters are given below for the heavy-hole model and, in parentheses, for a molecular model [4]. For the experimental conditions there are six distinct contributions to $P^{(3)}$, all of strength of the order of the probe field times the excitation intensity. We take the excitation field to be in the z-direction and the probe in the z(parallel) or x (orthogonal) direction. The background term originates from the gradual accumulation of carriers in the excited states. Self-diffraction terms are the sources for the experimental spikes.

Parallel Configuration:

All Contributions	Y_{zz} = 1/3 (1/3)	Y_{zzzz} = 2/15 (1/5)

Orthogonal Configuration:

Background (θ direction)	$Y_{xx} = Y_{zz}$ = 1/3 (1/3)	Y_{xxzz} = 1/10 (1/15)
Self-diffraction (θ)	$Y_{xz} = Y_{zx}$ = 0 (0)	Y_{xzzx} = 1/10 (1/15)
Self-diffraction (-θ)	Y_{xz} = 0 (0)	Y_{xzxz} = -1/15 (1/15)

In the orthogonal orientation, there is no grating-like modulation of the radiation intensity - the total concentration of excited-carriers is uniform in space. However the direction of the resultant polarization of the interfering beams does modulate. By analogy to the fact that for molecules in a liquid one would achieve greater excitation of those molecules aligned along the local polarization direction than normal to it, so also is there a preferred excitation in semiconductors. In this case, however, the nature of the heavy-hole eigenfunctions demands that those electrons with wave-vector $\underset{\sim}{k}$ normal to the polarization direction are preferentially excited. Consequently modulation of this $\underset{\sim}{k}$ orientation results and self-diffraction can still occur, as manifested by the nonzero Y_{xzzx} and Y_{xzxz} terms above. The presence of a signal in the $-\theta$ direction is testimony experimentally that a grating has been formed.

In the parallel configuration, reorientation plays only a small role. A concentration grating exists because the resultant radiation intensity modulates. In fact, band-filling as a consequence of free-carrier thermalization is expected to dominate the saturation behavior in this configuration [3]; the symmetry of the associated $P^{(3)}$ is of $Y_{zz}^{(1)}$, $Y_{zz}^{(1)}$ type. As a further point, we note that the response function $A(t-t')$ associated with band-filling is slowly varying so that the diffracted energy will reflect the coherence of the excitation pulse (~ 2 psec). However the anisotropic state-filling grating will decay rapidly-primarily through intraband deexcitation - and the diffracted energy will reflect the intensity correlation of the pulses (~ 10 psec).

In conclusion, additional detailed studies (not described above) indicate that the narrower spike in Fig.1a is indeed a result of a spatially-dependent dynamic Burstein-Moss shift of the absorption edge (band-filling). The corresponding concentration grating is formed at all intensities for which saturation of the direct absorption is observed, but only if the excitation and probe pulses have parallel electric field components. The grating has a measured lifetime of 50 psec, consistent with the expected decay by spatial diffusion.

By contrast, the broader spike has characteristics that are consistent only with an orientational grating formed by an anisotropic (in the Brillouin zone) filling of the optically-coupled states (state-filling). That is, a grating that diffracts light is formed regardless of the direction chosen for the polarization of the probe pulse. This grating is observed only at the very highest excitation levels, where the generation rate for optically-created carriers into energy states in a given $\underset{\sim}{k}$-direction might be expected to be of the same order as the scattering-out rate. The measured state-filling grating lifetime is short compared to the 8 psec pulsewidth. Furthermore, the polarization dependence of the scattered light is accurately predicted by the nonlinear polarization given above.

This work was supported by the Office of Naval Research and The Robert A. Welch Foundation.

References

1. C. J. Kennedy, J. C. Matter, A. L. Smirl, H. Weichel, F. A. Hopf, S. V. Pappu, and M. O. Scully, Phys. Rev. Lett. 32, 419 (1974).
2. C. V. Shank and D. H. Auston, Phys. Rev. Lett. 34, 479 (1975).
3. J. R. Lindle, S. C. Moss, and A. L. Smirl, Phys. Rev. B 20, 2401 (1979).
4. A. von Jena and H. E. Lessing, Appl. Phys. 19, 181 (1979).

A Picosecond Raman Technique with Resolution Four Times Better than Obtained by Spontaneous Raman Spectroscopy

W. Zinth, M.C. Nuss, and W. Kaiser

Physik Department der Technischen Universität München
D-8000 München, Fed. Rep. of Germany

A new Raman technique is presented which allows to observe a Raman transition with a bandwidth smaller than the common spontaneous Raman line-width.

This technique is based on short excitation and prolonged interrogation (SEPI) of molecular states /1,2/. During the short and transient excitation process the molecules are driven at the difference frequency $\nu_D = \nu_1 - \nu_2$ by two pulses of frequency ν_1 and ν_2. Raman transitions which are close to the frequency ν_D become coherently excited with amplitudes Q_i. This material excitation persists when the two pumping pulses have left the sample. After the excitation the molecules vibrate with their individual resonance frequencies and the coherent amplitudes Q_i decay exponentially with the time constants T_{2i}. A third delayed probe pulse interacts with the coherently vibrating molecules and generates a Stokes spectrum of the freely relaxing material excitation.

The crucial point of the transient excited Raman spectroscopy discussed here is the narrow Stokes spectrum produced by the long third pulse. Only molecules vibrating in phase contribute to the coherent Stokes light. Molecules which have suffered collisions are out of step and are not observed subsequently. For Gaussian shaped probing pulses the spectral width of the observed Stokes bands equals the width $\Delta\nu_L$ of the interrogating third pulse /1,2/. With long probing pulses of duration $t_p > 1.4\ T_{2i}$ the SEPI resolution will be better than the resolution of spontaneous Raman scattering.

The experiments on SEPI spectroscopy are performed using a picosecond Nd-glass laser system /2/. The second harmonic frequency ν_1 is used for one pumping and the probing pulse. The second pump frequency ν_2 is generated via transient stimulated Raman scattering in a generator cell.

Experimental results on liquid cyclohexane in the frequency range 2850 cm^{-1} to 2940 cm^{-1} are shown in Fig.1. Fig.1a shows the emission band-widths of the liquids employed to generate pulses at ν_2. The spontaneous polarized Raman spectrum of C_6H_{12} is shown in Fig.1b. Between the three strong CH-stretching modes one encounters a diffuse spectrum due to overlapping overtones or combination bands. Transition frequencies found in SEPI spectroscopy are marked with the vertical lines.

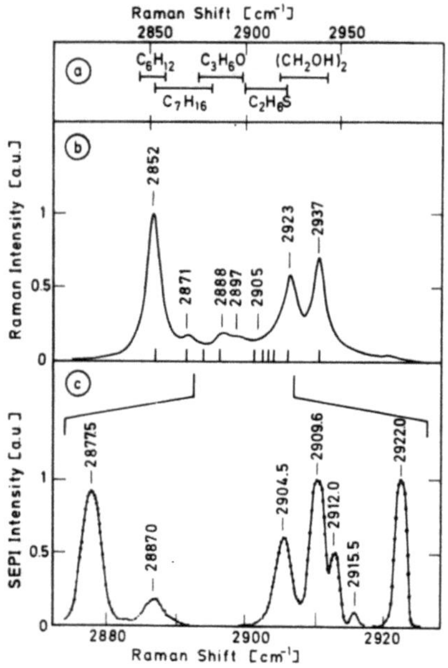

Fig.1 Experimental results of SEPI spectroscopy of C_6H_{12}. (a) Frequency ranges of the various Raman generators liquids used in the experiment. (b) Polarized spontaneous Raman spectrum of of C_6H_{12} recorded with a resolution of 1 cm^{-1}. The frequency positions of the resonances found in SEPI spectra are marked with vertical lines. (c) Three SEPI spectra taken with different generator liquids. New Raman lines are detected and the spectral resolution is improved. (Note, the frequency scale of (c) is 3.7 times larger than the one of (b).)

In Fig. 1c we show three SEPI spectra on an expanded scale (factor 3.7). Each spectrum was obtained by a single laser shot. On the r.h.s. we present the sharp SEPI band corresponding to the CH-stretching mode at 2923 cm^{-1}. We note that the SEPI band is considerably narrower than the corresponding band in the spontaneous Raman spectrum. The SEPI spectrum in the center shows four Raman transitions between 2905 cm^{-1} and 2916 cm^{-1}. Lines as close as 2.5 cm^{-1} are clearly resolved. In spontaneous Raman spectra the four transitions are hidden under the wing of the strong Raman band at 2923 cm^{-1} and cannot be detected. Fig.1c, l.h.s.,shows a SEPI spectrum of the frequency range 2875 cm^{-1} to 2890 cm^{-1}. We find two distinct Raman bands at 2877.5 cm^{-1} and 2887 cm^{-1}. The band at 2877.5 cm^{-1} has never been reported on previously. It is buried in the diffuse part of the conventional Raman spectrum.

The following points are relevant for the application of the SEPI technique: (i) The frequency positions of the observed Raman lines are independent of the excitation conditions since we observe freely relaxing molecules. (ii) In SEPI experiments the exciting and interrogating pulses should not overlap temporarily in order to avoid the generation of a coherent signal via the non-resonant four-photon parametric process. (iii) SEPI spectra taken for different delay times allow an estimate of the dephasing times T_{2i}. (iv) The frequency precision of the generated Stokes spectrum depends upon the frequency stability of the interrogating pulse. For highest accuracy the frequency ν_1 has to be measured simultaneously with the SEPI spectrum. (v) The scattering process may also be performed on the anti-Stokes part of the spectrum. The disturbing interference found in stationary CARS spectroscopy does not occur for the delayed probing used with SEPI spectroscopy.

The data presented here give convincing evidence of the potentiality of the short excitation and prolonged interrogation spectroscopy; new Raman lines are readily observed and vibrational energies are determined with improved accuracy.

References

1 W. Zinth, Optics Commun. 34 (1980) 479
2 W. Zinth, M.C. Nuss, and W. Kaiser, Chem. Phys. Lett. 88, (1982) 257

Broadband CARS Probe Using the Picosecond Continuum

L.S. Goldberg

Naval Research Laboratory, Washington, D.C. 20375, USA

1. Introduction

Coherent antistokes Raman scattering (CARS) provides a potentially useful diagnostic approach to the identification and study of transient molecular fragment species produced as the primary events in UV laser photolysis of molecules. GROSS, GUTHALS, and NIBLER [1] applied nanosecond dye laser techniques to obtain scanned as well as broadband single-shot CARS spectra of transient species from 266-nm photolysis of benzene vapor and derivatives. This work was extended to the picosecond time scale by HETHERINGTON III, KORENOWSKI, and EISENTHAL [2] who used optical parametric generation to provide tunable frequency Stokes pulses for a point-by-point probe of the photolysis spectrum. Earlier, GREEN, WEISMAN, and HOCHSTRASSER [3] had demonstrated single frequency picosecond CARS measurements in molecular nitrogen. In the present paper, development of a broadband picosecond CARS probe technique is reported. The method uses the picosecond white-light continuum [4] as Stokes light and enables an extensive antistokes spectrum to be obtained in a single 5-ps laser pulse.

2. Experimental

Figure 1 shows a schematic of the experimental arrangement. A recently developed modelocked Nd:phosphate glass laser system produces energetic pulses (25 mJ, 5 ps) of high beam quality at 1054 nm and harmonics, at a pulse repetition rate of 0.2 Hz [5]. The laser second harmonic at 527 nm serves as the pump frequency, ω_1, for the four-wave nonlinear CARS interaction. Its near transform-limited spectral width of ~4 cm^{-1} defines the spectral resolution of the measurements. A picosecond pulse continuum, extending throughout the visible and near IR spectrum, was produced by focusing the

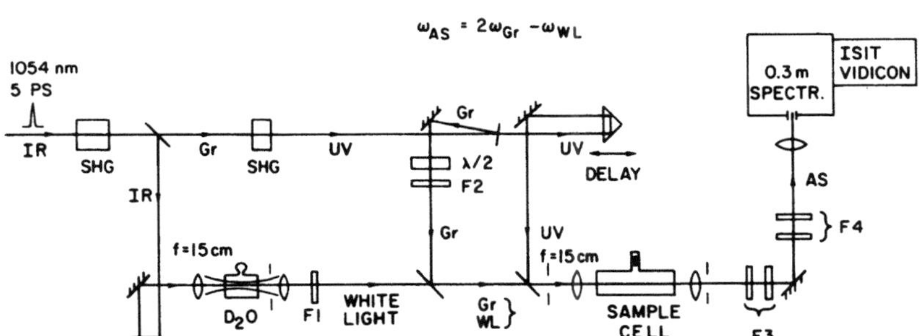

Fig.1 Schematic of the broadband CARS probe and photolysis experiment

Fig.2 Single-shot spectrum of white-light continuum in D_2O and in 50% D_3PO_4/D_2O mixture. The equivalent range of antistokes wavelengths are given in parentheses.

1054-nm fundamental into a 5-cm liquid D_2O cell (Fig. 2, upper trace). A 50% mixture of D_3PO_4/D_2O also has been used and produces a lower intensity, but spectrally more-uniform continuum (Fig.2, lower trace). The continuum beam is collimated and filtered to pass wavelengths >530 nm, thus providing a broad band of light at Stokes frequencies, ω_2. The ω_1 and ω_2 pulses are then combined spatially and temporally, and focused collinearly into a 22-cm gas sample cell. Relatively strong coherent antistokes signals are generated over the spectrum of frequencies, $\omega_3 = 2\omega_1 - \omega_2$, corresponding to Raman-active vibrational resonances, $\omega_1 - \omega_2$, in the third-order susceptibility $\chi^{(3)}$ of the system under study. The ω_3 beam is spectrally filtered using short-pass dielectric filters and focused into a 0.3-m grating spectrograph. The dispersed CARS spectrum is then recorded by an OMA II intensified vidicon system. For photolysis experiments, the fourth-harmonic beam at 264 nm is sent through an independent delay path, recombined collinearly with the probe pulse pair and focused into the sample cell.

3. Results

Figure 3 (lower trace) presents a CARS spectrum obtained at low resolution with a single 5-ps laser pulse from ground-state benzene vapor at 60 torr. The prominant narrow spectral features are identified as the fully symmetric 3070 cm^{-1} C-H stretch and 992 cm^{-1} C-C stretch modes of benzene. The vibrational frequency range encompassed in this measurement extends well over 2000 cm^{-1}. In direct estimates of signal strengths using the vidicon system, the CARS signal at 3070 cm^{-1} was determined to be ~10^{-4} of the corresponding white-light signal. A joulemeter measurement of the white-light pulse energy gave ~10 μJ from $550<\lambda<620$ nm, yielding an average of ~5 nJ/cm^{-1}. This corresponds to approximately 10^6-10^7 photons in the antistokes signal over a linewidth of 4 cm^{-1}.

Upon photolysis of benzene (Fig.3, upper trace), dramatic new spectral features appear in the region from 2300-2600 cm^{-1}. The sample was probed ~200 ps after arrival of the UV pulse. Figure 4 shows CARS spectra for ground-state and photolyzed toluene vapor at 20 torr. The new spectral lines appear the same for both molecules, although there are differences in line intensities that may be traceable to shot-to-shot variations in white-light intensity. There is also some correlation in the occurrence of intense green pulses with production of particularly strong fragment signals.

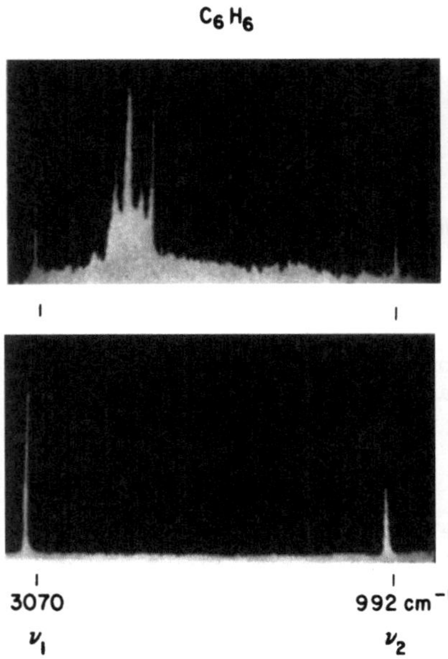

Fig.3 Single-shot CARS spectrum of benzene at 60 torr (bottom), after photolysis (top)

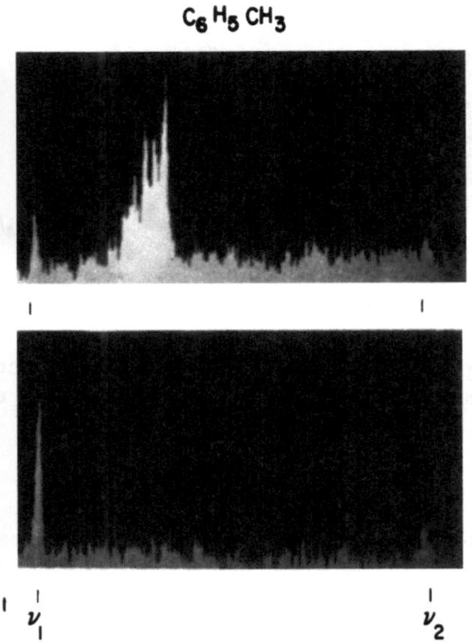

Fig.4 CARS spectrum of toluene at 20 torr (bottom), after photolysis (top)

4. Discussion

These measurements have shown an intense CARS product spectrum strikingly similar, albeit appearing in a distinctly different spectral region, to the complex spectrum attibuted by GROSS et al. [1] to the C_2 diradical. Their spectrum in the 2800-3000 cm^{-1} region was shown to have no hydrocarbon· component and was interpreted as arising from multiply resonantly-enhanced CARS signals of C_2. HETHERINGTON III et al. [2] observed the same spectral lines to be present within the 25 ps duration of their overlapping photolysis and probe pulses. BOESL et al. [6] and later HERING et al. [7], using mass spectrometry detection, have demonstrated that 2-photon UV ionization is the primary step in fragmentation of benzene yielding $C_6H_6^+$, and that subsequent photons, most effectively in the visible spectrum, lead to efficient further fragmentation of the parent ion species.

The strength of the new bands observed in the current experiments relative to the dominant ν_1 and ν_2 modes of benzene suggests that they also arise from strong multiple resonant enhancements [8] in the susceptibility of the product species. On the assumption that C_2 is the origin of these CARS signals, the difference in spectrum compared with that of [1] and [2] may result from the different laser wavelengths (527 nm vs. 532 nm) that define ω_1. The antistokes frequencies ω_3 then fall among the vibrational-rotational transitions in the $\Delta v = +1$ Swan series of C_2 ($d^3\Pi_g \leftarrow a^3\Pi_u$), while ω_2 lies in the region of $\Delta v = -2$, giving opportunity for several multiple resonances to occur.

References

1. K. P. Gross, D. M. Guthals, and J. W. Nibler, J. Chem. Phys. 70, 4673 (1979).
2. W. M. Hetherington III, G. M. Korenowski, and K. B. Eisenthal, Chem. Phys. Lett. 77, 275 (1981).
3. B. I. Greene, R. B. Weisman, and R. M. Hochstrasser, Chem. Phys. Lett. 59, 5 (1978).
4. The continuum has previously been applied to inverse Raman scattering; R. R. Alfano and S. L. Shapiro, Chem. Phys. Lett. 8, 631 (1971).
5. L. S. Goldberg, P. E. Schoen, and M. J. Marrone, Appl. Optics 21, 1474 (1982).
6. U. Boesl, H. J. Neusser, and E. W. Schlag, J. Chem. Phys. 72, 4327 (1980).
7. P. Hering, A. G. M Maaswinkel, and K. L. Kompa, Chem. Phys. Lett. 83, 222 (1981).
8. S. A. J. Druet, B. Attal, T. K. Gustafson, and J. P. E. Taran, Phys. Rev. A18, 1529 (1978).

Jitter-Free Streak Camera System

W. Knox, T.M. Nordlund, and G. Mourou

Laboratory for Laser Energetics, University of Rochester,
250 East River Road, Rochester, New York 14623, USA

Since the demonstration of a streak camera with 2 ps jitter,[1] we have evolved a versatile system for picosecond detection which facilitates study in a wide range of areas. The capabilities of the system are discussed with emphasis upon the advantage of signal averaging.

The detection system consists of a Photochron-II image-converter tube followed by an EMI 4-stage magnetically focused intensifier section and an OMA-II SIT vidicon detector. In this system, the deflection ramp voltage for the image-converter tube is generated with a picosecond high-power semiconductor switch: GaAs doped with chromium.[2] The switch operates in a linear photoconductive mode, resulting in a shot-to-shot jitter of only 2 ps, and complete absence of timing drift. Day-to-day reliability is ± 2 ps.

This enables the use of a particularly simple signal averaging technique: summation of successive laser shots (Nd^{+++}:YAG at 1 Hz). This results in an increase of signal-to-noise ratio in proportion to the square root of the number of shots averaged. Signal averaging allows the detection of very weak signals[3,4] and comparison of data to kinetic theory to a greater degree of accuracy[5] than single-shot results. Since the system has no internal drift, several experimental techniques become greatly simplified. Fluorescence lifetimes significantly less than the 30 ps laser pulsewidth can be measured by picosecond time-delay fluorimetry.[6] The time-resolved spectrum of luminescence can be acquired by using successive interference filters as opposed to a diffraction grating to record the entire spectrum on a single shot. The grating technique suffers from limited throughput and spectral and temporal resolution. Fluorescence depolarization measurements are simplified and we have achieved S/N > 15 in polarization signals by averaging up to 200 shots of fluorescence.

Samples are mounted in a closed-cycle helium circulator and can be maintained at any temperature between 8° K and 300° K. We have made detailed measurements of the fluorescence lifetime of purple membrane as a function of temperature. Long-term timing stability is necessary in detailed temperature cycling experiments.

The dynamic range of the streak camera is extended over single-shot values by signal averaging. In order to take advantage of the large dynamic range of the streak camera in this mode, it was necessary to adopt a new data acquisition software. Operating the OMA-II in the open system allows complete programming capability. Using this, we subtract a background signal immediately after each shot is acquired. This results in a cancellation of background drift which is a characteristic

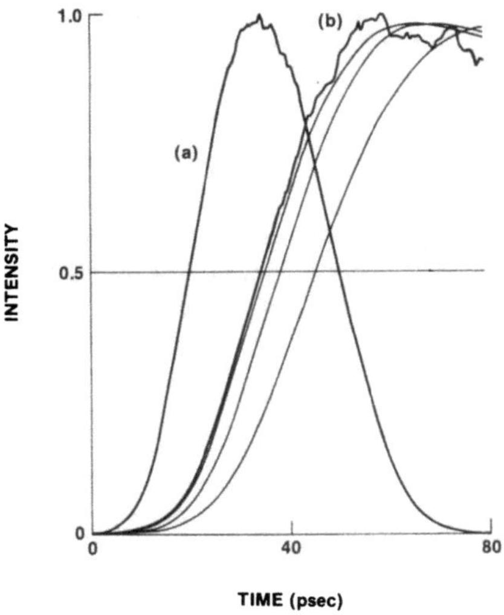

Figure 1 Timing of fluorescence rise of M centers in NaF. (a) 30 shots of scattered excitation pulse, (b) 30 shots of M center fluorescence, and calculated fluorescence response curves for response times of 0,3 and 12 psec. This indicates rapid vibrational relaxation.

of SIT vidicon tubes. Operating in the open system also allows full data-reduction and simulation routines to be run in parallel with data taking operations. This allows for an immediate comparison of streak camera data with kinetic models.

Figure 1 shows a measurement of fluorescence of color centers in an alkali-halide crystal at 300°K. The theoretical curves shown are numerical solutions to the following coupled differential equations which simulate a finite fluorescence response time τ:

$$\dot{N}_2 = \sigma I N_o - N_2/\tau$$

$$\dot{N}_1 = N_2/\tau - N_1/T \qquad \text{, where}$$

N_2 is the population density of a generalized intermediate state which relaxes to level 1 with rate constant $1/\tau$, and N_1 is the population density of the fluorescent state, with fluorescence lifetime T. (σ is the absorption cross-section, T is typically > 10 ns, and N_o is the ground state population density, assumed constant). Using curve (a) for I(t), solutions for $N_1(t)$ are generated for values of τ = 0, 3 and 12 ps. In order to best fit the fluorescence rise, it is necessary to use $\tau \sim 0$, however the accuracy of curves (a) and (b) is only ± ½ ps[1]. Therefore, it is concluded that the overall relaxation time of this center is less than 1 ps.

The jitter-free streak camera provides the basis for a powerful picosecond detection system. This system has found numerous applications in luminescence studies in solid-state physics, biophysics and chemistry. Use of signal averaging results in greatly improved data quality.

This work was supported by the Sponsors of the Laser Fusion Feasibility Project at the Laboratory for Laser Energetics of the University of Rochester and NSF grant #PCM-80-11819.

References

1. W. Knox and G. Mourou, Optics Comm. 37, 203 (1981).
2. G. Mourou and W. Knox, Appl. Phys. Lett. 35, 492 (1979).
3. R. W. Anderson and W. Knox, Journal of Luminescence 24/25, 647 (1981).
4. B. Weinstein, T. Orlowski, W. Knox, T. M. Nordlund and G. Mourou, APS Meeting, Dallas, Texas, March, 1982.
5. T. M. Nordlund and W. Knox, Biophysical Journal 36, 193 (1981).
6. M. Stavola, G. Mourou and W. Knox, Optics Comm. 34, 404 (1980).

Electrical Transient Sampling System with Two Picosecond Resolution

J.A. Valdmanis, G. Mourou, and C.W. Gabel

Laboratory for Laser Energetics, Institute of Optics,
University of Rochester, Rochester, New York 14623, USA

With the advent of picosecond photodetectors, photoconductive switches and other ultrafast devices, the need has arisen for a measurement system capable of characterizing small electrical signals with picosecond accuracy. Techniques for measuring ultrafast electrical signals to date have limitations in their use. Sampling oscilloscopes have temporal resolutions limited by their electronic sampling window. This is typically $\sim$ 25 ps. Recently, Auston[1] demonstrated a sampling technique in amorphous semiconductors that can resolve electrical transients as short as 5 to 10 ps. However, the ultimate resolution of that system is constrained by a material recovery time of approximately 10 ps.

We report the construction of a simple electrooptic sampling system based on the Pockels effect that avoids the fundamental limitations of these previous methods. The current system has demonstrated a temporal resolution of at least 2 ps (> 200 GHz bandwidth) with a sensitivity of less than 50 μV, and is believed to be limited by the test signal described later.

The system utilizes a Lithium Tantalate travelling wave Pockels cell as an ultrafast intensity modulator. A colliding pulse modelocked (CPM) laser[2] generating 120 fs pulses at 100 MHz is used to drive the electrical signal source and synchronously sample the electric field as it propagates across the crystal. Two detectors are employed to measure the intensities of both the transmitted and rejected beams of the analyzer. These signals are processed by a differential amplifier, lock-in amplifier, and signal averager. Optimum sensitivity is achieved with the modulator biased at its quarter wave point. The output signal is a linear, equivalent time representation of the electrical signal requiring no further processing.

The sampling system is tested by characterizing the impulse response of a Cr-doped GaAs photoconductive switch.[3] The figure shows the initial

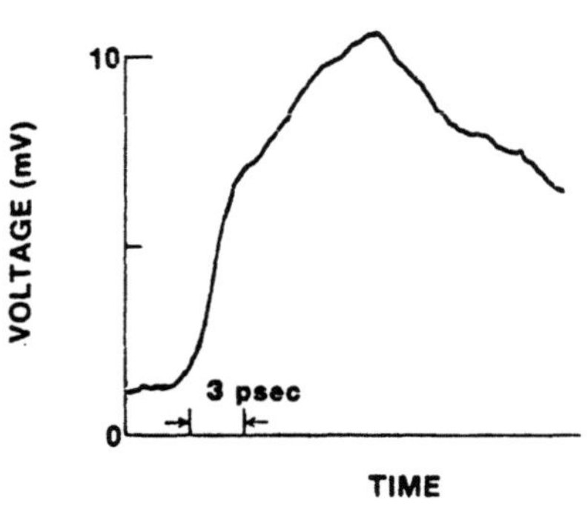

Figure 1: Response of the Cr-doped GaAs Photoconductive Switch

response of such a switch with a 30 μm gap, biased at 25 V when activated by a 120 fsec pulse of 0.1 nJ energy. Two components are clearly visible on the rising edge of the signal. The initial, faster component could be due to intervalley scattering or geometrical considerations of the gap and associated stripline.

In summary, we have developed a system capable of fully characterizing electrical transients with true picosecond resolution. This enables the possibility of analyzing ultrafast electrical processes such as those involved in photoconductive materials, photodetectors, and other picosecond electronic devices with the goal of understanding and improving their operation.

Acknowledgement

This work was partially supported by the following sponsors: Exxon Research and Engineering Company, General Electric Company, Northeast Utilities, New York State Energy Research and Development Authority, The Standard Oil Company (Ohio), The University of Rochester, and Empire State Electric Energy Research Corporation. Such support does not imply endorsement of the content by any of the above parties.

References

1. D.H.Auston, A.M.Johnson, P.R.Smith, and J.C.Bean; Appl. Phys. Lett. 37, 371, 1980.
2. R.L.Fork, B.I.Greene, and C.V.Shank; Appl. Phys. Lett. 38, 671, 1981.
3. Chi H. Lee; Appl. Phys. Lett. 30, 84, 1977.

High-Resolution Picosecond Modulation Spectrocopy of Near Interband Resonances in Semiconductors

S. Sugai, J.H. Harris, and A.V. Nurmikko

Division of Engineering, Brown University
Providence, R.I. 02912, USA

There are many methods of modulation spectroscopy which have been applied with success to the study of interband critical point structure in crystalline semiconductors under equilibrium conditions. We have used the sensitivity afforded by presently available wavelength tunable cw picosecond dye lasers to perform excite-probe spectroscopy of electronic excitations near lowest interband resonances in GaAs, $Ga_{1-x}In_xP$, and the semimagnetic semiconductor $Cd_{1-x}Mn_xTe$. From the time resolved optically modulated spectra new information has been obtained about the character and relaxation of free carriers, impurity and exciton states, and electronic spin-polarization.

The experimental arrangement employed by us is similar to that developed recently by Heritage, Levine, and co-workers in connection with Raman gain spectroscopy of transparent molecular systems (1). A pair of synchronously pumped, modelocked cw dye lasers provided the wavelength tunable excitation and probe radiation. The former was electro-optically modulated at an rf frequency (10 MHz) and the reflected or transmitted component of the latter (from a semiconductor sample) was synchronously detected. In our case, optically induced changes of less than 1×10^{-7} were detectable. Under conditions of low excitation, the changes in the transmission and reflection coefficients remained in most cases proportional to the corresponding changes in the absorption coefficient and the index of refraction, respectively.

We first illustrate the effect of free carriers in optically excited GaAs, probed in this instance within the $E_0 + \Delta_0$ interband resonance, thus giving sensitivity to the direct detection of changes in the electron occupation factor at the conduction band minimum.

Figure 1 shows the spectrum of modulated reflectance (ΔR), obtained from the free surface of a nominally undoped piece of crystalline bulk GaAs approximately 60 psec following the excitation. The strong spectral feature

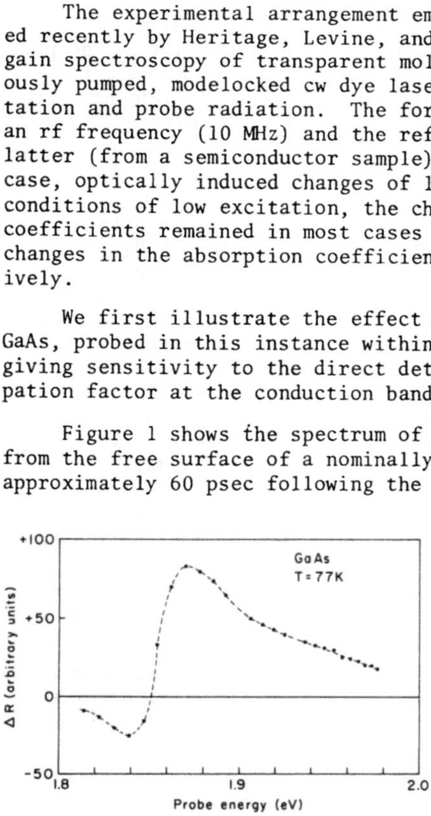

Fig. 1: Modulated reflection spectrum of bulk crystalline GaAs (undoped), measured 60 psec following excitation ($\hbar\omega_{ex}$ = 1.8 eV)

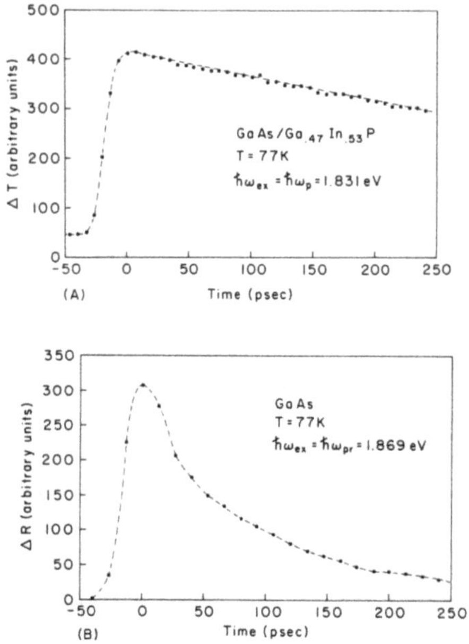

Fig. 2: Time resolved transmission and reflection of GaAs for the GaInP/GaAs/GaInP double heterojunction with electron concentration $n=5x10^{16}cm^{-3}$ (A), and the 'free' GaAs surface (B)

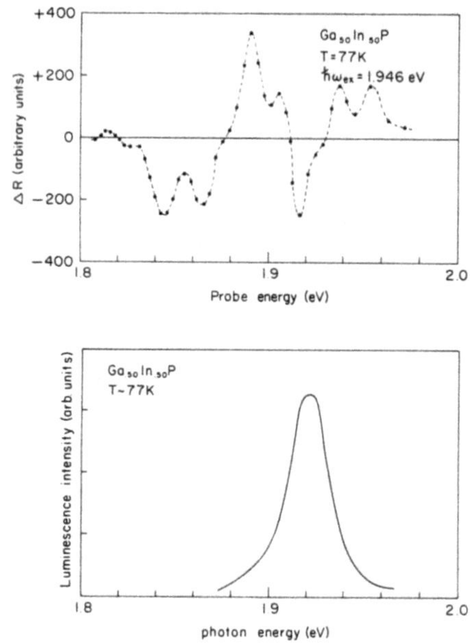

Fig. 3 : Modulated reflectivity (upper trace) in $Ga_{.50}In_{.50}P$ 60 psec following excitation at $\hbar\omega_{ex}$=1.946 eV. Lower trace shows photoluminescence spectra excited at λ=5145 A.

near the $E_0 + \Delta_0$ interband transition (∿ 1.850 eV) with its characteristic shape is also in good qualitative agreement with the measured changes in modulated transmission (ΔT) in a thin film of GaAs, when the two are compared through a Kramers-Kronig calculation. Figure 2 shows directly the time-resolved decay of excess free electron-hole pairs in GaAs with emphasis on demonstrating the influence of the interface environment on recombination. In part (A), the decay of modulated transmission is measured for a 0.18 μm thick GaAs layer in a MBE-grown double heterostructure, formed with the addition of surrounding layers of larger gap GaInP (x ∿ .50). In contrast, part (B) shows the considerably faster decay for a 'free' GaAs/air interface, now measured by modulated reflectance. Analysis of such data over a wider spectral region has permitted us to obtain direct information about the interface recombination for GaAs for different dielectric surroundings, as well as an indication of the carrier confinement in a heterojunction [2]. For example, consistent with recent work based on photoluminescence efficiencies [3] we find that the interface recombination velocity in this relatively strain free heterostructure is reduced to less than 1×10^4 cm/sec from the value of 5×10^5 cm/sec for a free GaAs surface.

At sub-bandgap photon energies the inherent sensitivity of our experimental method has lead to the observation of significant additional spectral structure in comparison with conventional steady-state photoluminescence data. This is illustrated in Figure 3 where modulated reflectance is shown for an epitaxial film of n-type $Ga_{.50}In_{.50}P$, obtained some 60 psec following the excitation at near bandgap energy (E_g ∿ 1.99 eV). In contrast, the steady state luminescence spectra on the same sample yields a single dominant feature in this energy range (bottom trace). A lineshape analysis of an optically modulated spectrum involving closely lying transitions is inherently difficult due to contributions from different effects (e.g. carrier occupancy and exciton screening). The fairly complex spectrum in Figure 3 may be decomposed into a superposition of five dispersive Gaussian resonance lineshape functions with comparable damping parameters of ∿20 meV and with adjustable amplitudes. The dispersive (as opposed to absorptive) part of the dielectric constant is expected to make the dominant direct contribution to the modulated reflectivity spectrum in a simple model where the optical transition probabilities for the probed resonances are affected by the excitation, without frequency shifts. In particular, the dispersive function with a zero crossing at 1.923 eV coincides with the peak of the measured luminescence and is identified as a 'fundamental' transition. The energetic position and behavior of this resonance strongly suggests an acceptor-like transition to be responsible in this compensated material. The energy spacing of the remaining lower energy dispersive functions relative to the primary resonance coincide closely with the 48 meV LO-phonon energy for GaInP (x = .50), as extrapolated from Raman data for GaP and InP. The relative amplitudes of these phonon sidebands in our spectrum may, however, be further affected by the relatively broad free electron contribution at the $E_0 + \Delta_0$ transition from the GaAs substrate. The modulated reflectance at the fundamental transition also shows a strong periodic dependence on the photon energy of excitation, with the periodicity matching the LO-phonon energy. This and additional evidence points to an acceptor-bound exciton with a strong polaron character as the main contributor to the spectrum in Figure 3.

Of the complicated time evolution in the modulated reflectivity in GaInP we show here only an example at the strongly phonon coupled resonance discussed above. Figure 4 shows the time dependence of the peak at $\hbar\omega_p$ = 1.940 eV for three different energies of excitation separated by approximately $\hbar\omega_{LO}/2$. The characteristic decay time is on the order of 200 psec for the three traces (amplitudes not normalized). At the highest excitation energy ($\hbar\omega_{ex}>E_g$) an initially faster decay is observed, probably due to an additional initial contribution from thermalizing and diffusing free carriers. The decay times, which showed only a weak temperature dependence in the range of 10-77 K, thus provide a direct measure of the electronic relaxation of the phonon coupled complex. The relative insensitivity of the relaxation time on $\hbar\omega_{ex}$ is not unexpected for the different states of a strongly phonon coupled transition. However, the observed time constants are quite short when compared e.g. with luminescence decay measured at low (liquid helium) temperatures for bound excitons in many III-V compound semiconductors. In the present instance the short lifetimes are not inconsistent with the very low steady-state photoluminescence yield in our samples [4], thereby indicating a rapid ionization or excitation transfer from the acceptor center. Add-

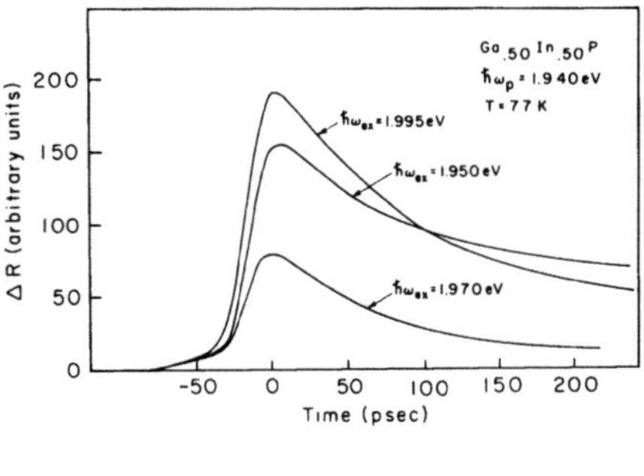

Fig. 4 : Time dependence of ΔR at $\hbar\omega_p$ = 1.940 eV for three energies of excitation.

itionally, a strong Auger process and diffusive transport of a mobile species (free exciton or free carrier) to the GaAs substrate may also contribute to lifetime shortening.

As a final example of our method, we have employed circularly polarized excitation/probe to study electron spin relaxation of free carriers and excitons. Of particular interest to us is the semimagnetic $Cd_{1-x}Mn_xTe$ (x = .30) which also shows strong excitonic subgap features in the modulated reflectance spectra. In particular, we have observed time-dependent spectra which appear to show a free electron/free exciton like component with a strongly temperature dependent spin relaxation. Such dependence may be influenced by the presence of a spin-glass transition in this alloy approached at low temperatures in our experiments. The details of this work will appear elsewhere [5].

Research supported by USAFOSR.

References

1. e.g. J. Heritage, Appl. Phys. Lett. 34, 470 (1979); B. F. Levine, C. V. Shank, and J. P. Heritage, IEEE J. Quant. Electr. QE-15, 1418 (1979)

2. G. B. Scott, G. Duggan, and J. S. Roberts, J. Appl. Phys. 52, 6312 (1981)

3. J. H. Harris, S. Sugai, and A. V. Nurmikko, Appl. Phys. Latt. 40, 885 (1982)

4. J. S. Roberts, G. B. Scott, and J. P. Growers, J. Appl. Phys. 52, 4018 (1981)

5. A. V. Nurmikko, S. Sugai, and A. V. Nurmikko, Proc. Int. Conf. Phys. Semiconductors (Montpellier) 1982, to be published

Electron Diffraction in the Picosecond Domain Steven Williamson and Gerard Mourou and Synchronous Amplification of 70 fsec Pulses Using a Frequency-Doubled Nd: YAG Pumping Source

J.D. Kafka, T. Sizer II, I.N. Duling, C.W. Gabel, and G. Mourou

Laboratory for Laser Energetics, University of Rochester,
250 East River Road, Rochester, New York 14623, USA

I. Electron Diffraction in the Picosecond Domain

With the exception of picosecond photoelectron switching recently demonstrated, streak camera tubes have been used exclusively as a fast optical and x-ray diagnostic tool. However, some of the most beautiful features of the image converter device used in the streak camera have been only partially exploited with this application. The image converter device produces a temporal and spatial monoenergetic photoelectron replica of the incident optical pulse. This replica is ultimately limited by the temporal and spatial resolution of the particular streak tube employed. Temporal and spatial resolution can be as good as subpicosecond and 100 μm respectively. It is also worth noting that this replica is accurately synchronized with the incident optical pulse. We have used this electron burst to generate an electron diffraction pattern, inferring that picosecond snap shots of laser induced structural changes in laser annealing or in the field of surface physics, can now be taken not only with a picosecond exposure, but also in picosecond synchronization with the laser induced kinetics. In the experiment (Fig.1) a demountable photochron II streak camera tube is used. The Al specimen of 150 Angstroms thickness is located in the drift space 1 cm behind the anode. The diffraction pattern is captured on a phosphor screen and photographed using an image intensifier with a gain of 3×10^4 lens coupled to the film. The optical burst used to generate the electron is the fourth harmonic of a Nd:YAG system. The pulsewidth is estimated to be about 15 psec. The optical energy on the Al photocathode of the streak camera is on the order of a few microjoules. The electron pulse width has been measured by using the camera in the normal streak mode and is found to be N 100 psec. This value departs significantly from the 15 psec pulse width expected. The pulse broadening is due to the space charge effect caused by the relatively high electron flux required to photograph the pattern with our present system. This electron flux level should decrease by a factor of over 100 by using a larger gain image in-

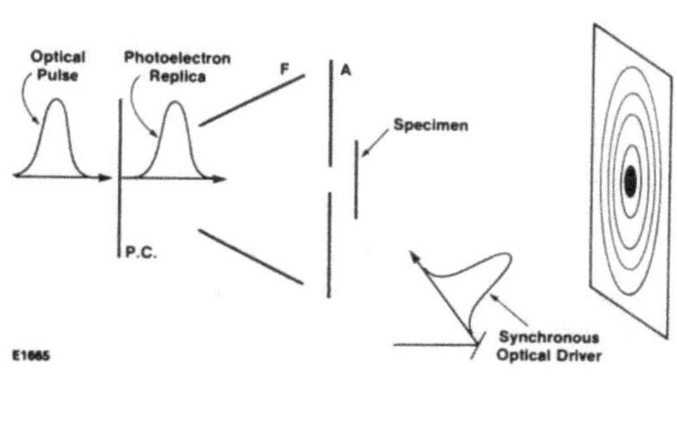

Figure 1 Picosecond diffraction experimental set up

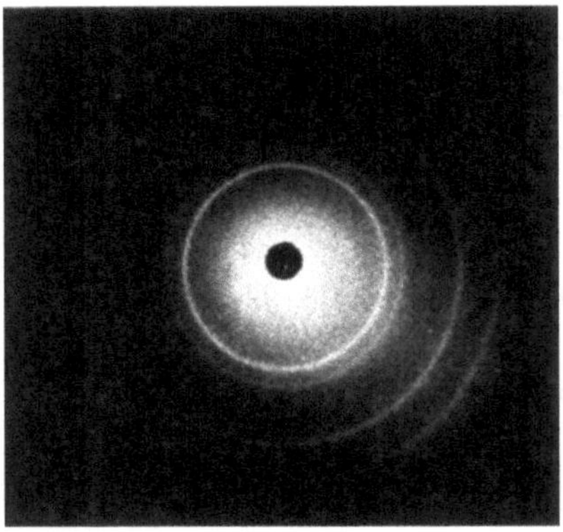

Figure 2 Transmission electron diffraction pattern of Al from 100 psec exposure of 20 keV electron

tensifier directly contacting the recording film. Additionally, for the investigation of single crystals diffracting singular spots, the electron flux is naturally relaxed still further. Fig. 2 represents the transmission electron diffraction pattern obtained with a single ∿ 100 psec electron burst passing through the Al specimen. The distinct rings correspond to the 111, 200, 220, and 311 planes. From the radius of each ring and the Miller indices, a lattice constant of 4.02 ± .04 Å is measured which agrees nicely with 4.05 Å previously reported.

In conclusion we have shown that electron diffraction can be used in the picosecond time domain using an electron burst generated by a streak camera tube. We believe that this technique will make possible the study of structural kinetics on a time scale that is currently of great interest.

II. Synchronous Amplification of 70 fsec Pulses

A. Optical Synchronization

Recently we reported on the generation and amplification of short dye laser pulses pumped by the frequency-doubled output of a CW modelocked Nd:YAG laser.[1] The advantages of this pumping source over others include a shorter pump pulse, increased component lifetime, and the relative ease of short pulse amplification. Pulses shorter than 70 fsec are presently generated by the use of a Rhodamine 6G-DQOCI dye mixture.[2] The autocorrelation of the pulse is shown in Figure 3.

The addition of a saturable absorber, DQOCI, to the same jet as the lasing medium, Rhodamine 6G, in a synchronously pumped dye laser combines the effects of active and passive modelocking to produce extremely short

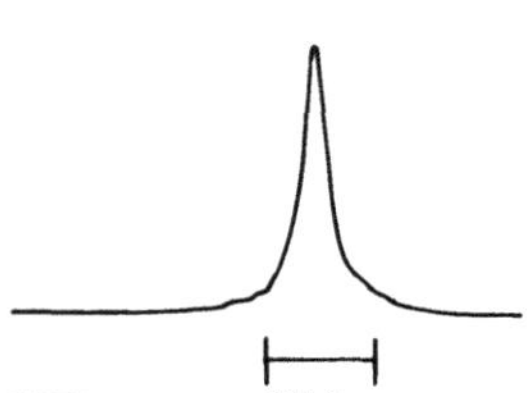

Figure 3 Autocorrelation trace of the 70 fsec pulse from the dye oscillator

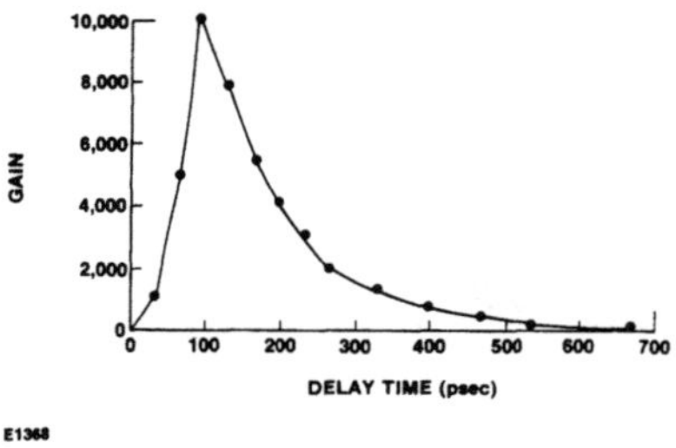

Figure 4 Timing diagram for the first amplifier stage

light pulses. Thirty milliwatts of power is obtained with 300 mW of 532 nm pump power. The shortest pulses are observed at 605 nm. Since the lasing bandwidth is 100 Å, greater than the fourier transform of a 70 fsec pulse, a frequency chirp is indicated which could be exploited, through compression techniques, to further reduce the pulse width.

The use of a synchronized 70 psec pump pulse for amplification reduces the problem of amplified spontaneous emission (ASE) and yields a more stable and efficient system with improved beam quality. The unconverted 1.06 μm radiation from the oscillator is used to seed a regenerative amplifier. The output is further amplified and then frequency doubled resulting in a 10 mj pump pulse precisely synchronized with the 70 fsec pulses from the dye laser. Three dye amplifier stages containing the dye Kiton Red are pumped longitudinally. This configuration preserves both the beam quality and the synchronization of the pulses. The evolution of the first stage gain with time is displayed in Figure 4.. The gain increase as the pump pulse energy is integrated by the dye, and then decreases as the ASE removes the stored energy. The input pulse must come within 50 psec of the correct time in order to access the optimal gain.

The temporal precision that is present in our optically synchronized system allows operation at the peak of the gain without having to constantly reinvert dye population. This elimination of wasted pump energy greatly reduces the ASE generated. In nanosecond pumped dye amplifier systems, to prevent the ASE from propagating down the amplifer chain, the stages must be isolated with saturable absorbers which degrade the spatial quality of the dye beam. Our picosecond pumped dye amplifier system no longer requires isolation and as a consequence we have removed the saturable absorbers from our system. The result is a trade off between pulse broadening and beam quality. With no saturable absorbers in the longitudinally pumped amplifier system, we obtain an output of 200 μj in less than 1 psec, with enhanced beam quality over systems requiring isolation. The efficiency of the system is 3%, which is much greater than comparable nanosecond pumped systems. Finally, the synchronization removes a major source of output fluctuations by eliminating temporal jitter between the pump and input pulses.

A diagram of the synchronously amplified subpicosecond system is shown in Figure 5. Advantages of this system include 70 fsec oscillator pulses, reduced ASE, and increased stability, efficiency and beam quality.

B. R. F. Synchronization

In the past we have produced subpicoseond pulses with low energy and high repetition rate (100 pJ, 100 MHz)[3] and high energy with low repetition rate (100 μJ, 10Hz).[1] To fill this gap a system has been built with a

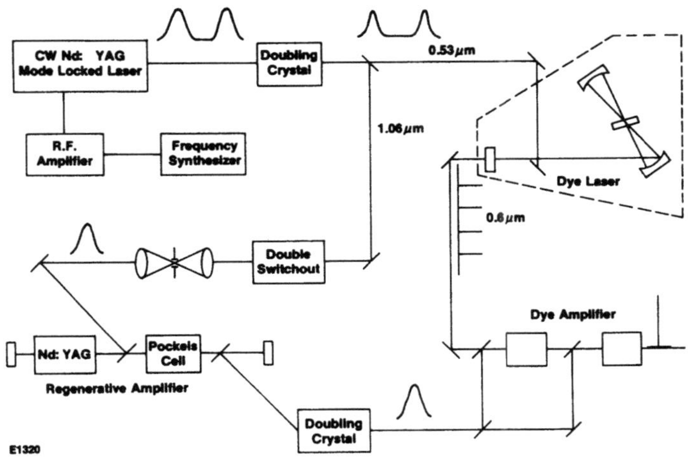

Figure 5 Layout of the synchronously amplified subpicosecond system

500 Hz repetition rate which would amplify pulses to a few microjoules. The system layout is shown in Figure 6.

The subpicosecond pulses from a synchronously pumped dye laser are amplified in a single pass two stage configuration. The amplifier is pumped by an actively Q switched and modelocked CW Nd:YAG oscillator which is frequency doubled in temperature tuned CD*A. The peak energy at 520 nm is 26 μJ. The dye laser pump and the amplifier pump are synchronized by using the same RF amplifier to drive both modelockers. In order to attain maximum efficiency from the amplifiers, the pump and dye pulses must be synchronized within 50 psec. Using autocorrelations of each pump laser and the cross correlation between the two, the maximum jitter was measured to be 40 psec. (Figure 7).

The gain depletion due to amplified spontaneous emission in the amplifier is shown by measuring the gain as a function of dye pulse delay. This measurement was made by varying the RF phase to the amplifier pump modelocker, eliminating the need for an optical delay line (Figure 8).

Currently pulse energies of 350 μJ have been obtained pumping with the entire train when thermal breakup is the limiting factor. This limit will

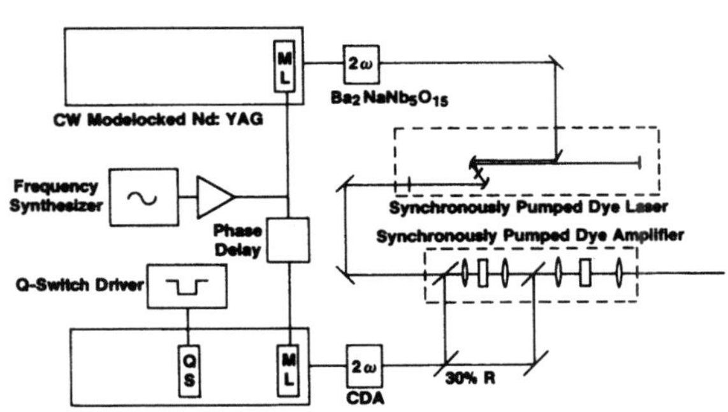

Figure 6 500 Hz Synchronous Amplifier for Subpicosecond Pulses

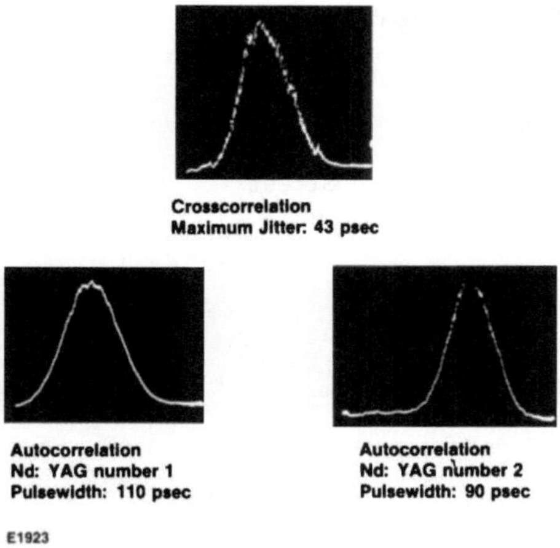

Figure 7 Correlation Measurements of Oscillator & Amplifier

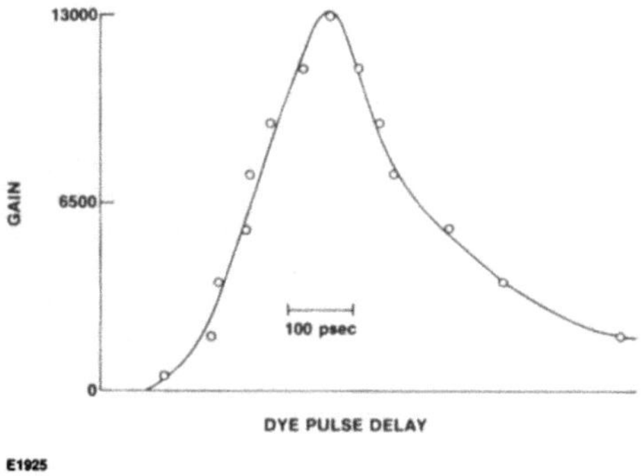

Figure 8 Amplifier Gain as a Function of Time Delay

be removed by an external switch out or an internal cavity dumper. It should be possible to use this technique to amplify a colliding pulse modelocked laser[4] if the amplifier pump modelocker RF is derived from the CPM laser output.

Acknowledgements

This work was partially supported by the following sponsors: Exxon Research and Engineering Company, General Electric Company, Northeast Utilities, New York State Energy Research and Development Authority, The Standard Oil Company (SOHIO), the University of Rochester, and Empire State Electric Energy Research Corporation. Such support does not imply endorsement of the content by any of the above parties.

A special ackowledgement goes to the Quantronix Corporation for their equipment support, and to Steven Zoeller for his assistance in data acquisition.

References

1. T. Sizer II, J. D. Kafka, A. Krisiloff and G. Mourou, Opt. Comm. 39, 259 (1981).
2. G. Mourou and T. Sizer II, Opt. Comm. 41, 47 (1982).
3. T. Sizer, G. Mourou and R. Rice, Optics Comm. 37 (1981) 207.
4. R. L. Fork, B. I. Greene and C.V. Shank, Appl. Phys. Lett. 38, 671 (1981).

Picosecond Time-Resolved Photoacoustic Spectroscopy

M. Bernstein, L.J. Rothberg, and K.S. Peters

Department of Chemistry, Harvard University, 12 Oxford Street
Cambridge, MA 02138, USA

1. Introduction

In this paper we report the development and application of photoacoustic spectroscopy (PAS) for the detection and characterization of transient intermediates with lifetimes on the picosecond timescale [1].

During the course of our investigations into photochemical reaction mechanisms we often have been frustrated by the insensitity of picosecond absorption spectroscopy. Current technology, based upon picosecond continuum generation and high quantum-yield vidicon and CCD detectors [2], tends to be both expensive and unreliable, especially since the very nonlinear process of continuum generation requires excellent laser stability and extensive signal integration. Sample fluorescence and phosphorescence can easily saturate these sensitive detectors, rendering emission species difficult or impossible to study. Even with considerable care and favorable conditions, transient absorbance levels lower than 0.01 A are difficult to study [3].

2. Picosecond Photoacoustic Detection [4]

In the picosecond photoacoustic experiment, a 25ps pump pulse (frequency ν_o, energy E_o) creates an ensemble of excited molecules which can be interrogated, after a time delay t, by a probe beam (frequency ν_p, flux F_p, duration t_p). For simplicity, we shall assume here that the unexcited sample is transparent to the probe beam, and absorbs the excitation beam with extinction coefficient ε. If the pump beam contains N_o photons, the number of transients initially created (assuming unit efficiency) will be

$$N = N_o(1 - 10^{-\varepsilon cl}). \tag{1}$$

The amplitude S_o of the photoacoustic signal rising from this excitation process is simply

$$S_o = KNh\nu_o \tag{2}$$

where the proportionality constant K is a function of sample geometry, transducer construction, and the quantum yield for radiationless conversion of the excitation energy.

If transients created by the excitation beam absorb the probe pulse with cross-section σ, they will, in turn, generate an additional photoacoustic contribution S_p:

$$S_p = KNh\nu_p[1 - \exp(-\sigma F_p t_p)]. \tag{3}$$

The changing composition and concentration of transient species may be studied by varying the delay t between excitation and probe pulses. Since the response time of the microphone is much longer than the temporal separation between the two laser pulses, the photoacoustic contributions are additive and the observed signal is $S(t) = S_o + S_p(t)$. The portion of the signal S_o due to the excitation beam can be estimated (2) from the measured energy of the excitation pulse. Hence, by measuring S(t) and E_o, one obtains the photoacoustic contribution $S_p(t)$ arising exclusively from photoproducts. In addition, analysis of the dependence of S_p on probe flux can provide a direct measure of the excited state absorption cross-section σ.

Note that the photoacoustic signal S_p can be comparable to S_o even though only a minute fraction of the probe beam is absorbed, providing only that the probe beam be sufficiently intense [4]. The incremental signal S_p/S_o is <u>independent of the concentration</u> of transients, depending only on the <u>fraction</u> of transients which absorb a probe photon. In contrast, direct absorption experiments measure the <u>number</u> of transients that absorb a probe photon, and must therefore fail at low transient concentrations.

A detailed description of the picosecond photoacoustic apparatus appears elsewhere [4]. We employ a Nd:YAG laser, using a 0.1 mJ 355 nm excitation beam and a 5mJ 530 nm probe beam. The excitation beam is weakly focused by a 1m lens to avoid multiphoton effect, while the probe is collimated with approximately 3[mm2] spot size. Increasing probe flux produces proportionately larger signals, limited only by saturation of the transient absorption or stimulated Raman scattering by the solvent.

3. Absorption in Diphenyl Polyenes

Diphenyl polyenes, long a focus of spectroscopic and theoretical interest [5], have recently gained significant technological importance as laser dyes and scintillators [6]. All except stilbene are strongly fluorescent, a property which gravely complicates conventional absorption studies. Photoacoustic detection, on the other hand, is ideally suited to such compounds since emission makes no contribution to the photoacoustic signal. Furthermore, photoacoustic detection affords substantially enhanced sensitivity in comparison to direct absorption measurements, and we routinely obtain adequate kinetic data on samples with peak absorbances of 0.001 A or less.

The photoacoustic absorption signal for the UV laser dye diphenyl stilbene is shown in Fig. 1. The measured lifetime (700 ± 100 ps: dioxane) is consistent with the fluorescence lifetimes observed in other solvents [7].

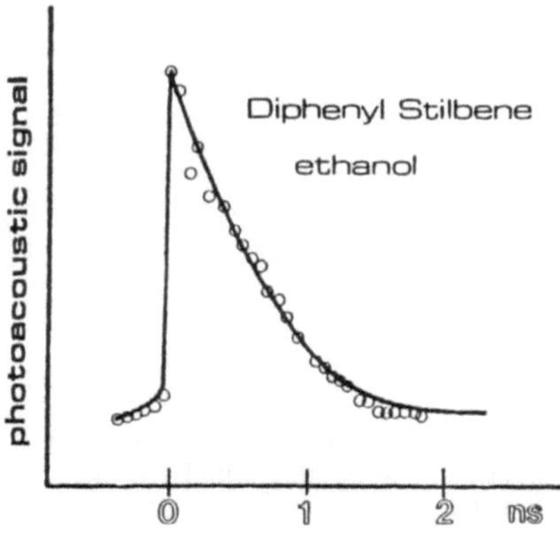

Fig. 1 Transient absorption decay kinetic of diphenyl stilbene (dioxane)

The photophysical properties of the diphenyl polyenes are strongly solvent dependent [5,8]. We have examined the transient absorption decay rate for this compound in a number of solvents, and have found that the lifetimes closely parallel observed fluorescence lifetimes [8]. This observation appears to preclude the possibility that diphenyl butadiene's short, solvent-dependent fluorescence lifetime arises from competition between emission and radiationless conversion to a non-emissive excited A_g state [8,9].

4. Stimulated Emission in Coumarin Dyes

We have also applied time-resolved photoacoustic spectroscopy to investigate excited state dynamics of laser dyes by direct observation of stimulated emission. In the absence of a probe beam, excited dye molecules partition their excitation energy between radiative and nonradiative decay channels, characterized by quantum yields Φ_r for emission and Φ_{nr} for radiationless conversion. Since Φ_{nr} is never identically zero, excitation always produces a finite photoacoustic signal.

If the excited sample is subject to an intense probe beam of appropriate wavelength, stimulated emission may become a significant decay mode. Since some molecules which undergo stimulated emission would otherwise relax nonradiatively, stimulated emission is observed as a net depletion of the photoacoustic signal.

Figure 2 presents the photoacoustic signal arising from an ethanol solution of Coumarin 485, an important laser dye [10], in the presence of diazobicyclo-[2.2.2]octane (DABCO), a potential electron donor.

Strong stimulated emission is observed initially, decaying with a time constant of 285 ± 50 ps, much longer than the approximately 5nsec lifetime [10] observed in the absence of DABCO. The 50 ps risetime of the stimulated emission signal is due entirely to the 25 ps pulsewidths. In fact, this measurement provides a uniquely simple and inexpensive method for characterizing picosecond laser pulses.

The stimulated emission decay, ascribed to formation of the coumarin radical anion, occurs at approximately the same rate as observed for electron transfer from DABCO to benzophenone in this solvent [11]. Although the Coumarin 485 anion appears to be transparent to the 530nm probe, we have observed strong absorption at this wavelength from the radical anions of related dyes (eg. Cou-

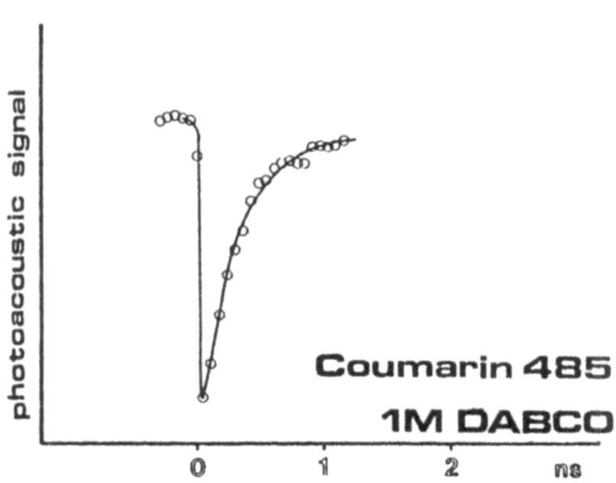

Fig. 2 Stimulated emission in Coumarin 485, quenched by 1M DABCO.

marin 540A). By resolving the temporal evolution of stimulated emission and absorption processes in dye solutions, we can thus obtain important information useful in designing new active media for dye lasers which can be optimized for stability, emission lifetime, wavelength, or other desirable properties.

References

1. Nanosecond transient absorption was detected photoacoustically, but without time resolution, by M. G. Rockley and J. P. Devlin, Applied Physics Letters 32, 24 (1977).
2. L. T. Netzel and P. M. Rentzepis, Chem. Phys. Lett. 29, 337 (1974).
3. B. I. Greene, R. M. Hochstrasser and R. B. Weisman, J. Chem. Phys. 70, 1247 (1979).
4. M. Bernstein, L. J. Rothberg and K. S. Peters, submitted to Chem. Phys. Lett.
5. B. S. Hudson, B. E. Kohler and Klaus Schulten in Excited States vol. 5, E. C. Lim, ed., in press.
6. K. H. Drexhage in Dye Lasers, F. P. Schafer, ed., Springer-Verlag, New York, 1973.
7. C. D. Amata, et. al., J. Chem. Phys. 48, 2374 (1968). I. B. Berleman, Handbook of Fluorescence Spectra of Aromatic Molecules, Academic Press, New York, 1971, p. 231.
8. S. P. Velsko and G. R. Fleming, J. Chem. Phys. 76, 3553 (1982). G. R. Fleming, personal communication. D. J. S. Birch and R. E. Imhof, Chem. Phys. Lett. 88, 243 (1982).
   Experiments with polarized pump and probe beams indicate that rotational diffusion makes only a minor contribution to the observed absorption decay rate.
9. J. R. Andrew and B. S. Hudson, J. Chem. Phys. 68, 4587 (1975).
10. While the fluorescence lifetime of Coumarin 485, a dimethylamine, has not to our knowledge been reported, it would be expected to resemble that of the diethyl analog, Coumarin 151, reported by G. Jones, W. R. Jackson and A. Halpern, Chem. Phys. Lett. 72, 399 (1980).
11. J. D. Simon and K S. Peters, J. Am. Chem. Soc., in press.

Subpicosecond Pulse Shape Measurement and Modeling of Passively Mode Locked Dye Lasers Including Saturation and Spatial Hole Burning

J.-C. Diels, I.C. McMichael
Center for Applied Quantum Electronics, North Texas State University
Denton, TX 76203, USA

J.J. Fontaine
ENSAIS and LSOCS, University Louis Pasteur, F-Strasbourg, France

C.Y. Wang
Tientsin University, Optical Department
Tientsin, The People's Republic of China

1. Introduction

We present a theoretical model of the linear and ring, passively mode locked dye lasers. This simple model gives a clear understanding of the role of the amplifier, the absorber, and the dynamics of the degenerate third order nonlinearity resulting from spatial hole burning [1].

The validity of the model is established by fitting the pulse shapes measured experimentally in the case of the linear laser. The model predicts a minimum pulse duration in the case of the ring laser which is within a factor of two of that obtained in experiments [2].

2. The Experimental Measurement of the Pulse Shape

In each roundtrip time for the linear cavity (4 ns) we observe a train of two pulses separated by 0.33 ns. The 0.33 ns corresponds exactly to the roundtrip time for pulses to travel from the dye jet, to the end mirror, and back to the jet. This indicates that the pulses must collide in the jet.

Two aspects of the linear laser output make it possible to extract information about each of the two pulses from the auto- and cross-correlations. First, since the peak intensity of the first pulse in the train of two is more than 15 times that of the second, the auto-correlation near zero delay is unaffected by the second pulse. Second, since the second pulse in the train of two is much broader than the first, the cross-correlation accurately portrays the shape of the second pulse.

The experimental measurements are shown in Fig.1. The front of the broad weak pulse has a slow rise followed by a fast decay. The auto-correlation of the sharp intense pulse scaled down to the peak amplitude of broad weak pulse is indicated by the dashed line.

3. The Theoretical Model

The theoretical model includes saturation of the gain and absorbing media, interaction of the pulses through spatial modulation of the populations of the active media, a finite jet thickness, and dispersion and bandwidth limitation in the cavity. The population difference W of either media will evolve according to the rate equation

$$\dot{W} = -\left\{\frac{E^2}{E_s^2}\ W + (W - W_{eq})\right\}/T_1 \quad . \tag{1}$$

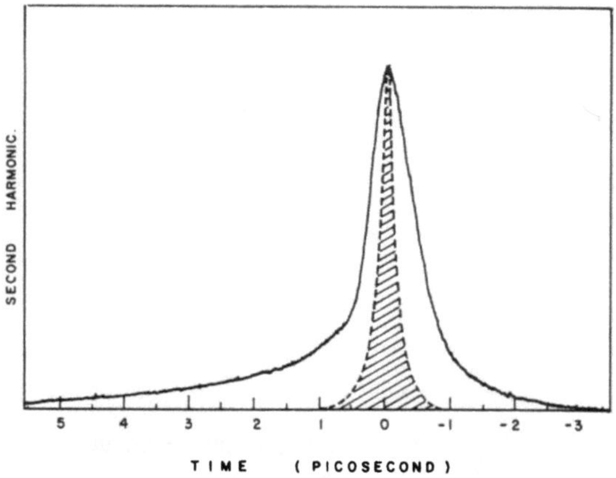

Fig.1 Cross-correlation: shape of the weak pulse

In Eq. 1 E_s is the saturation field, W_{eq} is the population difference in the absence of fields, and T_1 is the lifetime (corrected for the pumping rate in the case of the amplifying medium). With a driving field E composed of two counterpropagating components

$$E = \mathcal{E}_1 e^{i(\omega t - kz)} + \mathcal{E}_2 e^{i(\omega t + kz)}$$

the population differences will be of the form

$$W_j = W_{oj} + W_{2j} e^{2ikz} + W_{2j}^* e^{-2ikz}$$

with j = G for the gain medium, and j = A for the absorber.

For propagation in the dye jet the equations are written in advancing spatial coordinates for $\mathcal{E}_1$ and regressing spatial coordinates for $\mathcal{E}_2$.

$$\dot{\mathcal{E}}_1 = A_o \mathcal{E}_1 + A_2 \mathcal{E}_2$$
$$\dot{\mathcal{E}}_2 = A_o \mathcal{E}_2 + A_2^* \mathcal{E}_1$$

A_o includes contributions from the W_{oj}'s and A_2 includes contributions from the W_{2j}'s. One should not forget while looking at these equations that the A's are functions of the fields. The media comprising the dye jet are allowed to relax between pulse passages. At each cavity roundtrip the fields are modified to account for the spectral narrowing and dispersion introduced by the mirrors and solvent in a real cavity. Arbitrary shapes are assumed for $\mathcal{E}_1$ and $\mathcal{E}_2$ and then the calculation is cycled until a steady state is reached.

4. Discussion and Results

Figure 2 shows the calculated pulse shapes for the two pulses in the linear laser. The slow rise and rapid decline of the weak pulse appears as in the experiment. Inspection of the numerical values of $W_{2j}(t)$ indicate the

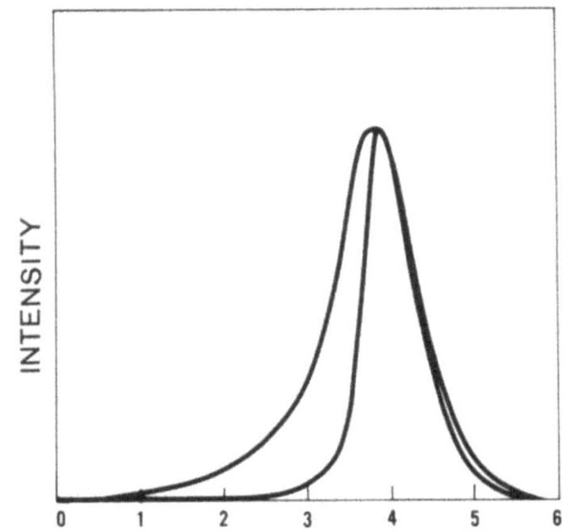

Fig.2 Calculated pulse shapes for the linear laser

formation of a strong grating during the slow rise of the weak pulse. As the peak of the strong pulse is approached, the population gratings are washed out by complete saturation. It should also be noted that the pulse shapes are nearly independent of the jet thickness, but strongly affected by dispersion.

This same model was used to describe colliding pulse mode locking in the ring laser. The only differences from the previous calculations are in the sequence of events and the fact that mutual interaction of the counter propagating pulses occurs only in the absorber jet. A steady state pulse shape is shown in Fig.3. Unlike the calculation for the linear laser, the calculation for the ring laser indicates that in the absence of dispersion the pulse duration is limited by the jet thickness. For this figure the medium thickness was 120 μm. The pulse duration corrected for medium dispersion is 0.4 psec in air. Scaling the thickness down to 10 μm yields a minimum duration of 35 ftsec. This is within a factor of two of the result obtained by FORK et al. [2].

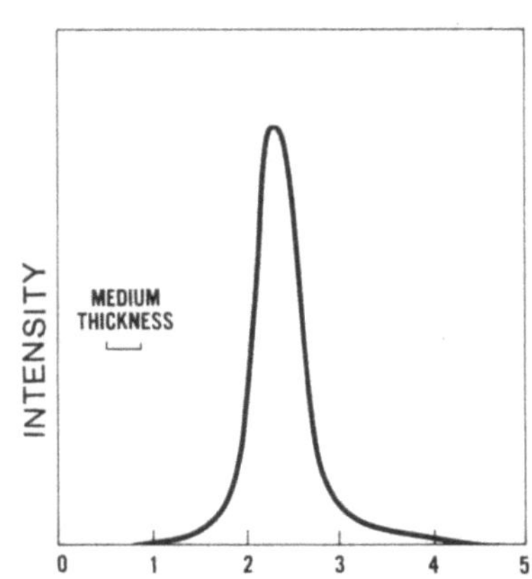

Fig.3 Calculation for the ring laser

5. Conclusions

A theoretical model has been made which describes the pulse forming mechanism in passively mode locked dye lasers. The model includes those parameters that are essential to the operation of the laser. The exact influence of each parameter has been evaluated by fitting the experimental data from a linear cavity.

In the case of the linear laser the model predicts that the pulse shape is independent of the dye jet thickness and that a population grating is formed in the leading edge of the broad weak pulse. In the case of the ring laser the model predicts that the pulse shape is dependent on jet thickness, that the induced grating yields pulse broadening, and that the stable regime results from a balance between pulse broadening and pulse compression by mutual saturation.

The agreement between these predictions and the results of experiment leads us to conclude that the proposed model accurately describes the colliding pulse mechanism in passively mode locked linear and ring lasers.

This work was supported by the National Science Foundation, Grant No. ECS-8119568, and the Office of Naval Research.

REFERENCES

1. J.-C. Diels, and W. C. Wang, Applied Phys. B 26, 105 (1981); R. L. Adams and R. C. Lind, Opt. Lett. 2, 95 (1978); J. B. Hambenne and M. Sargent, IEEE J. Quant. Elect. QE-11, 90 (1975).
2. R. L. Fork, B. I. Greene, and C. V. Shank, Applied Physics Letters 38, 671 (1981).

Experimental Demonstration of a new Technique to Measure Ultrashort Dephasing Times

J.-C. Diels
Center for Applied Quantum Electronics, North Texas State University, Denton, TX 76203, USA

W.C. Wang and P. Kumar
University of Southern California, Los Angeles, CA 90007, USA

R.K. Jain
Hughes Research Laboratory, 3011 Malibu Canyon Road
Malibu, CA 90265, USA

We demonstrate experimentally a new technique to measure ultrashort dephasing times. The method applies to the measurement of subpicosecond coherence memory time of dipole transitions. We perform demonstration experiments in metal vapors (sodium and lithium), using picosecond pulses. The techniques consists essentially in preparing the medium with a first pulse, and subsequently probing the atomic system with a second pulse. The second harmonic of the transmitted energy - as it is affected by successive absorption-reemission - is measured.

The basic instrument of our set up is an interferometric autocorrelator used previously to make accurate measurements on the coherence of subpicosecond pulses [1]. In the latter measurement, the beams issued from each arm of the interferometric delay line cease to interfere with each other when the relative delay exceeds the pulse coherence time. However, as seen in Fig.1, if the laser is tuned to resonance with an atomic vapor inserted between the interferometric delay line and the second harmonic detection, the interference region extends to delays beyond the coherence time (10 psec for the pulse used in Fig.1). Each vertical line corresponds to a scan through several interferences, for a coarse setting of the delay indicated on the abscissa. The envelope corresponding to destructive interferences starts from a small signal at minimal delays (destructive interference of overlapping pulses), then overlays the envelope of "constructive interferences" at larger delays, because of the larger transmission of "zero area pulse" sequences [2]. At large buffer gas pressures (1000 torr in the case of Fig.1) and with a pulse duration of 10 psec FWHM, Fig.1 can be accurately fitted by modeling the lithium as a pure two-level system with a phase relaxation time of T_2 = 15 psec. In this buffer gas

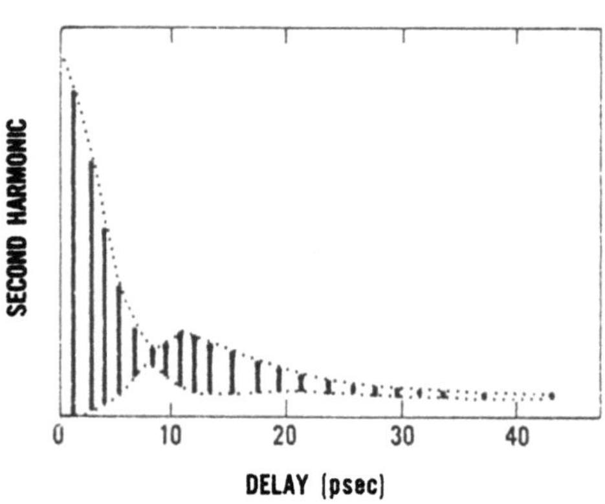

Fig.1 Interferometric second order autocorrelation through lithium vapor (buffer gas Ar at 1000 torr)

pressure range, the phase relaxation time can be measured directly as one exponential decay of the upper envelope in Fig.1 between 16 and 40 psec delay. We confirmed the value of 15 nsec-torr for the collisional relaxation time Li-Ar, by performing experiments of self-induced transparency in the pressure range of 0 to 150 torr of Ar. The value of the relaxation rate was deduced from the measurement of the delay [3] for 2 pulse transmission experiments. Our measurements in sodium yielded 16 nsec-torr for the Na-Ar collisional dephasing time.

The method demonstrated here applies ideally to discrete transitions, with a homogeneous broadening comparable to the pulse duration. With 50 fsec pulses, it should be possible to measure dephasing times in the 10 fsec to 1 psec range. We have shown [4] that the technique can also be extended to measurements of the dephasing time of multiphoton transitions. Care should be taken in trying to apply this method to dye solutions or systems with large inhomogeneous broadening. The use of weak pulses as proposed by CHO et al., [5] will result in a null effect in these systems [6]. With intensities far beyond the dye saturation intensity instead, one expects a recording qualitatively different from Fig.1, with no crossing of the lower and the upper envelopes (because there will be a larger absorption for out-of-phase sequences in media with large inhomogeneous broadening [6]).

Measurements performed without buffer gas, with shorter pulses, provide a completely different type of information (Fig.2). It enables one to remove the ambiguity present in an interferometric second order autocorrelation. Indeed, we have shown previously [7] that the latter measurement provides a very sensitive diagnostic for the pulse coherence, but cannot distinguish between a lack of "ensemble coherence" (the pulse of the train not being identical) or a lack of coherence (or phase modulation) in the individual pulses. In the measurement shown in Fig.2, the interference pattern would not decay if the pulses of the train were identical (chirped or unchirped). The 15 psec decay in the fringe pattern amplitude is representative of pulse to pulse bandwidth fluctuations in the train, or a figure of merit for the frequency stability of the pulse train.

The nodes seen in Fig.2 at 6 psec interval indicate a detuning of $(1/16\ \text{psec})^{-1}$ 60 GHz. It is thus possible to measure a detuning much smaller than the pulse bandwidth. This property can be used to measure

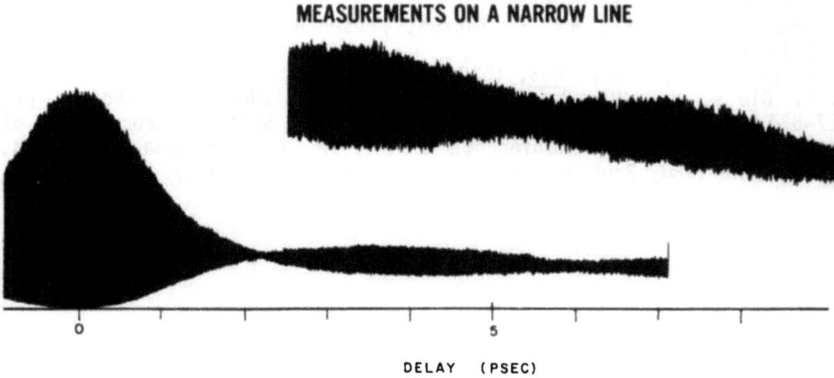

Fig.2 Interferometric second order autocorrelation through sodium vapor (no buffer gas)

pressure shifts of atomic transitions. The interference of one pulse with the polarization created by a previous pulse could be used inside or outside a laser cavity, to control the frequency of a laser with an accuracy much greater than the inverse pulsewidth. If used outside the cavity, the laser wavelength could be actively controlled to maintain a node at a predetermined delay setting (i.e. at a predetermined amount off resonance from an atomic line). The decay of fringe pattern amplitude (representative of the pulse train frequency stability) determines how close from line center a laser could be actively tuned by this procedure. Because the amplitude of the interference decreases when the laser is tuned away from line center, there is a maximum detuning of the order of one linewidth for which active wavelength control is possible.

It should be noted that there is an ambiguity in the interpretation of Fig.2 which can only be resolved by tuning the laser frequency across the resonance. Indeed, the beat note is identical above or below resonance. Therefore, if the laser carrier frequency has two components symmetrical with respect to the resonances, the same beat note will be observed as in Fig.2.

We have demonstrated a new method to determine dephasing times, ideally suited to the subpicosecond range with subpicosecond pulses. The method can be extended to multiphoton as well as to inhomogeneously broadened transitions, provided high intensities are used.

This work was supported by the National Science Foundation, under Grant No. ENG-7826209 and ECS-8119568.

REFERENCES

1. J.-C. Diels, E. W. Van Stryland, D. Gold, Proceedings of the 1st Int. Conf. on Psec. Phenomena, Hilton Head, South Carolina, Springer-Verlag, 117 (1978).
2. H. P. Grieneisen, J. Goldhar, N. A. Kurnit and A. Javan, Appl. Phys. Lett. 21, 559 (1972).
3. V. A. Alekseev and B. Ya Zel'dovich, Sov. J. of Quant. Electron. 5, 589 (1975).
4. J.-C. Diels, J. Stone, S. Besnainou, M. F. Goodman, and E. Thiele, Optics Commun. 37, 11 (1981).
5. Y. Cho, T. Kurobori, and Y. Matsuo, presented at the XIth IQEC, G6, June 1980.
6. J.-C. Diels, Phys. Rev. A13, 1520 (1976).
7. J.-C. Diels, UPS 80, Reinhardsbrunn (October 1980), Proceedings pp. 527-537 (1981); J.-C. Diels, J. Menders and H. Sallaba, proceedings of the 2nd International Conference on Picosecond Phenomena, Cape Cod, Mass., June 1980; Proceedings, Springer Verlag, 41 (1980).

Optical Pulse Compression with Reduced Wings

D. Grischkowsky and A.C. Balant

IBM Thomas J. Watson Research Center, P.O. Box 218
Yorktown Heights, New York 10598, USA

Most optical pulse compression schemes are based on the generation of a frequency swept pulse, followed by passage through a dispersive delay line. Since the group velocity of the light is determined by its instantaneous frequency, different portions of the frequency swept pulse travel at different speeds through the delay line. If the length of the line is adjusted so that the leading edge of the pulse is delayed by just the right amount to overlap the trailing edge at the output of the delay line, the output pulse can be as short as the reciprocal of the bandwidth of the frequency sweep.

The use of this scheme requires a method for producing a linear frequency swept (chirped) optical pulse. For picosecond optical pulses it is quite difficult, using electro-optic techniques, to obtain the required amount of frequency chirp. Consequently, all applications to date have used either the chirp produced by the mode-locked laser itself, or the chirp produced by the nonlinear process of self-phase modulation (SPM) in a nonlinear optical material. For this case, the instantaneous frequency is proportional to the time derivative of the optical pulseshape. Thus, for the case of the 6 psec, (hyperbolic secant)^2 pulse depicted in Fig.1a, only the central region of the pulse has the proper frequency chirp (Fig.1b) for pulse compression, while the chirp on the wings of the pulse will lead to temporal broadening of the wings.

We present here a new method to chirp intense psec pulses by propagating them through a single-mode optical fiber. During passage through the fiber the combined action of self-phase modulation and positive group velocity dispersion broadens both the pulseshape and the frequency bandwidth in such a manner that essentially the entire output pulse is positively chirped. This approach is based on a recent experimental result [1] made incidentally to a nonlinear pulse propagation experiment involving optical fibers. Our purpose here is to illustrate by a numerical calculation the importance of this method to pulse compression applications. For this illustration, we consider the example of a (hyperbolic secant)^2 input pulse (Fig.1a) with a peak power of 100W and a pulsewidth (FWHM) of 6 psec; the characteristics of the fiber are given in Ref. [1].

The results of our numerical integration of the nonlinear Schrodinger equation for this case are illustrated in Fig.2. As the pulse propagates through the fiber the calculated reshaping proceeds smoothly to the characteristic square pulseshape shown in Fig.2a, which occurs at 30 m. Simultaneously, the developing chirp grows in magnitude and extends over more and more of the pulse. For this example, the chirp shown in Fig.2b describes more than 95 percent of the total pulse energy, compared to only 59 percent for the SPM pulse.

The recompression of the output pulse of Fig.2a and the SPM pulse of Fig.1a are calculated as follows. The pulses are Fourier analyzed (Figs.1c and 2c) and are considered to pass through a dispersive delay line with a group velocity dispersion opposite to that of the optical fiber. At the output of the delay line the Fourier components are summed to give the recompressed pulses. The lengths of the delay lines are adjusted to obtain the shortest recompressed pulses shown in Figs.1d and 2d. For the fiber case the recompressed pulsewidth (FWHM) of 0.6 psec is to be compared to the input pulsewidth of 6 psec. The output intensity has correspondingly increased by approximately

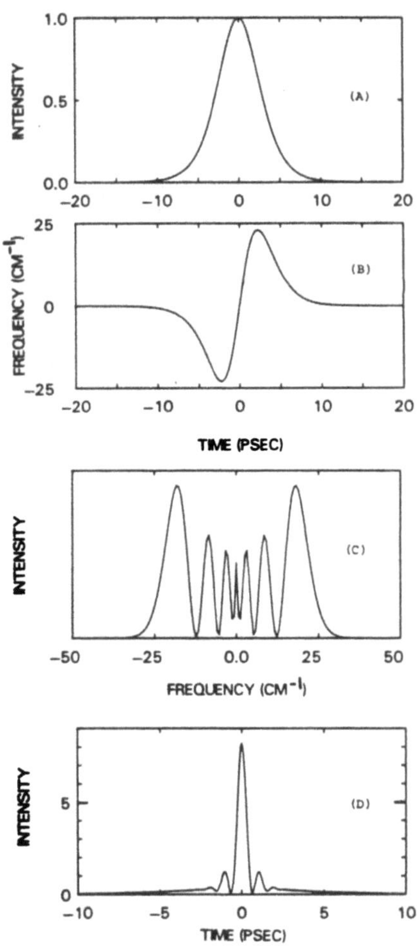

Figure 1.

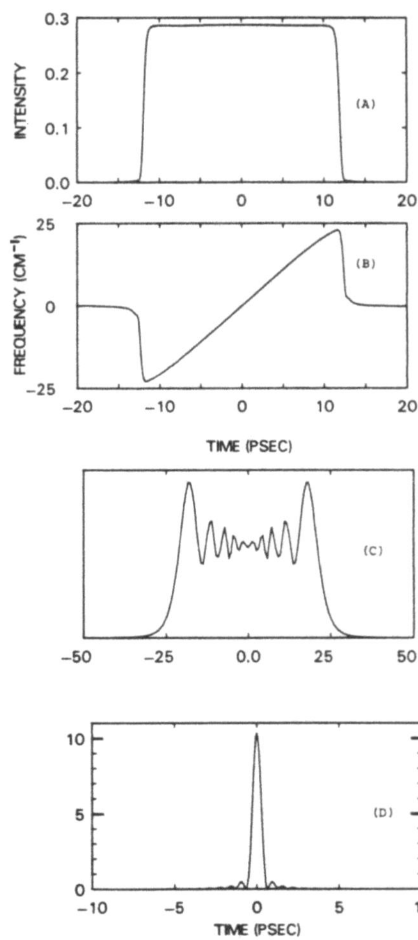

Figure 2.

10 times. The recompressed pulse of Fig.2d is more intense than the recompressed SPM pulse of Fig.1d, because of the fact that it contains more of the total energy of the input pulse. However, the most important feature is the fact that the recompressed fiber pulse has much less extensive wings than the SPM pulse. The relative energies in the wings of the two recompressed pulses, as measured from the first null next to the central peaks, are 42 percent for the SPM pulse compared to only 10.6 percent for the fiber pulse.

In addition, because of the single-mode propagation in the fiber, the entire output beam has the same chirp, i.e., the chirp is independent of the position on the output beam. The extensive frequency chirp enables the output pulses to be compressed to the frequency transform limit without any significant wings on the recompressed pulse, while the lack of any spatial effects in the frequency modulation allows the entire pulse to compress as a spatial unit.

Reference

1. H. Nakatsuka, D. Grischkowsky and A. C. Balant, Phys. Rev. Lett. 47, 910 (1981).

Polarition-Induced Compensation of Picosecond Pulse Broadening in Optical Fibers

G.W. Fehrenbach and M.M. Salour

Research Laboratory of Electronics and Department of Electrical Engineering and Computer Science, Massachusetts Institute of Technology, Cambridge, MA 02139, USA

During the past few years the use of optical fibers has grown because of their potential application as transmission media in long-distance high-bit-rate optical communication systems. Recently considerable attention has been given to the study of pulse distortions in single-mode optical fibers where the maximum data rate of signal transmission is limited by the group velocity dispersion (GVD). Distortion-free pulse propagation has been demonstrated in fused silica fibers at 1.3 μm and in the 1.3 - 1.7 μm spectral region [1,2]. Recent work showed the recompression of light pulses broadened by passage through optical fibers, using an atomic sodium-vapor delay line [3]. We report the compensation of the pulse broadening in optical fibers using the $n = 1$ exciton-polariton resonance of a thin direct-gap semiconductor.

The coulomb interaction of electron-hole pairs in semi-conductors causes discrete exciton resonances. Their coupling to the light field leads to the concept of the polariton, a mixed exciton-photon state [4]. In a cw absorption spectrum of the GaAs samples (see Fig 1(a)), the $n = 1$ and $n = 2$ exciton resonances are well resolved. Fig. 1(b) shows the predicted delay

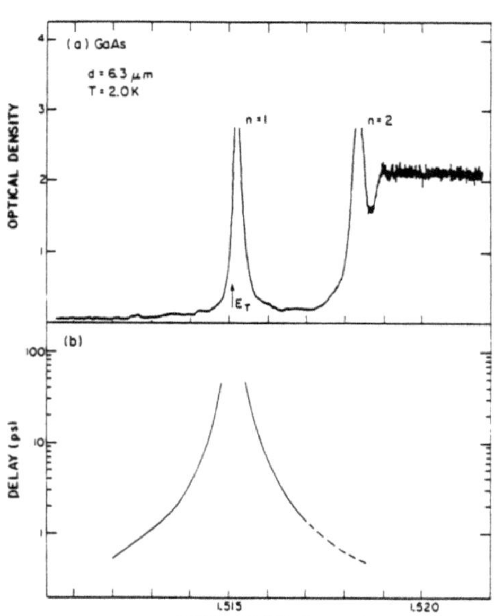

Fig.1 (a) Optical density spectrum of the 6.3 μm - thick GaAs sample (b) Temporal delay in this sample, determined by the group velocity $\partial\omega/\partial k$ for the single-exciton oscillator model

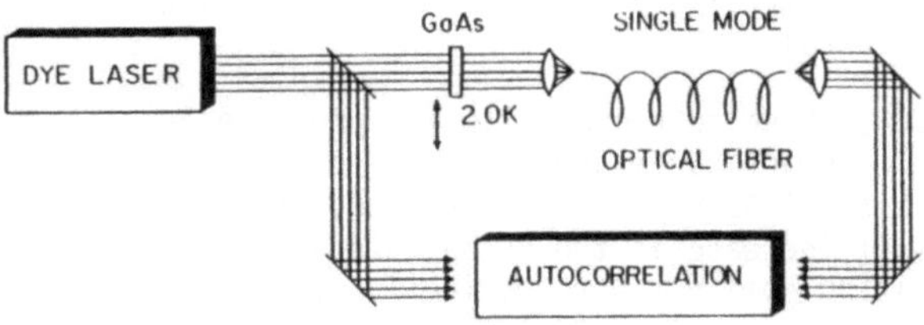

Fig.2 Experimental setup

near the n = 1 exciton-polariton resonance in the single-exciton oscillator model [4], using the values E_T = 1.5151 eV (transverse exciton energy), E_{LT} = 0.08 meV (longitudinal-transverse splitting), M_{ex}^h = 0.6 m_o (heavy exciton mass), and ε_b = 12.6 (background dielectric constant), known from resonant Brillouin scattering experiments [5]. This model was found to be in good agreement with the observation of slow-pulse propagation near this resonance [6].

Figure 2 shows the experimental arrangement: nearly band-width limited pulses with single-sided exponential pulse shape of 0.7-ps duration and 5.8 Å spectral width ($\Delta\nu\ \Delta t$ = 0.2) are generated with a synchronously mode-locked oxazine-750 dye laser. The laser beam was slightly focused into a 300 µm-diameter spot on the GaAs crystal, which was held in pumped liquid He at 2.0 K. The samples of controlled thickness d = 6.3 µm and ∿ 1 x 1 mm^2 size were prepared from high-purity (N_D, $N_A \leq 5.\ 10^{14}\ cm^{-3}$) liquid-phase epitaxy material by standard lapping and etching techniques.

The pulse shapes were measured by autocorrelation techniques, using background-free noncollinear second harmonic generation (SHG) in a 2-mm thick angle-tuned $LiIO_3$ crystal.

The distortion-free propagation of light pulses is demonstrated in Fig. 3. Fig. 3(a) shows the intensity autocorrelation traces of the 0.7-ps light pulses generated with the dye laser. The temporal dispersion D = -95 ps/nm/km of the optical fiber was directly measured using a Tektronix sampling scope

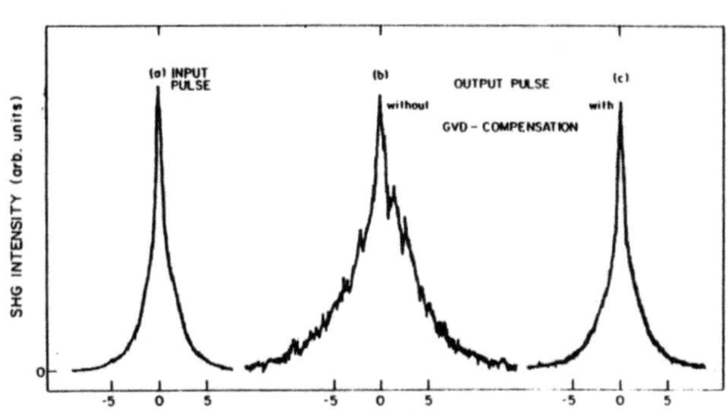

Fig. 3 Intensity-autocorrelation pulse-width measurements (a) 0.7-ps pulses with single-sided exponential pulse shape generated by the synchronously mode-locked dye laser, (b) 2.8-ps output pulses after transmission through the 100-m single-mode optical fiber, and (c) 0.8-ps output pulses after transmission through the optical fiber and the GaAs sample

and a fast photodiode. The pulses emerging from the optical fiber are broadened to 2.8 ps, as shown in Fig. 3(b).

At the wavelength of 8178.5 Å, which corresponds to the photon energy $\hbar\omega_0 = E_T + 0.83$ meV, we are able to achieve essentially distortion-free transmission in the optical fiber - GaAs crystal combination, as demonstrated in Fig. 3(c). However, the small increase in the pulse duration to 0.8 ps indicates that the GVD compensation does not work perfectly over the whole spectral width of the input pulses. The observed compensation energy is 0.5 meV closer to the resonance than the energy determined by the single-exciton oscillator model [4] and the condition of a vanishing GVD in the GaAs crystal - optical fiber combination. The internal transmission in the GaAs crystal at the matching wavelength was 30%, whereas the actual losses in our arrangement were higher due to reflection losses at the glass windows of the cryostat and the crystal surfaces.

In conclusion, we have demonstrated a new technique for achieving distortion-free pulse propagation through optical fibers at essentially any wavelength including those at which optical fibers have a large positive GVD. Our technique marks the first use of the anomalous dispersion near a discrete resonance in a solid to counterbalance the pulse broadening in optical fibers. The combination of a single-mode optical fiber and any direct-gap semiconductor with parameters similar to those used in our experiment should provide a useful tool for distortion-free pulse propagation in optical fibers at any wavelength.

This work was supported by the Air Force Office of Scientific Research.

References

1. D. M. Bloom, L. F. Mollenauer, Chinlon Lin, D. W. Taylor, and A. M. Delgaudio, Opt. Lett. 4, 297 (1979).
2. Chinlon Lin, H. Kogelnik, and L. G. Cohen, Opt. Lett. 5, 476 (1980).
3. H. Nakatsuka and D. Grischkowsky, Opt. Lett. 6, 13 (1981).
4. J. J. Hopfield, Phys. Rev. 182, 945 (1969).
5. R. G. Ulbrich and C. Weisbuch, Phys. Rev. Lett. 38, 865 (1977).
6. R. G. Ulbrich and G. W. Fehrenbach, Phys. Rev. Lett. 43, 963 (1979).

Part III

Advances in Optoelectronics

Generation and Pulsewidth Measurement of Amplified Ultrashort Ultraviolet Laser Pulses in Krypton Fluoride

P.H. Bucksbaum, J. Bokor, R.H. Storz, and J.W. White
Bell Telephone Laboratories, Holmdel, NJ 07733, USA

D.H. Auston
Bell Telephone Laboratories, Murray Hill, NJ 07974, USA

Rare gas halogen excimer laser amplifiers have long been considered attractive media for the production of high power ultraviolet picosecond light pulses. We have used standard nonlinear optical techniques to up-convert the output of a well mode-locked dye laser to 248 nm for amplification in KrF^*. Pulses with up to 20 mJ of energy and 17 psec in duration have been obtained at a repetition rate of 10 Hz, after a single pass through a KrF^* discharge. The output pulsewidth has been characterized using an electronic autocorrelator consisting of two ultrafast photoconducting detectors.

A block diagram of our apparatus is shown in Fig. 1. The ultrashort pulses were produced by a synchronously pumped mode-locked dye laser similar to the system described by Wokaun, et al [3]. 648 nm oscillator pulses were amplified, frequency doubled, and the 324 nm radiation which resulted was summed in KDP with a single 70 psec, 1.06 um pulse from the Nd:YAG pump laser to produce the desired wavelength. The amplified 648 nm laser pulse length has been measured by background-free second harmonic generation [4], yielding a second order autocorrelation FWHM of 20-30 psec, with the lower figure obtained when the oscillator is close to threshold. Assuming a gaussian pulse shape, this corresponds to an actual FWHM pulsewidth of 14-20 psec [4]. Up to 30 μj per pulse was available at 248 nm.

The bandwidth at 248 nm has been inferred from measurements of the 648 nm linewidth by observation of the interference fringes in transmission through a high finesse etalon. It was found that the bandwidth for each 648 nm laser pulse was essentially equal to the Fourier transform limit (1 cm^{-1}; however, the shot to shot frequency jitter was as high as 10 cm^{-1}. This jitter could be eliminated by reconfiguring the dye laser oscillator cavity at the expense of an increase in pulsewidth to about 30 psec.

Amplification of this laser pulse occurred in the 85 cm long discharge region of a Lambda Physics EMG 200 excimer laser with the mirrors removed and the windows tilted by approximately 20 degrees to eliminate feedback. However, even in the complete absence of optical feedback, the gain was sufficiently high that approximately 60 mJ of amplified spontaneous emission (ASE) was emitted from each end of the amplifier. A 1:4 cylindrical telescope was used to match the input beam to the 6 *mm* $\times$ 30 *mm* cross section of the discharge. Reflection losses here typically reduced the pulse energy at the amplifier input to 10 - 20 μJ. The discharge was approximately 15 nsec in duration.

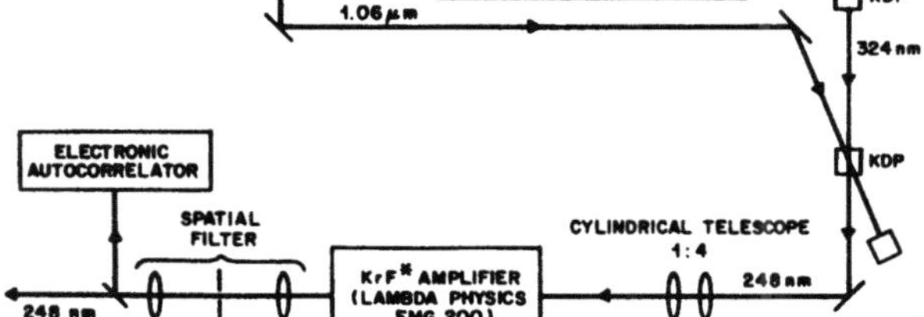

Fig. 1 Block diagram of the apparatus

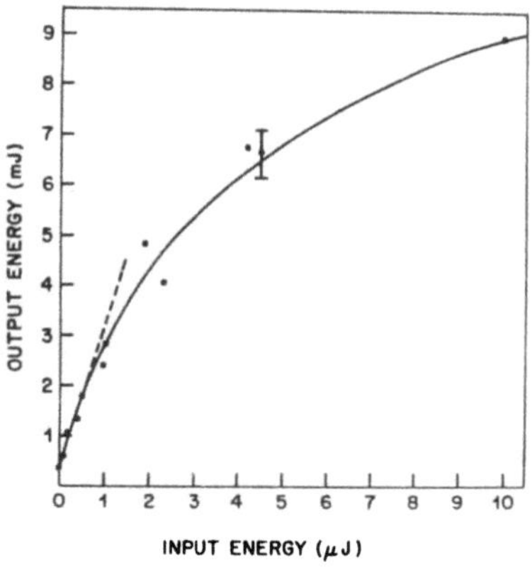

Fig. 2 KrF* amplifier gain curve

In Fig. 2, the input-output characteristics of the amplifier are shown together with a fit to the usual Franz-Nodvik formula [5] for a two-level system

$$E_{out} = E_{sat} \ln\left[1+e^{gl}\left(e^{E_{in}/E_{sat}}-1\right)\right] \tag{1}$$

where E_{out} and E_{in} are the output and input energy densities, respectively. The best fit corresponds to a small signal gain, $e^{gl} = 3500$, and a saturation energy density, $E_{sat} = 2.1\ mJ\ cm^2$. For this measurement, the input wavelength was tuned to the peak of the KrF* gain curve at 248.5 nm. The highest amplified energy observed was 20 mJ, obtained with approximately 20 uJ of input energy. In obtaining the data of Fig. 2 the strong ASE background was suppressed with a simple spatial filter consisting of a 500 mm focal length lens and a 0.5 mm aperture. This reduced the background to 0.4 mJ which appears in Fig. 2 at the nonzero intercept.

A somewhat larger value for E_{sat} was obtained from 2 nsec KrF* pulses in [1]. The discrepancy may be due several dynamical gain recovery mechanisms which are known to occur on 50 psec time scales [2]. Intensity inhomogeneities also tend to decrease the measured value for E_{sat}. The small signal gain g is also much smaller than values measured in shorted excimer discharge cavities [1]. This is probably due to gain saturation from the strong ASE output in our long cavity.

An important characteristic of the output is the pulse width, which we would like to keep as short was possible. However, under conditions of fairly strong gain saturation, it is in general very difficult to keep the output pulsewidth from broadening [6]. For this reason, it is extremely useful to have some means of monitoring the output pulsewidth. The usual technique of autocorrelation by second harmonic generation (SHG) is not applicable to these ultraviolet laser pulses since there presently exists no nonlinear crystal capable of generating the second harmonic of 248 nm. Streak cameras have been used to measure short pulses, and recently, ultraviolet short pulsewidth measurement by multiphoton ionization autocorrelation has been demonstrated [7].

We have developed a general method for measuring ultraviolet laser pulsewidths which is based on an electronic autocorrelator [8]. This technique is found to be quite comparable to SHG in terms of cost and simplicity. The device consists of two photoconducting switches connected in series by a terminated transmission line (Fig. 3). Ion-implanted silicon on sapphire is used as the photoconducting material. The signal from one photoconductor acts as a bias for the second photoconductor, which functions as a sampling gate. The photoconducting detectors are sensitive to all wavelengths shorter than the cutoff given by the material bandgap, which is in the near-infrared for silicon. Thus, this device is useful throughout the visible, ultraviolet and even soft X-ray spectral regions.

Pulsewidth measurements are performed via second order autocorrelation. The incident laser beam is split into two, with each beam directed at one of the photodetectors. By varying the relative time delay τ between the two beam lines, and simply measuring the total charge $Q(\tau)$ flowing through the second detector, we obtain the second order autocorrelator function

$$Q(\tau) \propto \int_{-\infty}^{\infty} I(t)I(t+\tau)dt \tag{2}$$

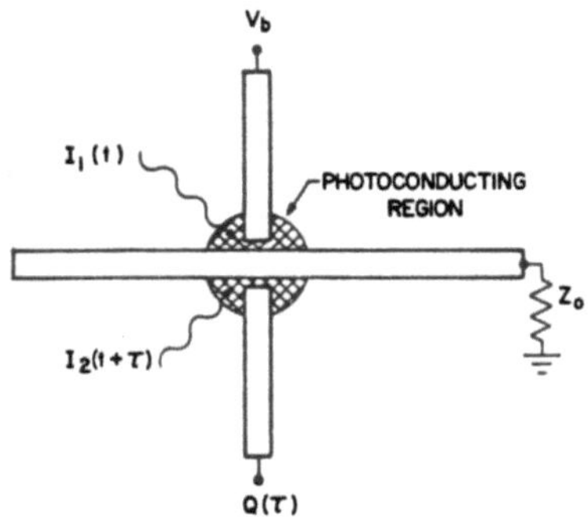

Fig. 3 (left) Electronic autocorrelator

where I(t) is the laser pulse waveform. Relation 2 holds when the laser pulse duration is much longer than the photoconductor and circuit response times. For laser pulse durations comparable to the device response time, a deconvolution procedure may be used. The device response time can be determined using visible laser pulses by comparing electronic and SHG autocorrelation measurements.

In Fig. 4 we show some results obtained using an electronic autocorrelator to characterize the amplified ultraviolet laser pulse. Figure 4(a) shows a SHG measurement of 20 psec 648 nm dye laser pulse. An electronic autocorrelation of the same pulse, using the same optical delay line is shown in Fig. 4(b). These may be compared to yield a response time for this particular device of 23 psec. Figure 4(c) shows an autocorrelation of 248 nm input pulses to the amplifier. If the same device response is assumed, this yields a pulselength of 14 ± 2 *psec* for a gaussian pulse shape. The precision of this measurement is limited by the response time, which is now longer than the light pulse. The output pulsewidth of the KrF* amplifier, shown in Fig. 4(d), is measured to be 17 ± 2 *psec*. Little, if any, broading is observed.

In conclusion, a system has been constructed for the generation of ultrashort uv laser pulses n the 15 – 30 *psec* range, with peak powers of over one gigawatt. Future improvements may include multiple passes through excimer amplifiers. In the regime of strong saturation, it may be possible to extract up to 100 mJ from this amplifier module. Careful pulse shaping will undoubtedly be required to accomplish this without significant pulse broadening.

We gratefully acknowledge helpful discussions with R. R. Freeman, J. P. Heritage, E. P. Ippen, C. V. Shank, and P. R. Smith, and the technical assistance of L. Eichner.

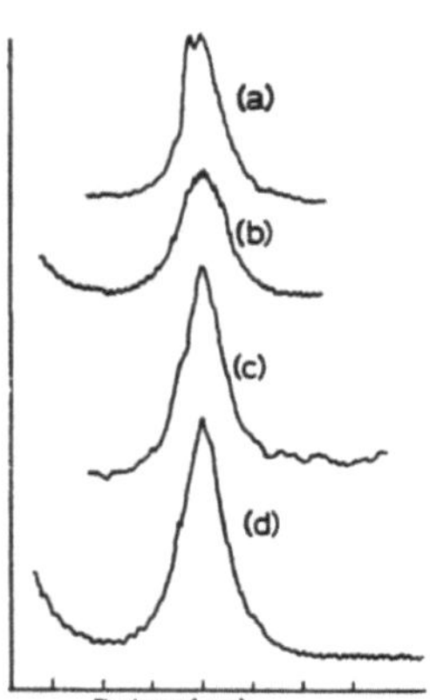

Fig. 4 (right) Autocorrelation traces for the KrF* amplifier system: (a) 648 nm laser pulse, using SHG; (b) 648 nm pulse, using electronic autocorrelator; (c) 248 nm input pulse to amplifier; (d) 248 nm output pulse from KrF* amplifier; the rising background to the left of traces (b) and (d) are due to an electronic reflection

References

[1] J. Banic, T. Efthimiopoulos, and B. P. Stoicheff Appl. Phys. Lett. *37*, 687 (1980).

[2] P. B. Corkum, and R. S. Taylor, IEEE J. Quantum Electron., (to be published).

[3] A. Wokaun, P. F. Liao, R. R. Freeman, and R. H. Storz, Opt. Lett. *7*, 13 (1982).

[4] E. P. Ippen and C. V. Shank, in *Ultrashort Light Pulses*, S. L. Shapiro, Ed., (Springer-Verlag, Berlin, 1977), p. 83.

[5] S. L. Shapiro, Ed., (Springer-Verlag, Berlin, 1977), p. 83. L. M. Franz and J. S. Nodvik, J. Appl. Phys. *34*, 2346 (1963).

[6] A. Migus, J. L. Martin, R. Astier, and A. Orszag, in *Picosecond Phenomena I*, R. Hochstrasser, W. Kaiser, and C. V. Shank, Eds., (Springer-Verlag, N. Y., 1980), p. 59; A. Migus, C. V. Shank, E. P. Ippen, and R. L. Fork, IEEE J. Quantum Electron., *QE-18*, 101 (1982).

[7] D. M. Rayner, P. A. Hackett, and C. Willis, Rev. Sci. Instrum.*53*, 537 (1982).

[8] P. R. Smith, D. H. Auston, A. M. Johnson, and W. M. Augustyniak, Appl. Phys. Lett. *38*, 47 (1981); D. H. Auston, A. M. Johnson, P. R. Smith, and J. C. Bean, Appl. Phys. Lett. *37*, 371 (1980).

Addressing and Control of High-Speed GaAs FET Logic Circuits with Picosecond Light Pulses

R.K. Jain, J.E. Brown, and D.E. Snyder

Hughes Research Laboratories
3011 Malibu Canyon Road, Malibu, CA 90265, USA

We demonstrate optically-addressed operation of high-speed GaAs FET logic circuits. More specifically, using picosecond light pulses from a cw mode-locked dye laser to illuminate specific FET's, we demonstrate complete logic level switching in NOR gates and inverters, and logic function control thereby. The latter is demonstrated by toggling a D-flip-flop in a divide-by-two mode with 76 MHz repetition rate pulses from the mode-locked dye laser. Besides their direct application to the optical addressing of ultrafast circuits, as might be required for data processing in GHz-rate optical communication links, such experiments show potential for contactless diagnostic procedures for test circuits, and for picosecond resolution measurements of the on-chip response times of specific logic gates via the use of optical sampling techniques.

Although other photosensitive circuit elements may be specially introduced into logic circuits, in depletion-mode FET logic circuits, metal-semiconductor junction field effect transistors (MESFETS) present themselves as natural circuit elements that are significantly photosensitive to light in the visible and the near infrared. At these wavelengths optical generation of carriers in the exposed GaAs results in photoconductivity in the source-gate and gate-drain regions. The effect of light is clearly more pronounced if the MESFET is biased in the pinch-off regions (i.e. the FET is non-conductive in the absence of illumination). Figure 1(a) and 1(b) show a magnified photograph and circuit diagram of a typical depletion-mode NOR gate. J_1 through J_5 are FETS and D_1 through D_3 are diodes. The logic levels are "0" ≡ - 2V±0.5V and "1" ≡ 0V±5V. J_3 and J_4 represent a nonlinear load and buffer amplifier stage, and do not exhibit strong photosensitivity. However, in the pinch-off mode, extreme sensitivity to illumination is exhibited by J_5, and by both of the input FET's (J_1 and J_2). With proper adjustment of the drain and source bias voltages (V_{DD}, V_{SS}), and by using ~5 ps pulses at $\lambda \simeq 6000$Å from a 76 MHz repetition rate synchronously mode-locked cw dye laser, we obtained complete logic level switching with only ~2 mW of average laser power (focused into a spot diameter of ~10 μm). When observed on an external 50Ω sampling scope, the "1" to "0" logic level switching of the NOR gate manifests itself as a negative going voltage pulse of ≈1.5 volt magnitude and of ~200 ps pulse duration. The shortness of the observed electrical signal was limited largely by inadequate high speed coupling (and impedance mismatch) between the circuit and the measuring instrumentation. Nevertheless, these results illustrate a secondary application of such standard logic circuits, viz. their usefulness as moderately sensitive high-speed photodetectors. Similar results were obtained with the use of an Inverter gate, which is essentially a NOR gate with only one input.

The logic level values of the optically-switched voltages were confirmed by using these electrical pulses to switch a second logic circuit. For this

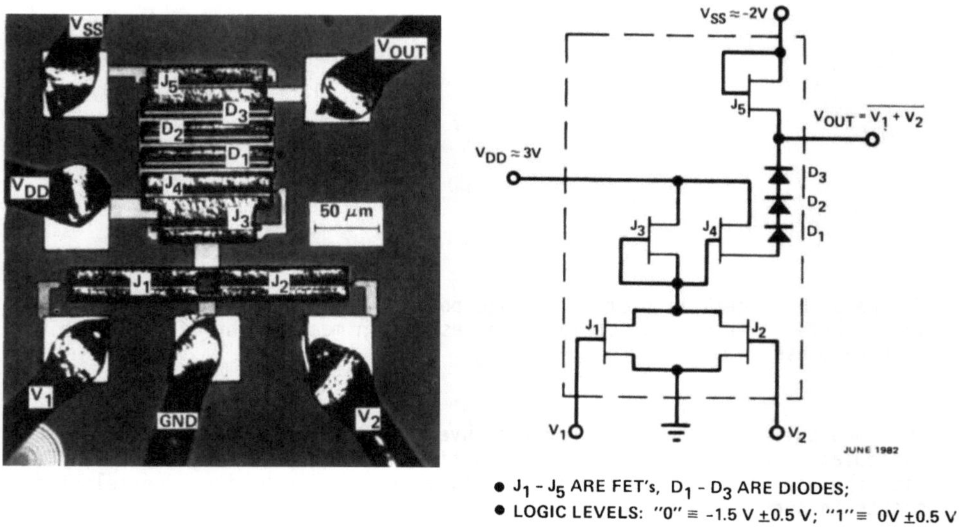

Figure 1 (a) Magnified photograph, and (b) Circuit Diagram of a typical depletion - mode NOR gate.

experiment, an inverter and D flip-flop were fabricated on the same GaAs chip, and the D flip-flop was bonded to operate in a divide-by-two mode, so that each logic level pulse at the flip-flop clock input results in a change in its logic state. The output of the inverter was then bonded to the flip-flop clock input, and a probe lead was also connected at this point, so as to monitor the waveform at the inverter output. To reduce loading problems, 1.1 K Ω series chip resistors were mounted on the printed circuit boards and electrical signals were fed in and out via a (10 GHz bandwidth) co-planar transmission line. Note that these series resistors result in a 23:1 voltage divider for observation of signals on the 50 Ω sampling scope.

As seen in this photograph, switching of the flip-flop output state occurs with each input pulse, confirming the logic level character of the optically

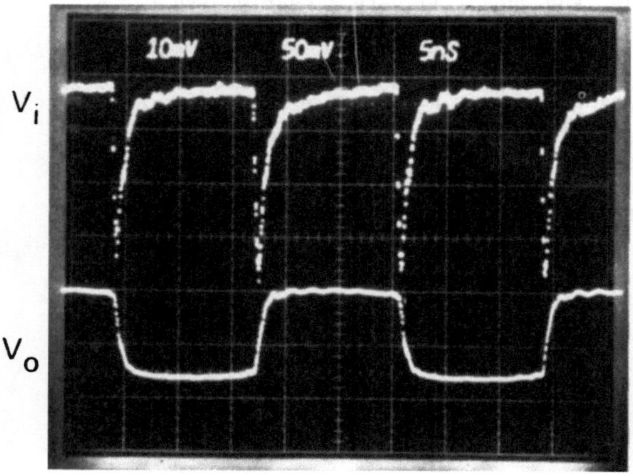

Figure 2 Electrical waveforms observed on a fast sampling oscilloscope at the outputs of the inverter (V_i), and the divide-by-two D-flip-flop (V_o). For this data, the output FET of the inverter was illuminated with dye laser pulses at a ~76 MHz repetition rate.

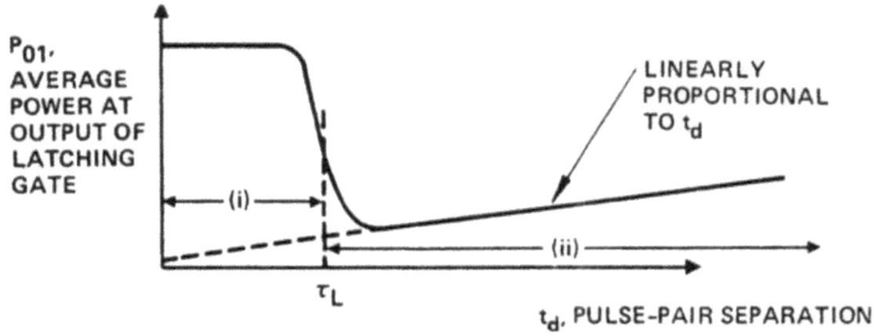

Figure 3 Hypothetical plot of average power P_{01} (at output of latching gate) versus pulse pair separation for high-resolution measurement of the latching gate response time.

switched pulses at the output of the inverter (top trace in Fig. 2). Clearly, manipulation of more complex logic functions is thus possible, especially with the use of more than one optical input, addressing various pre-selected points in such circuits.

The observed risetimes of the flip-flop output (~400 ps, as measured on expanded time scales) are still limited by loading problems. We describe here a novel technique for the measurement of picosecond risetimes of such circuits which uses essentially low speed (<100MHz) electronic instrumentation. In this technique, we split each optical pulse in our MHz rate pulse train into a pair of pulses with variable temporal delay t_d spanning the range of interest. Then, if $t_d > \tau_L$, where τ_L is the response time of the latching circuit (e.g. flip-flop), the output V_{01} of the latching circuit will be a rectangular pulse of width t_d. However, if the adjustable pulse pair separation t_d is less than the latching circuit response time (τ_L), then the latching gate will switch only once for each pair of pulses, and the output waveform will be a square waveform of period T given by the repetition period of the pulse-pair excitation. Thus, if one were to simply measure the "low-frequency" (MHz or slower) electrical power P_{01} (for instance, with an rms voltmeter) at the output of the flip-flop, for $t_d < \tau_L$ a plot of P_{01} vs. t_d would stay constant until $t_d \sim \tau_L$. At $t_d \sim \tau_L$, i.e. at temporal durations where the pulse pair can switch the flip-flop twice, the average power will drop (or rise) abruptly, and then will vary linearly with t_d, as shown schematically in Fig. 3.

Another very promising application of optical addressing of logic circuits is in the use of picosecond optical pulses for the contact-free diagnosis of complex high-speed logic circuits or test patterns. Specific locations in the FET circuits could be addressed optically, and one could look at the output or various outputs for predicted behavior. A high level of flexibility in the choice of address locations should help identify any failure points in the circuit.

Surface Metal-Oxide-Silicon-Oxide-Metal Picosecond Photodetector

S. Thaniyavarn and T.K. Gustafson

Electronics Research Laboratory, University of California
Berkeley, CA 94720, USA

1. Introduction

Recently there has been increased effort on the development of high speed photodetectors with ultimate response time down to the picosecond range. We have recently implemented a metal-oxide-silicon-oxide-metal junction photodetector having a surface structure as shown diagrammatically in Figure 1a [1]. The device has a relatively large circular optically sensitive area. A dark current of less than 1 nA at under 20 V bias, and a responsivity of the order 1/2 mA/mW at the cw He-Ne laser wavelength (6328 Å) has been observed. The device has a fast response time with a rise and fall time of about 50 psec.

2. Basic structure

A basic structure of the device is illustrated in Figure 1a. A very lightly doped, high resistivity p-type silicon wafer is used as a substrate. The thin uniform silicon dioxide surface layer is approximately 45-60 Å thick. The aluminum top layer of about 2000 Å thick is used to form both metal electrodes. A cross-section across the optically sensitive region is shown in Figure 1b. A photograph of the device displayed in Figure 2 shows the top two electrodes forming an interdigitated structure over the circular active area (150 microns in diameter). The interdigitated electrode fingers and the gap widths are both 5 microns .

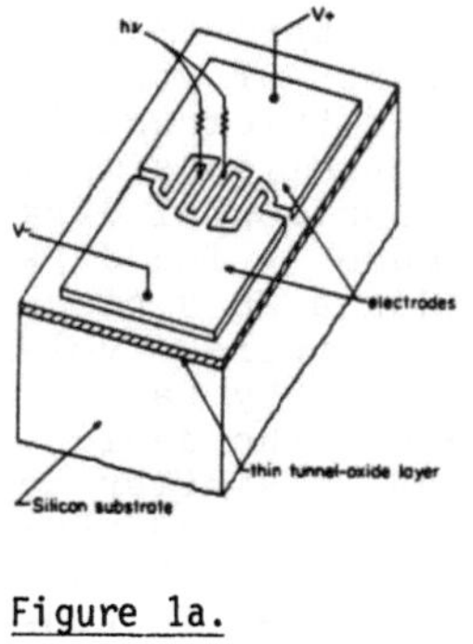

Figure 1a.

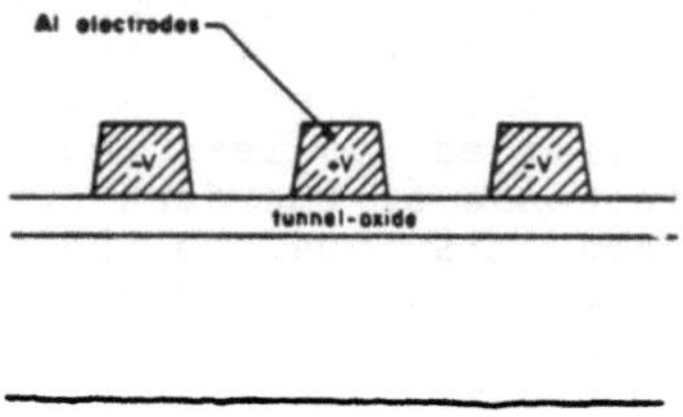

Figure 1b

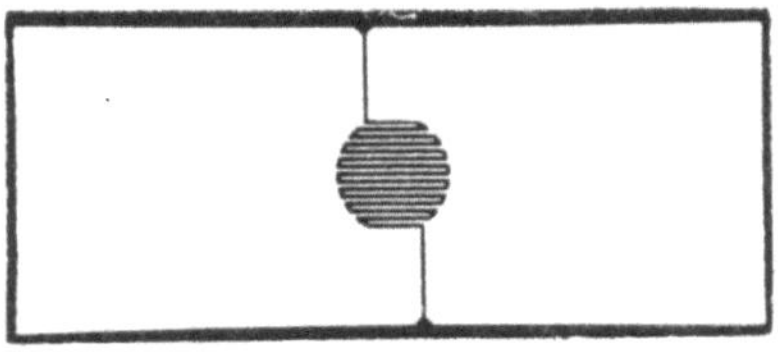

Figure 2

3. Fabrication

The silicon wafer is first stripped of its non-uniform native surface oxide layer. Then a thin very uniform surface oxide layer is regrown by thermal oxidation in a dry oxygen atmosphere at 700-850 °C for 10-15 minutes. This is followed by a thermal anneal in a dry nitrogen atmosphere for another 10-15 minutes at the same temperature. The top aluminum electrodes are subsequently formed by thermal evaporation and patterned by a photoresist lift-off technique. The fabrication process of the device is extremely simple. This is one of its major advantages. No semiconductor doping processes such as diffusion or ion implantation, necessary for the fabrication of most other detectors, are required here. Furthermore, for the basic structure, only one photomasking step is needed since both top electrodes are defined at the same time. There are no alignment steps. Thus the interdigitated electrodes and gap widths ,not being limited by the resolution of alignment, can be reduced to within approximately a micron. Submicron electrode fingers and gap widths, if needed, can be accomplished by techniques such as electron beam lithography.

4. Operations and Experimental results

The metal-oxide-silicon-oxide-metal junction is basically composed of two back-to-back M-I-S tunnel diodes. Since the semiconductor is very lightly doped, the active region immediately below the surface is totally depleted. The dark current flowing through the device is the sum of the minority electron and majority hole thermionic emission currents injected from the metal electrodes as in case of a completely depleted MSM structure [2]. The current is dominated by the minority thermionic emission current and saturates as the applied bias increases. This thermionic emission current is reduced by the presence of the tunnel-barrier layer. The barrier introduces a tunnel transmission factor which varies approximately exponentially with the product of the square root of tunnel barrier height in eV and the tunnel barrier thickness in angstroms [3]. Thus the dark current can be controlled simply by adjusting the tunnel-oxide thickness. An experimental dark current-voltage characteristics is shown in Figure 3. The dark current is limited to less than 1 nA for under a 20 V bias. The slight increase of the dark current, as the applied voltage increases, is due to the image force Schottky barrier lowering effect and the two dimensional nature of the device.

The photoresponse of the M-I-S-I-M junction is similar to that of an open-base bipolar NPN phototransistor. An aluminum/ tunnel-oxide/ p-type silicon forms a 'minority' M-I-S tunnel diode. The tunnel-oxide layer introduces asymmetries in tun-

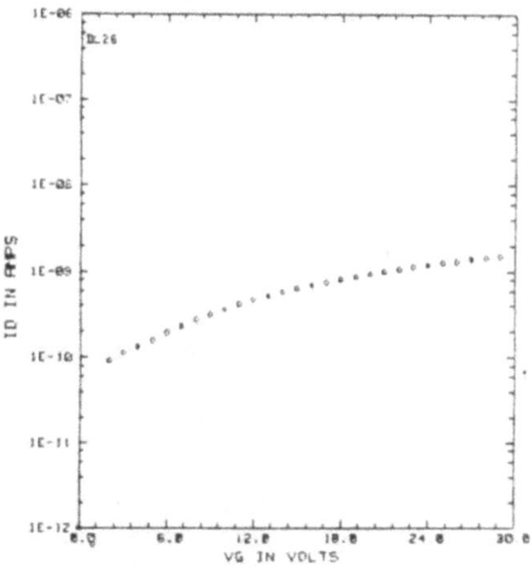

Figure 3

nel barriers for electrons and holes [4] in such a way that the junction current is dominated by the minority electrons rather than the majority holes as in the metal/ p-type silicon Schottky diode. Thus the minority M-I-S diode behaves similarly to a PN junction. It has been used in place of an N^+P junction in solar cell work [5], and has also been employed as an emitter in a bipolar transistor structure [6]. When a d.c. bias is applied to the M-I-S-I-M structure, a high electric field develops in the totally depleted base region. Photogenerated electron-hole pairs in this high field region will drift apart. The holes move towards the cathode, in effect, to forward bias the M-I-S junction. This induces more electrons to tunnel from the metal providing a current gain in a similar manner to an NPN phototransistor. Figure 4 shows the I-V characteristics of the device in response to a cw He-Ne (6328 Å) excitation with differing intensities of 0, 0.22, 0.45, 0.9 and 1.85 mW respectively. A responsivity of the order of 1/2 mA/mW corresponding to one carrier/ photon is observed.

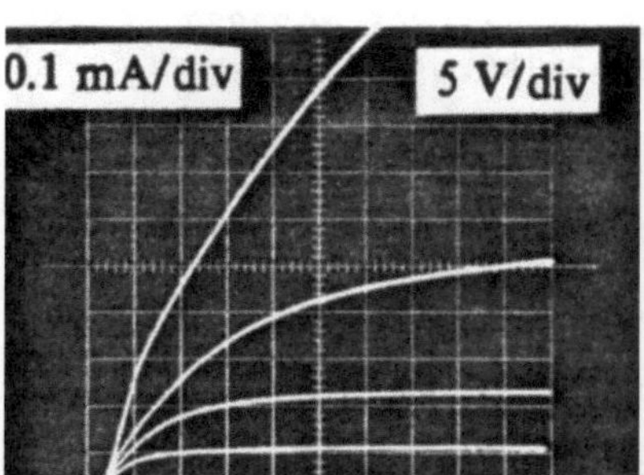

Figure 4

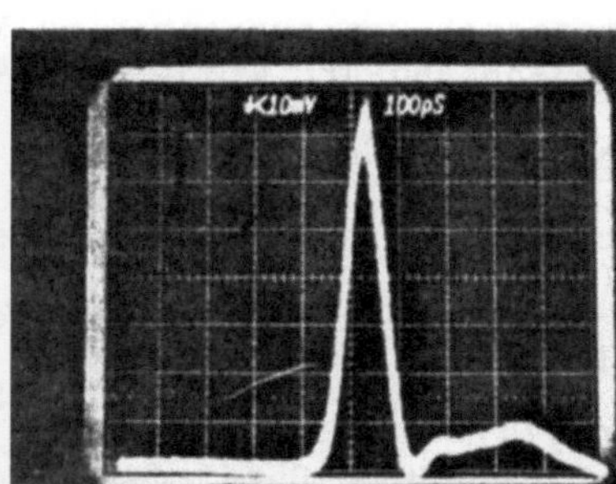

Figure 5

To test the response speed, a synchronously pumped mode-locked R6G dye laser providing <10 psec optical pulses at a wavelength of 6000 Å was used. The response in Figure 5 was taken with sampling oscilloscope with a limited rise time of 25 psec. The observed pulse response shows a symmetric fast rise and fall time of ~50 psec. The junction was biased to 20 Volts.

5. Advantages of the surface structure

Besides satisfying the basic requirements of high speed, low dark current and high responsivity, the device surface structure offers other advantages. Its planar structure makes it readily compatible with other microwave silicon integrated circuits. It can also be directly incorporated into a microwave microstrip transmission line. Moreover, the surface-type detector has a wider spectral response extending into the UV range. Since the biasing field is highest near the surface where most of the UV-generated carriers are created, no degradation in speed for UV radiation as in the case of a conventional vertical P-I-N diode is expected.

6. Conclusions

We have demonstrated that a simple silicon surface photodetector having a high response speed can be easily fabricated. The speed can be further increased by reducing the gap width, thus reducing the transit time limit. The silicon active layer thickness should also be reduced since long wavelength radiation can generate photo-carriers up to several microns deep. This depth must be limited if a higher speed is desired. This can be accomplished by using a thin slab of silicon. For example, a thin poly-silicon film of a micron or less in thickness can be deposited onto an insulating substrate such as sapphire or quartz. The film can then be recrystallized by annealing, and used as a substrate. In order to increase the detection sensitivity and reduce the dark leakage current further, the tunnel-oxide thickness should be optimized. An antireflection coating layer and a thick field oxide buffer layer should also be employed. A simple photodetector with a response time to within a few picosecond and a dark current of a few pico-amperes should be achievable ultimately utilizing this structure.

This research was supported by NSF grant ECS-7923877, NASA grant NAG3-88 and JPL grant 1-482427-26740.

References:

1. S. Thaniyavarn and T.K. Gustafson, Appl. Phys. Lett. 40(3), 255 (1982)
2. M.J. Malachowski and J. Stepniewski, Sol. St. Elec. 24, 381 (1981)
3. M.A. Green, F.D. King, and J. Shewchun, Sol. St. Elec. 17, 551 (1974)
4. K.K. Ng and H.C. Card, Jour. Appl. Phys. 51(4), 2153 (1980)
5. J. Shewchun, M.A. Green, and F.D. King, Sol. St. Elec. 17, 563 (1974)
6. H. Kisaki, Proc. IEEE, 61, 1053 (1973)

Solid-State Detector for Single-Photon Measurements of Fluorescence Decays with 100 Picosecond FWHM Resolution

A. Andreoni, S. Cova, R. Cubeddu, and A. Longoni

Centro Elettronica Quantistica e Strumentazione Elettronica, C.N.R., Istituto di Fisica del Politecnico, Piazza Leonardo da Vinci 32, I-20133 Milano, Italy

The single-photon timing or time-correlated single-photon counting technique for measurements of fluorescence lifetimes is well known [1,2] . It is a flexible technique, suitable to all cases where the emission is within the spectral sensitivity of single-photon detectors and a high repetition rate of the excitation pulse is available (> 1 kHz). It provides high linearity and accuracy over a wide dynamic range, extending down to very low intensity levels, and has better time-resolution than other techniques employing photodetectors. It is therefore very well suited and in many cases almost unique for physical, chemical and biological investigations that require measurements with high resolution and accuracy on fluorescence and scattering phenomena involving low detected light intensity.

After the development of laser systems for the generation of picosecond light pulses, the limit to the time-resolution in such measurements is set by the photodetector. The best resolution values so far reported are from 230 to 400 ps full-width at half-maximum (FWHM) for photomultiplier tubes with discrete dynodes. Values from 130 to 200 ps FWHM have been reported for PMTs with microchannel plate (MCP) multipliers [4,5]; the application of such MCP-PMTs, however, is still hampered by various problems, a major one being the degradation of the photocathode quantum efficiency, that results in limited working life of the device.

Single-photon detection can be obtained also with specially devised types of semiconductor photodiodes, operating in non-proportional avalanche multiplication [6,7,8] . These single-photon avalanche diodes (SPADs) have uniform breakdown over the junction area (diameters from 10 to 80 μm) and are biased above the breakdown voltage. With properly designed device structures, the carrier transit times in the junction can be as low as a few 10 ps, and other sources of time-jitter in the triggering of the avalanche current can be minimized. Theoretical evaluations, based on the physical phenomena involved, give for such devices expected resolution values in the range from 20 to 60 ps FWHM.

The avalanche-quenching necessary for generating single-photon pulses, however, was traditionally obtained by simple passive circuits. This passive quenching operation has features that degrade the detector performance for both photon-counting and photon-timing [9] . This fact, together with specific features of the devices and of the electronic equipment used, contributed to degrade to about one nanosecond the resolution actually observed by other experimenters [10] . The active-quenching method was devised by COVA and LONGONI [9] , in order to operate the SPADs with well controlled parameters and short deadtimes (less than 20 ns), thus overcoming the limitation of the passive-quenching. Preliminary experimental tests performed on active-quenched SPADs with laser pulses having durations down to 150 ps showed that the device resolution is better that this value [11] .

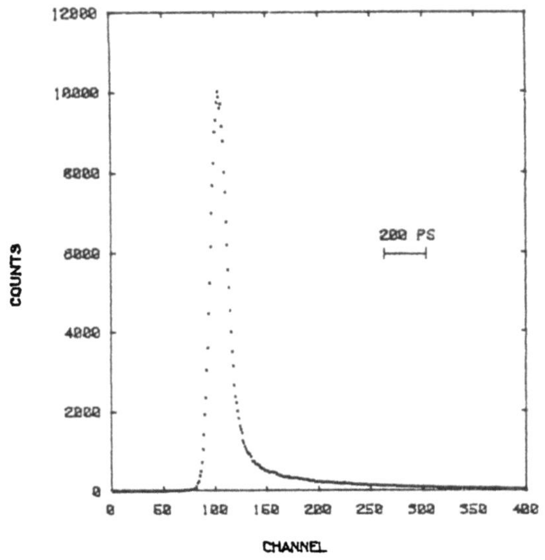

Fig. 1 Measurement of the laser pulse. Time scale: 5.16 ps/channel

The possibility of obtaining measurements of very low-intensity fluorescences with better than 100 ps resolution was deemed to be of the utmost interest and worthy of specific studies. Experiments were therefore undertaken by using a synchronously-pumped mode-locked Rhodamine 6G dye laser as the excitation source. The time duration of the dye laser pulses was measured by SHG autocorrelation function, and resulted to be less than 5 ps FWHM, assuming a gaussian shape. A time-to-pulse-height converter (TPHC) and a multichannel pulse-height analyzer (MCA) were used to measure the delay distribution of detected single photons with respect to reference start pulses, synchronized with the laser pulses. The start pulses were obtained by using a beam splitter and an ordinary fast photodiode, associated to fast electronic circuitry. Measurements were performed both with and without a pulse picker on the laser beam. In experiments without the pulse picker the repetition rate of the start pulses at the TPHC input was reduced to levels from 5 to 50 kHz by suitable demultiplying circuits. The use of the pulse picker allowed such repetition rates to be obtained directly for the light pulses. The synchronism jitter was preliminary checked by using a photodiode-circuit set-up identical to the start in the stop channel. The observed FWHM were about 20 ps with the pulse picker and between 35 and 70 ps without it, depending on various side effects.

Measurements of the laser pulse yielded FWHM from 90 to 100 ps; Fig. 1 shows a typical result. The slower tail is due to diffusion of the carriers generated in the neutral region beyond the depletion layer in the SPAD [11]. This effect increases with the wavelength λ of the detected radiation, due to the increase of the optical absorption length in the semiconductor. In the present SPAD structure, the tail is significant for $\lambda > 500$ nm and may complicate the analysis of fluorescence measurements, due to its marked wavelength dependence. However, the carrier-diffusion effect is found also in ordinary photodiodes and avalanche photodiodes and it is known that it can be strongly reduced by using suitably modified devices structures (see e.g. Ref. [12]).

Solutions of various dyes as DODCI, Erythrosin B, Rhodamine B and Rhodamine 6G, either in ethanol or in water, were used as tests for fluorescence decay

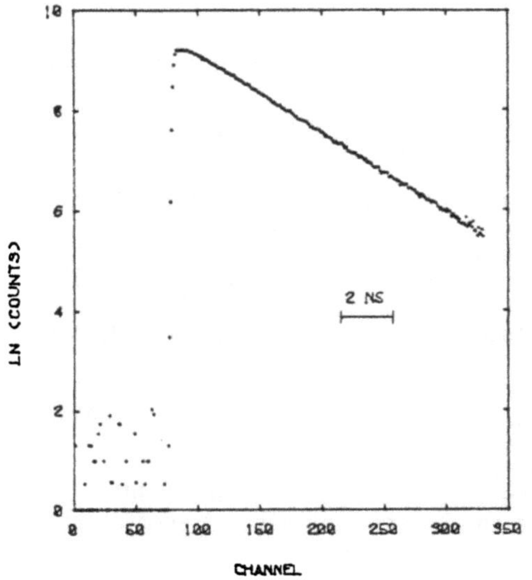

Fig. 2 Fluorescence decay of 10^{-5}M Rhodamine B (Rhodamine 610 Perchlorate-Excitation, Inc.) in ethanol. Time scale: 46.94 ps/channel. Measured decay time constant: 2.99 ns

measurements. The samples were placed in a quartz cuvette and the fluorescence was measured at 90° through cut-off filters by the detector placed about 1.5 cm far from the excitation beam. No optical device was used to improve the light collection efficiency on the sensitive area (about 35 μm diameter). The incident light intensity was adjusted so that the average number of detected photons per pulse was less than 0.1, as required for measurements of single-photon delay distribution [1,2] . A representative experimental result obtained with Rhodamine B 10^{-5}M in ethanol, is shown in Fig. 2. The fluorescence decay time was found to be 2.99 ns in agreement with the values reported in the literature [3] .

The results obtained demonstrate that SPADs can effectively be used for measurements of low-intensity fluorescence decays with the time-correlated single-photon counting technique, and that for such measurements they provide the highest time-resolution today available. In comparison with PMTs, besides the higher time resolution, SPADs have other advantages: the time-response curve is free from the small secondary peaks observed with almost all PMTs [13] ; the spectral sensitivity is more extended, in particular on the long wavelength side; the device is rugged and can easily be gated in short times [14] .

Other studies and experiments under way suggest that the resolution obtained in practice from the present devices can be improved towards the expected theoretical values. Furthermore, developments in the device design and technology may be expected to improve the detector performance in various respects (larger sensitive area, lower tails in the resolution profile, etc.). Developments in the fast circuits associated to the SPAD may also contribute to fully exploit the time resolution of the device.

Acknowledgements

The authors wish to acknowledge the assistance of S.De Silvestri and P. Laporta in the operation of the laser system and the fluorescence measurements and that of G.Ghielmi for the development of the electronic circuits.

References

1. W.R.Ware in Creation and Detection of the Excited State, ed. by A.Lamola (Marcel Dekker, New York, 1971) Vol. 1A
2. S.Cova, M.Bertolaccini and C.Bussolati, Phys. Stat. Sol. A18, 11 (1973)
3. V.J.Koester and R.M.Dowben, Rev. Sci.Instrum. 49, 1986 (1978) and V.J. Koester, Anal. Chem. 51, 458 (1979)
4. B.Leskovar and C.C.Lo, IEEE Trans.Nucl.Sci. NS25, 582 (1978) and NS26, 388 (1979)
5. T.Hayashi in Proc. 1981 Int. Symp. on Nucl. Rad. Detectors, K.Husimi and Y.Shida ed s, p. 259 (Univ. of Tokyo, Inst. for Nucl. Study, Tokyo, 1981)
6. R.H.Haitz, J. Appl. Phys. 35, 1370 (1964) and 36, 3123 (1965)
7. W.O.Oldham, R.R.Samuelson and P.Antognetti, IEEE Trans. Electron Devices ED-19, 1056 (1972)
8. P.P.Webb and R.J.Mc Intyre, Bull. Am. Phys. Soc. Ser II, 15, 813, June 1970
9. P.Antognetti, S.Cova and A.Longoni, Proc. 2nd Ispra Nuclear Electronics Symp., Euratom Publ. EUR 5370e, p. 453 (1975)
10. W.Fichtner and W.Haecker, Rev. Sci. Instrum. 47, 374 (1976)
11. S.Cova, A.Longoni and A.Andreoni, Rev. Sci. Instrum. 52, 408 (1981)
12. J.Müller in Advances in Electronics and Electron Physics (Academic Press, New York, 1981) Vol. 55 pp. 189-306
13. S.S.Stevens and J.W.Longworth, IEEE Trans. Nucl. Sci. NS-19, 356 (1972)
14. S.Cova, A.Longoni and G.Ripamonti, IEEE Trans. Nucl. Sci. NS29, 599 (1982)

Picosecond Optoelectronic Modulation of Millimeter-Waves in GaAs Waveguide

M.G. Li, V.K. Mathur, Wei-Lou Cao and Chi H. Lee

Department of Electrical Engineering, University of Maryland
College Park, MD 20742, USA

Optically controlled microwave or millimeter-wave devices have been a topic of great interest recently. Utilizing a laser induced electron-hole plasma in semiconductor waveguide to control the propagation of an RF signal, we have previously demonstrated the switching, gating and phase shifting of millimeter-wave signal in Si-waveguide with picosecond precision [1]. Phase shifts as large as 300°/cm at 94 GHz were observed. In the experiment involving switching and gating of RF waves, millimeter-wave pulses with pulsewidth as short as 1 ns and variable to tens of nanoseconds have also been generated. In these earlier experiments, high resistivity Si was used as the waveguide material. Since the carrier lifetime in pure silicon is in the millisecond range, to generate a short RF pulse, one generally requires two separate laser pulses, one to "turn on" and the other to "turn off" the millimeter-wave signal. Furthermore, the repetition rate of the device is limited by the carrier recombination rate to less than 10 KHz. In this work, we will report on our most recent study of this type of device by using Cr:doped GaAs as the waveguiding medium. Due to rapid carrier recombination, only a single picosecond optical pulse is needed to produce an ultrashort millimeter-wave pulse. This feature has been utilized to construct a high speed millimeter-wave modulator with a repetition rate well in excess of 1 GHz.

Optical control of RF-waves offers the following advantages: (a) near perfect isolation between the controlling and the controlled devices, (b) low static and dynamic insertion loss, (c) possibility of fast response with picosecond precision, and (d) high power handling capability.

The basic principle of opical control of millimeter waves is illustrated schematically in Fig. 1. The propagation constant, K_Z, in an interval ΔL of a rectangular GaAs waveguide (2.4 x 1.0 mm) is changed to K_Z' by illuminating the broadwall with optical radiation. The absorbed light generates an

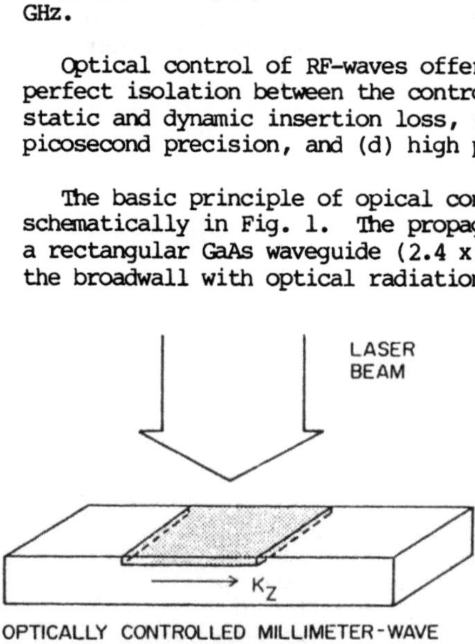

Fig. 1 Schematic diagram of an optically controlled phase shifter, k_z is the propagation vector in the waveguide.

electron-hole plasma resulting in a change of the complex index of refraction of the semiconductor thereby altering the boundary conditions of the waveguide and changing the propagation constant. A millimeter-wave launched into the waveguide experiences amplitude and/or phase modulation while propagating through the illuminated interval. The ratio of amplitude to phase modulation depends on the density and geometry of the plasma. For example, if the density yields a skin depth δ much less than the thickness of the plasma layer, the effect of the plasma is equivalent to an image guide. This yields a nearly pure phase shift, ϕ, given by the relationship

$$\phi = (k_z - k_z')\Delta L. \tag{1}$$

In general case, the transient response of the millimeter-wave depends upon the transport parameters of the optically induced carriers, such as carrier collision time, mobility, diffusion characteristics, etc. The mechanism for phase shift and attenuation can then be satisfactorily described in terms of a model developed in this work based on Marcatili's approximation [2].

An experiment was performed by using a millimeter-wave bridge similar to that used previously. The GaAs waveguide was inserted in one arm of the bridge. Initially, without laser-pulse illumination of the GaAs, the bridge was balanced by adjusting a mechanical attenuator and phase shifter in the other arm so that there was no signal at the output. When a single picosecond pulse of 0.53 μm extracted from a frequency doubled mode-locked Nd:YAG laser was illuminating the GaAs waveguide, the bridge became unbalanced and coherent signals appeared at the output of the bridge. Because the lifetime of the induced carriers is of the order of 100 picoseconds, the millimeter-wave signals rise and decay rapidly. If an optical pulse train is used to illuminate the waveguide, a millimeter-wave pulse train results mimicking the optical pulse train (see Fig. 3 of reference 3). This feature indicates that a modulation bandwidth approaching 1 GHz is attainable. The pulse width of the individual pulse is not resolvable since the combined response time of the detecting and display system is slower than the expected pulse width of about 200 picoseconds, wider than the predicted pulse width by a factor of three. This discrepancy can be resolved by realizing that the millimeter

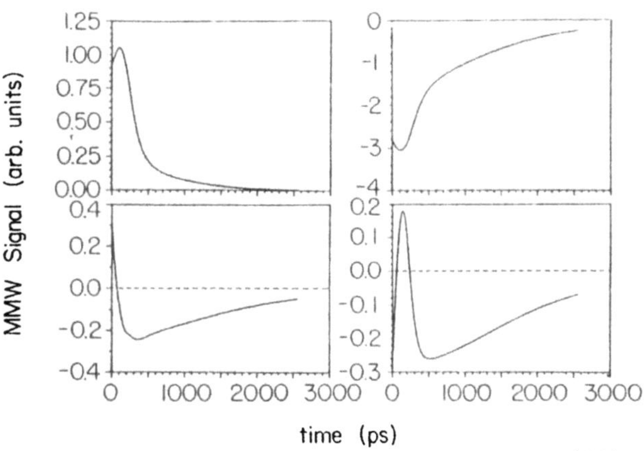

Fig. 2 Theoretical temporal profile of the millimeter-wave signals generated for the unbalanced bridge due to the decay of the optically induced carriers. The curves are plotted for different initial phase angles between the electric fields from two different arms. (a) 180°, the balanced case; (b) 0°; (c) 115°; and (d) 235°.

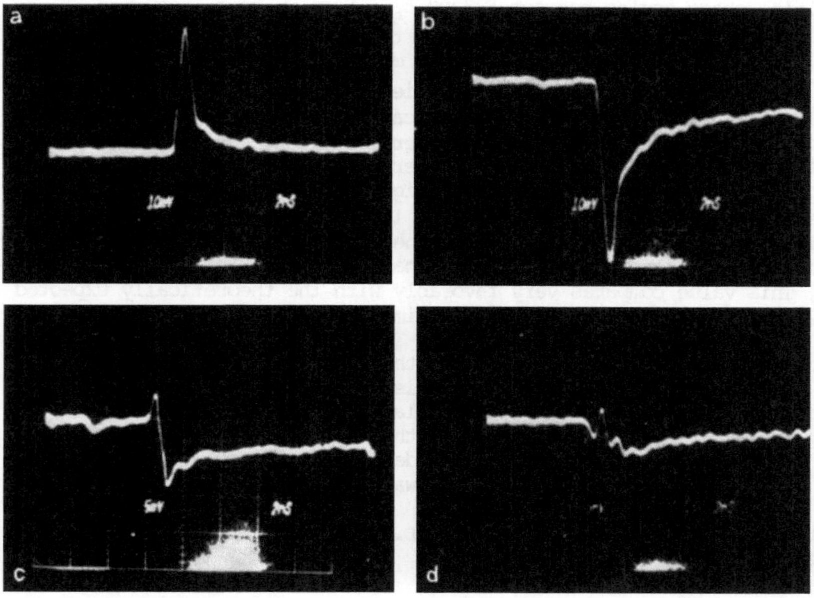

Fig. 3 Experimentally observed millimeter-wave signals corresponding to the theoretical ones depicted in Fig. 2 in the same cyclic order.

wave pulse is actually 'chirped' due to rapid phase modulation. Group velocity dispersion will broaden the 'chirped' millimeter-wave pulse when it propagates through a positively dispersive guiding structure. Mismatches between the dielectric and metallic waveguides will also contribute to some broadening.

Since the millimeter-wave signals are obtained by modulating the dielectric property of the Cr:doped GaAs waveguide, the pulse width of the signals are smaller than the combined response time of the detecting and displaying system. As a result the convenient calibration technique employed in the Si waveguide work [1] to obtain the values of phase shift and attenuation is not directly applicable here. We have, however, developed a dynamic bridge method to determine the values of the laser induced phase shift and attenuation. The output electric field of the bridge (see Fig. 2 of reference 3) is the sum of two phasors, E_A representing the electric field in the arm with dielectric waveguide, and E_B, the RF field in the other arm. E_A and E_B are linearly polarized in the same direction. When the bridge is balanced prior to laser illumination, $E_A = -E_B$ and the output is zero. Under laser illumination E_A is suddenly shifted to a new value and then relaxes back to its initial value as the laser induced carriers decay (represented by the rotation of the E_A phasor in a phasor diagram). The output waveform of the millimeter-wave detector is proportional to $|E_A + E_B|^2$. A positive pulse with a characteristic decay results. The amplitude as well as the detailed temporal profile of the pulse depends upon the initial density of the induced carriers and the material transport parameters. Based on the theoretically calculated curves of phase shift and attenuation as a function of carrier density [4] and assuming a certain decay characteristic of the excess carriers, we have calculated the temporal profile of the signal at the output of the detector. Figure 2 represents the results of these calculations with different initial phase angles between E_A and E_B. Here we have assumed a two component decay mechanism for the carriers with decay constants τ_1=100 ps and τ_2=1000 ps respectively. The mechanism for τ_1 may be due to efficient recombination at chromium impurities; while for τ_2, it may be due to ambipolar diffusion. The laser induced carrier density is estimated to be $2 \times 10^{18}/cm^3$, corresponding to the laser energy of 10μJ. The temporal profile is very sensitive to the initial carrier density from optical injection. Fig. 3 shows the observed millimeter-wave signals corresponding to the theoretical situations depicted in Fig. 2. It is apparent that there is a good qualitative agreement. By comparing the data with the theoretically calculated curve, one can conclude that a phase shift of 270° has been observed. This value compares very favorably with the theoretically expected value of 280° for a plasma column of 2 millimeters in length, or 1400°/cm [4].

In conclusion, we have demonstrated for the first time the generation of 'chirped' millimeter-wave pulses. Using this technique, the modulation of millimeter-wave signals at 94 GHz with modulation bandwidth in excess of 1 GHz is readily achievable. A dynamic bridge method has been developed to measure the phase shift and to monitor the carrier decay kinetics. A two component decay has been observed in a Cr:doped GaAs waveguide.

Acknowledgements are due to Professor C.D. Striffler and A.M. Vaucher for their contributions to these studies.

Work supported in part by the Harry Diamond Laboratory and by the Minta Martin Aeronautical Research Fund, College of Engineering, University of Maryland.

References

1. Chi H. Lee, P.S. Mak and A.P. DeFonzo, IEEE J. Quantum Electron. QE-16, 277 (1980).

2. E.A.J. Marcatili, Bell Syst. Tech. J. 48 2079 (1969).

3. M.G. Li, W.L. Cao, V.K. Mathur and Chi H. Lee, Electron. Lett. to be published.

4. A.M. Vaucher, C.D. Striffler and Chi H. Lee, unpublished.

Synchroscan Streak Camera Measurements of Mode-Propagation in Optical Fibers

J.P. Willson, W. Sibbett, and P.G. May

Optics Section, Blackett Laboratory, Imperial College,
Prince Consort Road, London SW7 2BZ, England

Introduction

The use of optical fibres in telecommunication and data processing links is a rapidly expanding field due to the potentially large information bandwidth ($\lesssim$10 Gbits/s) |1| and low transmission loss (<0.2 db/km at 1.55 μm) |2|. The combination of high data rates and long repeater spacing is now making optical systems very competitive with conventional copper cable systems and consequently the characterisation of optical fibres is of primary interest.

The maximum information bandwidth can be ensured by minimising the pulse dispersion through the choice of single-mode fibres to avoid modal dispersion and using light sources in the 1.3 - 1.5 μm spectral region where the chromatic dispersion is small |3|. Appropriate characterisation of manufactured fibres is usually performed by measuring the refractive index profile to determine the core radius and core/cladding index difference. From these data the normalised frequency parameter can be calculated and the cut-off wavelength for single-mode operation can therefore be established. However, this technique is indirect and can be rather inaccurate due to the substantial deviations of the refractive index away from the idealised step-index profile (refer to Fig. 1.).

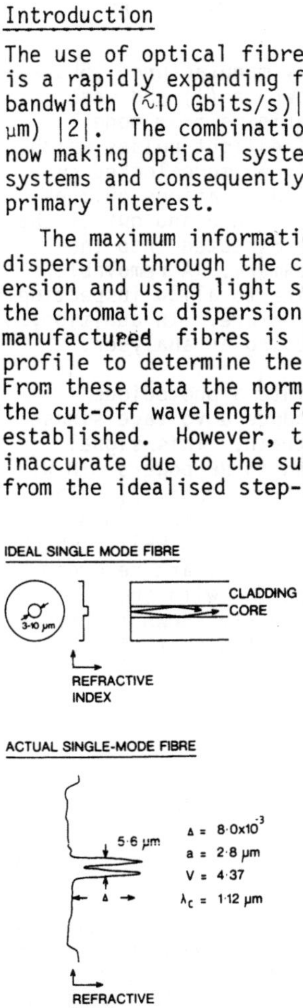

Fig.1 Ideal and actual refractive index profiles of optical fibres

A more direct measurement of the fibre characteristics can be made by exploiting the fact that a fibre which is single-mode at 1.3 μm is weakly multimode at wavelengths in the visible region. Measurement of the intermodal temporal dispersion then enables the effective core radius and index difference to be conveniently determined. Although a technique involving coherent optical filtering has been reported for the measurement of this dispersion |4|, it involved the use of a fibre interferometer with its associated criticality of alignment. In this paper we describe a convenient real-time intermodal dispersion measurement system involving a passively mode-locked cw ring dye laser used in conjunction with a synchronously-operating picosecond streak camera. From our results, we have been able to directly characterise the optical fibre.

EXPERIMENTAL TECHNIQUE

The experimental configuration is shown schematically in fig. 2.

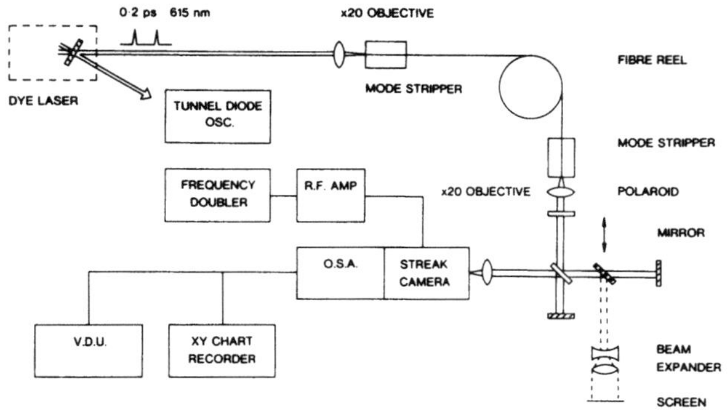

Fig.2 Experimental Configuration

The sample of optical fibre studied was a commercially manufactured |5| step-index fibre. It had the refractive index profile shown in (1) and was nominally single mode at 1.3 μm. To avoid problems arising from mode coupling, a short 28.5 m length of fibre was used. The light pulses which were propagated in the fibre were produced by a passively mode-locked cw Rhodamine 6G ring dye laser which generated pulses with durations ~0.2 ps at a wavelength of 615 nm |6|. The pulses were coupled into and out of the fibre by X20 microscope objectives and the output was directed into the optical delay calibrator for the Synchroscan streak camera. A removable mirror could be inserted into one arm of the delay line to allow inspection and photographic recording of the transverse mode structure. In addition, the polarisation state of the output was determined using an analyser.

The second output beam of the laser was directed into a photodiode/tunnel-diode oscillator microstripline circuit to produce a voltage sinusoid at 82 MHz synchronised to the mode-locked pulse train. This signal which was subsequently frequency doubled and amplified to RF powers ~10 W, provided the deflection voltage of a Photochron IIA streak tube |7|. The streak images were lens-coupled to a two-dimensional optical multichannel analyser and data processing console (B &M, Spektronic OSA 500-WP1/2) which gave a real-time display on a CRT and hard copy with a X-Y chart recorder. This Photochron IIA streak camera operating in conjunction with the passively mode-locked CW ring dye laser has been previously shown to have a time resolution ~1 ps and it has been directly demonstrated that the long-term jitter in the laser pulses is substantially less than 1 ps |6|.

RESULTS

The normalised frequency parameter for a step-index fibre is given by

$$V = \frac{2\sqrt{2}\pi a}{\lambda} (n\Delta)^{\frac{1}{2}}. \qquad (1)$$

where a is the core radius, n is the refractive index of the cladding, Δ is the core/cladding refractive index difference and λ is the free-space wavelength. In the approximation that the index difference is small (ie $\Delta \ll n$), the guided light will propagate along the fibre in linearly polarised (LP) mode groups |8|. The dependence of the normalised group velocity of the mode groups on the normalised frequency parameter is shown in fig.(3a), (3b). When the fibre dimensions (given in fig.1) are substituted into (1), then a value of V = 4.37 is obtained for λ = 615 nm. Reference to

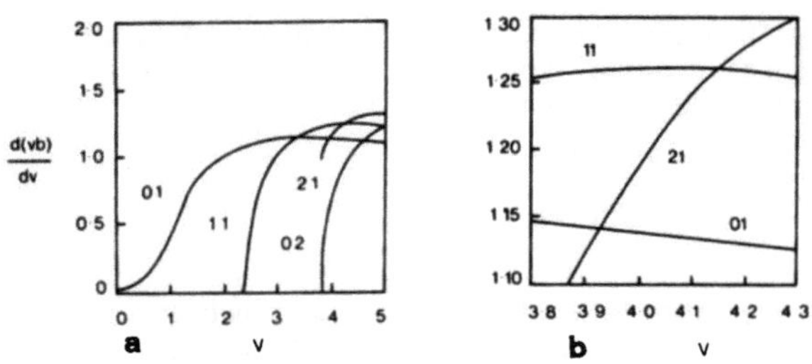

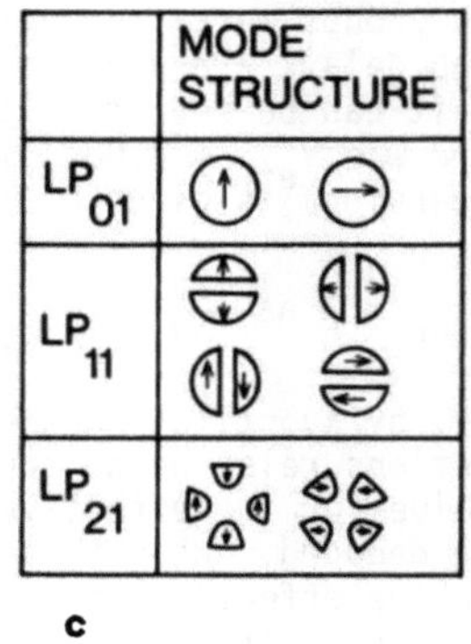

Fig.3 Dispersion curves and field distribution of LP mode groups

fig.3 would therefore indicate that up to five LP mode groups would be expected to be guided and the calculated field distributions for the three lowest order groups are reproduced in fig.3c.

In practice, only three mode groups were observed on the streak camera (refer to fig. 4). The individual mode groups were segregated by changing the launching conditions and polarisation axis of the analyser such that a single mode group was dominant at one particular time.

By simultaneously observing five "spatial" tracks along the streaked images and examining the transverse mode structure, it was possible to determine which mode groups were propagating in the fibre. The fundamental LP_{01} mode was characterised by a gaussian mode structure whereas Fig.5a shows the dominant LP_{11} two-lobe structure and the four-lobe LP_{21} pattern is shown in fig.5b. From these observations, it was concluded that the

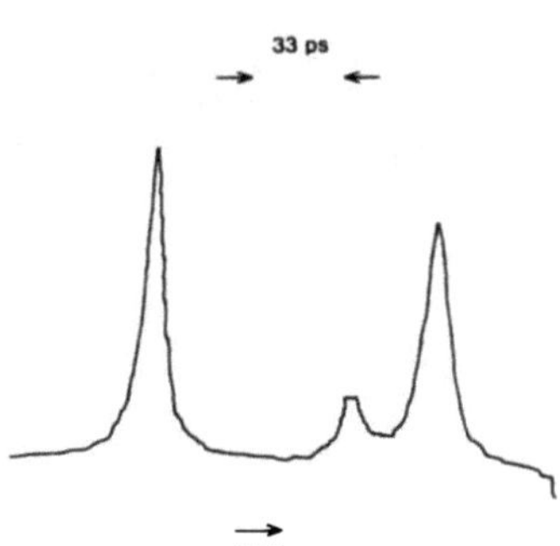

Fig.4 Observed Modal Delays

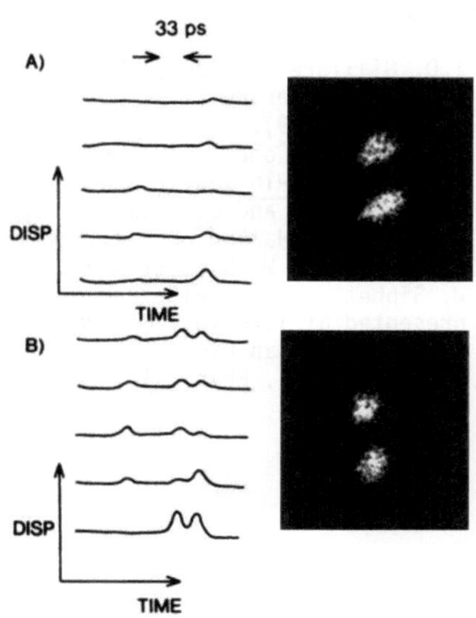

Fig. 5 Modal delays observed on 2-D synchroscan and photographs of associated field distributions

mode groups were propagating in the order LP_{01}, LP_{21}, LP_{11}. Their relative group delays are given directly from the streak image reproduced in fig.4 and it can be seen that the effective normalised frequency parameter of the fibre was $V_{eff} = 4.06$. The group delay of each mode group (neglecting material dispersion) is given by |7|

$$\tau_{gr} = \frac{L}{C} n\Delta \frac{d(vb)}{dv} \qquad (2)$$

Thus, by substituting the relevant dispersion data, two simultaneous equations relating to the three observed mode groups are obtained from which a value for the effective index difference $\Delta_{eff} = (6.0 \pm 0.2) \times 10^{-3}$ has been deduced. When these values for V_{eff} and $_{eff}$ are inserted into (1), then the effective core radius $a_{eff} = 3.0 \pm 0.1$ μm and consequently the actual cut-off wavelength of the fibre can be calculated to be 1.04 μm.

CONCLUSION

The mode-locked cw dye laser/Synchroscan streak camera has been demonstrated to be a powerful and convenient high temporal/spatial resolution technique with which to study the principle features of mode propagation in short lengths of step-index optical fibre. From the resulting time-dispersion data, the effective characteristics of the fibre have been directly determined.

We are glad to acknowledge V. Henderek and R. Epworth from STL Harlow for the supply of the optical fibre and useful discussions. The work was supported by the Science and Engineering Research Council.

REFERENCES

(1) T.D. Giallorenzi, Proc. IEEE 66, 744, (1978).
(2) T. Miya, Y. Terunuma, T. Hosaka and T. Miyashita, Elect. Lett., 15, 106, (1979).
(3) C. Lin, L.G. Cohen, W.G. French and H.M. Presby, IEEE J. Quant. Electron., QE-16, 33, (1980).
(4) A. Barthelemy and J. Piazecki, Opt. Lett., 6, 269, (1981).
(5) S.T.L. Ltd., Harlow, Essex, England.
(6) J.P. Willson, W. Sibbett and W.E. Sleat, Opt. Comm., (to be published).
(7) W. Sibbett, W.E. Sleat, J.R. Taylor and J.P. Willson, paper to be presented at 15th Int. Congress on High Speed Photography and Photonics, August 1982, San Diego, USA.
(8) D. Gloge, Appl. Phys., 10, 2252, (1971).

Part IV

Relaxation Phenomena in Molecular Physics

Picosecond Lifetimes and Efficient Decay Channels of Vibrational Models of Polyatomic Molecules in Liquids

C. Kolmeder, W. Zinth, and W. Kaiser

Physik Department der Technischen Universität München,
D-8000 München, Fed. Rep. of Germany

Convincing evidence is presented that anharmonic coupling between fundamental vibrational modes and overtones or combination modes is of major importance for the lifetime of vibrational states. The selection rules known to hold for Fermi resonance determine the decay channels. In a number of examples the decay pathways of vibrational energy were observed experimentally by measuring the population and depopulation of subsequent vibrations. Drastic variations of vibrational lifetimes were found for different vibrations of the same molecule.

Molecules are first excited by an ultrashort resonant infrared pulse and the instantaneous degree of excitation is monitored by observing the anti-Stokes Raman signal of a delayed probe pulse. Different vibrational modes are distinguished by their characteristic anti-Stokes frequency.

We have investigated numerous molecules and found widely varying values of the population life-times between 1 ps and 240 ps in polyatomic molecules at room temperature /1,2/. Special attention was paid to the CH-stretching modes in the frequency range of $3000\pm100\ \mathrm{cm}^{-1}$.

Vibrational energy is transferred from the CH-stretching modes ($\sim 3000\ \mathrm{cm}^{-1}$) via overtones and combination modes to lower energy states. Intramolecular anharmonic coupling, the Fermi resonance, manifests itself in the infrared and Raman spectra. Overtones and higher order combination modes borrow intensity from CH-stretching modes. We define as a measure of Fermi-resonance mixing the intensity ratio, R, between the final and initial state taken from the infrared or Raman spectrum. In a recent publication a formula was derived which allows to estimate the life time T_1 of vibrational states:

$$T_1 = N(1-R)^2 R^{-1} \exp(\omega/\Omega)^{2/3}\, T_2(f) \qquad (1)$$

N corresponds to the number of states initially excited, R is a measure of the Fermi resonance, and $T_2(f)$ stands for the dephasing time of the final state. $T_2(f)$ may be estimated from the Raman line-width $\Delta\tilde{\nu}$ as $T_2(f) = (2\pi c\Delta\tilde{\nu})^{-1}$ ($T_2(f)$ is equal to $T_2/2$ measured in coherent Raman experiments) . The frequency ω represents the energy difference between the initial and final state. Ω has a value close to $100\ \mathrm{cm}^{-1}$.

As an example for the importance of Fermi resonance we present data of the two molecules 1,1-dichloroethene and trans 1,2-

dichloroethene which are made up of the same atoms; only two atoms have exchanged their positions. The infrared and Raman spectra between 2950 cm^{-1} and 3250 cm^{-1} of both molecules are depicted in Fig.1. There are drastic differences in the anharmonic coupling of the CH-stretching modes. In Fig.1a we see strong Fermi resonance of $CH_2{=}CCl_2$ between the fundamental ν_1 and $\nu_2+\nu_3$, both of A_1 symmetry and between the fundamental ν_7 and $\nu_2+\nu_6+\nu_{11}$, both of B_1 symmetry /3/. This observation suggests that we have to consider at least two decay channels for the CH_2-stretching modes. For the decay $\nu_1 \rightarrow \nu_2+\nu_3$ we estimate the intensity ratio $R = 0.2\pm0.05$ and calculate $T_2(f) = 0.3$ ps from the Raman linewidth of $\Delta\tilde{\nu} = 17$ cm^{-1}. With $\tilde{N}=1$ and $\omega = 45$ cm^{-1} we calculate from (1) a value of $T_1 = 4\pm2$ ps. For the second decay channel $\nu_7 \rightarrow \nu_2+\nu_6+\nu_{11}$ we have to take a short dephasing time $T_2(f)$ of the combination mode. We estimate $T_2(\nu_2+\nu_6+\nu_{11}) \simeq 0.2$ ps from the observed line width. With $R = 0.6\pm0.1$, $\tilde{N}=1$ and $\omega = 45$ cm^{-1} we calculate $T_1 = 1.5$ ps.

According to Figs.1a and 1b the symmetric (ν_1) and asymmetric (ν_7) CH_2-stretching vibrations are separated by 100 cm^{-1}. We estimate an energy-transfer time between the CH_2-stretching modes of $T_1(\nu_1 \rightarrow \nu_7) \simeq 3.3$ ps using the formula $T_1(\omega_1 \rightarrow \omega_7) = T_2(\omega_1)\exp(\omega/\Omega)^{2/3}$, where $T_2(\omega_1)$ was taken from the Raman spectrum.

The estimates given here indicate that vibrational energy flows faster out of the ν_7 mode than it is supplied by the transfer $\nu_1 \rightarrow \nu_7$. For the excited and interrogated mode ν_1 we simply add the two decay rates for the two decay channels $\nu_1 \rightarrow \nu_7$ and $\nu_1 \rightarrow \nu_2+\nu_3$ and arrive at a lifetime $T_1(\nu_1) \simeq 2$ ps.

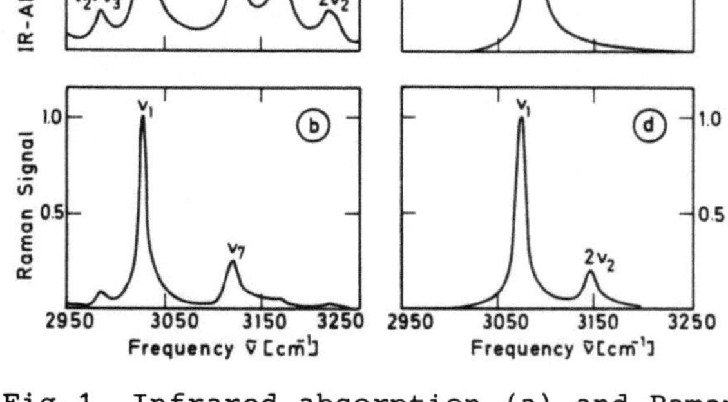

Fig.1 Infrared absorption (a) and Raman (b) spectra of CH_2CCl_2 between 2950 and 3250 cm^{-1}. Two combination tones are in strong Fermi resonance with the two CH-stretching modes ν_1 and ν_7. Infrared absorption (c) and Raman (d) spectra of trans CHClCHCl. There is less Fermi-resonance mixing than in CH_2CCl_2.

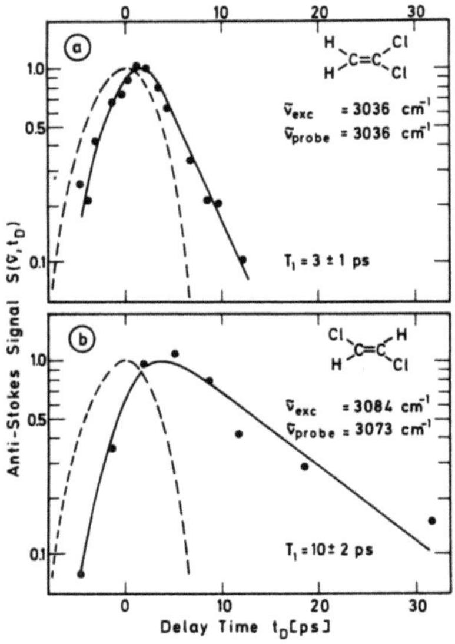

Fig.2 Anti-Stokes scattering signal versus delay time of the probing pulse. (a) CH_2CCl_2 in CCl_4 (c = 0.35 m.f.). The decay of the CH_2-stretching mode at 3036 cm^{-1} is shown. (b) trans CHClCHCl in CCl_4 (c = 0.35 m.f.). The CH-stretching mode at 3084 cm^{-1} is excited and the mode at 3073 cm^{-1} is monitored. The broken curves are the cross-correlation functions of the IR exciting and green probing pulses.

In Fig.2a we present experimental data of the direct determination of the T_1 value. The scattered Raman signal of the ν_1 mode rises to a slightly delayed maximum during the excitation process and decays with a relaxation time of T_1 discussed in the preceding paragraph. The broken curves in Fig.2 are cross-correlation curves of the excitation and probing pulse; they determine the zero point on the time axis and give a good indication of the time resolution of the experiment.

In Fig.1c we see the infrared active CH-stretching mode ν_9 and in Fig.1d the Raman active symmetric ν_1 mode of trans CHClCHCl. Here we find a considerably smaller Fermi resonance. The Raman spectrum of Fig.1d suggests some anharmonic coupling between ν_1 and $2\nu_2$, both of A_g symmetry /3/. With the values R = 0.15±0.02, T_2 = 0.3 ps, N=2 and ω = 80 cm^{-1} we calculate T_1 = 13 ps. It should be noted that there might be additional weak Fermi resonance between the ν_9 mode and higher combination modes (e.g. $\nu_2+\nu_5+\nu_{10}$) burried under the high frequency tail of the ν_9 fundamental. These additional decay channels may reduce somewhat the estimated T_1 value.

The time dependence of the CH-stretching modes of trans CHClCHCl is depicted in Fig.2b. The molecule is excited via the ν_9 mode at 3084 cm^{-1} and the population of the ν_1 mode at 3073 cm^{-1} is monitored by anti-Stokes Raman scattering. The rapid rise of the Raman signal, i.e. the fast population of the ν_1 mode, gives clear evidence of the quick energy exchange between the two CH fundamentals ν_1 and ν_9. The decay of the signal curve suggests a long lifetime of the two CH-stretching modes of T_1 = 10±2 ps. This number is in good agreement with the value estimated above. The small intramolecular coupling gives rise to the longer vibrational life time.

The vibrational states of acetylene are well documented in the literature /4/. Inspection of the energy-level system (see Fig.3) suggests the following interesting situations: (i) Energy in the high lying CH-stretching modes around 3200 cm^{-1} readily flows into several combination modes, all of which comprise the symmetric C≡C-stretching mode at ν_2 = 1968 cm^{-1}. (ii) The energy transfer from the ν_2 mode to neighboring combination modes is forbidden by symmetry selection rules. Thus we expect a long population life-time of the ν_2 mode.

Experimentally we investigated a solution of C_2H_2 in CCl_4. Acetylene molecules first are vibrationally excited via the infrared active CH-stretching mode ν_3 = 3287 cm^{-1} and the population and depopulation of the ν_2 mode at 1968 cm^{-1} is monitored. In Fig.4 we indeed see a rapid population of the ν_2 mode within ≲ 3 ps and a very slow depopulation with a time constant of 240 ps. The ν_2 mode in acetylene represents a bottle-neck state. It exhibits the longest relaxation time observed so far in a polyatomic molecule in the liquid state at room temperature.

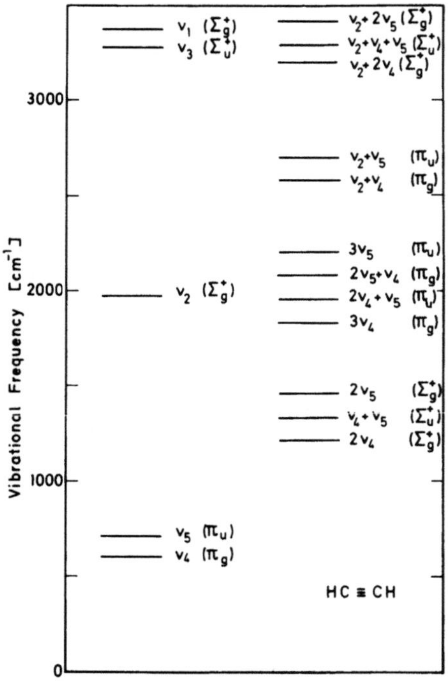

Fig.3 Energy-level diagram of Acetylene

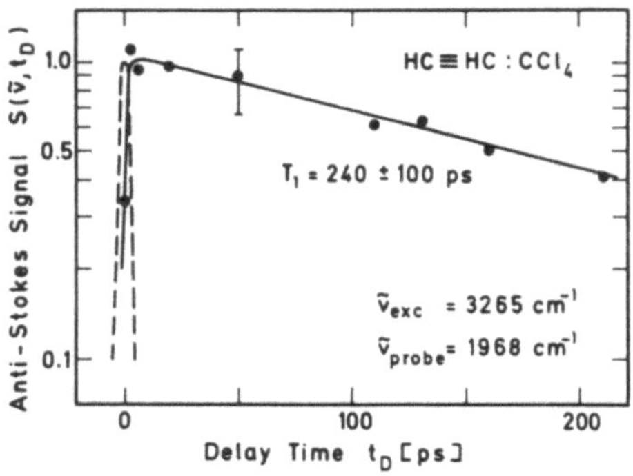

Fig. 4 Anti-Stokes scattering signal of C_2H_2 in CCl_4 versus delay time. Excitation frequency is 3265 cm^{-1}. The decay of the C≡C mode at 1968 cm^{-1} is monitored.

References

1 A. Fendt, S.F. Fischer, and W. Kaiser, Chem. Phys. 57, 55 (1981)
2 A. Fendt, S.F. Fischer, and W. Kaiser, Chem. Phys. Lett. 82, 350 (1981)
3 L.M. Sverdlov, M.A. Kovner, E.P. Krainov, Vibrational Spectra of Polyatomic Molecules, Wiley, New York, N.Y. 1974, and references therein
4 G. Herzberg, Infrared and Raman spectra of Polyatomic Molecules, von Nostrand, Princeton, N.J. 1945

Vibrational Population Decay and Dephasing of Small and Large Polyatomic Molecules in Liquids

H. Graener, D. Reiser, H.R. Telle, and A. Laubereau

Physikalisches Institut, Universität Bayreuth,
D-8580 Bayreuth, Fed. Rep. of Germany

Vibrational relaxation processes have received increasing interest in recent years. With picosecond light pulses and novel experimental techniques we are in the position to study vibronic relaxation in the excited electronic state of dye solutions and vibrational dynamics in the electronic ground state of simple liquids.

Ultrafast Polarization Spectroscopy of Dye Molecules

It has been shown in recent publications that valuable information on molecular rotation can be obtained investigating ultrashort pulse propagation in dye solutions under carefully selected polarization conditions. In these investigations enhanced pulse transmission was noted during and shortly after the occurrance of the excitation pulse. The time behaviour was interpreted as a coherent artifact via parametric 4-wave mixing of the excitation and probe pulses. We have carried out a detailed theoretical study of the processes which occur in ultrafast dichroism measurements. Our analysis reveals the importance of vibrational relaxation and of excited state absorption for a quantitative understanding of the signal transients at short delay time. Of particular interest is the limit of weak excitation intensity where the accurate knowledge of the pump pulse intensity is not required for the determination of the molecular relaxation times.

Fig. 1 presents experimental data on the system phenoxazone 9 in solid polystyrene obtained with a Nd:glass laser system.[1] The probe transmission behind a blocking polarizer is plotted versus delay time between pump and probe pulse. Equal probe and excitation frequencies, $\omega_{pr}=\omega_L=18990\ cm^{-1}$, are applied in Fig. 1a. We note a rapid signal rise to a maximum which occurs close to $t_D \simeq 0$. The curve then promptly decays and reaches a slow exponential asymptote. The rapid decrease for small t_D gives direct evidence that the vibrational relaxation time τ_V is shorter than the time resolution of the system, $\tau_V<2$ ps. Information on τ_V is provided by the signal enhancement $H = 2.5 \pm 0.3$. This value is significantly larger than the pure coherence peak of 1.4 for Gaussian pulses. Taking into account the excited state absorption we determine $\tau_V=0.7 \pm 0.2$ ps. Similar results were observed for phenoxazone 9 in dioxane and CCl_4. For Rhodamine 6G in ethanol our polarization technique yields $\tau_V=0.5 \pm 0.2$ ps in good agreement with earlier results by a nonlinear absorption technique.[2]

The data for the situation $\omega_{pr}<\omega_L$ are depicted in Fig. 1b. It is interesting to see the steep rise of the signal curve in

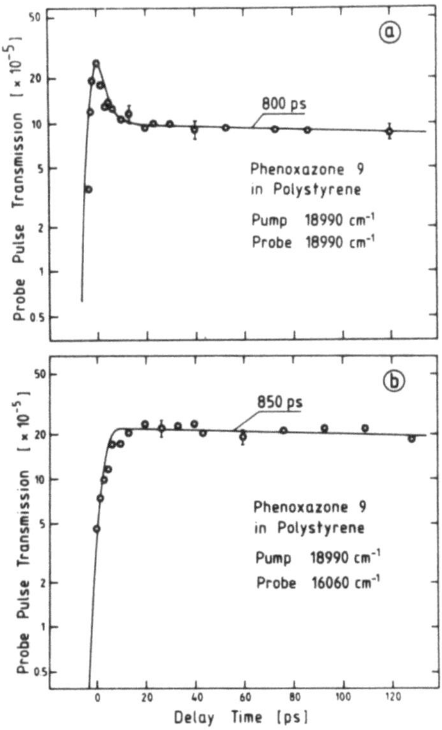

Fig. 1:

Probe pulse transmission via induced dichroism versus delay time between excitation and probing pulses; a) equal and b) different probe and excitation frequencies; theoretical curves.

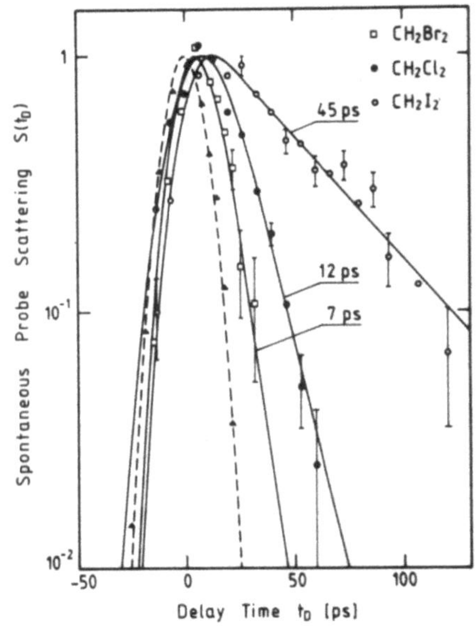

Fig. 2:

Spontaneous anti-Stokes probe scattering $S(t_D)$ versus delay time t_D for the symmetric CH_2-stretching mode of CH_2Br_2, CH_2Cl_2 and CH_2I_2; calculated curves. Broken line and full triangles represent the instrumental response of the CH_2Br_2 measurement.

the figure to a maximum value and a subsequent slow exponential decay. The rising part of the curve indicates a short vibrational relaxation time in accordance with the result of Fig. 2.

We emphasize the high sensitivity of the technique which allows to detect probe transmissions $\gtrsim 10^{-7}$ in single shot experiments. At this low excitation level disturbing nonlinear effects are clearly negligible.

Vibrational Population Lifetimes of Methylene Halides

We have devised a picosecond Raman spectrometer for the measurement of population lifetimes in liquids, which is based on a mode-locked YAG-laser with single picosecond pulses of 23 ps and a repetition rate of several Hz. The infrared excitation pulse is generated by a multiple step parametric generator - amplifier setup yielding tunable pulses of 10 ps and 200 µJ around 3 µm. The excited molecules are interrogated by a green probing pulse via spontaneous anti-Stokes Raman scattering.[3] We routinely control the time resolution of the setup measuring the cross correlation of excitation and probing pulses. The system provides values of the population lifetime $T_1 \gtrsim 4$ ps with enlarged dynamic range and increased measuring sensitivity as compared to previous experimental approaches. Examples for the observed signal curves are shown in Fig. 2. The scattering signal of the probe pulse from the symmetric CH_2-stretching mode of three methylene halides is plotted on a logarithmic scale versus delay time between excitation and probing pulses. The experimental points represent the accumulated information of 200 to 600 laser shots and extend over a factor of approximately 100. The open circles in the figure denote the simultaneously measured cross correlation function of excitation and probing pulses. The curves in the figure are calculated assuming Gaussian shape of the light pulses and using the population lifetime T_1 as fitting parameter. It is interesting to see that the delay of the signal maxima and the asymptotic decay are fully accounted for by the theoretical curves. Results on T_1 are listed in Table I, which indicates values of 6.5 to 50 ps for the investigated symmetric CH_2-stretching modes.[4] Data on $(CH_2ClCH_2)_2O$ are included to demonstrate the time resolution of the measuring system.

The population lifetimes are explained by vibrational energy transfer via anharmonic coupling to neighbouring combination bands and overtones. Numerical estimates have been performed considering Fermi resonance with the adjacent overtone of the CH_2-bending mode[5] yielding only fair agreement. It is concluded that higher order combination bands in the immediate neighbourhood of the symmetric stretching mode give a notable contribution to the observed population decay.

Subpicosecond Vibrational Dephasing in Liquids

In the past numerous investigations of the vibrational dephasing time T_2 in liquids have been performed. In these studies exponential decay of the coherent vibrational excitation has been observed on the time scale of several ps. No direct information is available on the dephasing in the subpicosecond time domain. For recent theoretical work on stimulated Raman scattering purely exponential time behaviour was assumed, which is not correct during very short times for general theoretical arguments. We

Table I: Measured population lifetimes

	concentration vol %	$\tilde{\nu}$ [cm^{-1}]	T_1 [ps]
CH_2Br_2	100	2987	7 ± 1
	30	2987	7 ± 1
CH_2Cl_2	100	2989	12 ± 2
	10	2989	12 ± 2
CH_2I_2	50	2967	45 ± 5
CH_2ClBr	30	2987	13 ± 2
CH_2ClI	30	2979	14 ± 2
$(CH_2Cl)_2$	10	2950	6.5 ± 1
$(CH_2ClCH_2)_2O$	5	∿ 2964	2 ± 2

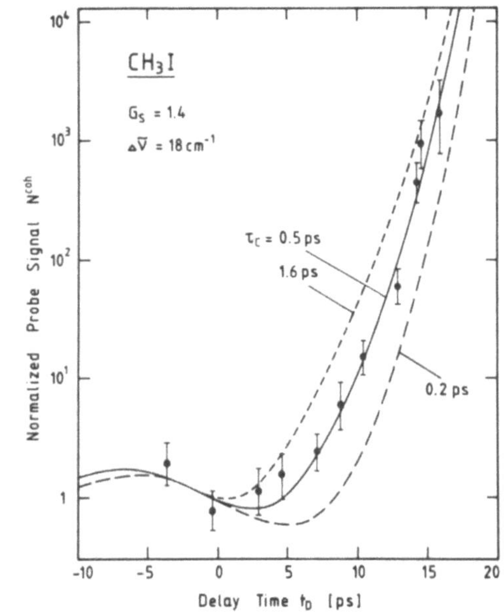

Fig. 3:
Ratio $N^{coh}(t_D)$ of coherent probe scattering signal to cross correlation signal versus delay time t_D for the symmetric CH_3-stretching mode of CH_3I; theoretical curves.

have generalized the theory of transient stimulated Raman scattering to account for non-exponential dephasing using band shape theory.[6] Our theoretical approach avoids the two-level approximation used previously and introduces the vibrational dynamics via the vibrational autocorrelation function. Using stochastic theory the following expression for this function may be derived:[7]

$$\phi(t) = \exp\{-\frac{|t|}{T_2} - \frac{\tau_c}{T_2}\left[\exp(-\frac{|t|}{\tau_c})-1\right]\} \tag{1}$$

Here the correlation time τ_c denotes the time scale of the interaction. The time constant T_2 represents exponential asymptotic dephasing. Pure exponential dephasing is contained in the present treatment for $\tau_c=0$. A flat top at t=0 results for $\tau_c>0$. Eq. 1 represents homogeneous dephasing for short values of $\tau_c \leq 1$ ps (T_2: a few ps). For large values of τ_c a Gaussian-like time dependence is predicted representing the inhomogeneous case.

We have directly observed the delayed onset of vibrational decay as predicted from Eq. 1 for finite τ_c using coherent Raman probing techniques. A Raman amplifier setup with small amplification factor ∿ 2, pumped by the second harmonic of a glass laser system and a suitable Stokes pulse, was used for the near-resonant excitation of CH-stretching modes. The coherent vibrational excitation was monitored by non-collinear coherent Stokes probing. As probe pulse we use part of the Stokes component of the pump process. Results for the symmetric CH_3-stretching mode of CH_3I are shown in Fig. 3.[6] For a reliable control of the time resolution of the measurement we simultaneously measure the cross correlation of the light pulses and plot the ratio of the coherent probing signal in the sample to the experimental cross correlation curve versus time delay between excitation and probing pulses. The signal points extend over three orders of ten and fully compare with computed theoretical curves. The result for the correlation time is τ_c=0.5 ps. Similar data have been obtained for other methane derivates. Our results on τ_c represent the first direct observation of non-exponential dephasing on the sub-picosecond time scale. Assuming dephasing via the repulsive part of the intermolecular potential to be predominant the constant τ_c denotes the time scale of the translational motion of the liquid molecules; i.e. the elastic collision time. Simple models of the liquid state suggest this time to be 10^{-12} to 10^{-13} s in satisfactory agreement with our experimental data.

References:

1. D. Reiser and A. Laubereau, Appl. Phys. B 27, 115 (1982); Optics Commun. to be published.
2. A. Penzkofer, W. Falkenstein and W. Kaiser, Chem. Phys. Lett. 44, 82 (1976).
3. K. Spanner, A. Laubereau and W. Kaiser, Chem. Phys. Lett. 44, 88 (1976).
4. H. Graener and A. Laubereau, to be published.
5. A. Fendt, S.F. Fischer and W. Kaiser, Chem. Phys. 57, 55 (1981).
6. H.R. Telle and A. Laubereau, to be published.
7. R. Kubo, Fluctuations, Relaxation and Resonance in Magnetic Systems, editor D. Ter Haar (Plenum, New York, 1962).

Mechanisms for Ultrafast Vibrational Energy Relaxation of Polyatomic Molecules

S.F. Fischer

Physik-Department der Technischen Universität München, D-8046 Garching, Fed. Rep. of Germany

In recent years ultra fast vibrational relaxation processes have been observed for molecules in a liquid[1]. The rate of energy depopulation of CH-stretching modes can vary by two orders of magnitude for different molecules like benzene(T_1=1ps) on the one side and acetylene (T_1=200ps)[1]on the other side. In order to understand the determining factors that influence the deactivation process one may argue on the basis of harmonic model calculations that the number of quanta exchanged in the transfer process should be small and the energy mismatch between the initial and the final state should also be small. These criteria, however, have to be specified further. For instance the medium or collision induced energy transfer between two CH-stretching modes, which may be localized predominantly at different sites R_n and R_m, will depend upon the local amplitude squared the mode localized at R_m has at the site R_n and vice versa. Only if both modes have essentially the same local amplitudes and only differ in the phases the energy relaxation is fast as the pure dephasing process of these modes. If the initial and the final state differ by many vibrational quanta the rate may still be fast provided these two states mix due to anharmonic interaction. If both states are actually in Fermi resonance the medium or collision induced transfer is again of the same magnitude as the corresponding dephasing process. Therefore, it is very essential to know the anharmonic coupling constants.

For the electronically excited state selective vibrational relaxation has been observed for several molecules in the gas phase[2]. It is found that the collision induced rates are usually faster than the corresponding rates in the ground state. This may partially depend upon the intermolecular forces but also upon the different intramolecular anharmonic interactions. We like to demonstrate here that vibronic interactions between different excited states can cause anharmonic interactions which may be much larger than those commonly known from anharmonic force fields in the ground state.

We consider two electronic states $|\psi_1\rangle$ and $|\psi_2\rangle$ and several harmonic oscillators. We denote the symmetric mode frequency by ω_s, those of the coupling modes by ω_p (promoting modes). The equilibrium position of the symmetric modes may differ for the two states by $\Delta R_s = (\hbar/M_s\omega_s)^{1/2}\, g_s$. Using creation and destruction operators b_s^+ and b_s or b_p^+ and b_p for the symmetric and the promoting mode respectively we obtain for the model Hamiltonian.

$$H = \left[\tfrac{1}{2}(E_2+E_1) + \sum_s \hbar\omega_s b_s^+ b_s + \sum_p \omega_p b_p^+ b_p\right]\left[|\psi_1\rangle\langle\psi_1| + |\psi_2\rangle\langle\psi_2|\right]$$

$$+ \left[\tfrac{1}{2}(E_2-E_1) + \sum_s g_s h\omega_s (b_s^+ + b_s)\right]\left[|\psi_2\rangle\langle\psi_2| - |\psi_1\rangle\langle\psi_1|\right]$$

$$+\left[\sum_p \lambda_p h\omega_p (b_p^+ + b_p)\right]\left[|\psi_1\rangle\langle\psi_2| + |\psi_2\rangle\langle\psi_1|\right] \tag{1}$$

This Hamiltonian contains a vibronic symmetry[3]and can be split into two parts by a proper canonical transformation

$$H = \hbar_+ \; |\psi_1\rangle\langle\psi_1| + \hbar_- \; |\psi_2\rangle\langle\psi_2| \tag{2}$$

The two vibrational Hamiltonians $\hbar_+$ and $\hbar_-$ can be further canonically transformed to make them very suitable for the final analysis[3].

$$\hbar_\pm = \hbar_\pm^o + \hbar_\pm^{anh} + \hbar_\pm^{int} \tag{3}$$

The zero order term reads to lowest order in $(\sum_p \lambda_p^2 h\omega_p)(E_2-E_1)^{-1}$:

$$\hbar_\pm^o = \tfrac{1}{2}(E_1+E_2) - \sum_s g_s^2 h\omega_s + \sum_s \hbar\omega_s b_s^+ b_s + \sum_p \hbar(\omega_p + \delta\omega_p) b_p^+ b_p \tag{4}$$

and the anharmonic interaction has the form

$$\hbar_\pm^{anh} = \mp \sum_{s,p,p'} g_s \hbar^2 \omega_s (\delta\omega_p \delta\omega_p)^{1/2} (E_2-E_1)^{-1} (b_s^+ + b_s)(b_p^+ + b_p)(b_{p'}^+ + b_{p'}) \tag{5}$$

with

$$\delta\omega_p = -2\lambda_p^2 \omega_p^2 (E_2-E_1)^{-1} \tag{6}$$

The interaction term is an odd function of the promoting mode coordinates. It is of higher order in $\sum_p \lambda_p^2 \hbar\omega_p (E_2-E_1)^{-1}$ and will not be discussed here.

As applications for vibronically induced anharmonic interactions we like to discuss benzene. Wunsch, Metz, Neusser, and Schlag [4] studied two photon spectra of benzene in the gas phase. They observed so called shadow bands. In benzene the transition $14_0^1 1_0^1$ is accompanied by a transition $14_0^1\ 16_0^1\ 17_0^1$ which is shifted by 29.5 cm^{-1} and has 13 % of the intensity from the $14_0^1\ 1_0^1$ transition. The relative intensity increases for the overtones of the v_1 mode. An even stronger effect is observed in deuterobenzene. Here the transition $18^1_0 1^1_0$ is followed by a transition $18^1_0 10^2_0$ which takes 50 % of the intensity. Both transitions show equal intensity for the combinations $19^1_0 1^2_0$ and $19^1_0 1^1_0$ and $19^1_0 1^1_0 10^2_0$. Fischer, Scharf, and Parmenter[5] discussed the effect of anharmonic couplings in benzene. They needed coupling constants of 11 cm^{-1} in order to interpret the intensity distribution between ν_1 and the overtone of ν_6. We can now estimate from our expression for $\hbar^{anh}_{\pm}$ (5) these constants.

For the coupling of ν_1 with the overtone of ν_6 we obtain the values $\delta\omega_p$ = 86 cm^{-1}, ω_s = 993 cm^{-1}, E_2-E_1 = 17600 cm^{-1} (E_2 is the $\pi\pi^*$ E_{1u}-state). The bond length change can be estimated from the bond order change and its emperical relation to bond length change. We find g = 0.4 and the anharmonic coupling constant is 19.5 cm^{-1}, which is even larger than the value assumed by Fischer et al.[5]. For the anharmonic coupling between ν_1 and ν_{16} + ν_{17} we need to consider a $\sigma\pi^*$ excitation. ν_{16} and ν_{17} are of e_{2u} symmetry, so E_2 must have E_{1g} symmetry. We suggest that the transition takes place from the highest occupied σ orbital[6] to the lowest unoccupied π^* orbital. There exist no good predictions on the energy location of this state. One may, however, locate it between 20000 and 30000 cm^{-1} above the $\pi\pi^*B_{2u}$ excited state of benzene. In this case g is larger since a σ bond is removed. Taking for g a value between 1 and 1.5 we obtain with ω_s = 993 cm^{-1}, $\delta\omega_{16}$ = 62 cm^{-1} and $\delta\omega_{17}$ = 255 cm^{-1} as approximate value $k_{1,16,17}$ = 12.4 cm^{-1}. This can well account for the observed intensity borrowing. In the case of the interaction of ν_1 with $2\nu_{10}$ in deuterobenzene again a σ-π^* excitation is required for the vibronically coupled state. We take $\delta\omega_p$=207cm^{-1}, ω_s = 925 cm^{-1}, E_2-E_1=25000cm^{-1}.

For g we obtain a value closer to two (g = 2) since the final π^*state has three nodal planes. This gives for the coupling constant $k_{1,10,10}$ = 15.3 cm^{-1}. This value gives just the right intensity ratio. The relative coupling increases by a factor of 1.4 if the overtone of the ν_1 mode is involved instead of the fundamental. Finally we like to mention that the strongest coupling is predicted between the overtone or higher overtones of the ν_4 mode and the ν_1 or ν_2 mode. This interaction does not show up in the frequency regime of the ν_1 mode, since the energy mismatch is too high. Higher order coupling terms, neglected in (5) couple the eighth overtone of ν_4 with ν_2. This effect may explain the sharp drop in the quantum yield of fluorescences in this energy regime, which has been described as the "channel three" problem of benzene[7].

Acknowledgement

The author likes to thank Prof. Kopelman and Dr. Hornburger for stimulating discussions.

References

1. A. Laubereau, S. F. Fischer, K. Spanner, and W. Kaiser, Chem. Phys. 3 (1978) 335; A. Fendt, S. F. Fischer, and W. Kaiser, ibid. 57 (1981) 55; see also contribution by W. Kaiser et al. in this Volume

2. C. S. Parmenter and K. Y. Tang, Chem. Phys. 27 (1978) 127
D. B. McDonald and S. A. Rice, J. Chem. Phys. 74 (1981) 4893, ibid. 4907
3. S. F. Fischer, Chem. Phys. Letters in press
4. L. Wunsch, F. Metz, H. J. Neusser, and E. W. Schlag, J. Chem. Phys. 66 (1977) 386
5. G. Fischer, B. Scharf, and C. S. Parmenter, Mol. Phys. 29 (1975) 1063
6. R. M. Stevens, E. Switkes, E. A. Laws, and W. N. Lipscomb, J. Amer. Chem. Soc. 93 (1977) 2603
7. H. Hornburger and J. Brand, Chem. Phys. Letters 88 (1982) 153

Studies of the Generation and Energy Relaxation in Chemical Intermediates-Divalent Carbon Molecules and Singlet Oxygen

E.V. Sitzmann, C. Dupuy, Y. Wang, and K.B. Eisenthal

Department of Chemistry, Columbia University, New York, NY 10027, USA

To understand the nature of complex chemical phenomena, it is necessary to describe more than the initial reactants and the identities and relative amounts of final chemical products. It is the sequence or sequences of competing elementary steps by which the initial molecules evolve through a variety of energy states, molecular structures, and short-lived chemical intermediates, which are necessary to explain the distribution of final product molecules. In recent years, there has been enormous activity aimed at identifying the transient molecular species and their structures using inert matrices at low temperatures ($\approx 4^{o}K$) as stabilizing environments. With picosecond laser methods, these unstable intermediates do not have to be "frozen" and stabilized but can be detected in their rapid evolution at room temperature in liquid systems where most of what we call chemistry occurs. Two chemical intermediates which we are studying are the divalent carbon fragments, R_1-$\ddot{C}$-R_2,(called carbenes)[1-11] implicated in an enormous variety of chemical reactions and singlet oxygen ($^1O_2$) [12-25] a prime species in materials degradation in photochemistry and photobiology. Little is known about the dynamics and competing pathways of their generation from precursor molecules such as diazo compounds and endoperoxides. There is also scant information about energy relaxation in carbene fragments, an issue which is of crucial importance since the chemistry is known to be state selective. With picosecond laser methods we can probe the energy states of the newborn chemical intermediates and thereby address the key issues of formation, energy relaxation and chemical reactivity of these transient species.

1. Generation of Singlet Oxygen, $^1O_2$, from Laser Induced Photoreactions

Photoexcitation of molecules which contain oxygen in the form of a peroxide bond generally results in the rupture of the O-O bond. For alkyl peroxides, R-O-O-R' where R and R' are groups of the type methyl (CH_3), ethyl (C_2H_5), etc., this is the mode of photofragmentation following UV excitation. However, for aromatic endoperoxides a new pathway for bond breaking becomes important, namely one which yields the aromatic moiety and oxygen in an excited singlet state, $^1O_2$. An example is the 1,4-endoperoxide of 1,4-dimethyl-9,10-diphenyl anthracene (DMDPA-O_2). We have been investigating[25] the nature of these processes which comprise an important class of excited state adiabatic photoreactions (i.e., producing excited state products) and which provide a convenient photochemical source for producing $^1O_2$ for further investigations as well.

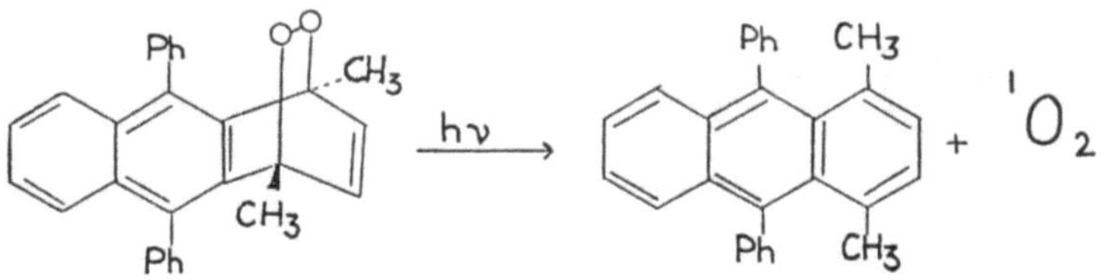

We have recently discovered[25] that one of the paths in the photodissociation of the anthracene endoperoxide, DMDPA-O_2, produces not only $^1O_2$, but also the electronically excited aromatic $^1DMDPA^*$. This represents one of the few cases of a photodissociation producing two electronically excited fragments. The experiment was carried out at a concentration of 1.5×10^{-4} M DMDPA-O_2 in acetonitrile at room temperature using the fourth harmonic of a Nd glass laser as the picosecond pulse excitation. The risetime of the fluorescence of the excited aromatic fragment $^1DMDPA^*$ was monitored with a streak camera and found to be less than 5 ps.

An important aspect of this newly discovered photodissociative route is the particular state in which oxygen is formed. By considering the energetics[20, 22] of the reaction, we have been able to determine that $^1O_2$ is produced in its $^1\Delta_g$ state for this pathway, the $^1\Sigma_g$ state being too high in energy to be reached (Fig.1). This finding represents the first time that the state of $^1O_2$ resulting from a photodissociation has been established.

To map out the photodissociation mechanisms leading to oxygen, we must consider other possible routes. The various spin-allowed processes are:

$$(1)\quad DMDPA\text{-}O_2 \xrightarrow{h\nu} {}^1DMDPA^* + {}^1O_2 \qquad (2)\quad DMDPA\text{-}O_2 \xrightarrow{h\nu} {}^3DMDPA^* + {}^3O_2$$

$$\text{and } (3)\quad DMDPA\text{-}O_2 \xrightarrow{h\nu} {}^1DMDPA + {}^1O_2$$

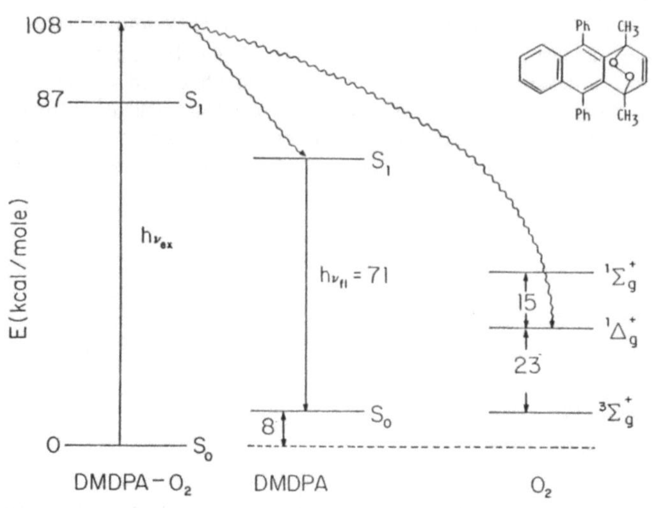

Fig.1 Energetics of endoperoxide photodissociation, λ_{ex}=266nm

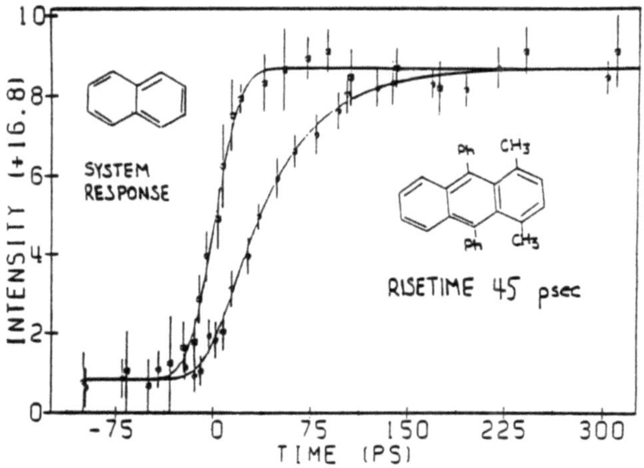

Fig.2 Best fit of risetime of ground state 1,4-dimethyl-9,10-diphenyl anthracene (^{1}DMDPA) following photoexcitation of endoperoxide at λ_{ex}=266nm

Using triplet-triplet absorption to probe for the appearance of ^{3}DMDPA* via route (2) we conclude that this is not an important decay path. The third channel, which produces the ground state aromatic fragment, was monitored by a laser induced fluorescence technique using a 355 nm Nd:YAG pulse which can excite ground state DMDPA but not the precursor DMDPA-O_2. From this study, we have determined that the dominant path for producing molecular oxygen is route (3) finding no evidence for route (2) and also that route (1) is roughly a factor of 10 less important than route (3). To obtain the kinetics of DMDPA formation the 355 nm pulse was time delayed using an optical delay line with respect to the 266 nm photodissociating pulse. Fig.2. We note that the risetime of DMDPA is 45±15 ps. If the routes (1) and (3) proceed from a common state of the photoexcited precursor ^{1}DMDPA-O_2^*, then they would have the same rise times. They do not, thus indicating that these dissociative channels do not arise from a commonly prepared state. The observed ground state DMDPA obtained via channel (3) does not proceed from the ^{1}DMDPA* since we find the decay time of ^{1}DMDPA* TO BE 0.5 ns. A possible interpretation of these results involves a biradical chemical intermediate living for 45 ps. A possible sequence is

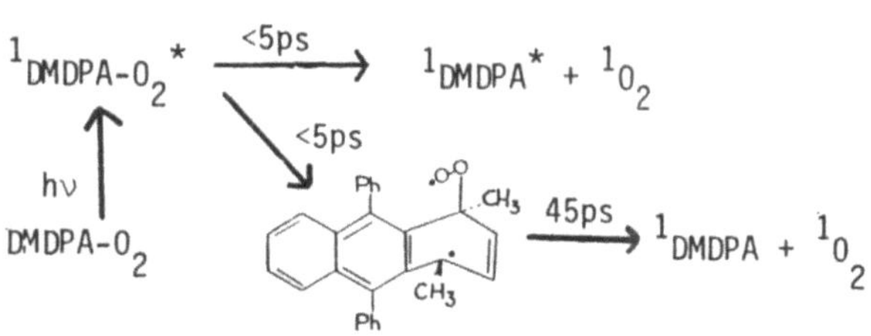

The chemical intermediate which we are proposing as a possibility and which we seek to identify results from the rupture of one of the C-O bonds which is

followed in 45 ps by breaking of the second C-O bond yielding $^1O_2$ and the aromatic fragment DMDPA. Further work is needed to delineate these various processes.

2. Excited State Photochemistry of Diphenylcarbene

The lowest singlet and ground triplet states of carbenes have different and distinctive chemistries[1-3]. For the case of diphenylcarbene (DPC), it is known that reactions with alcohols occur preferentially in its lowest singlet

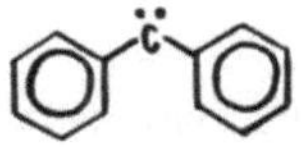

state whereas reactions with olefins such as isoprene, 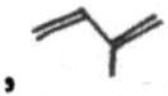, occur chiefly in its ground triplet state. Although the chemistry of these low-lying electronic states have been extensively studied, the nature of carbene chemistry in higher states remains unknown, in large part due to the reactivity of the thermally accessible singlet and ground triplet states. In order to examine the reactivity of excited triplet diphenylcarbene, $^3DPC^*$, with alcohols and isoprene, we have used picosecond laser methods and have found substantial differences between $^3DPC^*$ and ^3DPC.

The diphenylcarbene was generated by laser photolysis of the precursor molecule diphenyldiazoamethane (10^{-3} M) in a degassed acetonitrile solution. Excitation of the precursor molecule with a single pulse from a frequency quadrupled Nd:phosphate glass laser (264nm) produces two fragments, diphenylcarbene and N_2. The excited state triplet carbene was obtained by using a second picosecond pulse to photoexcite the ground triplet carbene to its excited triplet state, ($^3DPC \xrightarrow{h\nu} {^3DPC^*}$), or by using the small $^3DPC^*$ produced adiabatically from the photoexcited precursor molecule. The results obtained were found to be independent of the method used for generating the excited triplet carbene. The reactions of $^3DPC^*$ were monitored by measurement of the $^3DPC^*$ fluorescence lifetime, Fig.3, using a streak camera-optical multichannel analyzer system.

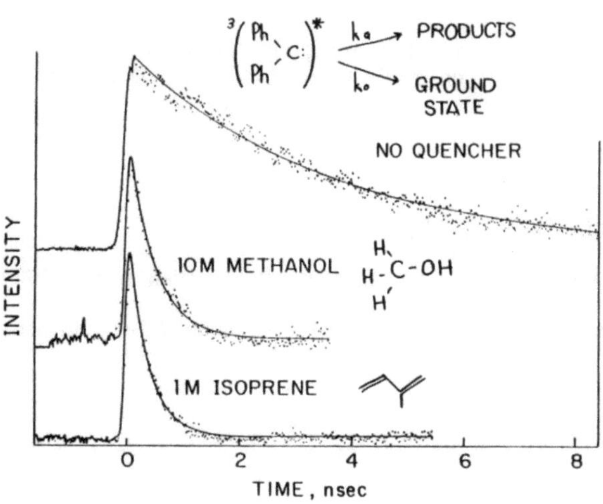

Fig.3 Quenching of $^3DPC^*$ fluorescence by reactions with methanol and isoprene

The lifetime of $^3DPC^*$ in acetonitrile with no quencher present was found to be 3.8 ns. Its decrease as a function of the added quencher concentration yielded the bimolecular reaction rate constant k.

$$^3DPC^* + \text{QUENCHER} \xrightarrow{k} \text{PRODUCTS.}$$

For isoprene, we find an increase in reactivity with $^3DPC^*$ relative to the ground triplet carbene ^3DPC [3,6] by roughly a factor of 10^4; with k for $^3DPC^*$ equal to $2 \times 10^9\ M^{-1}S^{-1}$. Whether this jump in reactivity is due to greater energy of $^3DPC^*$ (~58 kcal/mole) or the nature of the $^3DPC^*$ electronic state is not yet known.

With alcohol, we indeed do observe a rapid reaction with $^3DPC^*$ unlike the relatively inert ground triplet carbene. The bimolecular reaction rate constant k of $^3DPC^*$ with methanol (CH_3OH) is $3 \times 10^8\ M^{-1}S^{-1}$. A comparison of the rate constants we have measured for methanol, isopropanol and t-butanol shows a close parallel with the acidity of the alcohols. Combining this result with our observation of a kinetic isotope effect for the deuterated alcohols (ROD), we conclude that it is the O-H bond and not the C-H bond which is being attacked. A further point of interest is that the relative reactivities of $^3DPC^*$ with the alcohols is strikingly similar to that of ^1DPC reacting with the alcohols [5,9,11] pointing to an analogous reaction mechanism.

3. Summary

Using picosecond laser methods, we have investigated the dynamics of production and reactivities of significant chemical intermediates. The relative importance of various dissociative routes, their kinetics, and the possible intermediates in the generation of $^1O_2$ from a photoexcited endoperoxide are discussed. We have studied some aspects of the photophysics and photochemistry of diphenylcarbene as well, and have observed excited triplet state reactions with alcohols and isoprene and speculate on possible reaction pathways.

Acknowledgements

We thank the Air Force Office of Scientific Research, the National Science Foundation and the Joint Services Electronics Program (DAAG-79-C-0079) for their support.

References

1. Trozzolo, A.M., Accts. Chem. Res. 1, 329(1968).

2. Kirmse. W., "Carbene Chemistry", 2nd Ed.; Academic Press: New York, 1971.

3. Carbenes, eds., R.M. Moss and M. Jones, Jr., Vol. II (John Wiley and Sons, New York, 1975).

4. Roth, H.D., Accts. Chem. Res. 10, 85 (1977).

5. Closs, G.L., Rabinow, B.E., Ibid. 1976, 98, 8190.

6. P.P. Gaspar, B.L. Whitsel, M. Jones, Jr., and J.B. Lambert, JACS 102, 6108 (1980).

7. Zirpancic, J.J., P.B. Grasse, and G.B. Schuster, JACS 103, 2423 (1981).

8. Wong, P.C., D. Griller and J.C. Sciano, JACS 103, 5934 (1981).

9. Eisenthal, K.B., Turro, N.J., Aikawa, M., Butcher, J.A., Jr., Dupuy, C., Hefferon, G., Hetherington, W., Korenowski, G.M., McAuliffe, M.J., J.Am. Chem.Soc. 1980, 102, 6563.

10. Dupuy, C., Korenowski, G.M., McAuliffe, M., Hetherington, W., Eisenthal, K.B., Chem.Phys.Lett. 1981, 77, 272.

11. Hefferon, G.J., Ph.D. Thesis, Columbia University (1980).

12. Khan, A.U. and M. Kasha, J.Chem.Phys. 39, 2105 (1963).

13. Foote, C.S. and S. Wexler, JACS 86, 3879 (1964).

14. Corey, E.J. and W.C. Taylor, JACS 86, 3871 (1964).

15. Foote, C.S., Accts.Chem.Res. 1, 104 (1968).

16. Singlet Oxygen, eds. H. Wasserman and R.W. Murray (Academic Press, New York, 1979).

17. Kearns, D.R., Chem.Rev. 71, 395 (1971).

18. Srinvasan, R., K.H. Brown, J.A. Ors, L.S. White, JACS 101, 7424 (1979).

19. Wu, K.C., and Trozzolo, A.M., J.Phys.Chem. 83, 2823 (1979).

20. Turro, N.J., M.F. Chou and J. Rigaudy, JACS 101, 1300 (1979).

21. Drews, W., R. Schmidt and H.D. Brauer, Chem.Phys.Lett. 70, 84 (1980).

22. Olmsted, J.III, JACS 102, 66 (1980).

23. Harding, L.B. and W.A. Goddard III, JACS 102, 439 (1980).

24. Stevens, B., J.Phys.Chem. 85, 3357 (1981).

25. Hou, S.Y., C.G. Dupuy, M.J. McAuliffe, D.A. Hrovat and K.B. Eisenthal, JACS 103, 6982 (1981).

New Developments in Picosecond Time-Resolved Fluorescence Spectroscopy: Vibrational Relaxation Phenomena

B.P. Boczar and M.R. Topp

Department of Chemistry, University of Pennsylvania,
Philadelphia, PA 19104, USA

Problems associated with the determination of the pathways and rates of the relaxation of radiatively prepared vibronic states have occupied much attention in recent years. The subject has been approached from two main points of interest. First, the more significant advances have been made in the study of coherence loss phenomena associated not only with vibrational energy but also with "pure" dephasing contributions. This type of approach has some advantages in that it promises much higher spectral resolution and sensitivity since it is possible to work in the frequency domain as well as to make use of coherent amplification techniques including CARS spectroscopy and photon echo studies. On the other hand, the incoherent aspects of the relaxation phenomena are much harder to get at experimentally. Incoherent anti Stokes Raman [1] and two-step excitation of molecular fluorescence [2] have been used to probe vibrational relaxation in the ground state, while studies of the corresponding process in an electronically excited state have mostly used time-resolved fluorescence approaches. One exception has been the transient absorption spectroscopic work of HOCHSTRASSER and co-workers.[3]

We report here the successful utilisation of a tunable picosecond laser to study ultrafast-gated fluorescence spectra of aromatic hydrocarbons in solid solutions. Information available from these experiments allows the study not only of vibrational relaxation phenomena, but also of the time-evolution of molecular site distributions.

1. A Picosecond Fluorescence Spectrometer Using a Tunable Excitation Source

Although in principle the technology for the operation of tunable laser sources in the picosecond region has been developed for a decade, [4-6] they are not widely used in fluorescence spectroscopic applications. Furthermore, the scientific community has been slow to adopt optical gating techniques as opposed to electronic scanning techniques, even though the former have to date provided the only systematic studies of high-resolution, picosecond time-resolved fluorescence spectra. Having developed earlier a functional spectrometer for time resolved fluorescence spectroscopy, [7] we were particularly interested to incorporate a tunable excitation source into this instrument, of which the major considerations were reliability, narrow frequency bandwidth, short pulse-duration and relatively high power.

Experiments carried out in our laboratory have shown that the radiative rate constant is the dominant criterion in determining the strength of a fluorescence signal sampled by an ultrafast gate. Further, our experiments were extended to species such as benzophenone and its 2-hydroxy derivative, [8] of which the radiative lifetimes were measured to be in the range of several microseconds. Given this sensitivity, then even if only 10% of the irradiation intensity were available, still species with fluorescence radiative lifetimes in the range of several hundred nanoseconds would fall into range of our technique. Therefore, it was clear that aromatic hydrocarbons such as anthracene and perylene, having radiative lifetimes less than 10 nsec should be capable of study under high-resolution conditions.

One of the major limitations of the more usual sources for generating tunable radiation in the picosecond region is that the pulse-duration, output frequency and bandwidth are interdependent. Our approach has been to separate these functions as far as possible, thereby securing a higher reliability for the instrument as a whole.

The apparatus is depicted in Fig. 1. Single pulses from a Nd^{3+}-YAG laser oscillator are injected into a regenerative ring amplifier, [9], which produces well-defined pulses less than 10 psec in duration. Second harmonic pulses from this laser are used to pump a dye cell in a Hansch-type cavity, except that the cavity length is approximately the same as that of the regenerative amplifier. (Precise synchronous cavity adjustment is not necessary.) Normally, a diffraction grating would be expected to broaden the output pulses by imposing nearly 200 psec time-dispersion across the spatial profile of a 1 cm beam reflected from the grating. However, the gain in the transversely pumped dye cell is so large that gain saturation prevents the pulses reaching durations much greater than 50 psec. In the configuration used by us, the grating has the effect of broadening the bandwidth of the laser output close to 20 cm^{-1} as a result of a strong chirp.

For our present applications, we require ultraviolet pulses so that further efforts to reduce the pulse-durations and bandwidth available from the dye laser are unnecessary, since the process of up-converting the dye laser pulses into the ultraviolet region automatically produces shorter, narrow bandwidth pulses. The output pulses from the dye laser are carefully resynchronised with the short fundamental pulses from the regenerative amplifier and, after passing through a Type-II phase-matched crystal, the desired sum-frequency pulses in the near ultraviolet are generated. The pulse-duration of the sum-frequency pulses is slightly longer than that of the up-converting pulses because of a saturation of the efficiency of the up-converting process. The bandwidth is found to be much less than that of the chirped dye laser output.

Typical characteristics of the tunable ultraviolet laser output are durations down to about 12 psec, frequency bandwidths near 3 cm^{-1} and a power of from 10-20% of the energy of normal third harmonic pulses.

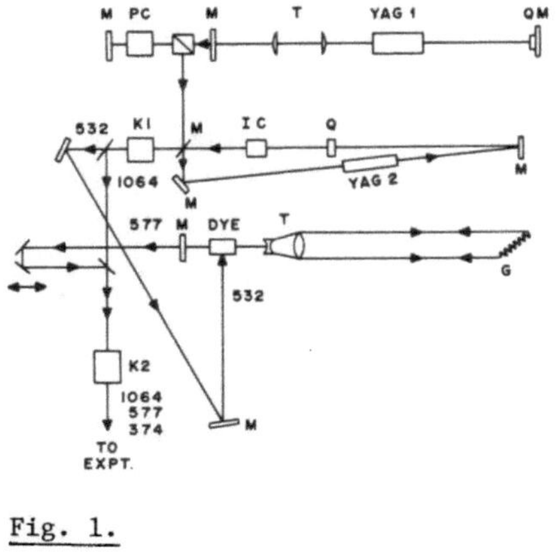

Fig. 1.

2. Fluorescence Line-Narrowing Effects in Solid Solutions of Anthracene

Studies of vibrational relaxation phenomena which involve electronically excited states in condensed media must always contend with inhomogeneous broadening effects, except at the very lowest temperatures in well-formed crystals. Therefore, a study of vibrational relaxation phenomena must involve the development of an understanding of effects such as fluorescence line-narrowing and molecular site relaxation.

That such effects can readily be observed using a tunable picosecond laser is shown in Fig. 2. A solution of anthracene in a 5:2 isopropanol: ether was frozen to 4.2°K and the fluorescence spectra measured using different irradiation wavelengths. Although the linewidths are broader than when a truly narrow-band laser is used, the spectra show typical features associated with the site selected molecular fluorescence (i.e. the fluorescence spectral shapes are strongly dependent on the excitation wavelength). This effect persists at 77°K in polymethylmethacrylate, as Fig. 3 shows.

Here, gated spectra are presented along with time-integrated spectra. One major advantage of using a picosecond-pulsed laser source for the measurement of such spectra is that the laser irradiation event may readily be removed from the spectrum by time-discrimination. Thus, while the fluorescence spectra excited at 373 nm and 374 nm are seen to be fairly similar in the presence of the intense scattered laser line, when this was

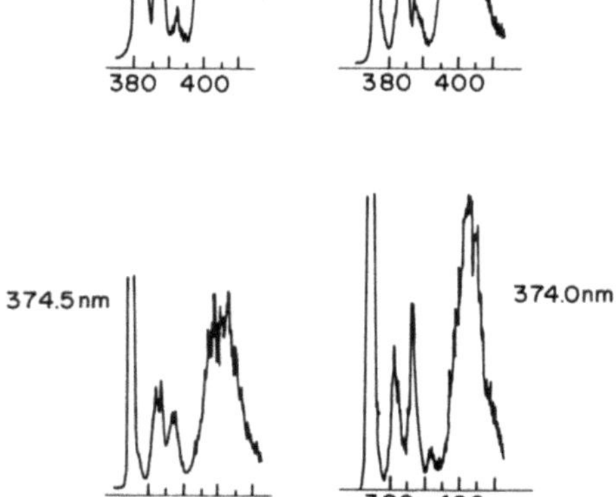

Fig. 2.

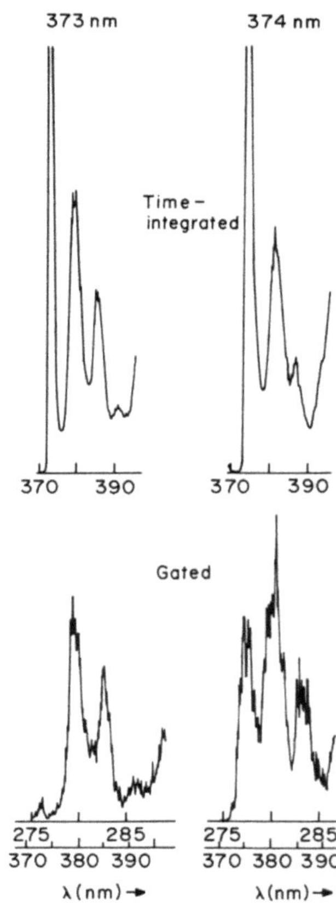

Fig. 3.

removed by time-gating, a substantial difference was revealed. Experiments carried out on the extreme red edge of the absorption band have shown the Franck-Condon progression in the 400 cm^{-1} mode of anthracene to exhibit an intensity ratio of approximately 1:0.5:0.25. Thus, the spectra excited by 373 nm and 374 nm are seen to represent different proportions of excitation of the anthracene molecules in different sites into v=0 and v=1 of this mode. Excitation into the v=1 level results in a spectrum the origin of which is shifted 400 cm^{-1} from the irradiation line, whereas excitation into v=0 gives rise to a fluorescence origin superimposed on the laser excitation wavelength.

An attempt was made to observe a persistence of resonance emission at 373 nm following laser excitation at that wavelength, but this was unsuccessful. This indicated that the v=1 level in the 400 cm^{-1} mode persisted for substantially less than the instrument resolution of 15 psec. More work is necessary, however, at even better resolution.

3. Evidence for the Interception of Vibrationally Unrelaxed Molecules

While the experiments near the origin of the anthracene first excited singlet state must be further developed in order to display evidence of emission from vibrationally hot molecules in identifiable states of excitation, it is a simpler matter to demonstrate the phenomenon of "hot" fluorescence if higher excitation energies are used. In fact, it is one of the characteristics of time-evolving fluorescence spectra, if the time-resolution is around 10 psec, that short-time components are observed, characteristic not only of vibrationally unrelaxed molecules, but also of higher states of electronic excitation. For example, we have reported such observations in the case of 3,4,9,10-dibenzpyrene [10] and diphenylpolyenes [11]. More recently, we have observed [12] emission from a pre-proton-transferred species in the intramolecularly hydrogen-bonded molecule 3-hydroxyflavone.

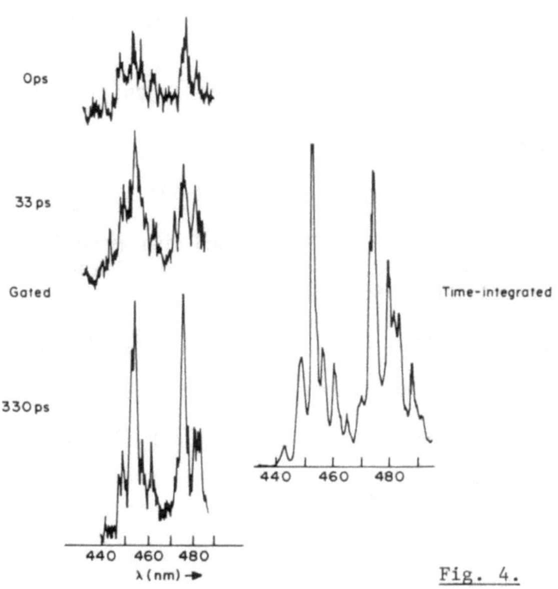

Fig. 4.

An even more recent example is shown in Fig. 4. This shows the time-evolution of the gated fluorescence spectrum of perylene in a heptane matrix near $4^{o}K$. Here, the molecule was excited with an energy equivalent to 5800 cm^{-1} above the origin of the first excited singlet state. The eventual structure shown in the time-integrated spectrum was found to evolve over a period of more than 50 psec. However, in this example, while one can indeed demonstrate that the molecule is fluorescing from unrelaxed vibrational energy levels, mechanistic information is not available without much greater control over the excitation level.

4. Summary

A picosecond time-resolved fluorescence spectrometer using a tunable laser has been shown to be capable of yielding important new information about vibrational relaxation dynamics of polynuclear aromatic hydrocarbons in solid solution. It has further been shown that, in addition to the use of well-characterised crystal matrices, the phenomenon of fluorescence line-narrowing can be used in a polymer even at temperatures above $77^{o}K$ to extend the useful range of these experiments.

We are grateful to the National Science Foundation for the support of this work through grants CHE 78-25312 and DMR 79-23647 (Materials Research Laboratory).

5. References

1. A. Laubereau, S. F. Fischer, K. Spanner and W. Kaiser, Chem. Phys. 31 335 (1978).

2. A Laubereau, A. Seilmeier and W. Kaiser, Chem. Phys. Lett. 36 232 (1975).

3. B. I. Greene, R. M. Hochstrasser and R. B. Weisman, J. Chem. Phys. 70 1247 (1979).

4. T. R. Royt, W. L. Faust, L. S. Goldberg and Chi H. Lee, Appl. Phys. Lett. 25 514 (1974).

5. L. S. Goldberg and C. A. Moore, Appl. Phy. Lett. 27 217 (1975).

6. C. A. Moore and L. S. Goldberg, Opt. Comm. 16 21 (1976).

7. L. A. Hallidy and M. R. Topp, Chem. Phys. Lett. 46 8 (1977).

8. K. J. Choi, L. A. Hallidy and M. R. Topp in "Picosecond Phenomena " II., eds. R. M. Hochstrasser, W. Kaiser and C. V. Shank, Springer Series in Chemical Physics, Vol. 14 (Springer, 1980) p 131.

9. K. J. Choi and M. R. Topp, J. Opt. Soc. Am. 70 607 (1980); 71 521 (1981).

10. K. J. Choi, B. P. Boczar and M. R. Topp, Chem. Phys. 57 415 (1981).

11. T. C. Felder, K. J. Choi and M. R. Topp, Chem. Phys. 64 175 (1982).

12. K. J. Choi and M. R. Topp (unpublished work).

Picosecond Photon Echo and Coherent Raman Scattering Studies of Dephasing in Mixed Molecular Crystals

K. Duppen, D.P. Weitekamp, and D.A. Wiersma

Picosecond Laser and Spectroscopy Laboratory, Department of Chemistry, University of Groningen, Nijenborgh 16, NL-9747 AG Groningen, The Netherlands

The advantage of time-delayed over dispersive four-wave mixing experiments is that in the former case power broadening effects are absent. This becomes particularly important when the exciting laser frequencies must be tuned into electronic resonances as is the case in the study of four-wave mixing of guest molecules in dilute mixed crystals.

We report here on the progress made on the theory of these pulsed four-wave mixing experiments and discuss some results obtained in psec delayed (vibrational) CARS experiments of pentacene in benzoic acid and naphthalene. First consider the level scheme of Fig.1. Here a is the ground state, b one of its vibrational levels, c an electronically excited state and d its vibrational level. When the exciting laser frequencies are close to the indicated resonances, it is sufficient in a calculation to only consider these four levels. In order to describe four-wave mixing effects in such a level system, what is quite often done in a description of dispersive coherence effects [1], is to calculate the third order susceptibility of the system using perturbation theory. This of course is a valid description as long as the coherent signals show the proper response with regards to the intensities of the incoming laser beams.

In pulsed high-power picosecond experiments this approximation breaks down and the predictions of the steady state theory may no longer be relevant. An intriguing example of this sort is found in the case of coherent Stokes Raman scattering (CSRS) in the four-level system of Fig.1, where perturbation theory predicts [2] only an ω_{cd} resonance damped by Γ_{cd} (dephasing rate) when pure dephasing (T_2^*) processes are present in the system. Nonperturbative calculations of the CSRS signal show [3] that, in time delayed CSRS experiments, the decay of the coherence between c and d (ρ_{cd}) can be probed even in the absence of pure dephasing processes.

When the vibrational discrepancy is large with respect to the Rabi-frequencies in the system: $(\omega_{ba} - \omega_{dc}) > \varepsilon_{ij}$ the relevant Hamiltonian,

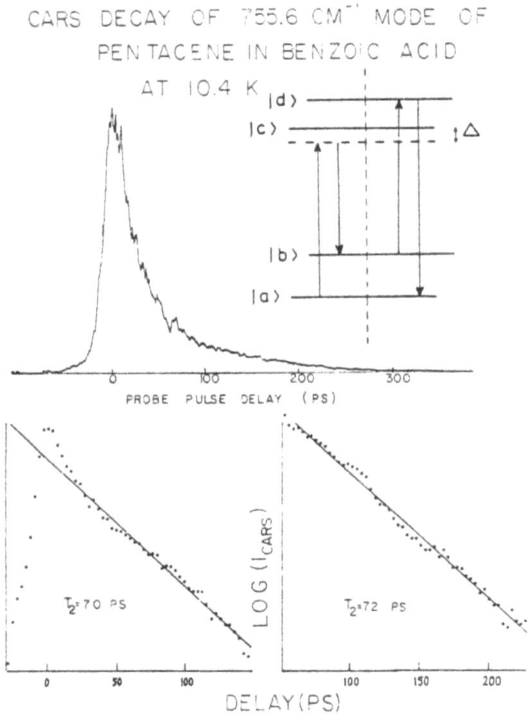

Fig.1 Upper right shows relevant level scheme for time-resolved coherent anti-Stokes Raman scattering experiments; $|a\rangle$ and $|c\rangle$ are purely electronic levels, $|b\rangle$ and $|d\rangle$ vibrational levels. Note that the probe pulse is resonant with the $\langle d| \longleftrightarrow |b\rangle$ transition. Upper left shows CARS FID decay of the 755.6 cm^{-1} mode of pentacene in benzoic acid at 10.4 K. Lower part of the figure shows that the vibrational coherence exhibits exponential decay, with lifetime of 35 ps ($\frac{1}{2}$ T_2), over three decades.

in an appropriate doubly rotating frame with the spatial dependence transformed away, is in frequency units [3]:

$$\mathcal{H} = \epsilon^{(1)}_{ac} I^{ac}_x + \epsilon^{(2)}_{bc} I^{bc}_x$$

$$\mathcal{H} = \epsilon(\cos\theta\ I^{ac}_x + \sin\theta\ I^{bc}_x) \tag{1}$$

with $\epsilon^{(\alpha)}_{ij} = \hbar^{-1}\langle i|E^{\alpha}_0 \cdot \mu|j\rangle$ and $I^{ij}_x = \frac{1}{2}(|i\rangle\langle j| + |j\rangle\langle i|)$. This three-level problem can be solved by hand and we obtain for the vibrational coherence at time τ:

$$\rho_{ba}(\tau) = \mathrm{Tr}(|a\rangle\langle b|\rho(\tau)) = \tfrac{1}{2}[\tfrac{1}{4}(3\cos 2\theta - 1)\sin 2\theta + \tfrac{1}{4}(1+\cos 2\theta)\sin 2\theta \cos\epsilon\tau - \sin 2\theta \cos 2\theta \cos(\epsilon\tau/2)] \quad (2)$$

with $\theta = \tan^{-1}(\epsilon^{(2)}_{bc}/\epsilon^{(1)}_{ac})$ and $\epsilon = ((\epsilon^{(1)}_{ac})^2 + (\epsilon^{(2)}_{bc})^2)^{1/2}$.
There are several interesting features worth noting of this function which are not evident from perturbative treatments. First the time evolution of the vibrational coherence during the preparative stage has Fourier components at zero, $\epsilon/2$ and ϵ. The relative weight of these components depends on the ratio of the applied fields. This ratio is important for the efficiency of the preparation process.

For short times $\epsilon\tau \ll 1$, the coherence amplitude is given by:

$$\rho_{ba}(\tau) = -(\epsilon^2\tau^2/16)\sin 2\theta \quad (3a)$$

which can also be written as

$$\rho_{ba}(\tau) = -\tau^2/8\ \epsilon^{(1)}_{ac}\epsilon^{(2)}_{bc} \quad (3b)$$

showing the relation with the perturbative approaches to the calculation of the vibrational coherence.

For large values of $\epsilon\tau$ the situation is very different. If an average over $\epsilon\tau$ is taken only the first term of (2) survives. This average vibrational coherence changes sign as the ratio of the fields is varied. Experimentally, averages over pulse angles can arise due to inhomogeneity of the beams across the sample.

The probing process has also been described to all orders in the field amplitude and can be found in [3] together with a complete account of this work. What still remains valid is that the decay of the delayed CARS FID signal is governed by the damping constant (homogeneous and inhomogeneous) of the vibrational transition.

We now turn to a short discussion of some of the results obtained in a study of the CARS FID decay of the 756 cm^{-1} vibrational mode of pentacene. In the actual experiments we have chosen the frequencies of the incoming beams such that the probing process was resonant with the $\langle d \leftrightarrow b\rangle$ transition. Note that this configuration, in principle, also allows measurement of the vibrational lifetime through observation of the probe-induced fluorescence. The upper part of Fig.1 shows the observed CARS decay signal at 10.4 K of the 756 cm^{-1} vibrational mode of pentacene in benzoic acid. The lower part shows a display of the logarithmic signal intensity versus probe pulse delay. First of all we conclude, from the exponential decay of the CARS FID, that the 756 cm^{-1} vibrational transition is homogeneously broadened and that the measured decay ($\tfrac{1}{2} T_2$ = 35 ± 2 ps) is due to vibrational relaxation. Note that this decay is a factor of three longer than previously [4] inferred from dispersive CARS experiments. Secondly, we find that the CARS FID decay, at least up to 20 K, is temperature independent. Very recently we have found the same result for the same mode of pentacene in naphthalene.

These results become very interesting in view of the fact that electronic dephasing of pentacene in both mixed crystals in the same temperature range is extremely rapidly increasing with temperature. It was previously shown, using psec photon echoes [5] that optical dephasing in these systems was basically due to uncorrelated phonon scattering processes in the ground and electronically excited state involving a specific guest librational level.

In the mixed crystal of pentacene in naphthalene [5] the frequency (ω) and lifetime (τ) of this libration were found to be rather different in the ground and electronically excited state. DE BREE and WIERSMA [6] showed that in such a four-level-coupled-to-a-bath system, optical Redfield relaxation theory should be used to interpret the homogeneous line shape. The pure dephasing constant $T_2^*(R)$ for such a four-level system takes the following form [6]:

$$1/T_2^*(R) = \tfrac{1}{2}[\tau_g^{-1}\exp(-\omega_g/kT) + \tau_e^{-1}\exp(-\omega_e/kT)] + \Gamma_{ge}^{(a)}, \qquad (4)$$

where the subscripts g and e refer to the ground and electronically excited state, respectively. $\Gamma_{ge}^{(a)}$ refers to the adiabatic [6] contribution to T_2^{-1}. For the pure electronic and vibronic transitions in pentacene this contribution to T_2^{-1}, up to 20 K, was shown [5] to be negligible.

In case of vibrational transitions one might expect [5] the ground and vibrationally excited state librations to be very close in frequency and lifetime, in which case, instead of (4), exchange theory [7] should be used to describe dephasing. The exchange-determined dephasing constant $T_2^*(E)$ is given by:

$$1/T_2^*(E) = [\delta^2\tau/(1 + \delta^2\tau^2)]\exp(-\omega/kT) + \Gamma_{ge}^{(a)}, \qquad (5)$$

where $\delta = |\omega_e - \omega_g|$, $\tau = (\tau_e + \tau_g)/2 \approx \tau_e$ and $\omega = (\omega_e + \omega_g)/2 \approx \omega_e$.

Note that exchange theory is only applicable when $\delta\tau \lesssim 1$. Furthermore in case of vibrational dephasing, in addition, τ should be much shorter than the vibrational relaxation time. For a thorough discussion of the limitations of exchange theory consult [6]. Comparing (4) and (5) it is clear that in case of vibrational exchange, the homogeneous line shape is narrowed compared to the no-exchange case. In the extreme case, where $\omega_g = \omega_e$, exchange-induced dephasing will vanish and the adiabatic contribution ($\Gamma_{ge}^{(a)}$) to the homogeneous line shape may become important.

The observation of a temperature-independent vibrational line shape (up to 20 K) suggests that we are in the extreme limit where $\omega_e = \omega_g$ and that the exchange-induced dephasing vanishes! Physically it means that the librational frequency is independent of vibrational excitation. Furthermore up to 20 K, the adiabatic contribution to vibrational dephasing is also absent.

We note that the results presented here (for pentacene in naphthalene) are in conflict with our recent (supposedly) resonance Raman scattering experiments [8]. We now believe that in those experiments we have, for unknown reasons yet, measured vibronic rather than vibrational dephasing.

References:

1. N. Bloembergen, H. Lotem, and R. T. Lynch, Jr., Indian J. Pure Appl. Phys. 16 (1978) 151.
2. A. Bogdan, M. Downer, and N. Bloembergen, Phys. Rev. A 24 (1981) 623; J. R. Andrews, R. M. Hochstrasser, and H. P. Trommsdorff, Chem. Phys. 62 (1981) 87.
3. D. Weitekamp, K. Duppen, and D. A. Wiersma, J. Chem. Phys., to be submitted.
4. J. R. Andrews and R. M. Hochstrasser, Chem. Phys. Lett. 83 (1981) 427.
5. W. H. Hesselink and D. A. Wiersma, J. Chem. Phys. 73 (1980) 648.
6. P. de Bree and D. A. Wiersma, J. Chem. Phys. 70 (1979) 790.
7. C. A. van't Hof and J. Schmidt, Chem. Phys. Lett. 36 (1975) 460; 42 (1976) 73. C. B. Harris, R. M. Shelby, and P. A. Cornelius, Phys. Rev. Lett. 38 (1977) 1415.
8. P. de Bree and D. A. Wiersma, Chem. Phys. Lett. 88 (1982) 17.

Picosecond Laser Spectroscopy of Molecules in Supersonic Jets: Vibrational Energy Redistribution and Quantum Beats

A.H. Zewail*

Arthur Amos Noyes Laboratory of Chemical Physics, California Institute of Technology, Pasadena, CA 91125, USA

I. Introductory Remarks

The dynamics of vibrational energy flow in large and isolated molecules following selective laser excitation is very interesting and challenging for many reasons. From a theoretical point of view, one would like to know how the coupling between bond vibrations influences the energy flow, and at what energy threshold does this flow or redistribution of energy from one mode-to-others occur. The energy region in which randomization or "chaotic behavior" dominates is important for mode-selective chemistry [1,2].

Our own interest in this problem has been focused on studying the time resolved dynamics by picosecond excitation of large isolated molecules (in S_1) in supersonic jets [3-7]. Here, we shall highlight the new findings we obtained from different studies (quantum beats, isomerization, selective energy transfer, etc.), with particular emphasis on the relevance of these observations to vibrational energy redistribution (statistical, RRKM-type vs nonstatistical) and dephasing in the isolated molecules.

II. The Picosecond Laser-Molecular Beam Apparatus

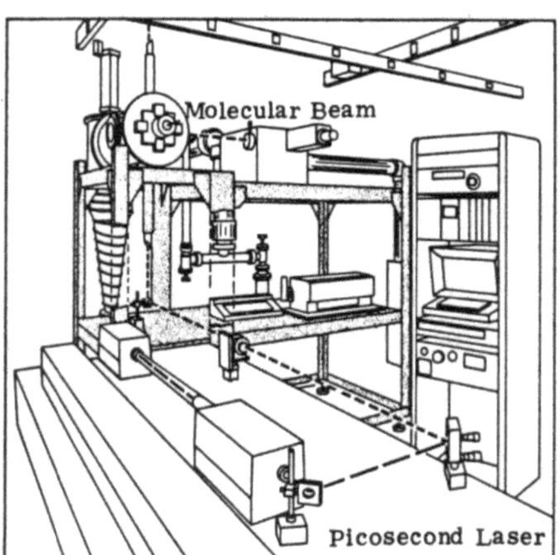

Fig. 1

Figure 1 displays the arrangement we used for interfacing a molecular beam (or free-jet) system to a picosecond tunable dye laser system. The latter is a synchronously pumped, cavity dumped dye laser. The laser was directed toward a spot down stream on the nozzle axis. The distance from, the temperature of, and the pressure behind the nozzle were varied depending on the experiment. The pressure in the beam chamber was 10^{-4} torr. For fluorescence measurements we dispersed the emission (after being collimated with f ~ 1 optics)

*Alfred P. Sloan Foundation Fellow and Camille & Henry Dreyfus Foundation Teacher-Scholar and used a fast photo-multiplier in connection with a time-correlated photon counting system interfaced to a PDP 11/23 computer. The coherent time and the actual laser band width were varied by using a combination of intrac vity filters and etalons. Finally, typical rotational and vibrational temperatures for these large molecules in the free-jet are 1 K and 20-40° respectively.

III. Applications

A. Coherent picosecond excitation and quantum beats in anthracene and stilbene

The emission spectra of anthracene in a gas bulb at ~480K reveal little intensity of quasi-sharp lines (I_s) and a large intensity of background that is very broad (I_b) or diffuse in nature. When anthracene is jet-cooled, the I_s/I_b varies dramatically depending on the excess vibrational energy in S_1. As shown in Fig. 2, this ratio varies from 100 or more at zero excess energy to 10^{-2} or less at 5600 cm^{-1}; the spectra in the high excess energy (E_x) region are very similar to the bulb spectrum.

At 1400 cm^{-1} of excess energy in anthracene, LAMBERT et al. [3,4] observed quantum beats on the dispersed fluorescence with a large modulation depth. The beats are very sensitive to the excess energy and to the fluorescence detection wavelength. On the other hand, the quantum beats in stilbene appeared at a number of different E_x. Clearly, the observation of beats in these large molecules must be related to the coherence of the vibrational states excited. From our results we conclude that the homogeneous width (dephasing) of the level (or bunch of levels) is definitely less than 0.1 cm^{-1}, and vibrational redistribution is in the intermediate coupling limit. The beats (due to rovibronic states) in anthracene are not expected to be sensitive to magnetic field in contrast with beats due to coupling between electronic states, one of them is a triplet state. In a

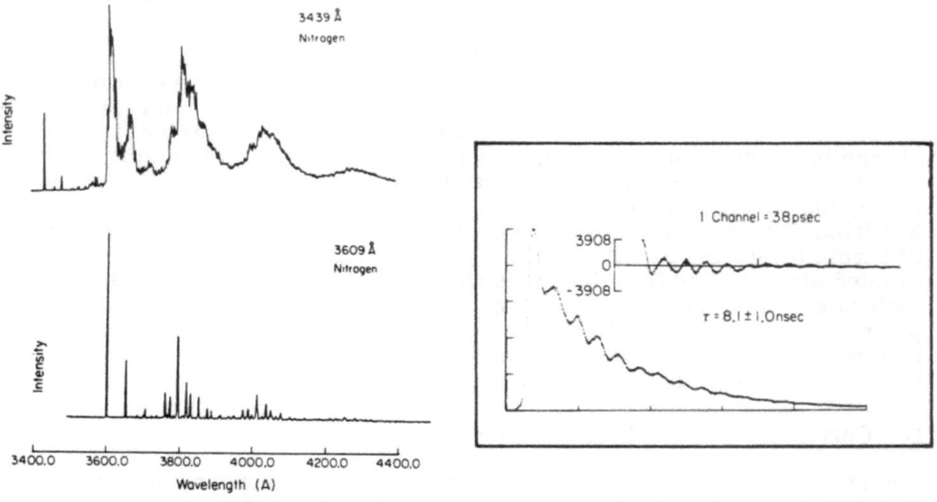

Fig. 2 Jet-cooled fluorescence of anthracene at two different excess energies (0 and ~1400 cm^{-1}). The quantum beats shown is obtained when the laser was tuned at 3439 Å.

recent work by FELKER et al. we reported on the Zeeman effect on these beats in pyrazine, and on the importance of beats to unravel coupling of singlet and triplet rotational levels of the isolated molecule (details are elsewhere).

B. Isomerization of trans-stilbene

The isomerization process of trans-cis stilbene involves some torsional vibrational modes for the twisting to finalize. This is a classical problem in chemistry, and much work has been done in solution and more recently under gas phase isolated molecule conditions [8]. SYAGE et al. [6] showed that the jet-cooled spectra displays the isolated molecule low-frequency modes which are optically active. But, perhaps, more importantly the observed rate as a function of excess energy exhibits a threshold at ~1200 cm^{-1}, which is in excellent agreement with the estimated barrier height for isomerization from S_1 in solution. The jet studies clearly indicate the absence of thermal energy influence on rates, and a restricted involvement of molecular modes in the redistribution. This latter point is now under examination in a collaborative effort with R. A. Marcus.

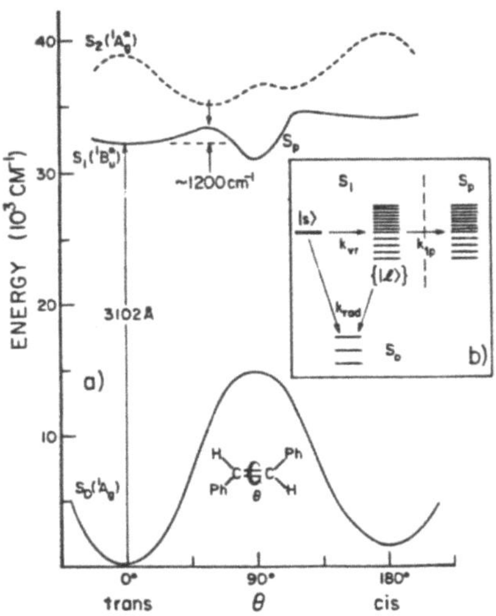

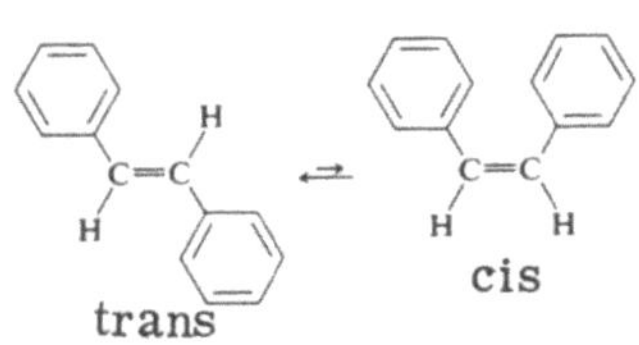

C. Hydrogen-bonded systems

With the same technique, FELKER et al. [5] examined molecules exhibiting (possible) proton transfer in the excited state. In other words, in these molecules there exist the possibility of "tunneling" between different configurations. The results for methyl salicylate in the jet show a drastic change in the equilibrium configuration upon excitation. The distorsion involves the OH, C=O, and a low-frequency mode of 180 cm^{-1}. Furthermore, the observed fluorescence rate exhibits again a marked change with E_x; from $(12nsec)^{-1}$ at zero E_x to $(160ps)^{-1}$ at E_x ~2000 cm^{-1}. We concluded from these studies that for the E_x=0, the distorted excited state is created simultaneously with the absorption of a photon in the isolated molecule. In the high excess energy limit, the low-frequency motions promote excited state electronic states coupling or configurational changes that are reminiscent of the stilbene case. Thus, the prevailing of a threshold effect (Fig. 3).

D. Chromophore-selective picosecond excitation

The idea of this experiment is to excite a single chromophore in a molecule (jet-cooled) and to observe real energy flow from the optically-pumped vibrations of the chromophore to other modes in the whole molecule. We chose the above molecule (I), which has been extensively studied by the Columbia group and by others [9]. This molecule (for

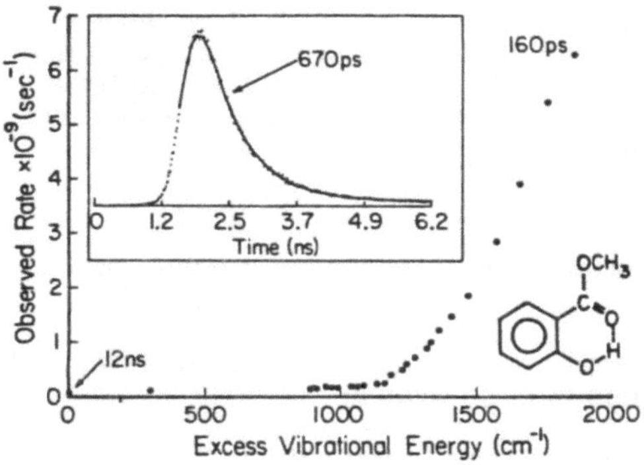

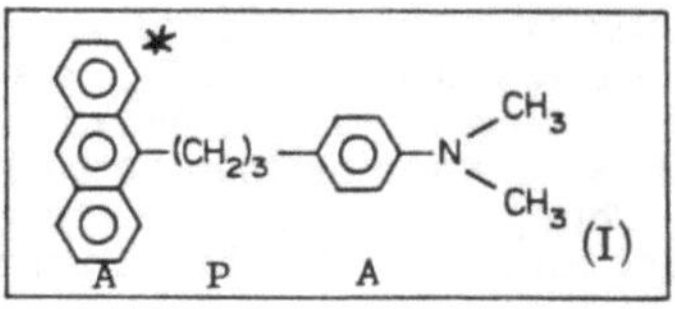

(I)

Fig. 3 The "threshold" effect in jet-cooled methyl salicylate.

brevity called APA), possesses an "optically active" chromophore--anthracene. Thus, in the jet we excite (picosecond) different modes in the anthracene moeity and follow the time-resolved dispersed fluorescence of anthracene-like emission and product emission (red shifted emission due to the folding of the aniline moeity towards anthracene: CT emission). FELKER et al. [7] observed a decay in the anthracene-like emission, and a buildup in the CT emission (Fig. 4). These rates were found to be very sensitive to the excess vibrational energy in the anthracene moeity.' This represents the first report on real time measurement of energy flow in an isolated molecule. Since the conformational change needed to give the red emission requires that the vibrational modes of the propyl linkage become populated, our results show that the excess vibrational energy in the anthracene moeity redistribute to vibrations (and "rotations") in the side chain, thus triggering the folding process which leads to a drastic change in the spectral and temporal behavior of the molecule. The threshold for this excess energy dependence of the product formation is $\simeq 1000\ cm^{-1}$ (3k cal/mole) consistent with a reaction barrier involving C–C type bonds. Finally, the decay and buildup time constants at $\sim 3000\ cm^{-1}$ of excess energy (~400 ps) are much different from the decay time constant of bare anthracene at similar excess energies (5. 7 ns). The question then; what does this rate in the isolated molecule mean and what is the role of the solvent?

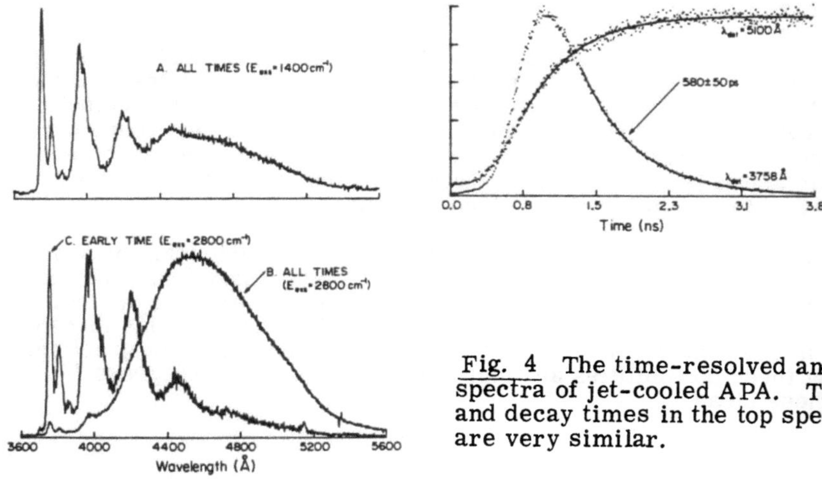

Fig. 4 The time-resolved and gated spectra of jet-cooled APA. The rise and decay times in the top spectrum are very similar.

Using an assumed set of vibrational frequencies for APA along with our jet lifetimes at different E_x, we have calculated the decay of the anthracene-like emission for a thermal distribution at 298K. The calculated lifetime (~600 ps) is shorter than actually observed in cyclohexane solution (1.4ns) [9]. Thus, it appears that the geometrical changes needed for the CT product formation are hindered by the solvent.

IV. Concluding remarks - Redistribution by low-frequency modes, the "threshold" effect, and coherence

From our studies of anthracene, stilbene, pyrazine, methyl salicylate, and APA in the jet the following conclusions may be drawn:

a) A threshold for an order of magnitude change in the rate at E_x between 1000 and 2000 cm^{-1}, depending on the molecule is present. In anthracene there is a leveling off of this rate at E_x ~1500 cm^{-1} - 3000 cm^{-1}. The effect of deuterium substitution depends on the excess energy.

b) The observed quantum beats in anthracene and stilbene are absent above the energy region for this abrupt change in the rate. Dephasing must be relatively slow at these excess energies.

c) The dispersed fluorescence around this threshold area displays two features--a sharp emission and a "diffuse" red shifted (from the 0,0 excitation) emission. The spectra become completely diffuse at high excess energies and resemble the trend found in other large molecules [10].

d) Energy flow at moderate excess energies in APA has a time constant of ~400 ps and not one or so ps.

An interesting question arises from these studies: What is the origin of this threshold effect? In what follows some ideas are discussed.

The initial temperature of the molecule is very low and certainly is much lower than $\hbar\omega/k$ of totally symmetric modes. We may divide the modes of the molecule into those which are optically active (predominately totally symmetric or "relevant" R) and those which are not excited directly by the laser (bath modes B including nontotally symmetric and torsion modes). This division of the system-bath interactions accounts for dephasing and energy relaxation by T_2 and T_1 time constants as discussed recently [2,11].

The redistribution of R excitation to B will have a threshold depending on the number of quanta that can be populated in B (density of states) and the degree of coupling. Invoking these low-frequency modes in the redistribution suggests the use of restricted density of states since high frequency modes are not efficiently populated. Further, coherence among some rovibronic states is expected since the modes density is not total, i.e., restricted. More details will be given elsewhere.

In conclusion, the measurements of rates vs. E_x in the beam provide a new way for obtaining the threshold energy ("solvent free") and the restricted number of modes. It is possible that B-type modes are the main constituent of the phase space for the redistribution, and that this threshold effect is almost universal. These studies raise interesting new questions regarding these possibilities and more experiments are in progress for testing these ideas.

Acknowledgments

This work was supported by grants from NSF. I would like to thank Profs. R. Marcus and J. Hopfield for very enlightening discussions. Without the efforts of Bill Lambert, Peter Felker and Jack Syage the story told here would not have been written.

References

1. R. A. Marcus, in this volume.

2. A. H. Zewail, Physics Today, **33**, 25 (1980).

3. W. R. Lambert, P. M. Felker, and A. H. Zewail, J. Chem. Phys. **75**, 5958 (1981).

4. A. H. Zewail, R. W. Lambert, P. M. Felker, J. W. Perry, W. S. Warren, J. Phys. Chem. **86**, 1184 (1981).

5. P. M. Felker, W. R. Lambert, A. H. Zewail, J. Chem. Phys. **77**, 1603 (1982).

6. J. A. Syage, W. R. Lambert, P. M. Felker, A. H. Zewail and R. M. Hochstrasser, Chem. Phys. Letts. **88**, 266 (1982).

7. P. M. Felker, J. A. Syage, W. R. Lambert and A. H. Zewail, submitted to J. Chem. Phys.

8. For a review see R. M. Hochstrasser, Pure and Appl. Chem. **52**, 2683 (1980).

9. M. Crawford, Y. Wang and K. B. Eisenthal, Chem. Phys. Letts. **79**, 529 (1981) and references therein.

10. For a review see C. Parmenter's paper in this volume.

11. See the article by N. Bloembergen and his coworkers in this volume.

Picosecond Studies of Intramolecular Vibrational Redistribution in S_1 *p*-Difluorobenzene Vapor

R.A. Coveleskie, D.A. Dolson, S.C. Muchak, C.S. Parmenter, and B.M. Stone

Department of Chemistry, Indiana University,
Bloomington, IN 47405, USA

Para-difluorobenzene (pDFB) is being used as a model aromatic molecule for detailed study of the picosecond dynamics of collision-free intramolecular vibrational redistribution (IVR). Three experimental methods allow the redistribution to be seen rather directly in the S_1 electronic state.

1. Collision-free room temperature fluorescence spectra after pumping various S_1 levels reveals the presence of IVR by congestion in the fluorescence spectrum that cannot otherwise be explained.

2. Line widths as determined from structure in high resolution ($S_1 \leftarrow S_0$) absorption band contours give a complementary view of IVR as one looks at rotational level widths within various S_1 vibrational levels. These data come from T.M. Dunn (University of Michigan).

3. The most direct view comes from picosecond time-resolved fluorescence spectra after various levels in S_1 pDFB are pumped by a tuned laser. By looking at the time-dependent disappearance of structure in the fluorescence spectrum, one can determine the actual kinetic characteristics of initial state decay by IVR. A special chemical timing technique is used to achieve simultaneously the fast timing and high fluorescence detection sensitivity that are required.

An overall view of redistribution as one climbs the S_1 vibrational ladder has emerged. Strong level mixing that is best described as ordinary Fermi resonance is seen as additional structure in fluorescence from levels as low as 819 cm^{-1}. The first secure indications of extensive level mixing are seen by picosecond timing when excitation reaches $\epsilon_{vib} \approx 1490\ cm^{-1}$. That mixing becomes sufficiently extensive by the time one reaches $\epsilon_{vib} \approx 2000\ cm^{-1}$ so that a severely congested background appears in the fluorescence spectrum. The background cannot be accounted for by other explanations. Both level widths and collision-free fluorescence spectra show that the level mixing associated with IVR persists for every accessible level above these thresholds.

The level widths correspond to lifetimes of tens of psec for the threshold region. The IVR lifetimes decrease to less than 5 psec for levels above 2500 cm^{-1}. While these lifetimes, as indicated by the level widths, show a trend to shorter values as vibrational energy increases, the trend is not completely monotonic. Some sensitivity to initial level identity seems to be present, but the data need refinement to press this issue further.

Chemical timing studies have been carried out for a variety of levels from 1250 cm^{-1} up. The method of picosecond fluorescence spectroscopy is clearly successful in displaying the collision-free time evolution of the S_1 vibrational identity.

As of this writing, detailed kinetic studies by picosecond timing have been completed only with excitation of an initial vibrational level lying in the threshold region near 2200 cm^{-1}. We are able to monitor the decay of the initial level and the build-up of the neighboring vibrational field by looking at the relative intensities of structured and background emissions as a function of time. The initial level decay is clearly non-exponential. If IVR is modelled by analogy with the theory of nonradiative transitions, the decay fits well the expectations of the intermediate case and not at all the statistical case. In such IVR studies, one learns only about the "short" decay of the intermediate case. It turns out to be about 35 psec.

The fit to intermediate case theory is the least surprising aspect of the results. With a density of vibrational levels of only about 40 per cm^{-1} (ungerade levels only for this case), such decay is perhaps to be expected. There is, however, another characteristic of the IVR that is more remarkable.

It concerns ergodicity. Our experiment is modelled by a "two-state" system comprised of the initial level and the field of levels as seen by structured and unstructured emission. The question is whether the entire field is needed to give the unstructured emission or could it be accounted for by a subset of available levels? If we seek an answer by computer simulated fluorescence spectra, we get a secure result. The full density of vibrational levels is not nearly <u>large</u> enough to account for the congested background. It is clear that some other source of level density is used by IVR. The only possibility appears to be rotational levels made accessible by Coriolis vibration-rotation coupling. One needs such coupling to explain also the appearance of IVR from lower levels. Thus IVR has another link with intermediate case radiationless transitions. Rotation-vibration coupling appears commonly used in both.

Direct Picosecond Resolving of Hot Luminescence Spectrum

J. Aaviksoo, A. Anijalg, A. Freiberg, M. Lepik, P. Saari, T. Tamm, and K. Timpmann

Institute of Physics, Estonian SSR Academy of Science, Tartu 202400, USSR

Hot luminescence |1| (HL) as a natural by-product of relaxation within excited electronic states enables relaxation kinetics to be investigated in detail without recourse to applying any additional (probing) action on the system, which may disturb the process under study. For systems possessing distinct-structured vibronic spectra (e.g. molecules in rigid matrices) the cw HL spectroscopy allows us to evaluate rather precisely vibrational relaxation times up to 10^{-13}s (see review |2|). Yet, by introducing direct picosecond time resolution, HL becomes a truly versatile tool in the study of transient states and relaxation in condensed matter.

On the other hand, the payment for the spontaneous nature of HL is its low intensity and therefore a need for a sophisticated signal-averaging picosecond set-up for a successful measurement of time-resolved HL spectra. As far as we know, the attempts

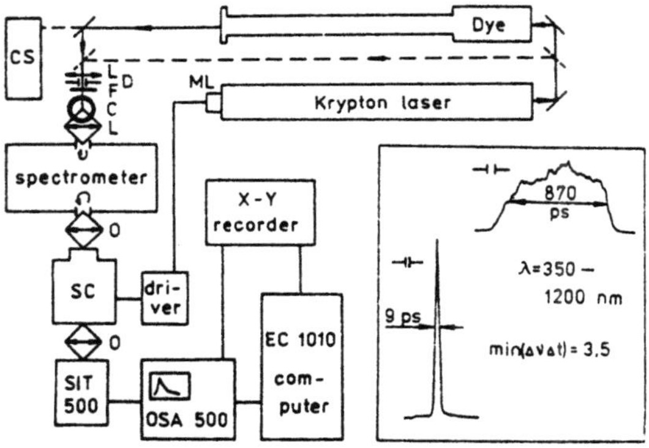

Fig. 1: Picosecond spectrochronograph for studying time-resolved emission spectra with uncertainty-principle-limited resolutions. For exciting the sample placed in a liquid He cryostat (C) 82 MHz pulses from an oxazine-dye laser are used, which is synchronously pumped by actively mode-locked krypton laser. The mode-locking voltage controls also the image scanning in the streak-camera (SC). The output image is recorded by a SIT-vidicon and an optical multichannel analyzer (OSA) and processed in a mini-computer. On the right the time response of the spectrochronograph is shown in comparison with an extremely poor response of a common Raman-quality spectrometer.

made up to now in order to observe time-resolved HL by single-shot streak-cameras have not given any reliable results. Moreover, if studying distinct-structured spectra an additional difficulty arises in connection with a considerable broadening of light pulses in a spectrograph with spectral resolution sufficient for displaying the vibronic structure. The reasons of such a broadening, far exceeding the uncertainty-principle-determined value, have been considered in |3|, where the solution to the problem - subtractive dispersion mount in the spectrometer, is also proposed. The scheme of a corresponding set-up is shown in Fig. 1 (a detailed description can be found in |4-6|). It should be pointed out that such a spectrochronograph has a resolution cell in frequency/time/intensity-space with volume close to the quantum-mechanical minimum value and therefore it opens for experimental study various interesting pecularities predicted for time-dependent emission spectra (see refs. in |3|).

In the present paper we discuss the first direct observation of the picosecond-domain temporal behaviour of vibronic lines in luminescence spectrum on an example of perylene molecule in paraffin lattice (preliminary results reported in |7|). From spectrochronograms obtained we derive lifetimes of vibrational relaxation in the S_1 state of the molecular system.

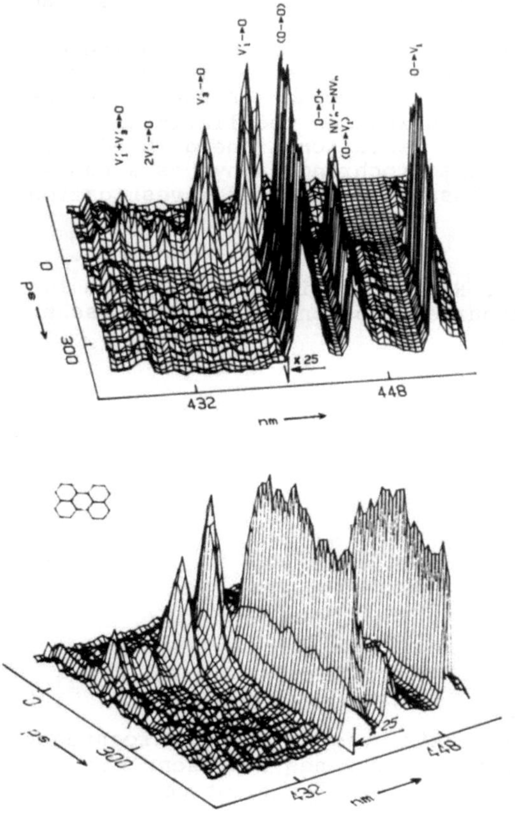

Fig. 2:

Two views on the spectrochronogram of the luminescence of the perilene molecules in the n -heptane matrix ($\sim 10^{-4}$ M/ℓ) at 4.2 K. The spectral slitwidth is 31 cm^{-1}. The lines are assigned by indicating the initial and final vibrational levels in the excited and ground electronic states, respectively. At t=0 excitation by frequency-doubled picosecond laser pulses (λ=396 nm) populates a selected high-lying intramolecular vibrational level of S_1 manifold.

The time-resolved HL spectrum is displayed in Fig. 2. The right hand side represents the slowly-decaying lines of ordinary luminescence. On the left one can see four short-living HL lines arising due to the radiative transitions from excited levels of two lowest totally symmetrical vibrational modes populated in the course of the relaxation process.

There exist two distinct series of ordinary luminescence lines which correspond to two - "red" and "blue" - inhomogeneity-caused types of centers. All marked HL lines belong to the "red" centers which have considerably higher density (the 0-0 line of these centers is suppressed by the reabsorption).

In Fig. 3 the time axis cuts of the spectrochronogram in the 0-0 and $\nu_1'\to 0$ transition wavelengths are given. If the shape of the last one is quite well approximated with the convolution of the spectrochronograph response function and the one-exponential decay, the shape of the first one is much more complicated. This is due to the overlapping of fastly decaying $N\nu'_n \to N\nu_n$-type hot transitions to the 0-0 transition.

The decay times of the excited state vibronic levels obtained by the one-exponential least-square fit of experimental curves are gathered in the Table. For comparison in the last column the decay times estimated from the steady-state HL spectrum |8| are also shown

From these results it is evident that despite the relative complexity of the perilene molecule and the existence of the thermal bath of the matrix the relaxation of the excitation energy is rather slow. This interesting aspect as well as peculiarities of relaxation of the different amounts of vibrational energy need further study and it seems that the spectrochronography has a number of advantages compared to other possible methods of investigation.

Acknowledgements

The authors are grateful to K. Rebane for introducing them into this field of study and numerous discussions and to J. Büscher for valuable advice and support.

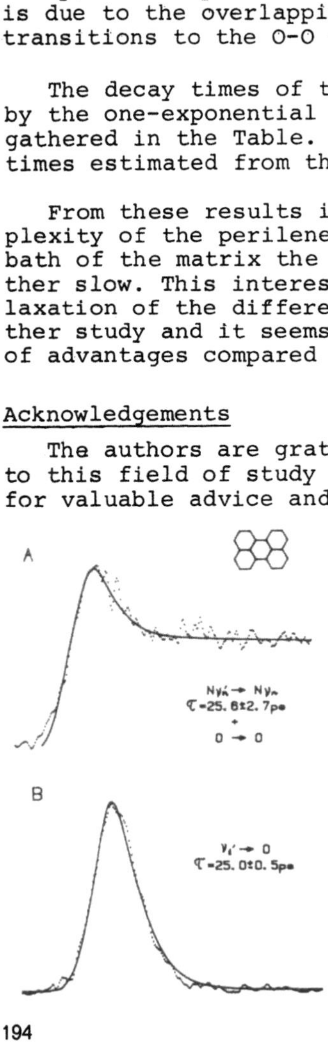

Fig. 3:

The time axis cuts of the spectrochronogram. The solid lines represent the least square fit of the dotted experimental curves.

Table: Lifetimes τ(ps) for lowest vibronic levels of perylene molecule in n-heptane at 4.2 K.

Vibrational energy (cm^{-1})	assignment to vibrational modes	τ (this work)	τ (from |8|)
1580	ν'_9		10 ± 3
1380	ν'_8		5 ± 2
1300	ν'_7		14 ± 3
900	$\nu'_1 + \nu'_3$	15.7 ± 2.5	
710	$2\nu'_1$	21.0 ± 1.4	
550	ν'_3	26.0 ± 2.2	
360	ν'_1	25.0 ± 0.5	35 ± 5

References

1. Saari P., Rebane K., Solid State Comm. 1969, 7, 887.
2. Rebane K., Saari P., Lumin. 1978, 16, 223.
3. Saari P., Aaviksoo J., Freiberg A., A. Timpmann K., Opt. Comm., 1981, 39, 94.
4. Freiberg A.M., Anijalg A.O., Kukk P.L., Mihkelsoo V.T., Saari P.M., Timpmann K.E. and Schultz E.A., in: Proc. 14 Intern. Congress on High speed photography and photonics, Moscow 1980, p. 97.
5. Anijalg A., Freiberg A., Kaarli R., Kukk P., Saari P. and Timpmann J., in: Ultrafast phenomena in spectroscopy, Proc. II, Intern. Symp. Reinhardsbrunn, GDR, 1980 (Academie der Wissenschaften der DDR) p. 95.
6. Anijalg A.O., Timpmann K.E. and Freiberg A.M., Pis'ma J. Tehn. Phys. (in Russian, to be published).
7. Saari P., Aaviksoo J., Freiberg A., Timpmann K., in: Proc. Int. Conf. "Lasers '81", Dec. 1981, New Orleans (to be published).
8. Tamm T., Saari P., Chem. Phys. 1979, 40, 311.

The Temperature Dependence of Homogeneous and Inhomogeneous Vibrational Linewidth Broadening Studies Using Coherent Picosecond Stokes Scattering

S.M. George, A.L. Harris, M. Berg, and C.B. Harris

Department of Chemistry, University of California and
Materials and Molecular Research Division, Lawrence Berkeley Laboratory,
Berkeley, CA 94720, USA

The theoretical interpretation of vibrational dephasing experiments has been dependent on several simplifying assumptions [1]. One important assumption has been low depletion (<5%) of the laser pump in the stimulated Raman excitation process. We have observed that, given exponential stimulated Stokes gain and pulse intensity fluctuations from passively mode-locked Nd:glass laser systems, maintaining stimulated Stokes conversions between 1-5% is a nearly impossible task. Figure 1 shows that immediately after the stimulated Raman scattering threshold is reached, the laser can be depleted >10% and depletion approaches ≈50% as the laser energy increases. Other studies measuring both laser and Stokes pulses after the Raman cell reveal coincident sharp thresholds for stimulated Stokes scattering and laser depletion [2]. Consequently, we believe that most vibrational dephasing experiments [3-6] have been performed in the high Stokes conversion regime.

A theoretical explanation for vibrational dephasing experiments conducted in the high Stokes conversion region is currently being developed. Solution of the coupled differential equations for transient stimulated Raman scattering reveals that energy oscillates between the laser and Stokes pulse via the

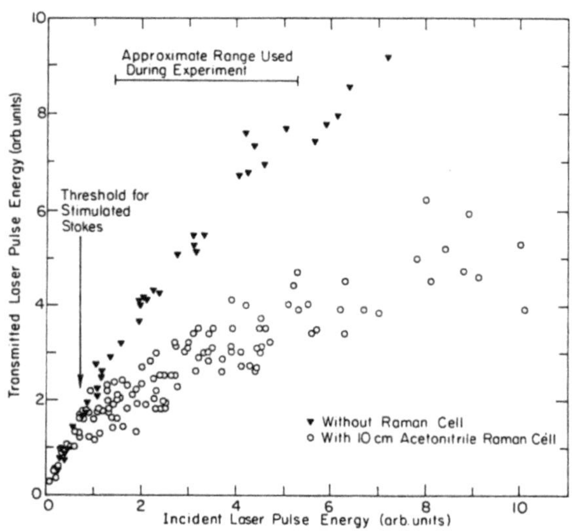

Figure 1
Transmitted laser pulse energy versus incident laser pulse energy with and without the 10 cm Raman cell using the experimental setup detailed in references 2 and 6.

coherent molecular polarization in the medium [7,8]. Calculations show that the position, width and slope of the stimulated Stokes pulse are all dependent upon T_2, the homogeneous dephasing time of the coherent molecular polarization. Because the vibrational amplitude closely follows the Stokes pulse, we believe that T_2 or a time associated with T_2 can be measured under conditions of high Stokes conversion using the selective coherent picosecond Stokes scattering technique [1,4,6]. The following experiment and discussion rely upon this interpretation.

In order to determine the temperature dependence of homogeneous and inhomogeneous vibrational linewidth broadening, the temperature dependent dephasing times were measured for the symmetric CH_3 stretch in liquid acetonitrile over its entire liquid range at P= 1 Atm [2]. The dephasing times from individual measurements, numbered according to their sequence, are shown in Figure 2 and the averages of this data are plotted in Figure 3.

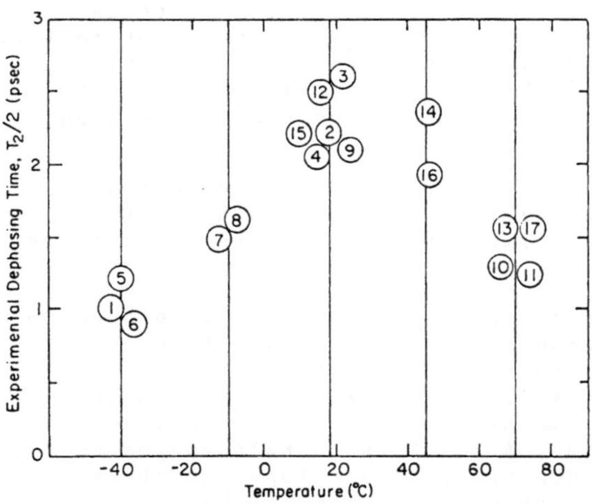

Figure 2
Experimental dephasing times from individual measurements versus temperature. The measurements are numbered according to the sequence in which they were performed.

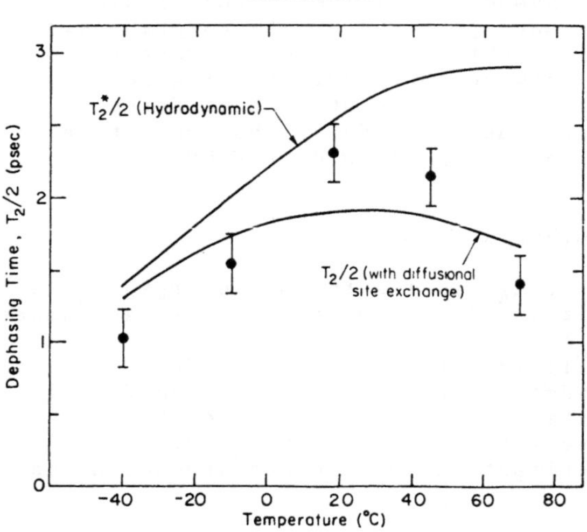

Figure 3
Average dephasing times versus temperature. Solid lines show $T_2/2$ and $T_2^*/2$ values which give the best fit to data according to Eqn. 3.

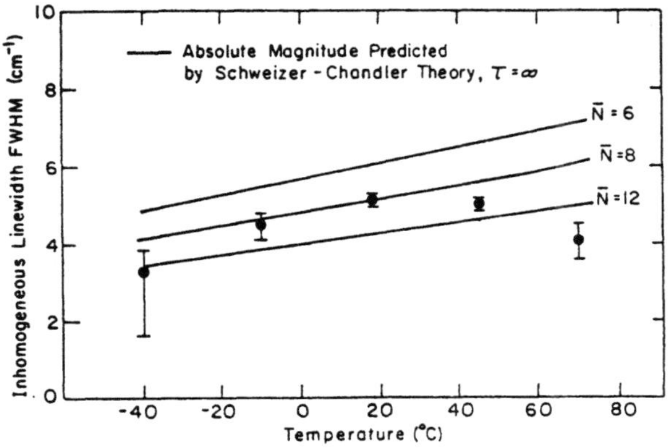

Figure 4
Measured inhomogeneous linewidths versus temperature. Solid lines show absolute magnitude of inhomogeneous linewidths predicted by Schweizer-Chandler theory.

The dephasing times at various temperatures are reproducible which demonstrates that they are not the result of day-to-day pulse fluctuations [9]. More importantly, these consistent dephasing times at various temperatures do not reflect the corresponding isotropic Raman linewidths which are nearly invariant with temperature [2,10]. Consequently, we believe these dephasing times are a measure of T_2 and the homogeneous linewidth.

The inhomogeneous broadening linewidths were obtained by deconvoluting the homogeneous linewidths determined by the dephasing times from the isotropic Raman linewidths [6] and are shown in Figure 4.

Many vibrational dephasing theories have been developed to explain homogeneous and inhomogeneous vibrational linewidth broadening caused by rapidly and slowly varying processes, respectively. The homogeneous linewidth broadening theories, based primarily on repulsive mechanisms, have been reviewed by OXTOBY [11]. Recently, GEORGE, AUWETER and HARRIS suggested that inhomogeneous linewidth broadening is proportional to the width of the distribution of local number densities in the liquid [6]. SCHWEIZER and CHANDLER have proposed a more comprehensive vibrational dephasing theory in which long range attractive forces provide the coupling between the slowly varying local density and the vibration [12]. When the attractive force correlation time, $T_A = \infty$, their theory leads to the prediction that LW, the inhomogeneous broadening linewidth (FWHM) is given by:

$$LW = 2(2\ \ln 2)^{1/2} / (\bar{N})^{1/2} * \langle \Omega_A \rangle (\bar{\rho}\ k_B T\ \kappa_T)^{1/2} \qquad (1)$$

where $\bar{\rho}$ is the number density, $k_B T$ is the thermal energy, κ_T is the isothermal compressibility, $\bar{N}$ is the estimated average number of nearest neighbors and $\langle \Omega_A \rangle$ is the attractive force contribution to the gas-to-liquid frequency shift. The inhomogeneous linewidths (FWHM) calculated according to Eqn. 1 for three different representative $\bar{N}$ values [12] are shown in Figure 4. These absolute predictions, based on $T_A = \infty$, are remarkably close to the experimentally determined inhomogeneous broadening linewidths.

The attractive force correlation time, T_A, is dependent upon the lifetime of the various local density sites, which is proportional to 1/D where D is the translational diffusion constant. Actual values of T_A are calculated by:

$$T_A > \sigma^2/(\pi^2 D) \qquad (2)$$

where σ is the hard core diameter [12]. Because of the temperature dependence of the translational diffusion constant, the attractive force correlation time will narrow the inhomogeneous linewidth as temperature increases and $T_A \rightarrow 0$. This motional narrowing occurs as translational diffusion causes interconversion of the various local density sites which define the inhomogeneous distribution of distinct vibrational frequencies. The narrowed inhomogeneous broadening linewidths observed at higher temperatures are consistent with this interpretation.

Likewise, as $T_A \rightarrow 0$, the attractive force correlation time will also broaden the homogeneous linewidth as translational diffusion causes population interconversion between the various local density sites at distinct vibrational frequencies. In analogy with T_1, this "population loss" effect on the T_2 time for an individual vibrational isochromat was modeled according to:

$$2/T_2 = 2/T_2^* + 1/\tau \qquad (3)$$

where T_2^* is the pure dephasing time and $\tau \propto 1/D$. The model's approximate fit to the measured T_2 times shown in Figure 3 was accomplished using the $T_2^*/2$ values scaled according to the hydrodynamic dephasing theory [13] given in Figure 3 and τ values scaled according to measured values of D [14] using τ= 4 psec at 70° C. We note that this τ is in excellent agreement with the calculated $T_A \geq 2.3$ psec at 70° C evaluated using Eqn. 2 with $\sigma = 4.3 \times 10^{-8}$ cm and D= 8.3×10^{-5} cm^2/sec.

References

1. A. Laubereau and W. Kaiser, Rev. Mod. Phys. 50, 607(1978).
2. S.M. George, A.L. Harris, M. Berg and C.B. Harris, to be published.
3. A. Laubereau, G. Wochner and W. Kaiser, Phys. Rev. 13A, 2212(1976).
4. A. Laubereau, G. Wochner and W. Kaiser, Chem. Phys. 28, 363(1978).
5. C.B. Harris, H. Auweter and S.M. George, Phy. Rev. Lett. 44, 737(1980).
6. S.M. George, H. Auweter and C.B. Harris, J. Chem. Phys. 73, 5573(1980).
7. G.I. Kachen, Ph.D. thesis, UCRL-51753, Lawrence Livermore Laboratory, Univ. of Calif., Livermore, CA 94550.
8. G.I. Kachen and W.H. Lowdermilk, Phys. Rev. 14A, 1472(1976).
9. W. Zinth, H.J. Polland, A. Laubereau and W. Kaiser, Appl. Phys. B26, 77(1981).
10. J. Yarwood, R. Ackroyd, K.E. Arnold, G. Doge and R. Arndt, Chem. Phys. Lett. 77, 239(1981).

11. D.W. Oxtoby, Adv. Chem. Phys. 40, 1(1979).
12. K.S. Schweizer and D. Chandler, J. Chem. Phys. 76, 2296(1982).
13. D.W. Oxtoby, J. Chem. Phys. 70, 2605(1979).
14. R. Landau and A. Wurflinger, Ber. Bunsenges. Phys. Chem. 84, 895(1980). κ_T values calculated using the Tait equation were obtained from A. Wurflinger.

A Picosecond CARS-Spectrometer Using Two Synchronously Mode-Locked CW Dye Lasers

J. Kuhl
Max-Planck-Institut für Festkörperfoschung,
D-7000 Stuttgart, Fed. Rep. of Germany

D. von der Linde
University of Essen,
D-4300 Essen, Fed. Rep. of Germany

Time resolved coherent anti-Stokes Raman spectroscopy has proven to be a powerful tool for studying the dephasing of molecular vibrations /1/. Excitation and delayed probing is usually performed with single high power picosecond pulses from Nd glass lasers. Up to now widespread application of this technique to study the decay of coherent optical phonons in solids was excluded by the lack of suitable sources for simultaneous generation of picosecond pulses with a material dependent difference frequency. In this paper a CARS-spectrometer with two synchronously mode-locked cw dye lasers parallely pumped by a mode-locked Ar^+ laser is described. It has been applied to explore the decay of coherent lattice vibrations in GaP and the dephasing of molecular vibrations in Toluene.

The experimental set up is as follows: The output of the ion laser (700 mW average power) is split into two equally intense pump beams. Both dye lasers are equipped with output couplers of 80 % reflectivity and two plate birefringent filters. Each laser generates a pulse train with 76.5 MHz repetition rate, average power of 30-40 mW and spectral band width of 6-8 cm^{-1}. The system permits synchronous generation of ps pulses at two wavelengths and continuous tuning of the difference frequency between 0 and about 4000 cm^{-1}, if Rhodamine 110, Rhodamine 6G and DCM are employed as laser dyes. Thus practically any Raman active mode in gases, liquids or solids can be coherently excited. In spite of their potential for time resolved spectroscopy experiments with such systems are rare and the time constants measured so far were several 10 ps long /2-5/. This may be explained by the considerable problems inherent in the synchronisation of two independently pumped lasers.

The dye laser pulses were analyzed measuring auto- and cross correlation functions by background free second harmonic and sum frequency generation. In addition phase matched four wave mixing of the two laser outputs (frequencies ν_1 and ν_2) according to $2\nu_1 - \nu_2 = \nu_3$ due to the third order nonlinear susceptibility $\chi^{(3)}$ in a GaP crystal was investigated. Fig. 1 shows a semilog plot of the autocorrelation, cross correlation and four wave mixing function versus time delay Δt. The signals can be traced over 6 orders of magnitude. The autocorrelations at the two different wavelengths (ν_1 = 17354 cm^{-1}, ν_2 = 17020 cm^{-1} for these measurements) have the same FWHM τ_p = 3.9 ps corresponding to a pulse duration t_p = 2.6 ps ($\tau_p/t_p \approx 1.5$) for the asymmetric profile described below. The almost noiseless cross correlation has a FWHM as short as 6.5 ps. This is about 20 % narrower than cross correlations obtained with similar systems so far /2/. Such a high synchronisation requires a highly stable pump source and careful optimisation and critical alignment of the dye lasers. Amplitude fluctuations of the pump pulse train, length variations of the dye laser cavities due to bubbles and thermal fluctuations in the jet, and mechanical instabilities of the cavity are the main sources of jitter. If the cavity length of one dye laser is changed by 1 µm the emit-

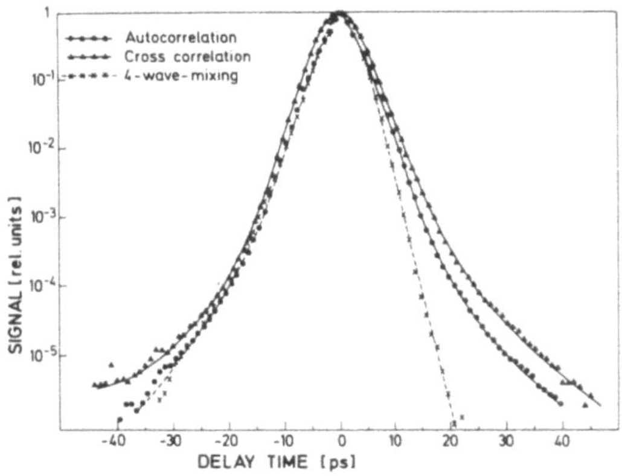

Fig. 1 Autocorrelation, cross correlation and four wave mixing traces of the dye laser pulses.

ted pulse train is shifted by 5-10 ps with respect to the train of the second laser. Similar shifts have been observed if the pump power for both lasers was varied by ± 10 %.

The four wave mixing signal exhibits a distinct asymmetry which reflects the different slopes of the rising and falling edges of the laser pulses. Deconvolution of this profile provides information about the pulse shape not attainable from second order correlation functions. Computer simulations have shown that the pulses are best described by a profile composed of hyperbolic secant functions with 0.5 - 1 ps and 4 - 5 ps FWHM for the leading and trailing edges of the pulse, respectively. The jitter has been considered in these calculations by assuming a normal probability distribution for the position of pulse ν_2 around the delay setting. The best fit to the experimental data is achieved with 4 ps FWHM for this distribution. The time bandwidth product for the dye laser pulses of ≈ 0.5 is close to the bandwidth limited case and indicates well mode-locked operation. The decay of the four wave mixing signal yields a lower limit of 1.3 ps for the resolution attainable with our present laser system in CARS experiments. Fig. 1 demonstrates that this is considerably shorter than the temporal resolution obtained with the same lasers in excite and probe experiments which depends on the shape of the cross correlation between the two pulses. The much faster decay of the CARS-signal is explained by the higher order nonlinearity and the asymmetry of the pulses.

For time resolved CARS-spectroscopy of a vibrational mode with frequency ν_0 and wavevector $\vec{q}$ the lasers have to be tuned to ν_L and $\nu_S = \nu_L - \nu_0$. The output of the laser operating at ν_L is split into a pump $(\nu_L, \vec{k}_L)$ and a probe beam $(\nu_L, \vec{k}_p)$ of comparable intensity. For coherent excitation the pump pulse is superimposed on the pulse of the second laser within the sample and the subsequent decay of the coherent vibrational amplitude is investigated by the delayed probe pulse. The beam geometry has to serve for spatial and temporal overlap in the excitation volume as well as for wavevector conservation $\vec{k}_{AS} = \vec{k}_L + \vec{k}_p - \vec{k}_S$. The CARS-signal at frequency ν_{AS} is recorded as a function of the delay time Δt between the probe pulse and the two pump pulses by a standard photon counting Raman spectrometer. In Fig. 2 the anti-Stokes count rate obtained from GaP at room temperature is displayed versus Δt for three different settings of ν_S. The trace marked by triangles is measured if $\nu_L - \nu_S$ is equal to the LO-phonon frequency ν_{LO}. The rapidly changing signal near $\Delta t = 0$ is mainly due to the nonresonant mixing of the three laser pulses by $\chi^{(3)}$ due to the electronic levels. This nonresonant part of the signal is very strong because the photon energies are close to the band gap of GaP.

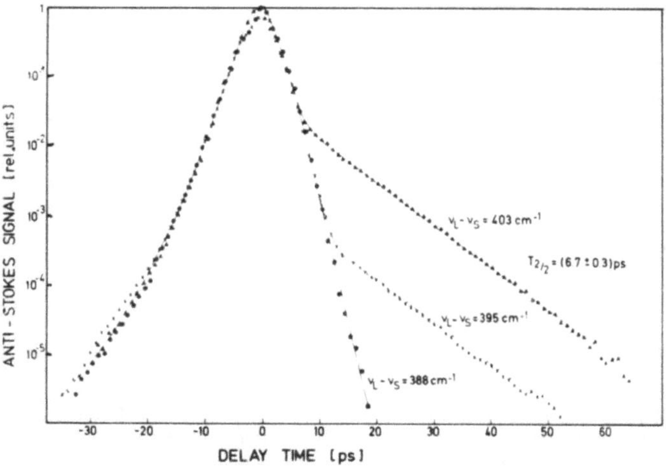

Fig. 2 CARS-signal measured in GaP at room temperature as a function of the delay time for different values of the difference frequency between the two pump lasers.

The exponential decay with a time constant $T_2/2 = (6.7 \pm 0.3)$ ps observed for $\Delta t > 10$ ps must be ascribed to the decay of coherently excited LO phonons. This interpretation is confirmed by the other two curves, which have been obtained for off resonant excitation. Detuning of ν_S by 8 cm^{-1} (curve 2, marked by crosses) leads to a smaller phonon population and consequently to a lower resonant signal. If $\nu_L - \nu_S$ deviates by more than 15 cm^{-1} from ν_{LO} no slow exponential decay is detectable (curve 3, marked by circles). When the sample is cooled to 5 K $T_2/2$ increases to (26.0 ± 2.5) ps. A temperature variation from 0 to 300 K should roughly double the relaxation rate if the decay of the LO phonon into two acoustic phonons of equal but opposite momentum is the dominant contribution to the finite phonon lifetime in GaP /6/. The present experiments give twice this value. Additional measurements of the energy relaxation time T_1 of the LO lattice mode by the technique which we have recently applied to GaAs /7/ and detailed investigations of the temperature dependence of T_2 are in progress in order to identify different processes contributing to the observed dephasing.

Further experiments have been performed to study the decay of coherent polaritons close to the pure TO lattice mode of GaP. For all investigated modes with energies between 338 and 365 cm^{-1} the time-resolved CARS-signal has the same shape and width as curve 3 in Fig. 2. Therefore we have to conclude that $T_2/2$ for these polaritons in GaP at 300 and 5 K is shorter than the resolution of 1.3 ps. This result agrees quite well with the damping parameters derived from line shape analysis of spontaneous Raman data /8,9/. In contrast a lifetime of (5.5 ± 0.5) ps was found for the ($\nu_0 = 361$ cm^{-1}, $\vec{q} = 2770$ cm^{-1}) - mode in an earlier CARS experiment /10/. This discrepancy is not understood at the moment. Accelerated decrease of the coherent signal due to the frequency distribution of the excited vibrational systems can be ruled out. For $\nu_0 = 361$ cm^{-1} our experimental configuration leads to an inhomogeneous broadening of 0.8 cm^{-1} which is not sufficient to explain the rapid decay.

Experimental results obtained for the dephasing times of several modes in liquids are summarized in Tab. 1. The dephasing of the 784 cm^{-1} and 1002 cm^{-1} modes of Toluene has been resolved for the first time. Within the experimental errors $T_2/2 = (2.6 \pm 0.4)$ ps is the same for both modes. This result agrees quite well with the value calculated from the width of the 1002 cm^{-1} Raman line.

Tab. 1: Measured dephasing times of several molecules and modes

	frequency [cm^{-1}]	$T_{2/2}$ [ps]
CCl_4	459	2.8±0.5
CS_2	656.6	10±0.5
C_7H_8	1002	2.6±0.4
C_7H_8	784	2.6±0.4
$CHCl_3$	360	<1.3
CH_3COCH_3	785	<1.3
$C_4H_8O_2$	830	<1.3

In summary, we have described a CARS-spectrometer with time resolution as short as 1.3 ps. It is distinguished by the independent broad tunability of the laser pulses and thus provides the ability to excite many different Raman active modes within the same material. In spite of the low peak power the quasi cw measurement techniques provide a dynamic range of 5 - 6 orders of magnitude. The performance has been demonstrated by the first measurement of the decay of coherent LO phonons in GaP and of the dephasing of two vibrational modes in Toluene.

We are indebted to N. Stath, Siemens AG, Regensburg, for supplying the GaP crystal. The expert technical assistance of A. Jonietz, W. König, H.J. Kühne, and R. Krause is gratefully acknowledged.

REFERENCES

/1/ A. Laubereau and W. Kaiser, Rev. of Modern Physics Vol. 50, 607 (1978).
/2/ J.P. Heritage, Appl. Phys. Lett. 34, 470 (1979).
/3/ N.P. Economou, R.R. Freeman, J.P. Heritage, and P.F. Liao, Appl. Phys. Lett. 36, 21 (1980).
/4/ R. Kaarli, Ya. Aaviksoo, A. Freiberg, and P. Saari, Toim. Eesti NSV Tead. Akad. Fuus. Mat. (USSR) Vol. 29, 181 (1980).
/5/ M.C. Sceats, F. Kamga, and D. Podolski in Picosecond Phenomena II, ed. by R.M. Hochstrasser, W. Kaiser, and C.V. Shank, Springer Verlag Heidelberg (1980).
/6/ P.G. Klemens, Phys. Rev. 148, 845 (1966).
/7/ D. von der Linde, J. Kuhl, and H. Klingenberg, Phys. Rev. Lett. 44, 1505 (1980).
/8/ A.S. Barker, Phys. Rev. 165, 917 (1968).
/9/ S. Ushioda and J.D. McMullen, Solid State Commun. 11, 299 (1972).
/10/ A. Laubereau, D. von der Linde, and W. Kaiser, Opt. Comm. 7, 173 (1973).

Picosecond Studies of Intramolecular Charge Transfer Processes in Excited A-D Molecules

H. Staerk, R. Mitzkus, W. Kühnle, and A. Weller

Max-Planck-Institut für biophysikalische Chemie,
D-3400 Göttingen, Fed. Rep. of Germany

1. Introduction

In compounds of the type A-$(CH_2)_n$-D (abbreviated AnD) photo-induced intramolecular charge transfer processes between electron donor (D) and acceptor (A^*) molecules produce in a picosecond to nanosecond time range the following high energy intermediates: exciplexes $^1(A^-nD^+)$, radical ion pairs $^2A_s^-n^2D_s^+$ and triplets $^3A^*nD$ [1-3].

The objective of the present work was to study the role of solvent and chain length on the dynamics of intramolecular electron transfer fluorescence quenching of the compounds shown in Fig.1.

2. Measurement Technique

We use a frequency-tripled (354 nm) picosecond laser pulse from a mode-locked Nd/YAG laser (Quantel) to excite the sample, and a modified GEAR streak camera with ITT streak tube to monitor the fluorescence of the sample (S) which is placed in front of an adjustable precision slit. The set-up is reproduced schematically in Fig.2. This also shows the function of an Auston-switch (Si) [4,5] made of intrinsic silicon which is used in some applications to gate the proximity-focused channel plate behind the photocathode [6]. The streak image is amplified by the channel-

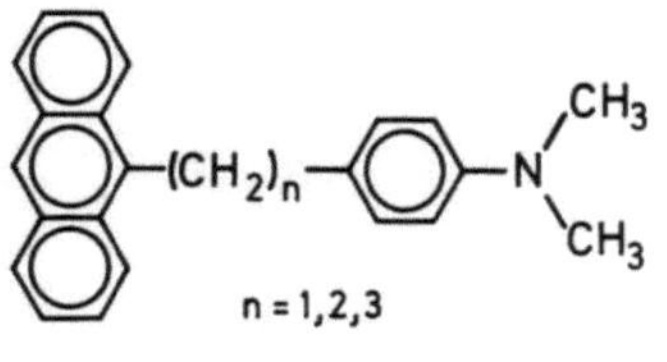

Fig.1 AnD, n = 1,2,3. Anthracene is the initially excited moiety

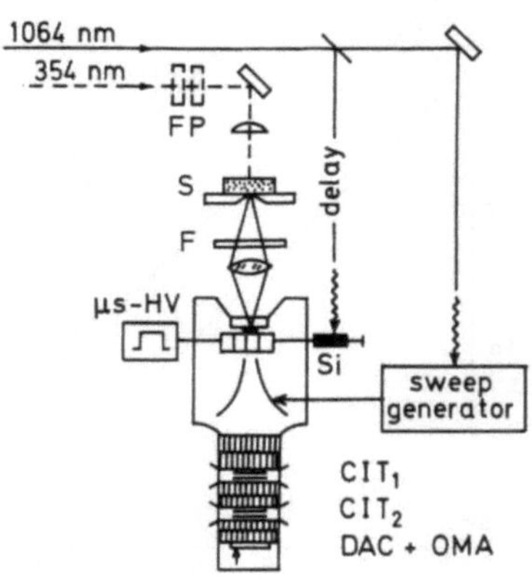

Fig.2 Streak camera arrangement with diode array camera (DAC + OMA)

plate intensifier tube CIT_1 (ITT) and the intensifier tube CIT_2 (Proxitronic) and is finally detected by a diode array camera DAC-OMA (B & M Spektronik) or a SIT-vidicon. The use of SIT-vidicon-OMA systems has been critically examined recently [7]. Streak camera signals have been digitally corrected with the flat-field function of the apparatus and with the sweep-time functions obtained by Fabry-Perot (FP) etalon calibration.

3. Results and Discussion

Fig.3 shows a log-log plot of the fluorescence quenching rate constant $k_q = (1/\tau - 1/\tau_o)$ against solvent viscosity η for the compounds $\overset{*}{A}nD$ dissolved in the linear alcohols $C_mH_{2m+1}OH$, $m = 1-8$. $\tau_o = 6-8$ ns is the lifetime of the unquenched acceptor viz. $^1\overset{*}{A}-CH_3$ or $^1\overset{*}{A}-(CH_2)_n-$⟨◎⟩ . τ is the measured $^1\overset{*}{A}nD$-fluorescence lifetime which, within experimental accuracy, decays exponentially down to pulse-width-limited decay times of about 10 ps.

The following facts are obvious from the experimental results:

(i) While keeping all other conditions constant, the quenching rate increases with decreasing chain length connecting the moieties A and D. (ii) The quenching rate constant k_q is inversely related to the solvent viscosity η. (iii) The ratio ϕ/τ of fluorescence quantum efficiency and lifetime are approximately of the order $5 \times 10^7\ s^{-1}$. (iv) No simple correlation is observed between the fluorescence lifetimes and the dielectric relaxation times as measured with microwave techniques [8].

In other experiments we compared the fluorescence decay times of AnD in propylene glycol ($\tau(^1\overset{*}{A}3D) = 110$ ps, $\tau(^1\overset{*}{A}2D) \approx 50$ ps, $\tau(^1\overset{*}{A}1D) \approx 15$ ps) with those of n-propanol ($\tau(^1\overset{*}{A}3D) = 130$ ps, $\tau(^1\overset{*}{A}2D) = 66$ ps, $\tau(^1\overset{*}{A}1D) = 22$ ps). Both solvent molecules have approximately the same size, but the macroscopic viscosities and dipole moments μ_s (prop-glyc) = 2.50 D, μ_s (n-PrOH) = 1.75 are very different.

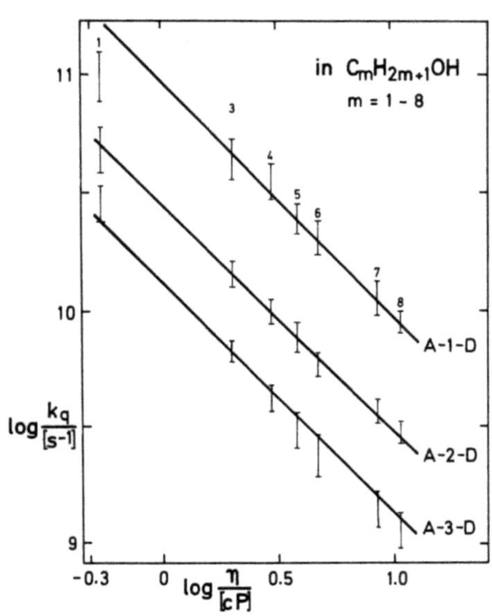

Fig.3 Variation of the observed fluorescence quenching rate constants k_q with viscosity η of alcohol solvents.
Solid lines: $d\log k_q / d\log \eta = -1$

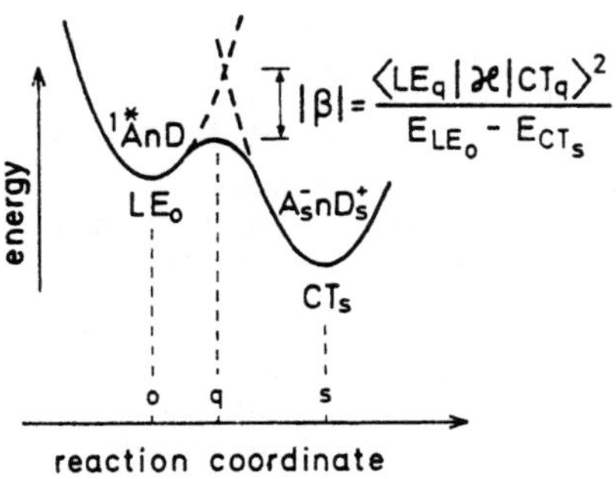

Fig.4 Transition between the locally excited state LE_o and the charge transfer state CT_s in charge transfer reactions.

All observations support a dynamic quenching mechanism to be predominant. The energetic interrelation between the locally excited state LE_o and the charge transfer state CT_s is indicated in Fig.4 (O, q, and s give the position along the reaction coordinate).

We interpret the observed viscosity dependence of the quenching rate constant as due to the dynamics of both solvent and solute, contributing to the development of a molecular configuration where charge transfer is possible.

While fluorescence decay functions were single exponentials for AnD in the alcohols (where $\Delta G \approx 0.5$ eV of free energy change is involved in the electron transfer reaction) biphasic fluorescence decays have been observed with A1D and A2D in the low polarity solvents toluene and diethylether. A change of free energy of only $\Delta G = 0.12$ eV and 0.08 eV respectively has been determined for exciplex formation in toluene [3,9]. The time-resolved picosecond experiments have been evaluated by fitting the standard solutions of the system of coupled differential equations describing the equilibrium situation, thus providing the forward and backward rate constants (k_c, k_c') associated with the charge transfer reaction [9].

A remarkable observation is the drastically enhanced intersystem crossing found with compound A1D, where the two moieties are close but not parallel. In the nonpolar solvents MCH and n-hexane, the exciplex is not stable, as a result of the Coulomb interaction, which is lower for A1D and A2D than for A3D because of the larger separation of the charges. The triplet yields in MCH (0.68) and n-hexane (0.63) are solely associated with intersystem crossing from the locally excited state $^1\overset{*}{A}1D$. The triplet yields found in the solvents toluene (0.85), diethylether (0.88), dimethoxyethane (0.88), tetrahydrofurane (0.93), n-propanol (0.43) and acetonitrile (0.30) are considered to be due to intersystem crossing from the exciplex, since it was verified [3] that the exciplex fluorescence decay time matches the rise time of the triplet absorption. A similar observation has been made recently by OKADA et al. [10] for (1-pyrenyl)-(CH_2)-(amines). It is important to note that even in the highly polar solvent acetonitrile, where the formation of solvated ion pairs is expected, the triplet yield is 30 per cent (while 1 per cent was found for A2D and <0.1 per cent for A3D). In the latter cases a triplet transition in the ion pair due to hyperfine modulation in the individual ions (as we usually find it in separated systems [11]) is ruled out because the two moieties in the $^1(^2A^- n ^2D^+)$ molecule with overall singlet multiplicity cannot separate far enough to yield the very low exchange interaction necessary for the dephasing process of the precessing free electron spins to occur.

Finally, we wish to report briefly on the kinetic work we have done on A3D in a larger solvent polarity range. In contrast to the results published by others [12] we find a fluorescence lifetime decrease by a factor of ≈150 on going from n-hexane to acetonitrile. There is a drastic increase of the quenching rate constant in the medium polarity range. This can be readily correlated with the free energy change ΔG of the states involved [2,9], where the reaction is assumed to proceed in two steps $^1\overset{*}{A}3D \xrightarrow{1} {}^2A^- 3 {}^2D^+ \xrightarrow{2} {}^1(A^- 3D^+)$, i.e. the ion pair formation is followed by the Coulomb attraction of A^- and D^+ to yield the sandwich exciplex [2]. We call this process "harpooning".

References

1 M. Schulz, H. Staerk and A. Weller in "Excited States of Biol. Molecules", ed. J.B. Birks, Proceedings Internat. Conference, Lisbon (1974), J. Wiley & Sons, London.
2 M. Schulz, Doctoral Thesis, Universität Göttingen (1974).
3 J. Fuhrmann, Doctoral Thesis, Universität Göttingen (1981).
4 D.H. Auston, Appl.Phys.Lett. 26 (1975) 101.
5 G. Mourou and W. Knox, Appl.Phys.Lett. 35 (1979) 492.
6 R. Mitzkus, H. Meyer and H. Staerk, Rev.Sci.Instr., to be published.
7 H. Staerk, R. Mitzkus and H. Meyer, Appl.Optics 20 (1981) 471.
8 S.K. Garg and C.P. Smyth, J.Phys.Chem. 69 (1965) 1294.
9 J. Fuhrmann, M. Schulz, H. Staerk and A. Weller, to be published.
10 T. Okada, I. Karaki, E. Matsuzawa, N. Mataga, Y. Sakata and S. Misumi, J.Phys.Chem. 85 (1981) 3957.
11 K. Schulten, H. Staerk, A. Weller, H.-J. Werner, B. Nickel, Z.Phys.Chem. N.F. 101 (1976) 371.
H.J. Werner, H. Staerk and A. Weller, J.Chem.Phys. 68 (1978) 2419.
12 T. Okada, T. Fujita, M. Kubota, S. Masaki, N. Mataga, R. Ide, Y. Sakata and S. Misumi, Chem.Phys.Lett. 14 (1972) 563.

Femtosecond Transient Birefringence in CS_2

B.I. Greene and R.C. Farrow

Bell Laboratories, Murray Hill, NJ 07974, USA

With the technological achievement of stable well characterized optical pulses in the 100 femtosecond time regime, [1] it seems likely that many previously studied systems will now be reexamined with heightened time resolution. Optimistically, such efforts will go beyond tracing previously determined response functions closer to their time origins. With fully an order of magnitude increase in time resolution over most previous picosecond studies, qualitatively new results bearing on ultrafast relaxation mechanisms in solids, liquids and gases should be expected.

Recently, with a time resolved interferometric technique utilizing ultrashort light pulses from a CPM laser, a measurement of transient refractive index changes in CS_2 has revealed a novel 0.33 ps response in addition to the well known ca. 2.0 ps relaxation [2]. If this new result proves correct, it might have significant bearing on an understanding of the physical nature and origin of nonlinear molecular polarizabilities as well as provide information pertinent to the dynamical structure of complex liquids. It is for this reason that we have undertaken, by an alternate technique, to independently verify the results of HALBOUT and TANG.

In the present study we repeat a by now classic picosecond measurement of the relaxation rate of an optically pumped CS_2 Kerr gate [3,4,5]. We utilize the Kerr gate configuration rather than the interferometric method of HALBOUT and TANG as the two methods should yield consistent results. Particular care is taken to establish a precise time delay origin. This is of critical importance when one wishes to discuss the possible contribution of a "coherent artifact" [6] in the measured sample response.

A schematic of the experimental optical apparatus is shown in Fig.1. Linearly polarized pulses (0°) derived from a passively model-locked RGG/DODCI CPM laser [1] enter the diagram in the upper right hand corner. The pulses are 150 fs in duration as measured by dividing the autocorrelation function FWHM by a factor of 1.5, contain approximately 0.5×10^{-9} J, and occur at a repetition rate of 110 MHz. The pulse stream is divided, one being mechanically chopped at 1 KHz and sent down a variable delay line, the other being passed through a polarizer oriented at 45° with respect to the original plane of polarization. At the sample, a 1 mm pathlength fused silica cuvette containing spectrophotometric quality CS_2, the chopped delay beam (pump) has an average power of 6 mw, which has been adjusted to be a factor of 35 greater than the probe beam. Both beams are focussed through a single 5.0 cm focal length lens into the sample. After the sample, the pump beam is blocked, while the probe beam is recollimated and passed through a quarter waveplate and a polarizer that has been oriented at an angle of -45° with respect to the laser output polarization (0°). The waveplate is adjusted to circularly polarize the probe beam before it is polarization analyzed and detected

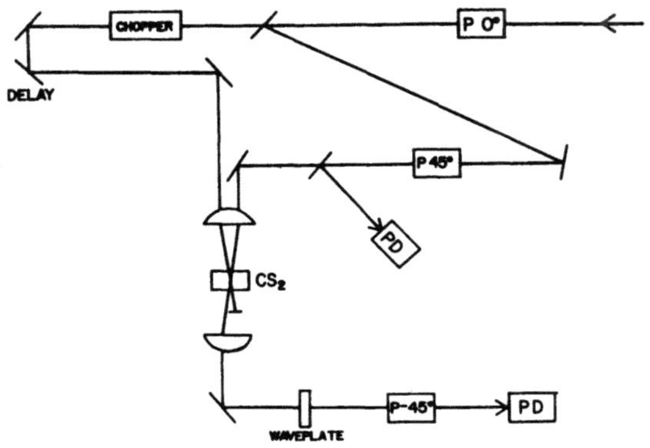

Fig.1 Schematic of Kerr gate apparatus: P, polarizer; PD, photodiode. 150 fs pulses enter from upper right hand corner.

by a photodiode. A lock-in amplifier is utilized in the differential mode, subtracting signals from two identical photodiodes looking at the probe beam before and after it passes through the sample. Peak signals corresponding to modulations of the probe beam of approximately 0.05% were observed. The delay line was driven by an encoded motorized micrometer and controlled by an Apple II computer, which simultaneously digitized the analogue signal. A waveplate is utilized to bias open the Kerr gate in order to ensure, in the weak signal limit, a linear signal response. The average of four single scan data traces is shown in the top half of Fig.2.

To establish the delay time corresponding to precise pump and probe pulse overlap, an angle tuned KDP crystal was situated in the sample position. Absolutely no change in the beam configuration before the sample was made. A background free crosscorrelation trace appears in the bottom half of Fig.2, and its peaks defines the t=o position.

Of key concern in an evaluation of the above data is the question of to what extent, if any, is a coherent artifact present in the signal. In general, if such an effect is present to an appreciable degree, it must be subtracted out before evaluation of a sample response can be made. This holds particularly true in the present case of CS_2, where distortion to a fast sample response (comparable to the pulse width) would be severe.

Our data argues against the presence of such an artifactual signal. As has been well established, [6,7] the coherent artifact signal will always appear symetrically disposed about the time origin, and to first simple approximation have the functional form of the crosscorrelation of the two pulses. When we attempt to algebraically subtract out of the data even small fractions (ca. 10%) of the normalized experimentally derived crosscorrelation, a clearly nonphysical derived rise in the response curve results. Additionally, such a small fractional subtraction is seen to have a negligible effect on the detailed shape of the decay curve.

We conclude therefore that the relaxation of induced birefringence in CS_2 exhibits a fast 0.24±.02 ps component and a slower 2.16±.1 ps component. The data can be fit to the functional form $Ae^{-t}/\tau_1 + Be^{-t}/\tau_2$. The fast relaxation only contributes significantly to the first 30% of the response curve (see Fig. 2). Therefore, by fitting the tail of the response curve (t>1 ps) to a single exponential and subtracting the result from the original data, a reli-

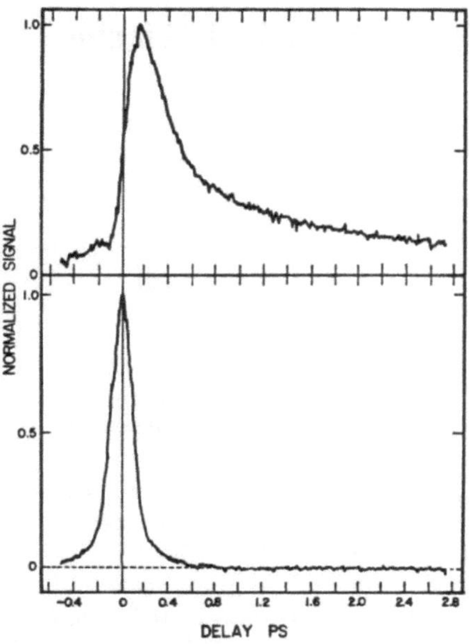

Fig.2 Demodulated signals. (top): pump induced transmission through Kerr gate vs pump/probe delay. (bottom): pump/probe crosscorrelation function, 225 fs FWHM

able fit of the faster response is obtained. The values of A and B were determined to be 0.65 (fast component) and 0.35 (slow component) respectively. The data was not deconvoluted to correct for the finite laser pulse width. It should be noted that the detailed sample response will depend both on the laser pulse width and the material damping factors. Our results are therefore considered to be in good agreement with those of Halbout and Tang [2].

It is generally accepted that the origin of the 2.1 ps response is rotational diffusion in the liquid. The origin of the ca 0.2 ps response is by no means certain. This timescale however is close to that of collisional events in liquids. It seems likely therefore that future investigation and speculation will explore relationships between the detailed microscopic nature of CS_2 liquid and the presently observed ultrafast birefringent response.

References

1. R. L. Fork, B. I. Greene, and C. V. Shank, Appl. Phys. Lett. 38, 671 (1981).
2. J. M. Halbout and C. L. Tang, Appl. Phys. Lett. 40, 765 (1982).
3. M. A. Duguay and J. W. Hansen, Appl. Phys. Lett. 15, 192 (1969).
4. E. P. Ippen and C. V. Shank, Appl. Phys. Lett. 26, 92 (1975).
5. P. P. Ho and R. R. Alfano, Phys. Rev. A 20, 2170 (1979).
6. E. P. Ippen and C. V. Shank, "Ultrashort Light Pulses" ed. by S. L. Shapiro (Springer, Berlin, Heidelberg, New York) p. 83.
7. Z. Vardeny and J. Tauc, Opt. Com. 39, 396 (1981).

Time-Resolved Observation of Molecular Dynamics in Liquids by Femtosecond Interferometry

C.L. Tang and J.M. Halbout

Cornell University, Ithaca, NY 14853, USA

Results on the relaxation of molecular rotations in various organic liquids obtained using subpicosecond interferometry [1] are discussed in this talk.

The laser used was a colliding-pulse mode-locked [2] rhodamine 6G ring dye laser which produces a train of highly stable pulses. The shortest pulse achieved with the laser was 60 fs (Fig. 1) with a peak power of 5 KW and a pulse repetition rate of 10^8 s^{-1}.

The first liquid we studied was CS_2 since it is a relatively well-known simple linear molecule. Until now most of the information on the relaxation of molecular rotations in the liquid state has come from Rayleigh scattering or electron spin resonance studies. However, the interpretation of the data is complicated because the relaxation process is not governed by a single exponential relaxation process. The Rayleigh line shape, for example, is known to consist of three regions [3]. Nearest to the incident line, the shape is Lorentzian and the width is determined by the Debye relaxation time. Further out in the wings, the intensity first flattens out and then drops off rapidly again in a third region. The Debye relaxation time for CS_2 is known to be on the order of 2 ps. In the large-frequency wing region, because the intensity is low and the different regions merge continuously into each other, it is difficult to obtain accurate quantitative information from the Rayleigh scattering data. It is only known qualitatively that the boundary of the first and second regions corresponds approximately to 0.5 to 1 ps and the third region begins at approximately 70 fs [3]. Since the large-frequency region of the Rayleigh-wing corresponds to the short-time domain in transient studies where the signal is strong, for sub-picosecond dynamics time-domain transient studies are more suitable.

With the femtosecond lasers, it is now possible to suddenly realign the molecules through the optical Kerr effect and then observe the molecular

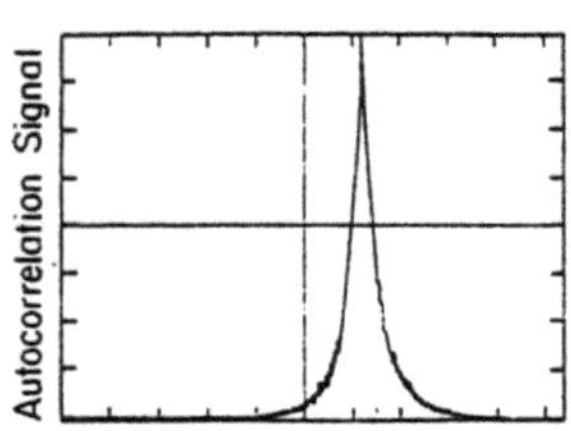

Figure 1 Autocorrelation trace of 60 fs (FWHM) mode-locked laser pulse

Work supported by the National Science Foundation through the Materials Science Center of Cornell University, Ithaca, NY 14853.

relaxation. Realignment of the molecules leads to an anisotropy in the refractive index of the liquid proportional to the averaged angular rotations of the molecules. By monitoring the time evolution of the index-change following a short pulse excitation, one can then follow the relaxation of the molecular rotation back to equilibrium. In our experiment, a Mach-Zehnder interferometer was used with the sample in one of the arms as described in [1]. Although the scheme is conceptually very simple, because the light-induced index-change is relatively small, the interferometer must be stabilized by feedback control. The minimum phase shift $\Delta\phi$ measurable in our initial system is approximately 10^{-4} rad. At this level, the sensitivity of the system is comparable to that of the optical Kerr gate [4]-[6] type of measurement. However, the sensitivity of the interferometer can still be improved by perhaps two orders of magnitude and is, therefore, potentially more sensitive than the Kerr gate for this type of studies.

Figure 2 gives the measured relative phase-shift of the CS_2 sample as a function of the delay between the pump and probe pulses. Since the details of the experiment on CS_2 have already been given in [1], here we concentrate on the interpretation of the results. First to be noticed is that the decay of the phase-shift deviates from that of the cross-correlation trace almost from the peak indicating very little electronic contribution to the third order nonlinearity of CS_2 consistent with Hellwarth's suggestion [5]. Also because the pulse length (70 fs for this trace) was much shorter than the known vibrational relaxation time of CS_2 (on the order of 20 ps), the excitation pulse was too short to cause any distortion of the molecules via the electronic polarizability; the corresponding vibrational relaxation component is not present in the decay as expected. The observed relaxation of the light-induced index-change is, therefore, primarily due to molecular rotations. There are, however, clearly two rotational relaxation components. Numeri-

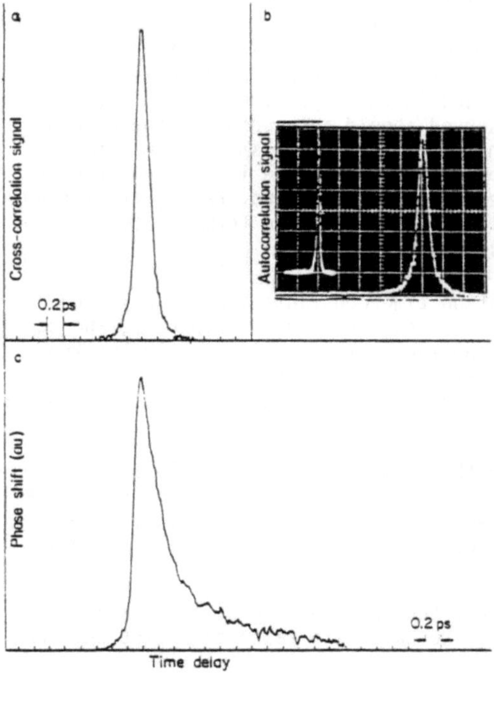

Figure 2 a) Cross-correlation between the pump and probe pulses; FWHM 190 fs. b) Auto-correlation of the laser pulse. Upper trace: 1 ps/div; lower trace: 0.2 ps/div. FWHM 110 fs. c) Light-induced phase-shift vs. delay between pump and probe pulses.

cally, the observed decay outside of the range covered by the cross-correlation curve can be fitted very well by a 4-parameter double-exponential function of the form:

$$A \exp(-t/\tau_a) + B \exp(-t/\tau_b) \quad . \tag{1}$$

The best-fit numerical values of the parameters are, without taking into account the finite pulse width, $\tau_a = 0.36$ ps, $\tau_b = 2.1$ ps, A = 0.86, and B = 0.14. The standard deviation in the fitting is comparable to the experimental error due to the finite step-size ($\sim$17 fs) of the pulse delay.

Before considering the meaning of the results, let us consider first whether it is meaningful to attempt to fit the observed decay curve to a double-exponential function of the form (1). The physics of molecular relaxation in the liquid state is very complicated and not well understood. According to the model developed by Fabelinskii, Starunov, and co-workers [3] the relaxation of molecular rotations is characterized by three time regions. In the large-time region, the molecular relaxation process can be described as a rotational or anisotropy diffusion process resulting from the average of random hops by the molecules from one local orientational equilibrium position to another as in Brownian motion. The characteristic decay time is the Debye relaxation time which is proportional to the viscosity of the liquid. For times shorter than the mean hopping time, the molecule librates in a potential well centered around a local equilibrium, but this libration is damped by a frequency-dependent internal friction due to collisions with other molecules in the liquid. For times shorter than the mean hopping time but longer than the mean collision time or the relaxation time for the internal friction, the molecular libration is damped. For times still shorter, the friction is absent and the molecule librates around the local equilibrium position. However, because the local environment for each molecule is expected to be quite different in the liquid state, the librational frequencies for different molecules will be different and it is unlikely any discernable oscillations will actually be seen even in the shortest time domain for the liquid as a whole. For CS_2, the mean hopping time is estimated [3] to be about 500 fs and the relaxation time for the internal friction is about 70 fs [3]. Since the pulse length used in the experiment was 70 fs, the rotational relaxation in CS_2 we observed should correspond to the average of damped librational motions of the molecules in the short-time domain (70 fs to 0.5 ps) and the anisotropy relaxation in the long-time (greater than 0.5 ps) domain. Both are of exponential forms and, hence, it is meaningful to attempt to fit our data to a function of the form (1) even in the short-time (but greater than 70 fs) domain.

As pointed out above, two relaxation time constants are clearly seen. After deconvolution taking into account the finite pulse width, τ_a and τ_b are, respectively, 0.33 and 2 ps. The long component is clearly the known Debye relaxation time. The fast relaxation component has not been seen before. It is possible that it corresponds to the damped librational motion as discussed above.

Because the excitation pulse length is shorter than the relaxation times τ_a and τ_b, the amplitude constants A and B in Eq.(1) depend upon the pulse length because the rotations of molecules do not reach the steady-state within the pulse duration. With longer excitation pulse, B will increase relative to A.

To remove any remaining doubts that the observed signal and relaxation characteristics are intrinsic to CS_2, we have also taken extensive data on

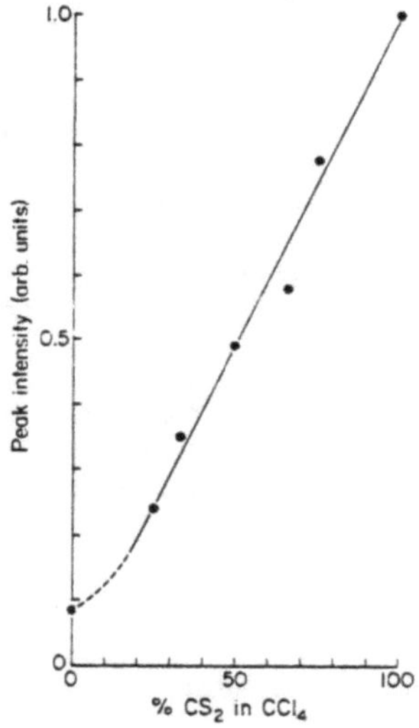

Figure 3 Measured peak signal vs. CS_2 concentration in mixtures of CS_2 and CCl_4

other liquids. The details will be reported elsewhere. Here we give a summary of the relevant results. In Fig. 3, we show similar results obtained with mixtures of CS_2 and CCl_4. The light induced index-change in CCl_4 is nearly an order of magnitude smaller than that in CS_2 for a 70 fs excitation pulse. The measured peak signal varies essentially linearly with the CS_2 concentration in the mixtures, indicating that the signal observed did come from CS_2. In Fig. 4, we show similar results obtained in nitrobenzene.

The viscosity of nitrobenzene is nearly four times larger than that of CS_2; thus, the long relaxation component should be much longer than the excitation pulse width and the corresponding amplitude coefficient B should be much smaller relative to A in nitrobenzene. The observed decay for nitrobenzene shown in Fig. 4 has indeed a very small long relaxation component. It is interesting to note that the short component is even faster than that for CS_2 indicating that the fast component observed in CS_2 is intrinsic to the liquid and not some spurious results due to the laser pulse. Because the fast relaxation component in nitrobenzene is not too much longer than the pulse width, one must be careful in assigning any specific numerical value

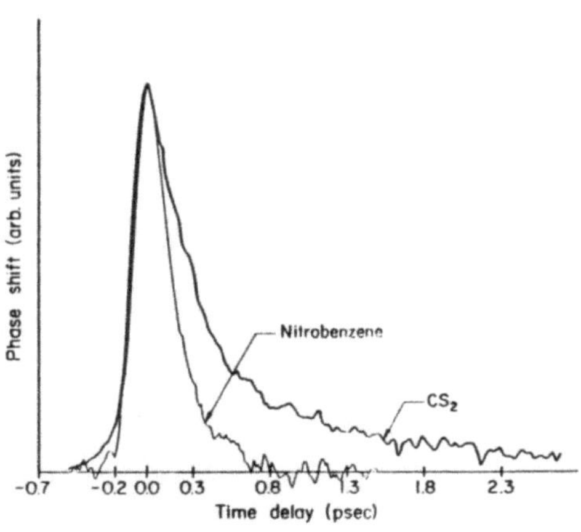

Figure 4 Measured light-induced phase-shifts in nitrobenzene and in CS_2

to the relaxation time. A careful numerical analysis is being carried out. Qualitatively, τ_a for nitrobenzene is on the order of 150 fs, which is one of the fastest relaxation times ever observed.

In conclusion, we have reported here the first direct time-domain observation of the relaxation of molecular rotations in the liquid state. Two relaxation components are observed. The long component corresponds to the Debye relaxation time. The short component is tentatively identified as due to the damped librations of the molecules within the potential well around the local orientational equilibrium. With further progress in the reduction of laser pulse width, a decay due to the superposition of undamped librational motions of the molecules may become observable. But due to the inhomogeneously broadened nature of the molecular librations in the liquid state, it is unlikely that any coherent oscillations will be seen even in this short-time domain due to the rapid dephasing of the librations of the molecules in different local environments in the liquid.

References

1. J. M. Halbout and C. L. Tang, App. Phys. Lett. 40, 765 (1982).
2. R. L. Fork, B. I. Green, and C. V. Shank, App. Phys. Lett. 38, 671(1981)
3. I. L. Fabelinskii, Molecular Scattering of Light, (Plenum Press, N.Y., 1968); V. S. Starunov, Optical Studies in Liquids and Solids, Proc. of the P. N. Lebedev Phys. Inst. (Consultants Bureau, NY 1969) 39, 147.
4. G. Mayer and F. Gires, C. R. Akad. Sci. Paris 267, 54 (1968); M. Dugay and J. Hanson, App. Phys. Lett. 15, 192 (1969).
5. R. W. Hellwarth, Prog. in Quant. Elect. (Pergamon, NY 1977), 5, p. 1.
6. B. I. Greene and R. C. Farrow, Proceedings of this Conference, have reported at this conference that they have confirmed the essential features of our initial results [1] using the independent optical Kerr measurements.

Time Resolved Measurement of Nonlinear Susceptibilities by Optical Kerr Effect

J. Etchepare, G. Grillon, R. Astier, J.L. Martin, C. Bruneau, and A. Antonetti

Laboratoire d'Optique Appliquée, Ecole Polytechnique - E N S T A,
F-91120 Palaiseau, France

With the advent of powerful and ultrafast optical pulses, many physical phenomena, non linear and time dependent in the picosecond and subpicosecond time scale, become potentially directly accessible to study.

Here, we are interested in the direct time resolved measurement of the non linear refractive index changes in liquid media. The method, a kinetics study of the Kerr effect, can be advantageously used to extract directly the different physical mechanisms responsible for the induced transient anisotropy, namely the elements of the third order optical susceptibility tensor $\chi^{(3)}_{ijk\ell}(-\nu,\nu,\omega,-\omega)$.

Our laser source [1] creates a high peak power pulse that serves three functions:1. obtain a significant transmission of the Kerr cell(1 to 10 percent range);2. simultaneously, part of the pulse can be converted to a continuum of light where we extract a narrow spectral band pulse;3. a pulse width in the subpicosecond range in order to discriminate between time dependent processes in the early picosecond region.

The experimental setup is shown in fig.1. The pump pulse (ω) is linearly polarized(polarizer P). White light continuum probe pulses are generated in the cell C_1, filled with H_2O. One of the probe pulses passes through the

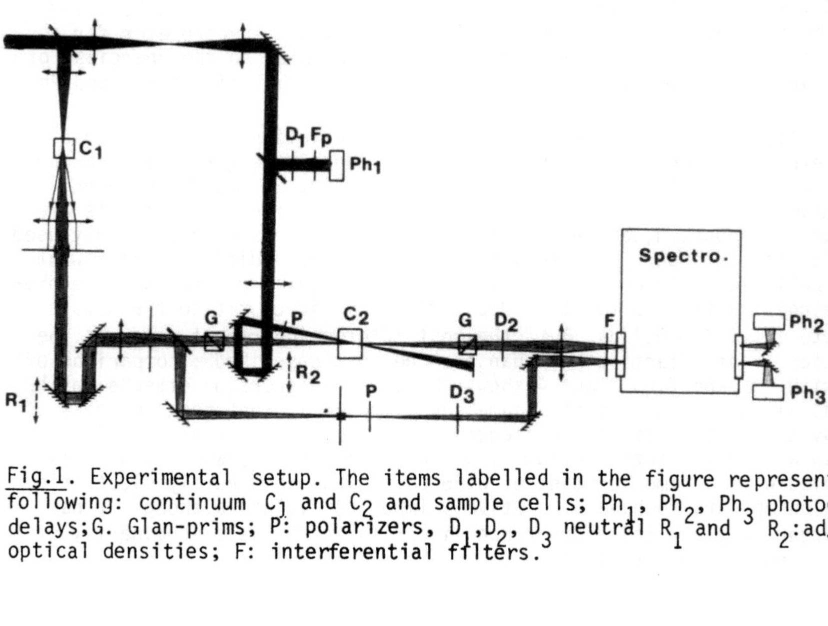

Fig.1. Experimental setup. The items labelled in the figure represent the following: continuum C_1 and C_2 and sample cells; Ph_1, Ph_2, Ph_3 photodiodes; delays;G. Glan-prims; P: polarizers, D_1,D_2, D_3 neutral R_1 and R_2:adjustable optical densities; F: interferential filters.

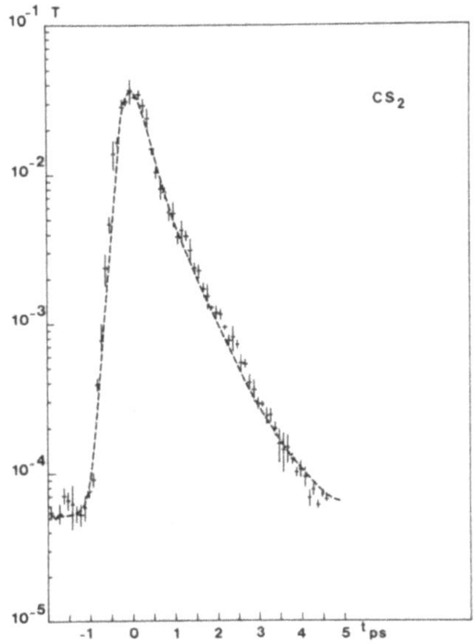

Fig. 2a
Transmitted signal (T) through a 1 cm long CS_2 cell, at various delay times t(ps).

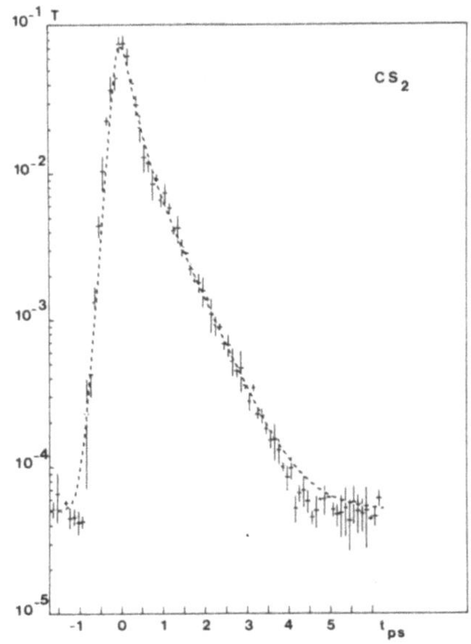

Fig. 2b
Transmitted signal through a 5 mm long CS_2 cell.

Each experimental point is an average (error bars are represented) of 3 times 5 laser shots; their spacing is of 0.1 ps. The dashed lines are numerically calculated assuming biexponential pulse shapes (see text).

sample cell C_2, located between two crossed good quality Glan-Air polarizing prisms (G); if properly aligned, the probe beam can be attenuated by a 10^5 factor. The transmission of the cell is measured, via a spectrometer for any given probe wavelength (ν), and normalized with respect to the energies, of the other probe and the pump beams measured with the photodiodes Ph_3 and Ph_1.

The results presented below seem to us characteristics of liquid media. They concern experiments always using the same two wavelengths: ω=615 nm for the pump pulse and ν=650 nm for the probe pulse. Fig. 2a and 2b represent time-resolved transmission of liquid CS_2, for which third order effects are particularly pronounced, and which has been for this reason, extensively used as a calibration standard. The results are surprisingly different from what has been published before. The reorientational molecular relaxation time process is significantly shorter than 2 ps [2]; from direct fit to the data, we estimate $\tau_{OR}=1.4 \pm 0.1$ ps. But, the most interesting result concerns the contribution of effects other than the one just described : comparison of fig. 2b with 2 a and figs 3 and 4 shows that, when dispersion effects may be neglected, all kinetics present a same superimposed instantaneous process, limited by the pulse width and consequently of purely electronic origin. There also we are in contradiction with recent results [3] where a process was observed with a time constant of 0.36 ps. Third, this purely electronic contribution appears to have a relatively important value, although it has been inferred that the third order non linearity in CS_2 is predominantly of nuclear (or molecular) origin [4,5].

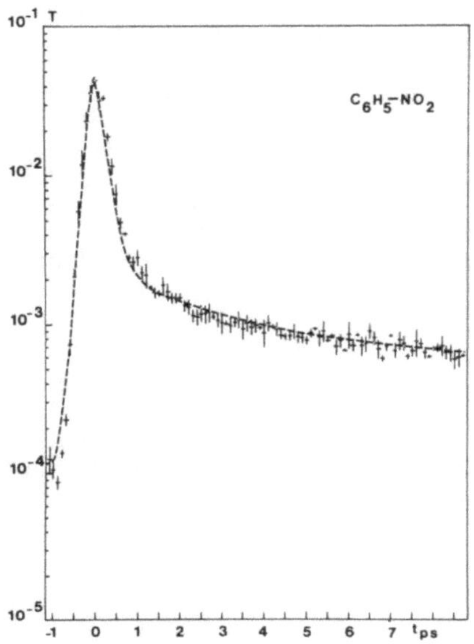

Fig.3 : transmitted signal through 1 cm long nitrobenzene cell.

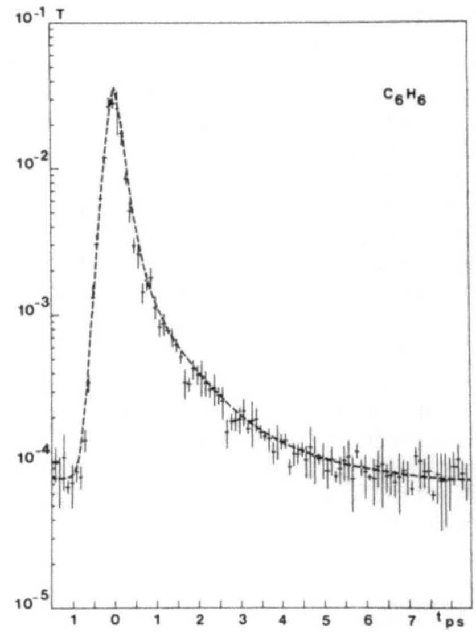

Fig. 4: transmitted signal through 1 cm long benzene cell.

To fit the data of benzene and nitrobenzene we had to take into account three different time dependent processes. For nitrobenzene, τ_1=4 ps and τ_2=42 ps; for benzene, τ_1=1.8 ps, τ_2= 10 ps; in all cases, the instantaneous process (τ_o < 0.2 ps)is of the same order of magnitude. We verified that the " Kerr constant " associated to molecular reorientation on benzene is about one order smaller than its equivalent in CS_2.We can conclude, that C_6H_6 is for subpicosecond pulses experiments a good candidate for an ultra fast shutter based on the optical Kerr effect.

At this stage, it is possible to give more precise informations on the pulse width and shape. If we assume that the probe and pump pulses have the same characteristics, the response of an optical Kerr shutter is a third order correlation of the pulse itself and thus contains specific informations on the shape of the pulse which initiates it, and especially about any possible assymetry. If, generally, a severe limitation of such measurements results from non-instantaneous Kerr processes, the use of very short pulses on compounds where exists an important instantaneous process, enables us to obtain an accurate representation of the pulse. We found our rapid responses to be well described by a biexponential pulse model $I(t) = \exp - |\frac{t}{T_i}|$, with the subsequent parameters T_1= .18 ± .01 ps(t<0) and T_2=.10 ± 01 ps (t>o), which results in a FWHM of 200 fs around 620 nm.

The formalism of non linear optical susceptibilities, which is employed to describe induced refractive index changes, shows that Kerr effect experiments lead to the measurement of its non null tensor elements [5.6]. In a configuration where the pump and probe pulses are linearly polarized, at 45° from each other, and where the induced polarization is measured at 90° to the direction of polarization of the probe pulse, the intensity which passes through crossed polarizers is :

$$I_T \alpha \mid \chi^{(3)}_{1212} (-\nu,\nu,\omega,-\omega) + \chi^{(3)}_{1221} (-\nu,\nu,\omega,-\omega) \mid^2$$

So, we can measure directly, by this way, the contributions of different origin to the induced polarization, third order in the electric field strength. From the results obtained, we have seen that the $\chi^{(3)}_{ijk\ell}$'s values of electronic origin are of the same order of magnitude for all the compounds studied and perhaps surprisingly high in the case of CS_2. One of the possible enhancement of this process of electronic origin may be a two photon absorption resonance [7,8]. The most attractive experiment to study resonant process- two photon absorption effects as well as Raman types resonances - is the configuration presented first by D. Heiman et al.[9]which consists in the use of a circular polarization of the pump beam. There, the transmitted intensity through crossed polarizers is :

$$I_T \propto | \chi^{(3)}_{1212} - \chi^{(3)}_{1221} |^2$$

As for non resonant processes, $\chi_{1212} = \chi_{1221}$, one can measure directly the difference between resonant susceptibility tensor elements by simply adjusting the wavelength of the pump(or probe)beam or, directly by adjusting the spectrometer to the wavelength wanted.Experiments are in progress in this direction.

References

1.a. J.L. Martin,C.Poyart,A. Migus,Y.Lecarpentier,R.Astier and J.P.Chambaret, this issue

b. A. Migus, J.L. Martin, R. Astier, A.Antonetti, A. Orszag, this issue.

2. E.P. Ippen and C.V. Shank, Appl. Phys. Lett. 26, 92, 1975.

3. J.M. Halbout and C.L. Tang, Appl. Phys. Lett. 40, 765, 1982.

4. G. Hauchecorne, F. Kerhervé and G. Mayer, Journal de Physique 32,47,1971

5. R.W. Hellwarth, Progress in Quantum Electronics, 5,1, 1977.

6. A. Owyoung, Ph D Thesis - Pasadena -1971

7. D.A. Ramsy, Determination of Organic structures by Physical Methods, New York 1962.

8. H. Lotem and R.T. Lynch Jr, Phys. Rev. Lett, 37, 334, 1976.

9. D. Heiman, R.W. Hellwarth, M.D. Levenson and G. Martin, Phys. Rev. Lett. 36, 189, 1976.

Subpicosecond Laser Spectroscopy: Pulse Diagnostics and Molecular Dynamics in Liquids

C. Kalpouzos, G.A. Kenney-Wallace, P.M. Kroger, E. Quitevis, and S.C. Wallace

Department of Chemistry, University of Toronto,
Toronto, Canada M5S 1A1

Molecular Dynamics and Memory in Liquids

The advent of picosecond (ps) [1,2] and more recently the first femtosecond (fs) [3] laser pulses now offers access to a time regime in simple liquids which approaches an interval corresponding perhaps to less than one hundred collisions. Yet, within this period, intermolecular potentials are sensitive to local density and configurational fluctuations, and the molecular dynamics to spatial correlations, which may persist over hundreds of collisions. This is the glimpse of memory in liquids. To what extent the early time behaviour of a molecular system at liquid densities can be adequately described by hydrodynamic models, developed from generalised Langevin equations, or by kinetically motivated models emphasizing collisional behaviour, has been a topic of considerable theoretical interest[4] in recent years and a long standing goal of our research in picosecond and nonlinear laser spectroscopy[5]. However, because the temporal profile of the laser pulse is often of finite width with respect to the time-evolution of the molecular response and its decay, the properties of the pump and probe laser pulses become a significant and even deterministic part of the problem when exploring such memory in liquids.

We first report here selected aspects of recent pulse diagnostic measurements that are part of a comprehensive study of laser pulse profiles in this time domain. Then we outline new results from a tunable, four-wave mixing phase conjugation study of the picosecond molecular dynamics of CS_2 diluted in various host liquids. This is the first time such measurements have been reported and demonstrate the potential for nonlinear spectroscopy in probing molecular dynamics and polarization memory in liquids.

The ps and fs dynamical behaviour can be linked to theory through the time-correlation functions (TCF) of the optical induced properties of the system, probed during and after the laser interaction. To fully describe the system of N particles we need to define the properties of the laser pulse, its interaction ($|i\rangle$, $|f\rangle$) with a molecule N_i in a reservoir R at temperature T, the coupling of N_i to R, and the TCF for all these interactions, while preserving explicit links for the interaction probability (W_{if}) to measurable quantities such as dipole moments, polarizabilities and so forth. Of particular interest to us has been the time-dependence of the rise and fall of the ps optical Kerr (OKE) transients[5,6]. These should reveal a temporal asymmetry in the case that the perturbation is far above kT and laser-field induced anisotropy carries the system into a non-stochastic regime, that is, one in which the intrinsic assumptions of the fluctuation-dissipation theorem are no longer valid[5].

A further fundamental consideration concerns the transition between many particle (Γ) and single particle (τ) relaxation behaviour. By analogy with depolarized Rayleigh light scattering: this can be written as $\Gamma^{-1} = (g_2/j_2)\tau_2$, where g_2 and j_2 are the static and dynamic orientational pair correlation functions, respectively [4]. The j_2 term is frequency-dependent, and at high frequency contains information on collision-induced events. In a weakly anisotropic system, collision may dominate diffusive effects. In these phase conjugation studies we set out to investigate the nature of the relaxation behaviour in CS_2 below the intense field regime.

Pulse Diagnostics

The synchronously mode-locked (SML) dye laser oscillator in Fig.1 is presently operating in a linear extended cavity configuration and can generate pulses at 82 MHz of ≥ 0.8 ps with an average power of 150 mW, tunable from 580 - 620 nm. Full details of its operation and the three-stage amplification to generate trains of ~800 μJ/pulse at 10 Hz in the visible have been given elsewhere[6].

We have measured the temporal profiles for both average of the pulse ensemble and a single pulse selected from the ensemble. Autocorrelation traces are inherently of an ensemble-averaged nature. We employ a real-time (speaker) autocorrelator system[7] as a permanent on-line monitor to measure the background free, second-order autocurrelation $G_0^{(2)}(\tau)$ function of the unamplified pulses at the output of the dye laser. A step-by-step measurement of $G_0^{(2)}(\tau)$ is taken for the amplified pulses using translation stages capable of electronically-controlled $\pm 1\,\mu$ precision to determine the delay time (τ) in one arm of the correlator.

In order to determine the true single pulse profiles, we have taken unamplified and amplified pulse measurements using an Imacon 500 (Hadland Co.) streak camera coupled to a 20/30 intensifier, with a minimum slit width of 25 μ, maximum streak speed of 20 ps/mm and an intrinsic technical response time of 0.75 ps for a threshold flux of $> 10^5$ photons/event at the slits. In practice, before this maximum response of the camera can be fully evaluated, significant electronic synchronization has been necessary to coordinate the multiple functions of this laser oscillator-amplifier system to the camera and the SIT vidicon detector and OMA II (PAR Corp). Details will be published elsewhere.

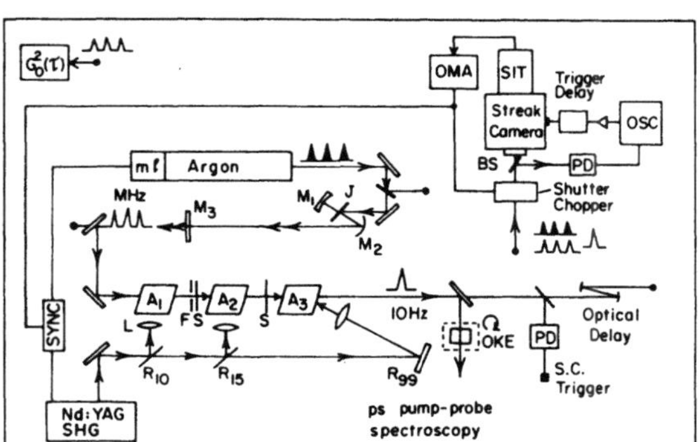

Fig.1 Schematic of SML oscillator-amplifier dye laser system for picosecond absorption and emission studies

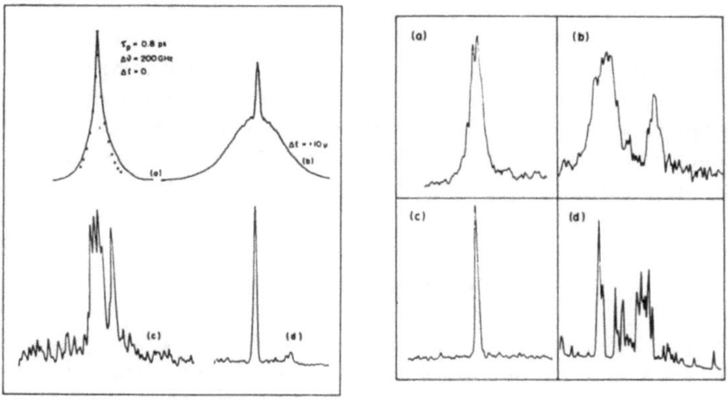

Fig.2 Left: Comparison of typical autocorrelation traces (a, b) and streak camera traces (c, d) for SML dye laser pulses, with cavity detuning of $\Delta \ell \geq 0$. Right: Dependence of argon ion (a, b) and dye laser (c, d) pulse profiles on pumping and cavity parameters. See text for details.

Figure 2 illustrates the general effect of dye laser cavity detuning by 10 μ as revealed by the autocorrelation trace versus the streak camera trace of the oscillator pulse, and the influence of the argon ion pulse on multiple pulse structures in the dye output. The effects of detuning on measurements of $G_0^{(2)}$ for SML dye laser systems have been reported earlier by us[7] and by others[8-10], but we are aware of only two other reports[9,10] which have compared these effects with streak camera traces in the single shot mode. Reproduced here to indicate that indeed apparent "single" pulse $G_0^{(2)}$ traces can actually include a major, if not dominant, contribution from multiple pulse structure, are selected data. Of the four profiles on the left of Fig.2, trace (a) is obtained under optimum conditions, can be fitted yielding $\tau_p = 0.8$ ps ($\Delta\nu = 200$ GHz). In fact, the additional intensity in the wings of (a) above the theoretical fit is indeed a signature of some broader pulses within the ensemble. The pulse burst of (c)has a FWHM of 75.2 ps, with many narrower components. The analogous trace after detuning in (d) shows only single pulses with negligible satellite structures corresponding to the FWHM base of (b), namely 10.3 ps. The four profiles on the right of Fig.2 illustrate (a) a streak camera trace of a single Ar^+ pulse (FWHM = 117 ps) lasing at 514 nm at 30 amps, just above threshold, average power 650 mW and $\nu_{mL} = 41.1215$ MHz, and (b) the double Ar^+ pulses which appear at 31 amps, 800 mW and $\nu_{mL} = 41.1225$ MHz, with major peaks separated by 259 ps. The pulse in (a) leads to only single dye pulses as in (c) when cavity matching is attained and, not surprisingly, pulse structure in (b) generates mostly pulse bursts (FWHM = 168 ps) in the dye laser seen in (d).

There are three important conclusions from our studies so far. First, SML systems can generate multiple structure dye pulses with rather small changes in laser operating parameters. Single pulse operation is a very sensitive function of many variables. Secondly, these pulse bunches cannot be obviously recognized through autocorrelation traces, which average over millions of pulses. This is because $G_0^{(2)}$ autocorrelation techniques based on second harmonic generation, while quite capable of indicating ultrafast subpicosecond and femtosecond components, are nevertheless fated to be symmetric in temporal profile and such pulse bunching appears as a somewhat dispersed contribution to the wings of the signal. Thirdly, streak camera traces reveal pulse asymmetry and pulse bunching or multiple structure very clearly, but nevertheless cannot at the present time readily break the subpicosecond barrier for a single shot on several technical accounts. These conclusions support earlier observations[9,10], but indicate single pulse performance to be a much more sensitive function of SML cavity parameters than previously supposed.

Phase Conjugation Studies of Molecular Dynamics

An isolated molecule in the gas phase has an intrinsic linear polarizability (α), which is modified upon transition to the liquid phase to give an interaction-induced polarizability, $\delta\alpha_i$, which takes into account the effect of intermolecular attractions and orientational correlations. Raman and depolarized Rayleigh scattering monitor fluctuations in the polarizability density (ρ_α) that arise from modulations in $\delta\alpha_i$ as a consequence of molecular motion and collision-induced perturbations. If, in addition to these fluctuating polarizabilities determined by internal fields, we impose an orientational torque, through the interaction of an external laser field, then the response of the molecular system will carry information not only on the polarization anisotropy $\Delta\alpha_i$ but on fluctuations in ρ_α ($\Delta\rho_\alpha$) as well.

Nonlinear optical responses can thus be used to monitor certain types of molecular relaxations through studies in the time and frequency domain, via both the relaxation of the laser-induced anisotropy and the time-dependence of $\chi^{(3)}_{ijkl}$, which must explicitly contain damping terms in the denominator of the susceptibility expression[11].

Both phase conjugation using four-wave mixing[12] and the optical Kerr effect [13-15] operate through a nonlinear polarizability

$$P^{(3)}_{NL} = \chi^{(3)}_{ijkl} \vec{E}\cdot\vec{E}\cdot\vec{E}$$

third-order in the electric field. The optical Kerr effect can reveal on a subpicosecond time scale the temporal evolution of the molecular response and its subsequent decay in terms of the components $\chi^{(3)}_{elec}$, $\chi^{(3)}_{vib}$ and $\chi^{(3)}_{rot}$ [13,14].

While clearly there is a temporal separation between electronic and orientational contributions to $\chi^{(3)}$, it should be noted that the intrinsic damping terms in $\chi^{(3)}_{elec}$ will be apparent as a finite relaxation time for this electronic contribution whenever resonance enhancement effects are present[16]. Thus the electronic nonlinear process, whose decay is often intuitively described as instantaneous with the radiation field, may not be so in the proximity of resonance effects[16]. With femtosecond duration pulses, we anticipate that such ultrafast relaxation times will be readily observable, since only these electronic contributions will have an instantaneous response to the laser field while the orientational contribution will exhibit a $\delta\alpha_i$- and field-dependent lag[5].

We present here new phase conjugation data, which demonstrate for the first time the intrinsic ability of a four-wave mixing interation to yield dynamical information concerning the single particle and many particle interactions in liquids.

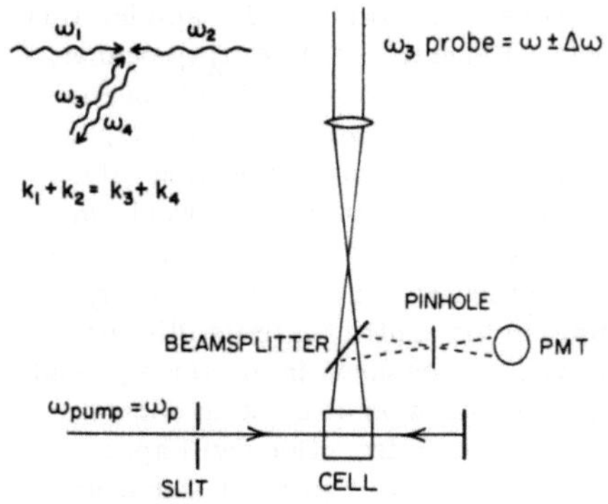

Fig.3 Schematic of phase conjugation and degenerate four-wave mixing experiment, using tunable dye laser spectroscopy and 90° detection geometry for ω_4.

The experimental arrangement, based on a Nd:YAG pumped, oscillator-amplifier tunable dye laser, is shown in Fig.3. We employ degenerate four-wave mixing to measure the frequency-dependence of $P^{(3)}_{NL}$ for the case

$$P^{(3)}_{(\omega_4 = \omega_1 + \omega_2 - \omega_3)} = \chi^{(3)} A_1^{(\omega_1)} A_2^{(\omega_2)} A_3^{*(\omega_3)}$$

where $\omega_1 = \omega_2$, $k_1 = -k_2$ and ω_3 is tunable. The intensity, linewidth, frequency and polarization of the conjugate wave carry the dynamical information on the molecular relaxation mechanisms in the nonlinear medium. The intensity of the conjugate wave I_{ω_4} is proportional to $(I_{\omega_2} I_{\omega_2})$ and is recorded at 90° to the path of the object wave, using a fast photomultiplier (PMT) and boxcar integrator coupled to a computer. For the data reported here, $\omega_1 = \omega_2$ at $\lambda = 5321.8$ Å, $\Delta\bar{\nu} = 0.1$ cm^{-1}, and the power density in ω_3 (tunable) was ~ 0.1 MWcm^{-2}. Cylindrical lenses cofocused ω_1 and ω_2 to a vertical line of $\leq 80\,\mu$ width in the sample cell, which contained the liquids at room temperature.

A strong signal was readily observable from pure CS_2 and its mixtures, and shown in Fig.4 are pure CS_2 and CS_2 in n-pentane (20% vol fraction). Note the change in intensity, linewidth and frequency maximum of ω_4 upon dilution of CS_2.

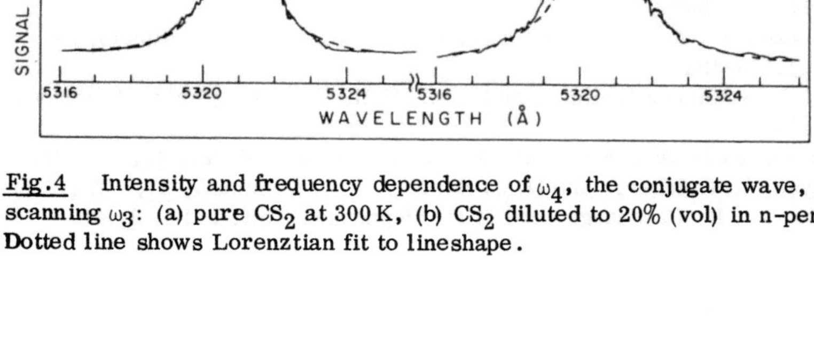

Fig.4 Intensity and frequency dependence of ω_4, the conjugate wave, observed by scanning ω_3: (a) pure CS_2 at 300 K, (b) CS_2 diluted to 20% (vol) in n-pentane. Dotted line shows Lorenztian fit to lineshape.

Spectra exhibited good signal: noise ratios and were fitted to single Lorentzian line shapes (dotted curve) over a wide dynamic range. Since n_2 $(CS_2) \ggg n_2$ (pentane), we conclude that even in dilute CS_2 the latter molecules are the dominant source of the conjugate wave. The Lorentzian line shapes imply lifetime broadening, a single exponential decay for the relaxation and hence that we are observing diffusive motion. Transit time limitations due to $\leq 80\,\mu$ interaction width would mask phenomena faster than ~150 fs. In pure CS_2, τ_{pc} value is 2.2 ± 0.1 ps, which decreases in subpicosecond intervals as a linear function of dilution and viscosity, until for 10% CS_2 in pentane $\tau_{pc} = 1.0 \pm 0.2$ ps. In the limit of infinite dilution, $\tau_{pc} < 1$ ps. The τ_{pc} is in good agreement with τ_{Kerr} measured in previous ps and subpicosecond OKE values[13,14,17] and a fuller discussion of these in comparison to data from other techniques and the solute-solvent interactions will appear elsewhere[16]. In brief, we attribute the decreasing τ_{pc} to a gradual transition from τ_{or} dominated by many-particle collective effects

$$\sum_{i \neq j} \langle A_i(t) A_j(0) \rangle$$

to one which in essence is approaching the single particle reorientation time $\langle A_i(t) A_i(0) \rangle$. It is unlikely that T_2 vibrational contributions are significant because of the narrow frequency range. Also, dephasing of the ν_1(Raman) mode[4] is ~20 ps, long compared to correlation times in the liquid and indicative (via $T_2^{-1} \propto \{\langle V^2 \rangle \tau_{corr}\}$) that CS_2 molecules interact weakly, further supporting our conclusion that the loss of the phase grating is primarily via a diffusive, orientational motion. We are presently pursuing these experiments in both frequency and (ps, fs) time domain to explore the details of the dilute cases for CS_2 and other "symmetric rotor" liquids, in which polarization memory can reveal novel dynamical effects and probe the onset of collisional regimes.

The financial support of the Natural Sciences and Engineering Research Council of Canada, the Connaught Foundation, the U.S. Office of Naval Research, and the Research Corporation is gratefully acknowledged.

References

1. See Picosecond Lasers and Applications, ed., L. Goldberg, Proc. SPIE, vol. 322 (1982).
2. Picosecond Phenomena I, eds., C. V. Shank, E. P. Ippen and S. L. Shapiro, Springer-Verlag, New York, 1978.
3. R. Fork, B. Greene, and C. V. Shank, Appl. Phys. Lett., 38, 671 (1981); C. V. Shank, R. L. Fork, R. Yen, and J. Thomlinson, in press (1982).
4. D. Kivelson and P. A. Madden, Ann. Rev. Phys. Chem., 31, 523 (1980).
5. G. A. Kenney-Wallace, Proc. R. Soc. (London), Ser. A, 299, 309 (1980); and in ref. 2, p.208.
6. C. Kalpouzos, G. A. Kenney-Wallace, P. M. Kroger, and E. Quitevis, in ref. 1, pp.188-198.
7. K. Sala, G. A. Kenney-Wallace, and G. E. Hall, IEEE J. Quantum Electron., 16, 980 (1981).
8. For a recent summary, see G. Fleming in Adv. Chem. Phys., 49, 1 (1982)
9. J. P. Ryan, L. Goldberg, and D. J. Bradley, Opt. Commun., 27, 127 (1978).
10. S. L. Shapiro, R. Cavanaugh, and J. C. Stephenson, Opt. Lett., 6, 470 (1981).

11. T. K. Yee and T. K. Gustafson, Phys. Rev. A, 18, 1597 (1978).
12. D. M. Bloom, C. V. Shank, R. L. Fork, and O. Tescke, in ref. 2, p.96.
13. J. Etchepare, G. A. Kenney-Wallace, G. Grillon, A. Migus, and J.-P. Chambaret, IEEE J. Quantum Electron., in press (1982).
14. J. M. Halbout and C. L. Tang, Appl. Phys. Lett., 40, 765 (1982).
15. R. Y. Chiao, P. L. Kelley, and E. Garmire, Phys. Rev. Lett., 17, 1158 (1966).
16. S. C. Wallace, to be published.
17. E. P. Ippen and C. V. Shank, Appl. Phys. Lett., 26, 92 (1975).

Viscosity-Dependent Internal Rotation in Polymethine Dyes Measured by Picosecond Fluorescence Spectroscopy

A.C. Winkworth, A.D. Osborne, and G. Porter

Davy Faraday Research Laboratory, The Royal Institution,
21 Albemarle Street, London W1X 4BS, U.K.

Introduction

Polymethine dyes are known to undergo ultrafast internal conversion, the rate of which depends on solvent viscosity (1), and this is thought to be due to a torsional motion of the heterocyclic quinolyl rings which allows the excited state energy to be dissipated by intramolecular vibration and rotation. The radiationless relaxation rate has been found to obey an empirical law of the form $\tau = \text{const.}\eta^{\alpha}$, where τ is the relaxation lifetime, and τ, for a particular dye molecule, is strongly dependent on the type of solvent used. In ethanol/glycerol mixtures, it is found that the power dependence varies between 0.35 and 0.70 (2). A recent study (3), using both mixed solvents and a homologous series of unbranched alcohols, established that α approaches 0.5 in the former, but is close to unity in the latter.

We have measured the fluorescence lifetimes (τ_{FM}) of a number of carbocyanine dyes, selected for their varying degrees of steric hindrance and size of substituent groups, with a view to gaining a fuller understanding of the mechanism of the viscosity dependent electronic relaxation.

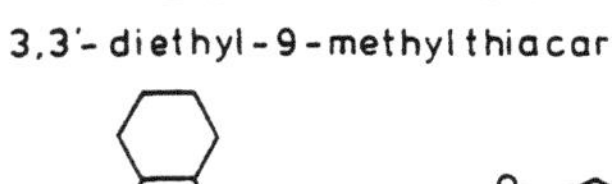

Br^- IV

3,3'-diethyl-9-methylthiacarbocyanine bromide

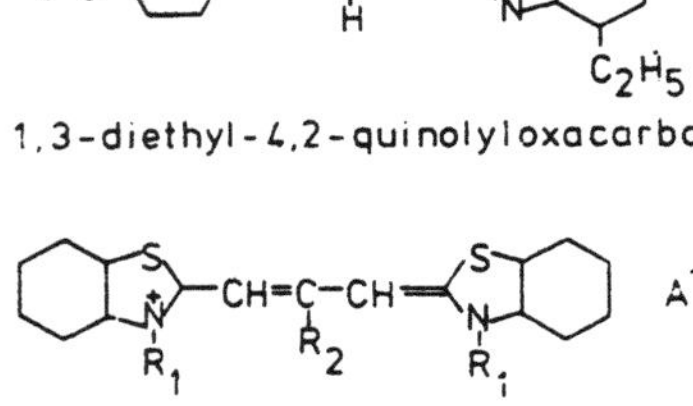

I^- V

1,3-diethyl-4,2-quinolyloxacarbocyanine iodide

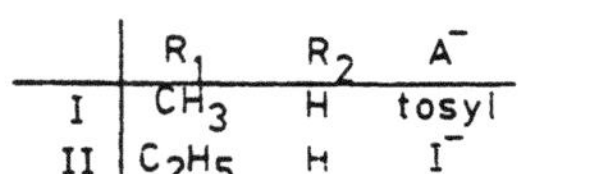

A^-

	R_1	R_2	A^-
I	CH_3	H	tosyl
II	C_2H_5	H	I^-
III	C_3H_5	H	Br^-
IV	C_2H_5	CH_3	Br^-

Figure 1

Experimental

A complete description of the picosecond laser system used may be found elsewhere (4). A frequency doubled mode-locked neodymium glass laser is used to generate the 6 ps duration 530 nm excitation pulse. Fluorescence emitted by the sample is time-resolved by an Imacon 600 Streak Camera with an S20 cathode. The fluorescence decay curves are digitised by a vidicon optical multi-channel analyser (OMA, Princeton Applied Research), and transferred to a computer for analysis. Up to ten fluorescence traces were recorded for each sample and summed within the computer. The measured decays were all found to fit a single exponential convoluted with the instrument response function. In general, concentrations of dye were selected to give an optical density <0.5 in a 1 mm cuvette. To preclude self-aggregation, self-absorption of fluorescence or stimulated emission, dye concentrations were kept below 10^{-4}M. Figure 1 shows the structures of the dyes whose fluorescence lifetimes are reported here. I, II and III were chosen as a series of varying size of N-alkyl group and of nature of counterion. IV was chosen as an example of the effect of methyl substitution in the polymethine chain. V is an example of an unsymmetrical dye.

Results and Discussion

Table 1 shows the results of measurements of τ_{FM} in a variety of solvents. These data are plotted in Fig. 2 on log-log diagrams whose slopes give the value of the viscosity power dependence α. From the data we note the following :

TABLE I - fluorescence lifetimes (ps) viscosity (cP)

solvent	viscosity	I	II	III	IV	V
methanol	0.55	92	125	137	12	11
ethanol	1.2	149	173	227	19	20
propan-1-ol	2.0	182	261	270	22	33
propan-2-ol	2.4	156	262	261	25	35
butan-1-ol	2.6	254	302	354	32	49
pentan-1-ol	4.0	381	371	415	43	73
hexan-1-ol	5.4	353	482	545	63	111
heptan-1-ol	7.0	406	524	644	79	145
octan-1-ol	8.8	448	627	742	125	196
nonan-1-ol	10.3	524	641	692	157	247
decan-1-ol	12.3	553	756	814	175	316
undecan-1-ol	14.2	575	798	808	214	389
cyclohexanol	33	562	884	889	171	367
ethan-1,2-diol	20	390	603	526	104	137
propan-1,2-diol	40	480	811	840	155	224
pentan-1,5-diol	98	-	-	-	257	470
pentan-1,4-diol	163	-	-	-	336	654
glycerol	900	-	-	-	886	1104

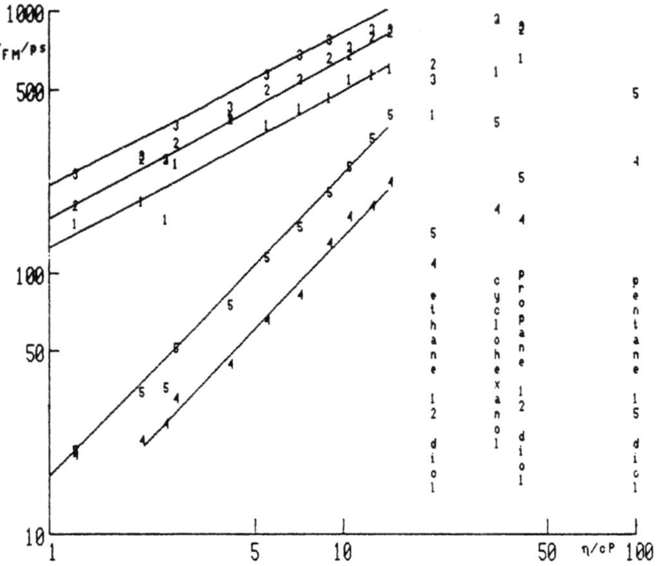

Fig 2

(a) Internal conversion (k_{GM}) is anomalously fast in the diols and in glycerol as was found for N-aryl substituted rhodamine dyes (5,7). However, the anomaly is not nearly so great, τ_{FM} in ethane-1,2-diol (η = 20cP) being similar to that in the C_6 - C_8 mon-ols (η = 5-9cP) whereas in the aryl substituted rhodamines, τ_{FM} in ethane-1,2-diol was similar to that in propanol (η = 2.4cP).

(b) τ_{FM} in the group I, II, III are similar indicating that the size of the N-alkyl groups and the nature of the counterion do not have a significant effect upon k_{GM} (excluding the unlikely event that the size of the group and the nature of the counterion have effects which fortuitously cancel).

(c) α was lower in the diols than that in the mon-ols. An attempt was made to fit the results to KRAMERS (9) equation in a manner similar to that described by McCASKILL and GILBERT (7) but without success. This equation gives the rate of passage over a torsional angle dependent energy barrier as a function of η, the nature of the barrier and the diameter of moiety (taken to be spherical) undergoing the torsional motion. The equation was used (8) in the form :

$$K_{GM} = \frac{Q}{Q+\Delta}^{\frac{1}{2}}(2\pi)^{-1} \quad \frac{1}{4} C^2 d^2 \eta^2 + 66(Q+\Delta)^{\frac{1}{2}} - \tfrac{1}{2} Cd\eta \quad \exp(-Q/kT)$$

where C is related to the reduced moment of inertia (I) of the moiety, d is the diameter and Q and Δ define the geometry of the barrier. It was found that, despite the complexity of the equation, it was approximately equivalent to τ_{FM} = const.η^{α} with $\alpha \sim 1$ under widely varying values of the parameters d, C, Δ and Q. WILHELMI (10) has also recently considered this model and concluded that low values of α are due to a viscosity independent channel for internal conversion and OSBORNE (8) has shown that, for the rhodamine dye, Fast Acid Violet 2R, a value of α of 2/3 can be explained if it is assumed that solvent attachment occurs so that, within a solvent homologous series, I and hence C increases with viscosity. The presence of an η-

independent component of k_{GM} must not be discounted but does not explain why for the same dye α is smaller in mixed solvents than in pure solvents. It seems likely that this is connected with the anomalous behaviour of the dyes in the diols and glycerol so that mixtures of these with mon-ols give rise to a smaller η-dependence. To test this hypothesis τ_{FM} measurements of I in mixtures of heptan-1-ol (η = 7cP) and propane-1,2,diol (η = 40cP) were made. These solvents were chosen because τ_{FM} is similar ($\sim$ 400 ps) in both. It was found that τ_{FM} was approximately constant in such mixtures, i.e. $\alpha \sim 0$.

(d) Dyes IV and V both had significantly smaller τ_{FM} than I, II and III. This is taken to be due to relatively weaker conjugation in the polymethine chain due respectively to steric hindrance (11) and to asymmetry (1). Both were found to have $\alpha > 1$ in mon-ols which would be explained on the basis of KRAMERS equation if there were more solvent attachment in the lower members of the homologous series of mon-ols than in the higher members.

Acknowledgement

We acknowledge financial support from the SERC and helpful discussions with T. Doust.

References

1. C.J. Tredwell and C.M. Keary, Chem. Phys., **43** (1979) 307
2. W. Sibbett, J.R. Taylor and D. Welford, IEEEJ. Quantum Electronics QE-**17:4** (1981) 500
3. V. Sundström, T. Gillbro, Chem. Phys., **61** (1981) 257
4. W.J. Getting and E. Wyn-Jones (Eds.) Techniques and Applications of Fast Reactions in Solutions, NATO Advanced Study Institutes, Series C, D. Reidel, Holland **50** (1979) 157
5. C.J. Tredwell and A.D. Osborne, JCS Faraday II, **76** (1980) 1627
6. A.D. Osborne and A.C. Winkworth, Chem. Phys. Lett., **85** (1982) 513
7. J.S. McCaskill and R.G. Gilbert, Chem. Phys., **44** (1979) 389
8. A.D. Osborne, JCS Faraday II, **76** (1980) 1638
9. H.A. Kramers, Physica, (1940) 284
10. B. Wilhelmi, Chem. Phys., 66 (1982) 351
11. L.G.S. Brooker, F.L. White, R.H. Sprague, S.G. Dent, Jr., G. van Zandt, Chem. Rev., (1947) 325.

Rotational Diffusion in Mixed Solvents Measured by Picosecond Fluorescence Anisotropy

T.Doust and G.S. Beddard*

Davy Faraday Research Laboratory, The Royal Institution,
21 Albemarle Street, London W1X 4BS, U.K.

Introduction

The influence of the microstructural changes of a mixed solvent system, as its composition is varied, on the rotational dynamics of a solute chromophore has been observed by picosecond fluorescence anisotropy measurements. This effect is mediated by the preferential interaction of one component of the solvent with the chromophore. From our observations we infer that there are long-lived (> 10^{-10}s) structural features in the solvent systems used.

Experimental

The solute used was cresyl violet chloride (Lambda Physik) and perchlorate (Exciton), both laser grade. Analar grade alcohols and DMSO and doubly distilled deionized water were used to prepare the solvent mixtures. The fluorescence depolarisation measurements were made using a modification of the upconversion spectrometer described in (2). The fluorescence was detected at 670 nm and excited at 590 nm with $\sim$5 ps pulses at 75 MH_z from a Coherent CR12/CR590 synchronously pumped laser system.

Results and Discussion

Preliminary measurements of this behaviour were made in ethanol-water mixtures (1) but this behaviour now seems to be more general and appears in aqueous solutions of MeOH, EtOH, 1-PrOH and DMSO. The data for a series of 1-propanol-water mixtures is plotted in fig. 1, which shows the rotational relaxation time τ_R plotted against viscosity relative to water. As may be clearly seen, at a given solvent viscosity, two very different rotation times are observed, depending on composition.

There are two distinct regimes in the data; the crossover from one to the other occurs approximately at the maximum viscosity which is around 0.2 mole fraction of alcohol. In the low alcohol regime we observe a linear dependence of τ_R on the macroscopic solvent viscosity. The slope of this line follows closely that predicted for the unsolvated dye molecule by Debye-Stokes-Einstein (DSE) theory at the stick boundary condition. We have shown previously (2) that in the pure alcohols, solvent interaction (presumably H-bonding) to the amino groups results in anomalously high rotational relaxation times; normal DSE behaviour is observed in water. In contrast the related dye, oxazine-1, in which the amino groups are fully ethylated, shows no anomalous rotational behaviour in either alcohols or alcohol water mixtures (3).

---

* Present address: Department of Chemistry, University of Manchester, Oxford Road, Manchester M13 9PL, U.K.

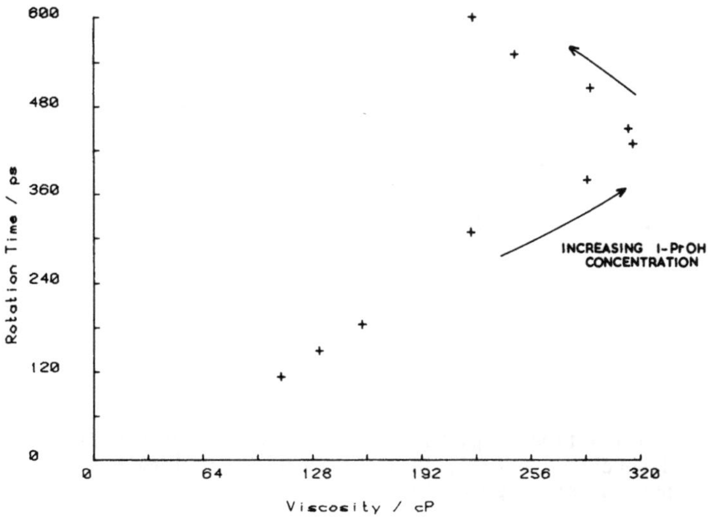

Fig. 1
Rotational relaxation times (τ_R) vs relative viscosity for cresyl violet chloride in a series of 1-propanol-water mixtures ranging from pure water to pure 1-propanol.

Thus, in the low alcohol regime where the predicted diffusional behaviour is observed, the alcohol must be fully solvated by the water and is unable to interact with the solute chromophore. These water "cages" surrounding the alcohol must last for longer than the rotation time of either the dye or the alcohol and water molecules ($\sim 10^{-11}$s).

At viscosities above the maximum the rotation time changes with decreasing viscosity in a very non-linear fashion with values much larger than predicted by DSE theory; this is clearly due to the availability of free alcohol molecules to interact with the dye. These results are consistent with models for alcohol-water mixtures derived from nmr, thermodynamic and other measurements (4).

The possibility of the formation of ion-pairs has been considered. Preliminary conductivity measurements indicated that the dye was at least partly dissociated but we observe different rotation times for the perchlorate and chloride over the whole composition range for the ethanol water system (2) except at very low ethanol concentration; no difference is observed between chloride and perchlorate in pure DMSO. It is possible that these differences are due to ion pair formation but if this is the case the deviations from DSE behaviour are not sensitive to whether the rotor is the cationic chromophore or an ion pair.

It appears that this type of composition dependent rotational behaviour appears to be quite general when the solvent undergoes microstructural changes with varying composition and the solute is able to interact selectively with one component of the solvent system.

Acknowledgement

We thank Professor Sir George Porter for his advice and encouragement in this work; we would also like to thank J. Hudales for his help with some of the measurements and the SERC for financial support.

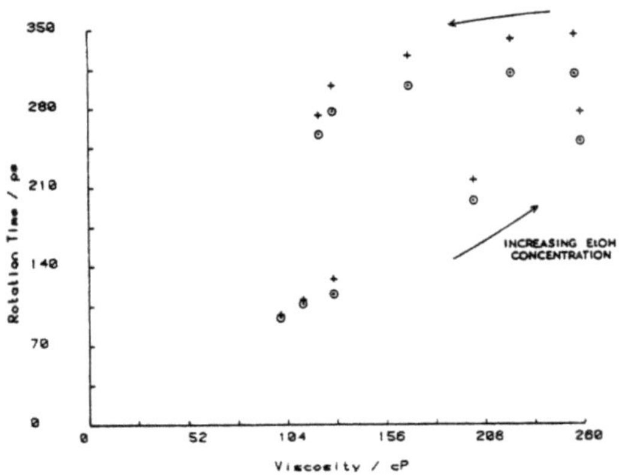

Fig. 2 Rotational relaxation times (τ_R) vs relative viscosity of cresyl violet chloride (+) and perchlorate (⊙) in a series of ethanol-water mixtures ranging from pure water to pure ethanol

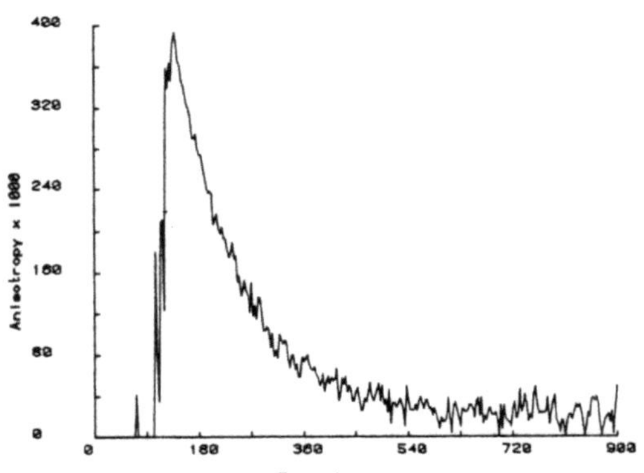

Fig. 3 Decay of fluorescence anisotropy of cresyl violet perchlorate in 2% 1-propanol calculated from the measured parallel and perpendicularly polarized fluorescence decays (τ_R = 106 ps)

References

1. G. Beddard, T. Doust and J. Hudales, Nature, 294, (1981), 145
2. G. Beddard, T. Doust and G. Porter, Chem. Phys., 61, (1981), 17
3. G. Fleming, D. Waldeck and G. Beddard, Il Nuovo Cimento, 63B, (1981), 151.
4. F. Franks (ed.), Water - A Comprehensive Treatise, Vol. 2, Plenum, N.Y. (1975).

Investigation of Level Kinetics and Reorientation by Means of Double Pulse Excited Fluorescence

D. Schubert, J. Schwarz, H. Wabnitz, and B. Wilhelmi

Friedrich-Schiller-Universität Jena, Sektion Physik,
GDR-6900 Jena, German Dem. Rep.

Fluorescence spectroscopy is advantageously applied in order to measure excited state lifetime and reorientation of the transition moment after excitation (see, e. g., [1], [2]). Applying ultrashort light pulses the time resolution is limited in most cases by the signal detection system. MAHR et al. [3] proposed a method in which an excite-and-probe scheme is combined with fluorescence observation. The dependence of the integrated fluorescence signal on the relative delay between two successive exciting light pulses is measured. That dependence arises under the condition that the second pulse views a ground state population depleted by the first pulse. The time resolution is only limited by the pulse duration and hence values in the sub-ps-range can be achieved. The method proposed in [3] represents a modification of that given in [4] where the fluorescence from a higher singlet level has been measured after a two step excitation $S_0 \rightarrow S_1 \rightarrow S_x$, in which one of the exciting pulse trains has been modulated.

Under the condition that the pulse energies are small in comparison with the saturation energy the modulated part of the fluorescence signal is proportional to the product of the energies of the first and second exciting pulses. For that reason the detection sensitivity is only comparable with that of excite-and-probe spectroscopy. The main advantage of the fluorescence technique consists in the possibility to perform measurements not only in well transmitting samples but also in inhomogeneous and opaque ones. With this aim the fluorescence has to be observed in the backward direction.

In [3] a special modulation of the high repetion rate pulse trains has been applied using a third reference pulse train with a fixed delay being large in comparison with the signal relaxation time. A very precise adjustment of the pulse intensities in the different channels is needed which has to be independent of the delay. To avoid this requirement MAHR et al. [5] and the authors together with KIRMSSE [6] modified the method by modulating two exciting beams with frequencies ω_1 and ω_2, respectively, and measuring the fluorescence signal at one of the frequencies $|\omega_1 \pm \omega_2|$. Regarding this method care has to be taken to minimize any nonlinearities in the whole detection system.

Another possibility considered by us consists in the modulation of the polarization direction of one of the pulse trains which is a well known method in cw-polarization spectroscopy. Concerning this method we shall discuss here the switching of the polarization direction by an angle of 90°. An analogous technique has been applied by VON JENA and LESSING [7] in probe beam spectroscopy.

These methods have been investigated by us with respect to the measurement of level kinetics and reorientation of the transition moment after excitation. The reorientation influences the absorption of the second pulse and the emission of light due to the first and second pulses. The time-integrated signal depends on the polarization directions of the second pulse with respect to the first one and on the polarization chosen in the detection channel. Assuming isotropic rotational diffusion (coefficient $D = 1/6\tau_{or}$) in both the ground and excited states, single exponential decay of the excited level (lifetime τ) and weak δ-pulse excitation the following delay time dependence of the signal under the polarization geometry specified in fig. 1 results:

$$S(t) \sim L(t) \left\{ \left[1 + \frac{2}{5} \frac{1}{1 + \tau/\tau_{or}} c_1 \right] + \left[\frac{2}{5} (3\cos^2\Psi - 1) + \frac{2}{5} \frac{1}{1 + \tau/\tau_{or}} c_2 \right] R(t) \right\}$$

where

$$c_1 = \frac{3}{2}\sin^2\Psi \sin^2\Theta_F \cos 2\varphi_F + \frac{3}{2}\sin 2\Psi \sin 2\Theta_F \cos\varphi_F + \frac{1}{2}(3\cos^2\Psi - 1)(3\cos^2\Theta_F - 1)$$

$$c_2 = -\frac{6}{7}\sin^2\Psi \sin^2\Theta_F \cos 2\varphi_F + \frac{3}{7}\sin 2\Psi \sin 2\Theta_F \cos\varphi_F + \frac{1}{7}(6\cos^2\Psi + 5)(3\cos^2\Theta_F - 1)$$

$$L(t) = \exp(-t/\tau)$$

$$R(t) = \exp(-t/\tau_{or})$$

This signal is a complex superposition of level kinetics and reorientation such that certain arrangements of the polarization directions are necessary in order to separate these two processes. If one does not take into account this complex dependence, experimental results might be misinterpreted (cf., e. g., the reevaluation of results of [3] in [8]).Another inference is that under consideration of reorientation the signal is in general no longer independent of the sequence of the two pulses [5] .

With regard to the three pulse method [3] and the double modulation method [5, 6] the influence of reorientation can be eliminated

- in the case of observation under 90° ($\varphi_F = 90°$) for $\Psi = 54.7°$, $\Theta_F = 65.2°$
- in the case of observation in the backward direction ($\varphi_F = 0°$) for $\Psi = 54.7°$, $\Theta_F = 54.7°$.

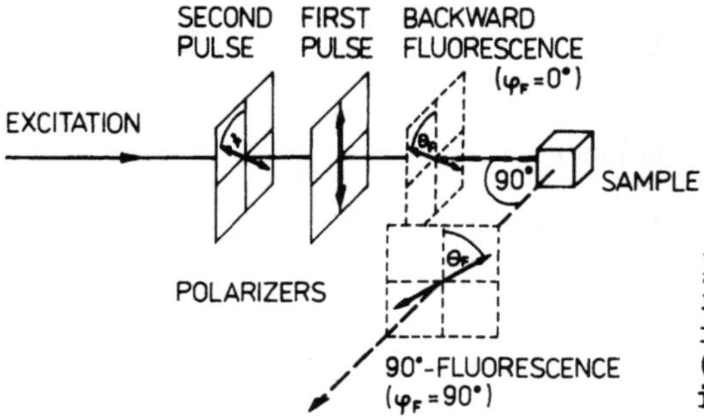

Fig. 1 Geometrical configuration of two-pulse excited fluorescence measurements (Polarization directions are indicated by ⟷)

Then, τ can be extracted from $S(t) \sim L(t)$. On the other hand, determination of τ_{or} would require the formation of certain linear combinations of signals with different polarization directions.

A simpler possibility for obtaining τ_{or} is the modulation of the polarization direction of the second pulse. If Ψ is switched between $\Psi_1 = 0°$ and $\Psi_2 = 90°$ and the fluorescence is observed, e. g., in the backward direction under $\Theta_F = 45°$, the fluorescence excited by each of the two pulse trains alone does not show any modulation. In this background free method a modulation signal arises only if the second pulse views an anisotropic orientational distribution of ground state molecules. This signal decays with reorientation as well as with excited state relaxation during the delay time t:

$$[S(t;\Psi_1) - S(t;\Psi_2)] \sim L(t)\cdot \dot{R}(t) \; .$$

Thus for known τ the value of τ_{or} can be determined.

References

1 Ultrashort Light Pulses, Topics in Applied Physics, vol. 18 (ed. S. L. Shapiro), Springer Berlin, Heidelberg, New York, 1977

2 Picosecond Phenomena II, Springer Series in Chemical Physics, vol. 14 (eds. R. M. Hochstrasser, W. Kaiser, C. V. Shank), Springer Berlin, Heidelberg, New York, 1980

3 H. Mahr, A. G. Sagan, C. P. Hemenway and N. J. Frigo Chem. Phys. Lett. 79 (1981) 3, 503-505

4 E. P. Ippen, C. V. Shank, R. L. Woerner Chem. Phys. Lett. 46 (1977), 20-23

5 H. Mahr, A. G. Sagan Opt. Commun. 39 (1981) 4, 269-271

6 H. Kirmsse, Diploma Thesis, Jena 1982

7 A. von Jena, H. E. Lessing Ber. Bunsenges. 83 (1979) 3, 181-191

8 H. Wabnitz Chem. Phys. Lett., in press

Dynamics of Photoisomerization

G.R. Fleming, S.P. Velsko, and D.H. Waldeck

Department of Chemistry and James Franck Institute,
The University of Chicago, Chicago, IL 60637, USA

Introduction

There has recently been a good deal of theoretical [1-3] and experimental [4-7] interest in activated barrier crossing processes in liquids. The best known model is that of Kramers [8] who derived expressions for the escape of particles over a barrier in various limiting cases. Combining Kramers ideas with recent kinetic theory calculations [1-3] leads to the following general picture. At low friction (viscosity) the rate of barrier crossing increases with viscosity (inertial region) until it turns over and begins to decrease with increasing friction (intermediate friction). Finally at high friction values the rate decreases linearly with increasing friction (Smoluchowski limit region). In terms of Kramers Eq. (1)

$$k = \frac{\omega}{4\pi\omega'\tau_v}\{[1+(2\omega'\tau_v)^2]^{1/2}-1\}\exp(-E_0/RT) \qquad (1)$$

the transition from intermediate friction to high friction is determined by the magnitude of the barrier curvature (ω') times the velocity relaxation time for the relevant motion (τ_v). If $\omega'\tau_v \ll 1$ (1) simplifies to the Smoluchowski limit and with the hydrodynamic approximation $\tau_v \propto \eta^{-1}$ where η is the solvent viscosity, the barrier crossing rate becomes inversely proportional to η.

Photochemical isomerization dynamics should provide a valuable testing ground for the theoretical approaches, and in the following we present a summary of our recent studies.

Results

We have already presented data for the excited state isomerization of diphenyl butadiene (DPB) in alkane solvents [7], and of DODCI in alcohol solvents [6]. Recently we have extended these studies to polar solvents for DPB [9] and to return of the isomer to the normal form on the ground state surface of DODCI [10]. These measurements enable us to investigate the influence of barrier height and of the characteristic frequency (pre-exponential factor in (1) [7]) on the barrier crossing dynamics.

Our analysis involves extracting the internal barrier (E_0 in (1)) by means of isoviscosity plots (see Refs. [6,7] for details). We find that the observed activation energy can be decomposed into 'molecular' and solvent contributions in a consistent and reliable way. The results are listed in Table I along with the value of the effective frequency and the product $\omega'\tau_v$ evaluated at 1 cp. These values are obtained by fitting a plot of k* vs viscosity to the hydrodynamic Kramers expression

Table I Summary of barrier crossing data

System	Barrier Height	Effective Frequency at 1 cp	$\omega'\tau_v$ at 1 cp	a ($k \propto \frac{1}{\eta^a}$) value
DODCI GS	13.7 kcal	4×10^{12} s^{-1}	7.5	0.26
ES	2.7	8×10^{11}	0.8	0.43
[Rotation]	0	10^7	--	0.97
DPB ES alkanes	4.7	1.5×10^{12}	0.71	0.59
ES alcohols	∿0.5	1.5×10^{10}	0.13	1.07
FAV 2R ES alcohols (Tredwell & Osborne)	∿0.2	9×10^{10}	0.02	0.98

$$k^* = k\,e^{E_0/RT} = A/(B/\eta)\{\,[1+(B/\eta)^2]^{1/2}-1\,\} \qquad (2)$$

where $A = \omega/2\pi$ and $B/\eta = 2\omega'\tau_v$. Figure 1 shows such a plot for the return isomerization on the ground state surface of DODCI. As can be seen the Kramers expression does not reproduce the curvature of the data well. The same qualitative shape is obtained for DODCI excited state and for DPB in the alkane solvents, but not for DPB in the alcohol solvents. Also shown in Table I is the value of a obtained by fitting the data to $k = B/\eta^a$ (3) Note that $a = 1$ corresponds to the Smoluchowski limit (2). Equation (3) always fits our data extremely well although the value of a ranges from 0.26 to 1.07.

For both DODCI data sets and for DPB in alkanes the value of $\omega'\tau_v$ corresponds to intermediate friction. In all these cases the barrier is larger than kT. Theoretical studies of diphenyl polyenes have suggested that the twisted form reached after barrier crossing has significant charge transfer character [11]. Stabilization of the twisted form may then change the shape of the potential surface along the twisting coordinate and influence both the sharpness and height of the barrier. The data for DPB in alcohols show that the barrier height and curvature are greatly reduced and the the twisting dynamics are now in the high friction or Smoluchowski regime (Table I). The reason for the failure of Kramers equation for high (sharp?) barriers may lie in the necessity to consider the frequency dependence of the medium response at the effective reaction frequency [1]. We are currently investigating this aspect of the problem.

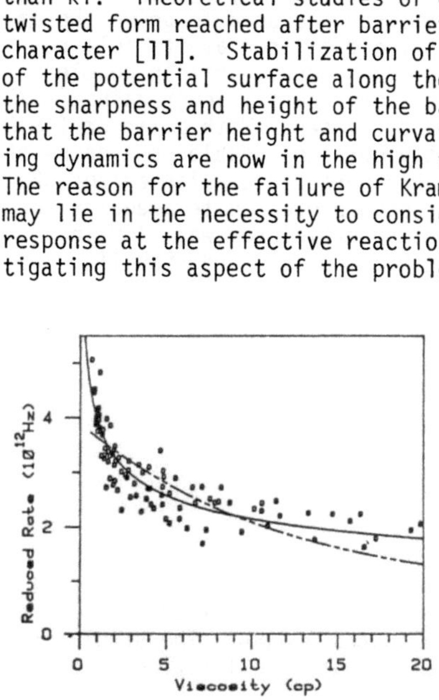

Fig.1 Plot of k* vs η for DODCI ground state isomerization. The dashed line is the best fit of Eq. (2) and the solid line is the best fit of Eq. (3).

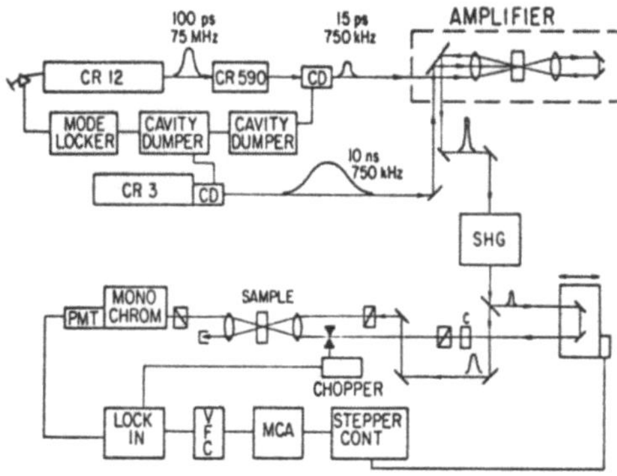

Fig.2 Experimental arrangement (CD: Cavity Dumper; C: compensator; VFC: voltage to frequency converter; MCA: multichannel analyzer). See Ref.14 for further details.

Rotational Motion and Isomerization

The use of (2) rather than (1) involves the assumption of hydrodynamic friction. We have tested this assumption by using the rotational diffusion times for the same molecule in the same solvent as a measure of the solvent friction. Equation (1) is rewritten in terms of the reduced reorientation time obtained via the Hubbard relation (see [10] for details.) We have completed this study for DODCI [10] but the fit to Kramers expression is not improved as might be expected from the close adherence of the rotational diffusion data to Stokes-Einstein theory [12]. (Note $a=1$ in Table I).

A High Repetition Rate Amplifier

To obtain the rotational diffusion times for DPB we have extended the anisotropic absorption technique [13,14] into the ultraviolet by use of a high repetition rate ($\lesssim$5 MHz) amplifier. The system is shown in Fig. 2 and consists of a cavity dumped argon laser synchronously amplifying pulses from a synchronously pumped dye laser. A 3W argon laser was used to drive the amplifier. Two configurations were used with the dye laser at its full repetition rate or also cavity dumped. In the first configuration double pass gains of 30 times were obtained, while significantly lower gains (2.75 times) were obtained with the cavity dumped dye laser. With a larger (18 W) pump laser T. Gustafson of Sohio has recently obtained double pass gains of 7 times for a cavity dumped laser. This corresponds to a pulse energy 32 nJ. However, the gains obtained in our initial setup proved quite adequate for UV pump-probe spectroscopy. Frequency doubling efficiency was ∿15% in $LiIO_3$ and 7.6% in ADA. The ADA gave better beam quality and stability and was used for the experiments.

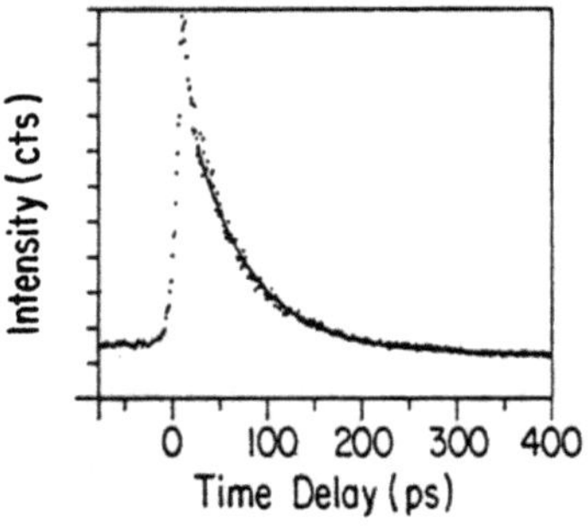

Fig.3 Anisotropic absorption decay of DPB in tetradecane. Excitation and probing wavelengths 303 nm. The single exponential fit gives a lifetime of 62ps.

Figure (3) shows the anisotropic absorption signal from DPB in tetradecane. The excitation and probe wavelengths were 303 nm. Using the measured decay time and the fluorescence lifetime in (4) gives τ_{or} 66ps/cp. The calculated values for stick and slip boundary conditions are 127ps/cp and 62ps/cp. Thus the result agrees well with slip hydrodynamics. The reorientation time is obtained from the measured time by (4)

$$\tau_M^{-1} = 2(\tau_{or}^{-1} + \tau^{-1}) \qquad (4)$$

When the excited state is depopulated by a process that does not reform the ground state on the timescale of the measurement (4) is inadequate. The general form of the signal is

$$T(t) \propto [\frac{2}{5} N'(o)((1-\phi)e^{-6\theta t} + \phi e^{-(k_f + 6\theta)t})]^2 \qquad (5)$$

where ϕ is the sum of the yields of processes $S_1 \rightarrow S_0$, $k_f = k_r + k_{nr}$, and θ is the rotational diffusion coefficient. If (5) applies to DPB in ethanol where the fluorescence yield is very small (0.04) then the anisotropic absorption signal decays as 2 τ_{or}. We measure 25 ± 3ps giving τ_{or} 50 ps in reasonable agreement with the above value.

Acknowledgement

This work was supported by grants from NSF and NIH. We thank W.T. Lotshaw and D.B. McDonald for their assistance with the amplifier and K. Keery and Dr. T. Gustafson (Sohio) for access to their unpublished results.

References

1. R.F. Grote and J.T. Hynes, J. Chem. Phys. 74, 4465 (1981).
2. J.A. Montgomery, D. Chandler and B.J. Berne, J. Chem. Phys. 70, 4056 (1979).
3. J.L. Skinner and P.G. Wolynes, J. Chem. Phys. 72, 4913 (1980).
4. R.M. Hochstrasser in Picosecond Phenomena II, 259 (1980).
5. C.J. Tredwell and A.D. Osborne, J. Chem. Soc. Faraday II, 76, 1627 (1980).
6. S.P. Velsko and G.R. Fleming, Chem. Phys. 65, 59 (1982).
7. S.P. Velsko and G.R. Fleming, J. Chem. Phys. 76, 3553 (1982).
8. H.A. Kramers, Physica 7, 284 (1940).
9. K.M. Keery and G.R. Fleming, in preparation.
10. S.P. Velsko, D.H. Waldeck and G.R. Fleming, submitted J. Chem. Phys.
11. I. Baraldi, et al., Chem. Phys. 52, 415 (1980).
12. D.H. Waldeck and G.R. Fleming, J. Phys. Chem. 85, 2614 (1981).
13. G.R. Fleming, et al., Proc. Lasers '81, in press.
14. D.H. Waldeck, W.T. Lotshaw, D.B. McDonald and G.R. Fleming, Chem. Phys. Lett. 88, 297 (1982).

Evidence for the Existence of a Short-Lived Twisted Electronic State in Triphenylmethane Dyes

V. Sundström, T. Gillbro, and H. Bergström

Division of Physical Chemistry, University of Umeå,
S-901 87 Umeå, Sweden

Introduction

The viscosity dependence of the radiationless relaxation rate of triphenylmethane (TPM) dyes has lately been studied in numerous works [1-10]. Time-resolved absorption [1-4] and emission [5-7] techniques as well as more conventional fluorescence quantum yield measurements [8-10] have been employed in these studies. The results have often been interpreted in terms of the Förster-Hoffman theory [8] which predicts a magic $\eta^{2/3}$ dependence of the fluorescence quantum yield. Due to the rapidity of the studied processes in solvents of low viscosity, most experiments have been performed in glycerol or solvents of even higher viscosity. Here we want to present some new results on the relaxation of TPM dyes in n-alcohols that support the previously assumed twisting mechanism for the relaxation process, and what is more important also suggest the existence of a highly unstable intermediate. We tentatively identify this species with a photolytically formed short-lived geometrical isomer. In addition to these results the presented data show the necessity of performing picosecond absorption studies at several wavelengths in order to extract all information available.

Experimental

Absorption recovery rates after picosecond pulse excitation of the TPM dyes crystalviolet (CV) and ethylviolet (EV) in n-alcohols were studied using a synchronously mode-locked and cavity-dumped dye laser [11]. Pulselengths and pulse repetition rate were ca 2-8 ps and 82 kHz respectively. A flow cell system using a peristaltic pump was employed to circulate the sample through a 0.5 mm cell at room temperature. Below room temperature the sample was thermostated in an Oxford Instruments cryostat.

Results and Discussion

The time resolved absorption recovery of CV and EV was studied in several n-alcohols as different wavelengths and temperatures. The measured kinetics of EV/EtOH at λ = 625 nm at various temperatures display interesting features. At room temperature the recovery signal is biphasic consisting of a positive (bleaching) and negative (absorption) part. As the temperature is lowered the relaxation rates of both processes slow down and the intensity of the negative part of the signal decreases. Within a narrow temperature range ($\sim \pm$ 5 °C) around ca -90 °C the appearance of the signal is drastically changed, as the negative part of the signal continously transforms into a positive bleaching signal, see Fig. 1a-b. The variation in recovery signal of EV and CV when going through the n-alcohol series methanol-decanol is similar to that observed in ethanol upon changing temperature. Measured kinetic and spectral data ov EV and CV in alcohols are

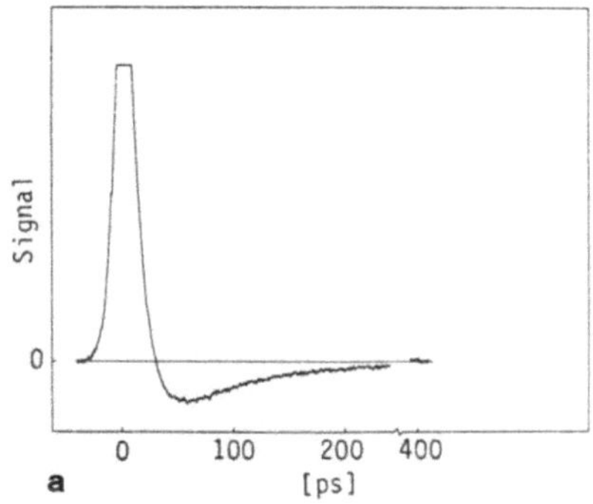

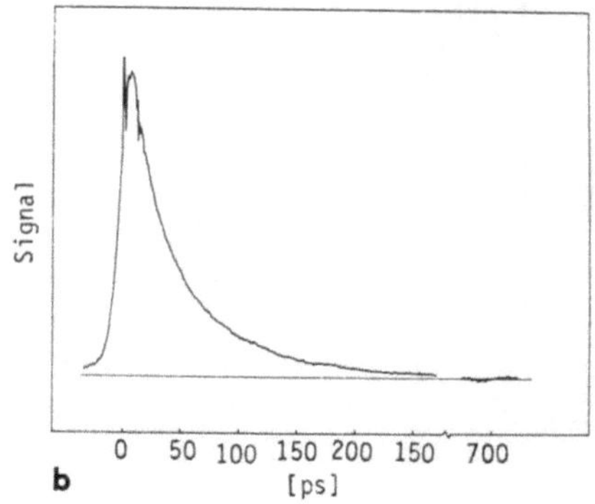

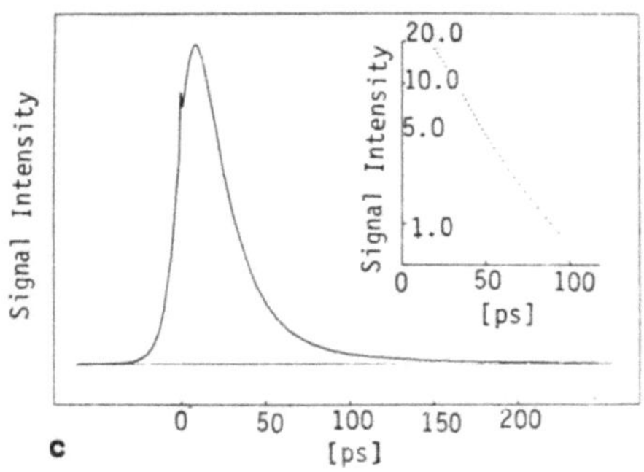

Fig. 1
a EV/EtOH, 625 nm, - 59 °C.
b EV/EtOH, 625 nm, - 96 °C
c CV/HexOH, 585 nm, r.t.

summarized in table I. From these it is evident that there exists a wavelength for each dye/solvent system where a single exponential recovery is obtained (the isosbestic wavelength). At wavelengths shorter than this the recovery signal again consists of a double exponential, now with two positive contributions. At ≈ 585 nm the two recovery rates are ca two times slower than at λ = 625 nm. The isosbestic wavelength changes from ca 600 nm in met-

Table I EV- and CV/n-alcohols, kinetic parameters at different wavelengths and room temperature 23 ± 1 °C

ROH	λ = 625 nm τ_1 [ps]	τ_2 [ps]	A_1/A_2	λ_{iso} [nm]	τ_{iso} [ps]	λ = 585 nm τ_1 [ps]	τ_2 [ps]
EV:							
Methanol			-	599	4.1	-	-
Ethanol	6.8	10.7	-	605	5.9	10.5	-
Butanol	13.5	18.0	- 1.5	611	16.1	19	-
Hexanol	15.2	63.0	- 7.2	613	19.0	32	123
Octanol	22.3	128.0	-17.0	615	30.0	48	176
Decanol	31.3	330.0	-25.0	617	42.0	62	-
CV:							
Ethanol	-	-	- 1	599	3.7	9.0	-
Butanol	7.6	13.3	- 2.3	603	7.2	13.8	-
Hexanol	9.3	29.6	- 4.0	606	12.0	22	-
Octanol	11.5	50.0	- 8.3	610	17.4	28	114
Decanol	14.4	105	-15	612	19.0	33	212

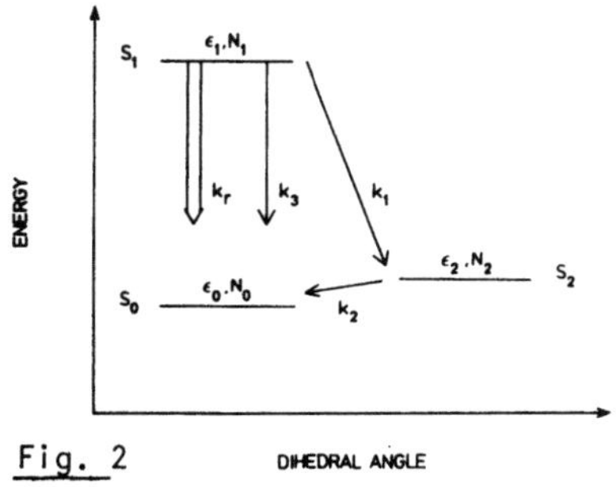

Fig. 2

hanol and ethanol solutions to ca 615 nm in decanol. The observations described above can satisfactory be explained using the energy level scheme of Fig. 2. Assuming a delta-pulse excitation the kinetic equations (1) - (3)

$$\frac{dN_0}{dt} = N_1k_3 + N_2k_2 \tag{1}$$

$$\frac{dN_1}{dt} = -N_1k_1 - N_1k_3 \tag{2}$$

$$\frac{dN_2}{dt} = N_1k_1 - N_2k_2 \tag{3}$$

$$N_0 = 0 \text{ at } t = 0$$

give the expression (4) for the induced transmission change ΔT in the sample

$$\Delta T \sim A_1 \exp(-k_1t) + A_2 \exp(-k_2t) \tag{4}$$

where

$$A_1 = (k_2 - k_1\varepsilon')/(k_2 - k_1), \quad A_2 = k_1(\varepsilon' - 1)/(k_2 - k_1) \text{ and } \varepsilon' = \varepsilon_2/\varepsilon_0$$

The rest of the symbols are explained in Fig. 2. For simplicity $k_3 = 0$. With this scheme the measured quantity ΔT is predicted to obey a double exponential decay. The amplitudes and signs of the individual components of the recovery signal depend on the relative magnitudes of k_1, k_2 and $\varepsilon_0, \varepsilon_2$. Thus, the observed recovery signals at $\lambda \approx 625$ nm see Fig. 1a correspond to the case when $k_1 > k_2$ and $\varepsilon' > 1$. At the isosbestic wavelength $\varepsilon' = 1$ and consequently $A_2 = 0$ which reduces the signal to a single exponential with lifetime $\tau_1 = 1/k_1$. At wavelengths shorter than λ_{iso}, $k_2/k_1 < \varepsilon' < 1$ result in a double exponential decay with two positive contributions as shown in Fig. 1c. At wavelengths shorter than ca 590 nm the situation is further complicated by the appearance of another isomer [3,12,13]. This is evidenced by a factor of two slower relaxation rates at these wavelengths. In fact, even at λ_{iso} the measured lifetimes are slightly affected by the presence of this species, see Table I. Both the rate of formation (k_1) and decay (k_2) of the intermediate state S_2 is seen to be viscosity dependent (Table I). k_1 has approximately a $\eta^{-2/3}$ dependence wheras k_2 displays a much stronger viscosity dependence, $k_2 \sim \eta^{-1.5}$. Since both the formation and decay of S_2 is viscosity dependent it is near at hand to consider a twisting motion of the phenyl rings (or possibly the NR_2-groups) as responsible for the observed relaxation. This is also the view traditionally assumed to explain the viscosity dependence of the fluorescence quantum yield [8-10]. From the picosecond absorption measurements it is not possible to determine if S_2 is an excited state or a ground state. To settle this point time-resolved emission experiments on a few of the systems in Table I was performed [14]. Only the fast decay (k_1) was observed in these experiments, no long-lived emission corresponding to k_2 of table I could be detected within the experimental error limits, $A_2/A_1 < 10^{-3}$. This result indicates that S_2 probably is an unstable ground state species.

Finally it remains to relate the observations in EV/EtOH upon changing the temperature, Fig. 1a-b, to the proposed model. The measurements were performed at a fixed wavelength, 625 nm. When the temperature is lowered λ_{iso} gradually moves towards longer wavelengths (cf. table I in the case of n-alcohols), consequently $\varepsilon' \rightarrow 1$. At a particular temperature it is expected that $\varepsilon' = 1$ and thus $\lambda_{iso} = 625$ nm. Evidently this happens at T $\sim$ - 90 ^{o}C for EV/EtOH. Below this temperature a double exponential decay composed of two positive terms is observed, Fig. 1b.

In conclusion, we have observed a twisted intermediate state in the relaxation pathway of the TPM dyes EV and CV in n-alcohols. We have also confirmed earlier proposals concerning the existence of an equilibrium between two different conformers in CV and EV. What at first sight was a puzzling variation in absorption recovery with changing wavelength, temperature and viscosity could be interpreted according to a simple model after performing detailed absorption measurements with both picosecond and wavelength resolution.

References

1. D. Magde, M.W. Windsor: Chem. Phys. Lett. *24*, 144 (1974)
2. E.P. Ippen, C.V. Shank, A. Bergman: Chem. Phys. Lett. *38*, 611 (1976)
3. J.M. Grzybowski, S.E. Sugamori, D.F. Williams, R.W. Yip: Chem. Phys. Lett. *65*, 456 (1979)
4. D.A. Cremers, M.W. Windsor: Chem. Phys. Lett. *71*, 27 (1980)
5. W. Yu, F. Pellegrino, M. Grant, R.R. Alfano: J. Chem. Phys. *67*, 1766 (1977)
6. M.D. Hirsch, H. Mahr: Chem. Phys. Lett. *60*, 299 (1979)
7. G.S. Beddard, T. Doust, M.W. Windsor: Picosecond Phenomena II, eds. R.M. Hochstrasser, W. Kaiser, C.V. Shank, p. 167 (1980)
8. T. Förster, G. Hoffmann: Z. Phys. Chemie N.F. Bd *75*, 63 (1971)
9. L.A. Brey, G.B. Schuster, H.G. Brickamer: J. Chem. Phys. *67*, 2648 (1971)
10. C.J. Mastrangelo, H.W. Offen: Chem. Phys. Lett. *46*, 588 (1977)
11. V. Sundström, T. Gillbro: Appl. Phys. *24*, 233 (1981)
12. G.N. Lewis, T.T. Magel, B. Lipkin: J. Am. Chem. Soc. *64*, 1774 (1942)
13. V. Sundström, T. Gillbro, H. Bergström: Submitted to Chem. Phys.
14. V. Sundström, A. Holzwarth, T. Gillbro: to be published.

Kinetics of Stimulated and Spontaneous Emission of Dye Solutions Under Picosecond Excitation

B.A. Bushuk, A.N. Rubinov, A.A. Murav'ov, and A.P. Stupak

Institute of Physics, BSSR Academy of Sciences,
Minsk 220602, USSR

As was demonstrated in [1], ultrafast superfluorescence pulses can be developed in a dye solution under pumping by a powerful picosecond mode-locked solid-state laser, though normally superfluorescence emission of dye has a rather broad spectrum and its frequency tuning is comparatively inconvenient (it requires changing the dye concentration in a solution). The extreme simplicity of the method makes it quite attractive and useful for some applications in picosecond spectroscopy.

However, the mechanism of ultrashort superfluorescence pulse formation in dye solution is not quite clear. Two different cases have been observed experimentally: (1) dye emission synchronized with a pumping pulse, and (2) superfluorescence pulses delayed about 10 ps with respect to the pumping pulse. The reason for such different behaviour is not completely understood.

In this paper we present some new experimental data on the time development of ultrashort superfluorescence in dye solution. These data show the importance of self-focusing phenomena in this process. In addition, a new technique for time-resolved measurements of low-intensity light signals is demonstrated. The experiments were carried out with a mode-locked phosphate glass neodymium laser. The second harmonic of a single 5-ps pulse extracted from the oscillator and amplified in a two-stage amplifier was used for the excitation of rhodamine 6G in alcohol. The time evolution of dye superfluorescence radiation was investigated using the up-conversion method [2]. Frequency mixing of the dye emission with a 1.054-μ picosecond pulse was observed in a KDP crystal. Selected by a monochromator, the sum radiation was registered with a photomultiplier at different time delays between pumping and 1.054-μ pulses. At the same time, the spectrum and the far field distribution of the superfluorescence were detected at each shot.

The following dependence of superfluorescence parameters on the pumping power density was observed in the experiments: when the pumping power was less than or equal to 10^9 W/cm^2, the superfluorescence pulse had a duration

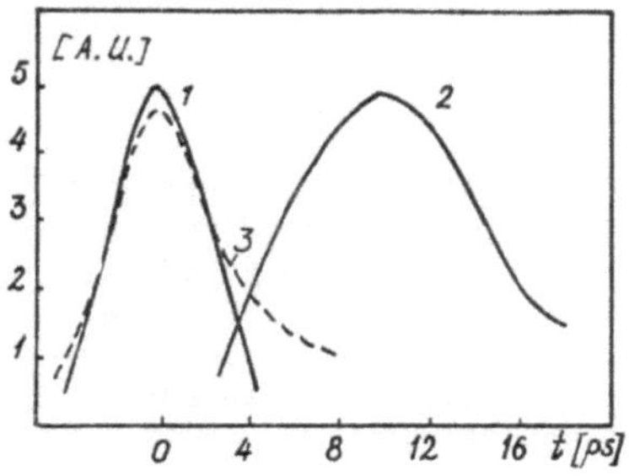

Fig.1. Kinetics of the superfluorescence maximum of rhodamine 6G in ethanol at low (2) and high (3) levels of pumping. (1) Pumping pulse

of ~15 ps and was delayed for ~10 ps in respect to the pumping pulse. The correlation function of the up-conversion of superfluorescence is shown by curve 2 in Fig.1, while curve 1 corresponds to the third harmonic generation of the pumping pulse. The diffused structureless spectrum and the smooth intensity distribution in a far field were characteristic features of the superfluorescence emission in this case.

A drastic change in the superfluorescence behaviour was observed when the pumping power density was increased up to 10^{10} W/cm^2. Under these conditions a periodic structure showed up in the spectrum, and characteristic "hot spots" appeared in a far field picture of superfluorescence. Such behaviour gives a clear indication of the formation of self-focusing filaments in a solution induced by the pumping beam. At the same time, a change in the temporal evolution of the superfluorescence was observed (Fig.1, curve 3). The superfluorescence pulse was shortened to the duration of the pumping pulse and coincided with it in time.

The correlation between the temporal behaviour change and the change in the spectral and spatial intensity distribution gives direct evidence of the important role of self-focusing phenomena in the formation of the ultrashort superfluorescence pulse synchronized with the pumping one. An investigation of the evolution of dye superfluorescence in different solvents and direct observation of filament dynamics give additional information on this point. We observed a spectral change in the superfluorescence of dye after it passed through pure benzene in which the self-focusing filaments were synchronously induced by additional radiation. After the benzene cell the initially smooth spectrum acquires the characteristic periodic structure. Our experiments showed that with self-focusing of the pumping beam, the duration of dye superfluorescence is close to the filament lifetime. An increase in superfluorescence duration is observed when a solvent with a longer Kerr relaxation time is used. The ultrashort superfluorescence pulses proved to be useful in the investigation of orientational relaxation processes in dye solutions under excitation with different wavelengths.

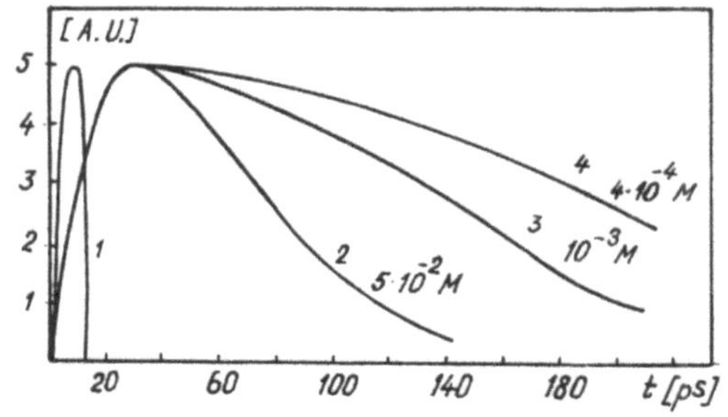

Fig.2. Kinetics of 5-mm base dye laser pumping with 7-ps pulse of the second harmonic of a Nd-laser. (1) Pumping pulse, (2,3,4) dye laser pulses at different concentrations of dye

In some cases, the frequency-tunable light pulses of variable duration may be of practical use. Our experiments showed that such pulses can be generated in a dye solution placed in a resonator with a short base. While pumped with 5-ps duration pulses, such a system generates pulses of 200-50 ps; the duration of the pulses can be varied by changing the dye concentration or the resonator length. Figure 2 shows the kinetics of the emission of an ethanol solution of rhodamine 6G in the 5-mm-long resonator at different concentrations of dye molecules.

The possibility of ultrafast kinetic measurements of low-intensity light pulses using picosecond CARS spectroscopy is also demonstrated in this paper.

The second harmonic of a single 5-ps Nd-laser pulse was superimposed with a dye superfluorescence pulse in a benzene cell. When the emission spectrum of the dye was overlapped with the first Stokes component of Raman scattering of the pumping frequency in benzene, and the angle between two interacting beams was kept within 1°-2°, anti-Stokes scattering of pumping light and amplification of the superfluorescence at the first Stokes component were observed. The duration of dye emission was found by measuring the intensity of the anti-Stokes scattering versus time delay between pumping and probing pulses. The measurements were easily carried out with a probing pulse having an intensity three orders of magnitude lower than the pumping pulse.

The orientational relaxation of solvent molecules in a dye solution not only influences the duration of the dye superfluorescence via the self-focusing mechanism, but also causes a temporal change in the spontaneous emission spectrum. It is known that the excitation of a dye molecule in polar solution leads to an orientational redistribution of solvent molecules incorporated into the solvate, the process being followed by a temporal shift in the fluorescence (or gain) spectrum of the solution. If the solvent molecules are rotationally anisotropic, one can expect at least two different relaxation times to appear in the process. The shortest time character-

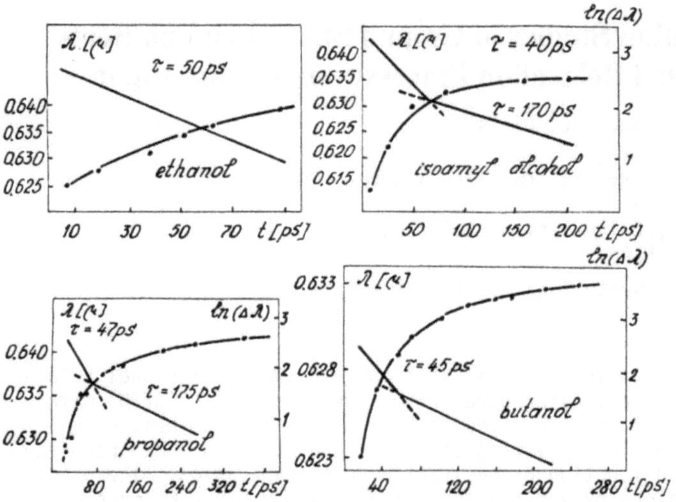

Fig.3. Kinetics of gain band maximum of oxazin 17 dye in different solvents. Picosecond (5-7 ps) excitation at $\lambda = 0.527\mu$

izes the fastest component of rotation, i.e., the rotation around the axis perpendicular to the smallest cross-section of the molecule, while the longer time corresponds to rotation in a perpendicular plane.

To observe experimentally the existence of the two relaxation times mentioned, we investigated the spectral kinetics of the gain of the oxazin 17 dye in different solvents using the conventional picosecond continuum probe technique. The results presented in Fig.3 show that different relaxation times (40-45 ps and 120-170 ps) exist in all investigated alcohols except ethanol. It is quite probable that ethanol also has two different relaxation times, but that the shorter one is too short to have been observed in our experiments. This hypothesis is in accordance with the fast relaxation of self-focusing filaments observed for ethanol in the experiment with superfluorescence kinetics discussed above (Fig.1).

References

1 A.N. Rubinov, M.C. Richardson, K. Sala, A.J. Alcock: Appl. Phys. Lett. *27*, 358 (1975)

2 L.A. Hallidy, M.R. Topp: Chem. Phys. Lett. *46*, 8 (1977)

Picosecond Resolution Studies of Ground State Quantum Beats and Rapid Collisional Relaxation Processes in Sodium Vapor

R.K. Jain, H.W.K. Tom[1], and J.C. Diels[2]

Hughes Research Laboratories
3011 Malibu Canyon Road, Malibu, California 90265, USA

[1]Department of Physics, University of California, Berkeley, California, USA

[2]Department of Physics, North Texas State University, Denton, Texas, USA

We report experimental studies of coherent transients and collisional relaxation processes in Na vapor, using degenerate four-wave mixing (DFWM) with picosecond pulses. These include investigation of the quantum beat modulation (due to the hyperfine splitting of the ground state) of the intensity of the DFWM signal as a function of excitation pulse separation. The picosecond resolution obtained by varying the pulse delay in our DFWM experiments has resulted in the first observation of harmonic structure in the quantum beat modulation, and in the observation of an anomalously rapid (e^{-1} decay time <50 psec) collisional relaxation process at increased temperatures.

In our experiments, we used the standard DFWM optical arrangement with forward (f) and counter-propagating backward (b) pump pulses, and a probe (p) pulse arriving at the Na cell at times: t_f, t_b, and t_p, respectively. Optical delay lines allowed us to vary both t_p and t_b independently with respect to t_f, whose instant of arrival is defined here as the zero reference of time. For our excitation source, we used ∿35 psec pulses ($\Delta\nu$ = 9.5 GHz) at 5890Å (D_2-line) from a cw Rh 6G dye laser synchronously-pumped by a mode-locked Ar^+ laser (repetition rate ∿80 MHz). DFWM signals were easily observed and the average power in the signal wave (i.e., the energy in the signal pulse) was monitored with a slow photodiode and phase-sensitive (~1 KHz) detection electronics. In contrast with other related work [1-3], no cavity dumping or amplification of the output of the cw mode-locked dye laser was used for these experiments, and pulse areas were typically less than π/100. A Faraday isolator was used to prevent feedback into the dye laser.

In Fig. 1, we show the signal intensity as a function of t_p, with t_b fixed at values of 1670 psec and 370 psec. The signal intensity is plotted on a logarithmic scale, to emphasize the dynamic range of the data in the region of fast decays, such as $t_p < 0$. The modulation of the signal in both Figs. 1a and 1b at both the fundamental and second harmonic of the ground state hyperfine frequency (ν_{12} = 1.77 GHz, T_{12} = 565 psec) is easily observable. In the special case in which pulses are incident in the sequence f,p,b (i.e., $0 < t_p < t_b$), the four-wave mixing signal has been identified as a photon echo (called the backward-wave echo) with emission occurring at $t=t_p+t_b$[1,4]. No echo occurs for $t_p < 0$, or for $t_p > t_b$. Quantum beat modulation in the echo itself has been discussed by Nakatsuka, et al., but this is the first experimental study of quantum beat modulation of the pulse energy as a function of excitation pulse separation times. Characterizing this quantum beat modulation is essential before ultrashort pulses can be used for DFWM studies of rapid relaxation processes, such as for the study of the effect of dephasing or rephasing collisions with a large number of external perturbers.

The rapid decay of the signal in Fig. 1b, at t_p ∿ 370 psec, is due to the change in the pulse sequence. This rapid decay is consistent with the absence

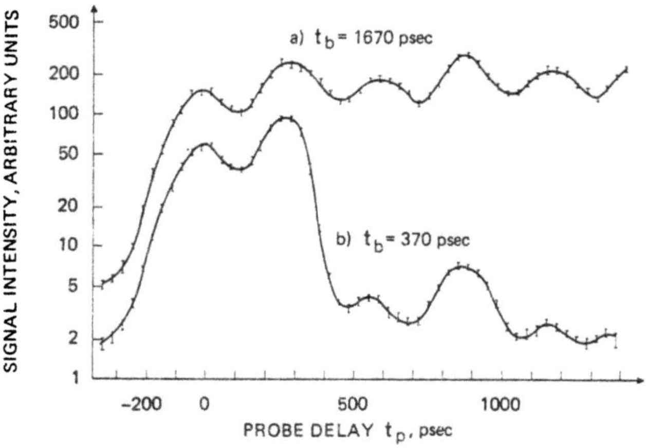

Fig.1 Signal intensity vs. t_p at two values of the delay t_d and at fixed cell temperature. The relative magnitude of the curves is not significant.

of the photon echo, and was verified by noting rapid decays at the appropriate time delays t_p for various other choices of t_b. The modulation in the signal for $t_p > t_b$ is attributed to subsequent backward pulses (from the high repetition rate laser) scattering off the ν_{12} modulated spatial grating formed by an initial f-p pulse pair.

It is well established that transient four-wave mixing in the backward-wave echo geometry is a useful probe of relaxation processes affecting single states and coherent superpositions of states [1,4]. In Fig. 2, we show evidence of an ultrafast collisional relaxation process that we observed at increased temperatures. 250 mTorr of He buffer gas was used for the data of Fig. 2(a). Such a strong temperature dependence and anomalously rapid decay is not directly attributable to known cross sections of Na-He and Na-Na collisions. In the absence of He buffer gas (see Fig. 2(b)), similar decays were observed at slightly higher (∿15°C) temperatures, excluding the possibility of relaxation via Na-He collisions. One possible source of such behavior is an unknown impurity, perhaps due to outgassing from the specific cells used; we are presently unaware of an impurity with such a large Na-perturber cross section. This rapid decay is presently a topic of further investigation [6].

In our talk, we will compare our experimental observations with theory. The relation between quantum beats observed in our backward-wave echo experiments and quantum beats observed in fluorescence [5] and transmission [2] experiments will be discussed.

References

1. M. Fujita, H. Nakatsuka, H. Nakanishi and M. Matsuoka, Phys. Rev. Lett. 42, 974 (1979).
2. H. Harde, H. Burggraf, J. Mlynek and W. Lange, Optics Lett. 6, 290 (1981).
3. H. Nakatsuka, M. Fujita and M. Matsuoka, Optics Comm. 36, 234 (1981).
4. T. Mossberg, A. Flusberg, R. Kachru and S.R. Hartmann, Phys. Rev. Lett. 42, 1665 (1979).

a

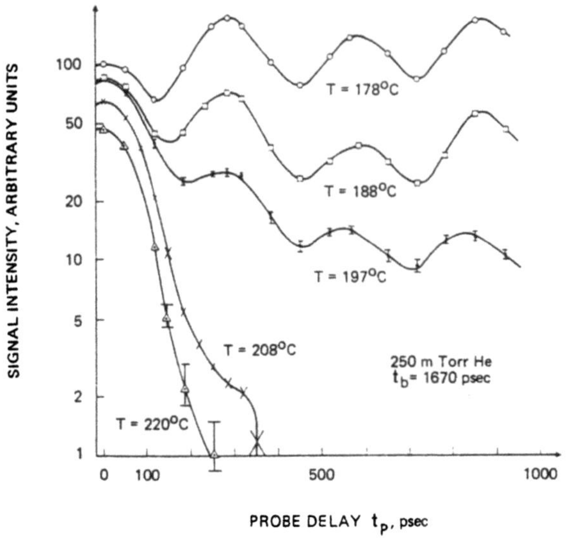

b

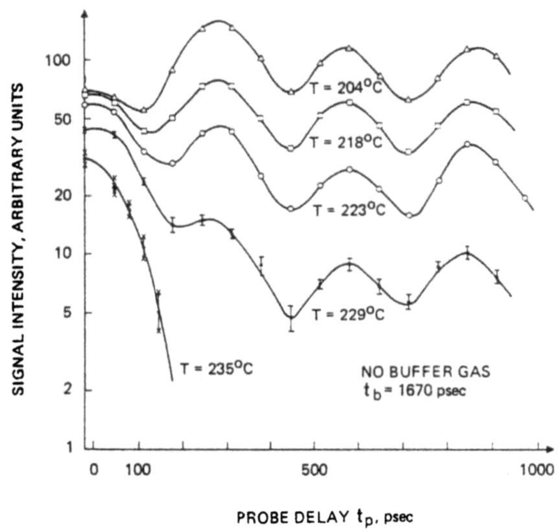

Fig.2 Signal intensity vs. t_p at fixed t_b and different values of the cell temperature. In Fig. 2(a), 250 mTorr of the buffer gas was also present in the cell.

5. S. Haroche, in High-Resolution Laser Spectroscopy, ed. K. Shimoda, Springer-Verlag, New York (1976).
6. Another possible mechanism for this rapid decay is the associative ionization of the sodium from the laser-excited $3P_{3/2}$ states. For instance, see: F. Roussel, B. Carré, P. Breger, and G. Spiess, J. Phys. B14, L313 (1981); A. de Jong and F. van der Valk, J. Phys. B12, L561 (1979).

Part V

Picosecond Chemical Processes

Unimolecular Processes and Vibrational Energy Randomization

R.A. Marcus

A.A. Noyes Laboratory of Chemical Physics, California Institute of Technology, Pasadena, CA 91125, USA

Intramolecular vibrational energy transfer is the subject of much current experimental and theoretical research. Some aspects of this problem are reviewed in this talk, together with a few new additions. Inasmuch as extensive references have been given in a recent review,[1] they are omitted in this extended abstract in the interests of brevity. Topics discussed include (1) experimental work, (2) some theoretical concepts on "regular" and "chaotic" motion in the classical and quantum mechanics of anharmonic systems, and (3) the question of whether experiments can distinguish regular from chaotic behavior. By way of introduction we begin first with a review of the unimolecular rate constant of reactions occurring in the gas phase.

Unimolecular Reaction Rate Constant

The study of intramolecular energy transfer began with investigations of unimolecular reactions in the 1920's and 1930's, and was resumed especially in the post 1950's. The current interpretation of these data via a Lindemann mechanism for a reaction A → B involves activation by collision with a molecule M to form a vibrationally-hot molecule A*, which can either be deactivated by subsequent collisions or form a reaction product B:

$$A + M \rightleftharpoons A^* + M, \qquad A^* \xrightarrow{k(E)} B \tag{1}$$

Current interpretation of experimental data in these reactions is typically via a statistical theory - RRKM theory - in which a microcanonical distribution of states is assumed for each A* of energy E and for the transition state of the second step in (1). Transition state theory is then used to calculate the rate constant k(E). One finds, in RRKM theory,

$$k(E) = N^{\dagger}(E)/h\rho(E), \tag{2}$$

where $N^{\dagger}(E)$ is the number of quantum states of the transition state of energy less than or equal to E, h is Planck's constant and ρ(E) is the number of states per unit energy for molecule A*.

This expression for k(E) is then multiplied by the steady state distribution function $\rho_s(E)$ dE for the probability of finding an A* in the energy range (E, E + dE), and integrated over all E. k(E) may also depend on other quantum numbers, such as the angular momentum of A*, J, e.g., via the centrifugal potential, and then one integrates the corresponding $k(E,J)\ \rho_s(E,J)$ dEdJ over all E and all J.

Much of the current experimental research is designed, in effect, to avoid the need for this convoluting of k(E) by preparing molecules A* with a narrower energy distribution, namely by some way other than the collisional path in eq.(1). Before considering some of these methods, we first give another form of eq.(2) for k(E), derived in the Appendix. It is somewhat more revealing of its content and appears to be new:

$$k(E) = \nu_c N^\dagger(E)/N_1(E), \tag{3}$$

where ν_c is any classical vibration of A* and $N_1(E)$ is the number of states of A* with energy equal to or less than E and with this one degree of freedom removed. Thus, k(E) equals this frequency ν_c multiplied by the quantum equivalent of a ratio of phase space available to the transition state to that available to a molecule with the same number of degrees of freedom. The geometry and vibration frequencies may differ for $N^\dagger$ and N_1, and $N^\dagger(E)$ is zero until E exceeds some critical E_0, which in turn is related to the thermal activation energy. As a function of E, k(E) has a threshold at $E = E_0$, then rises and tends to "saturate" at high E.

Some Experimental Studies

Among the studies that have been made are those of (1) unimolecular reactions, (2) chemical activation (in bulk and in molecular beams), (3) photochemical excitation followed by internal conversion, (4) infrared multiphoton dissociation (in bulk and in beams), (5) high overtone induced reactions, (6) dissociation of van der Waals' complexes, (7) unimolecular reactions in beams at low energies, (8) infrared excitation followed by the study of subsequent fluorescence or absorption, (9) vibronic excitation followed by dispersed and other fluorescence, and (10) high resolution spectroscopy. Several of these are considered below:

One of the first methods used to secure a narrower range of E's for A* involves its formation via a chemical activation step, e.g., the addition of an atom or free radical to an olefinic double bond, or the insertion of a CH_2 group into a CH bond,

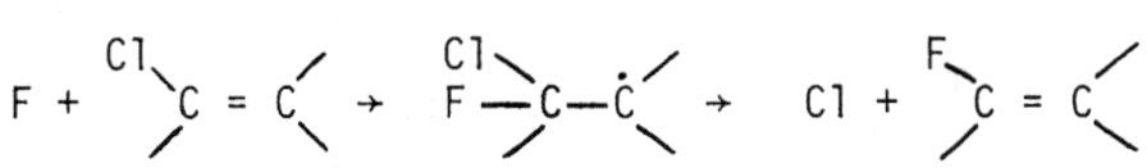

(4)

The intervening free radical or molecule is vibrationally hot. Its energy distribution is mainly due to the thermal distribution of the initial substate. Chemical activation has been studied in bulk and in molecular beams. In the bulk case k(E) has been inferred via measurements of products in competition with the frequency of deactivating collisions, and E has been varied by varying the initial reaction leading to the formation of a hot molecule. A classic example of the use of the former to study the rate of vibrational energy randomization has been the measurement of ratios of products in carefully selected reactions by Rabinovitch _et al_. Typically, the intramolecular energy redistribution time thus inferred was about 1 ps at the high vibrational energies involved.

The products' energy distribution for chemical activation in beams has been used to infer that of the transition state and, thereby, whether or not the microcanonical assumption used for A* in RRKM theory is valid. Such measurements are unambiguous when there are no exit channel effects (no subsequent translational-vibrational energy exchange in the exit channel), namely when the exit transition state is "loose". Reaction (4) displayed statistical behavior. An interesting example of nonstatistical behavior occurred, instead, in the reaction of Cl addition to a bromoolefin, with Br subsequently leaving the hot free radical. The estimated lifetime (< 1 ps) was presumably too short for statistical intramolecular randomization to occur.

Another method for preparing vibrationally hot molecules is a photochemical one in which the molecule is excited with light to form an electronically-excited state. In some systems the resultant excited molecule undergoes internal conversion to form a vibrationally hot ground state and its subsequent unimolecular reaction can be studied.

Vibrationally hot A*'s have also been prepared by selectively pumping a vibrational mode of a molecule using infrared multiphoton absorption. Such studies provide interesting insight into different properties of the states of the molecule in the lower energy range and in the higher one (the so-called quasi-continuum). To be sure, the system yields a distribution of E's of the reactive A*'s, rather than a nearly monochromatic one. A number of the studies have tested the RRKM form of k(E) indirectly, via measurement of the energy distribution of the reaction products, usually measuring the translational distribution but on some occasions a vibrational one. Another test has been via measurement of branching ratios, e.g., in the measurement of the HF/HBr ratio in a reaction in which CF mode was pumped:

$$CH_2FCH_2Br \xrightarrow{nh\nu} CH_2 = CHBr + HF \quad \text{or} \quad CHCF = CH_2 + HBr \tag{5}$$

The observed ratio was found to be the statistical one computed from RRKM theory.

By and large, the behavior of the dissociating or isomerizing A* prepared by the above methods has been interpreted using RRKM theory. More recently, vibrationally-hot molecules have been prepared by excitation of a high CH overtone in a molecule. The lifetime has been measured indirectly as a function of the vibrational energy, via competition with deactivating collisions, and, subsequently, directly using studies in real time. In the case of $CH_3NC \rightarrow CH_3CN$, the results are in ballpark consistency with those inferred from RRKM theory and unimolecular studies, while data on the unimolecular studies is needed for comparison in the other cases. Some argument regarding nonrandom behavior has been made from apparent ±50% fluctuations from a monotonic k(E) vs E plot in one reaction. Such effects if real would probably go unnoticed in unimolecular plots, but the main question is rather whether or not the k(E) measured in such selectively-prepared molecules is or is not in rough agreement with that inferred from statistical (e.g., RRKM) theory.

Another interesting source of information has been initially cold van der Waals' complexes (such as I_2He) in which the I_2 is vibrationally-excited via lasers. The subsequent lifetime has been inferred indirectly from line width studies. Typically, one has $I_2(v)X \rightarrow I_2(v-1) + X$ and RRKM theory does not apply to such systems. In the language used below, their motion is much too regular, rather than chaotic.

Other more recent interesting and relevant studies include the infrared emission from a molecule (methyl formate) in which a CH band is excited (one photon) with an infrared laser and emission from other bands is observed, (McDonald), the observation of vibrational quantum beats (Zewail), the pump-probe work and vibrational quantum beats reported here by Bloembergen, the study of k(E) versus E for molecules in which the energy barrier for reaction is very low - $\sim$ 1500 cm^{-1} (reported here by Zewail) instead of the more usual 15,000 cm^{-1} barrier, the study of modal intramolecular relaxation via dispersed fluorescence in substituted benzenes, (Parmenter at this conference, Smalley) and the measurement of highly resolved spectral sequences in small molecules.

Some Theoretical Concepts

We turn next to current theoretical work on regular and chaotic motions within a molecule. An anharmonic molecule, viewed classically, is an example of a nonlinear mechanical system. Such a system has been extensively studied in recent years, particularly in astronomical journals. There is a relatively recent theorem (Kolmogorov-Arnold-Moser) which shows that at low energies the motion of such a system is for most initial conditions highly "regular" (i.e., quasi-periodic): In an N-coordinate system such a motion has N constant "action variables" and so moves on an N-dimensional surface in a 2N dimensional phase space, a torus, rather than on a 2N-1 constant energy surface. It is thus quite highly restricted in its motion. An example of this regularity is given for N=2 by the classical trajectory in coordinate space (e.g., Figs.1-4 below). Such motion is highly nonstatistical.

At higher energies the motion tends, for an increasing fraction of the initial conditions, to be chaotic rather than regular, as for example in Fig.5. The spectrum of the classical trajectory, determined from the Fourier transform of an autocorrelation function of the trajectory, changes correspondingly. It is "regular" for a quasi-periodic trajectory (N fundamental frequencies, plus overtones and combinations). In the chaotic case it is more diffuse, centered around principal frequencies usually red-shifted from those at lower E.

The above dynamical behavior is consistent with one current view in which statistical theory is appropriate for interpreting rate processes at higher energies; at lower energies, a quasi-periodic type of theory should be used, e.g., by treating a molecule as possessing independent vibrational modes, using some version of the usual radiationless transition theory and its Franck-Condon type matrix elements to treat rate processes. Such a treatment would, in a sense, be a quantum modification of N.B. Slater's treatment of unimolecular reactions. In that treatment, which was classical, a special case of quasi-periodic motion was assumed, namely, independent harmonic oscillations. The drastic effect of even small anharmonic coupling on the motion in the presence of internal resonances was neglected but can be included in any modified theory. The "transition state" (using transition state terminology) was assumed to be a hyperplane in the coordinate space.

The question arises as to whether this classical mechanical transition from quasi-periodic to chaotic motion has quantum implications. According to our recent findings in a numerical study, the former may be a necessary but not a sufficient condition for quantum "chaotic" behavior (as judged by the spectrum). Sufficient conditions have been postulated, both by Kay and by the writer, based on Chirikov's theory for the onset of chaotic motion ("overlap" of internal resonances). Our sufficient conditions for quantum chaos include one where the overlap region in phase space should (when divided by h^N) exceed unity, i.e., contain at least one quantum state.

One limitation of the Chirikov theory for classical chaos is that it is basis-set dependent, i.e., dependent on the choice of the unperturbed Hamiltonian, H_0. Another approach, which I am suggesting and which we are currently testing, is a basis-set independent extension of the Chirikov resonance, namely we examine the trajectories and find the various domain or domains in phase space where a trajectory of given shape dominates. One

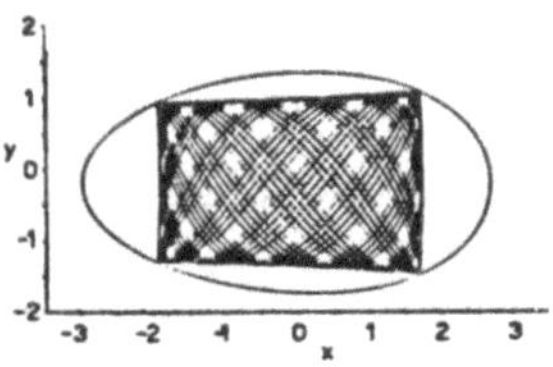

Fig.1

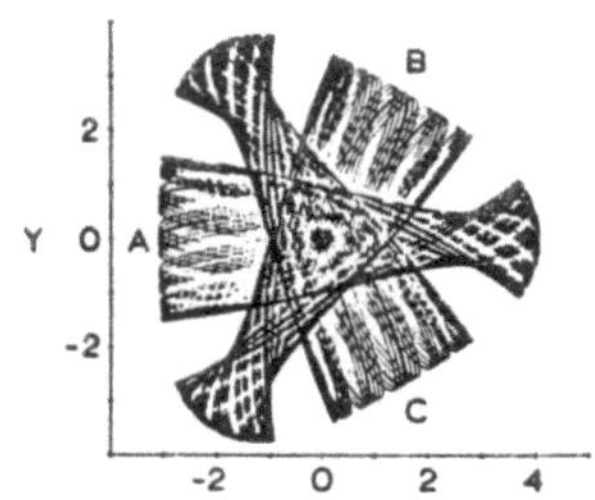

Fig.2

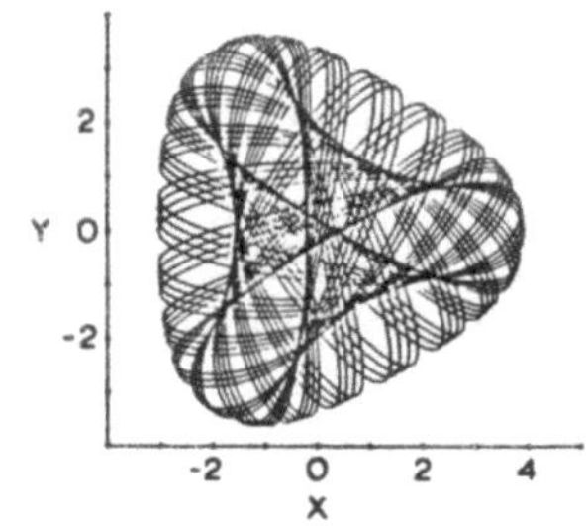

Fig.3

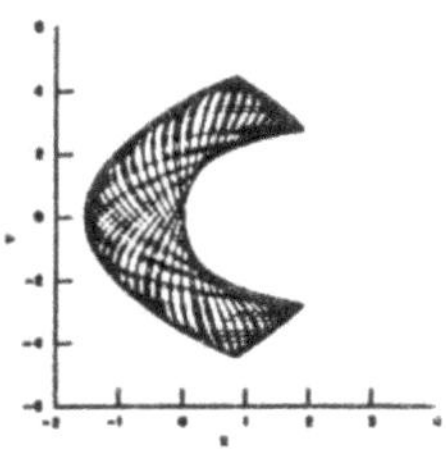

Fig.4

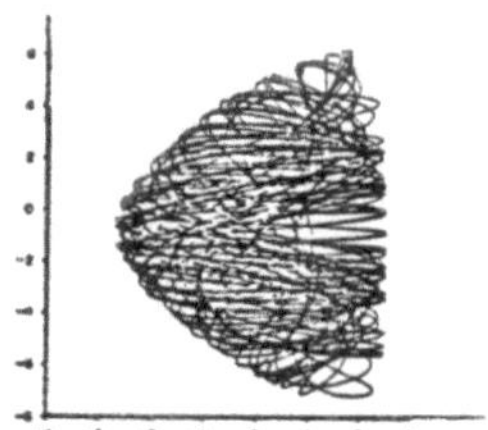

Fig.5

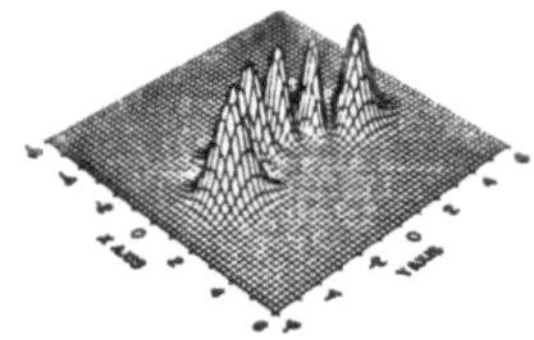

Fig.6

Legend to Figures 1-6

Classical trajectories y(t) vs x(t) for the Hamiltonian $H = \frac{1}{2}\left(p_x^2 + p_y^2 + \omega_x^2 x^2 + \omega_y^2 x^2\right) + \lambda x(y^2 + \eta x^2)$, for several cases: Fig.1 ω_x and ω_y not commensurable; Fig.2, $\omega_x = \omega_y$ (3 librating ellipse-like trajectories); Fig.3, $\omega_x = \omega_y$ (1 precessing ellipse-like trajectory); Fig.4, $\omega_x = 2\omega_y$ (1 librating figure-eight like trajectory); and Fig.5, $\omega_x = 2\omega_y$ (chaotic trajectory). In Fig.6 is the wave function for the system of Fig.4

would then try to determine if a chaotic region corresponds to an overlap of such domains. Further, the size of that overlap domain relative to h^N would indicate, according to this model, whether or not a local "quantum chaos" exists. A trajectory of a given shape corresponds, incidentally, to a given resonance (cf Figs.2-4).

Another approach to "quantum chaos" that we have suggested, again basis-set independent, involves examination of plots of vibrational energy eigenvalue vs a perturbation parameter, and seeing whether avoided crossings and, more particularly, whether overlapping avoided crossings occur. Overlapping of avoided crossings will result in destruction of regular shapes of wave functions, so that fairly regular nodal patterns no longer occur. (A regular-shaped wave function is given in Fig.6.) The relation between the "overlap of different trajectory shapes" method and the overlapping avoided crossings method will be discussed elsewhere.

Can Experiment Distinguish Chaotic from Regular Behavior?

Not considered above is the relation, if any, between experiment and the classification of states as regular or chaotic. The more well-defined the "preparation of the initial state" in an experiment the more one expects a difference between regular and chaotic behavior. For example, if one examines the vibrational spectrum of a molecule at ultra high resolution and finds regular sequences of eigenvalues, the corresponding quantum states of the molecule are apt to be "regular": An isolated avoided crossing at the given value of a perturbation parameter will cause some mixing of two nodal patterns of the wave function, and some perturbation of the regular eigenvalues' sequence. [E.g., the eigenvalue sequence in a pendulum problem (libration-rotation), which is a model for an isolated resonance, is not as simple as that of, say, a Morse oscillator.] Overlapping avoided crossings will produce even more irregularity in the spectrum.

In reactions, the difference in rate constant of A* in the two models - quasi-periodic or chaotic - can be large, depending on the preparation of the initial state. (Strictly speaking, a dissociating molecule is never quasi-periodic, at most only approximately so, but we'll neglect this detail.) If A* is prepared via a collision in eq.(1) to form a nearly microcanonical distribution of states, or a distribution of states centered about the microcanonical, the energy dependence of the rate constant might be similar in the regular and chaotic cases. If, on the other hand, the state of preparation is less coarse-grained the experiment may distinguish between the two cases. Relaxation by itself is no indication: For example, it has been shown that even an integrable system (no classical chaos at any energy) shows relaxation of a classical state if the initial excitation involves the excitation of many KAM tori at once, by exciting some zeroth order mode, rather than by exciting motion on a single KAM torus.

This field is in its infancy, both experimentally and theoretically, and it will be interesting to see the growing relation between these new experiments and the new theories.

Appendix. Derivation of Eq.(3)

The $\rho(E)$ is $dN(E)/dE$, where $N(E)$ is the number of states of A* with energy equal to or less than E. $N(E)$ can be written, at least approximately, as $N_1(E-E')$, the number of states of an A* with one degree of freedom removed and with energy equal or less than E-E', times the number of states $\rho'(E')dE'$ of the missing degree of freedom having the energy in E', E' + dE', integrated over E' from 0 to E. If for this extra degree of freedom we select a vibration of A* whose frequency ν_c is low enough to be regarded as classical and which can also be treated as a harmonic oscillator, then $\rho'(E')$ equals $1/h\nu_c$. One then finds that $\rho(E)$ in eq.(2) equals $N_1(E)$ [using $dN_1(E-E')/dE = -dN(E-E')/dE'$]. From these results and eq.(2) eq.(3) follows.

Acknowledgement

This work has involved an active and stimulating collaboration with former students, particularly with Drs. Noid and Koszykowski, my co-authors in [1].

This research was supported by a grant from the National Science Foundation.

[1] D. W. Noid, M. L. Koszykowski, and R. A. Marcus, Ann. Rev. Phys. Chem. 32 267 (1981).

Picosecond Dynamics of I_2 Photodissociation

P. Bado, P.H. Berens, J.P. Bergsma, S.B. Wilson and K.R. Wilson
Department of Chemistry, University of California, San Diego,
La Jolla, CA 92093 USA

E.J. Heller
Theoretical Division, T-12, Los Alamos National Laboratory,
Los Alamos, NM 87545 USA

1. Introduction

While liquid solution reactions are much more important in chemistry, gas phase reactions are much better understood. Given the central importance of solution reactions to inorganic, organic, industrial and biochemistry, it is rather surprising that, as yet, there is not a single such reaction whose molecular dynamics are understood in detail. Theoretical and experimental evidence already makes clear that much of the important molecular dynamic action in solution reactions occurs on the picosecond and subpicosecond time scales. The dihalogen photodissociation and recombination reactions, $X_2 + h\nu \rightarrow X + X \rightarrow X_2$, involving the simplest possible molecular reactants and products, diatomics, and in rare gas solution involving only two elements, seem excellent candidates for study.

2. Theory

The first deterministic theoretical study of the molecular dynamics of reactions was by BUNKER and JACOBSON[1], who computed the classical trajectories for I_2 in CCl_4 solvent represented by 26 spherical, structureless particles in a specular cube. MURREL, STACE and DAMMEL[2] modelled the photodissociation of I_2 in dense inert gases, I_2 plus 22 gas atoms in a spherical, soft-walled container. We have similarly modelled I_2 plus 50 Xe atoms at liquid density in truncated octahedral periodic boundary conditions[3, 4], computing the photodissociation, solvent caging, atomic recombination and vibrational energy decay to the solvent from the new I_2 molecule, as shown in Fig. 1. The conclusion of all three molecular dynamic studies is that geminate recombination is usually a very fast process, over within a few picoseconds. An important caveat, and a weakness in these theoretical studies, is that the process whereby the I atoms dissociating on an excited state potential surface refind the ground state surface on which they recombine is not well understood, and is therefore handled in these calculations by arbitrary assumptions which may be incorrect. If so, the real time for geminate recombination may be longer than the few picoseconds calculated.

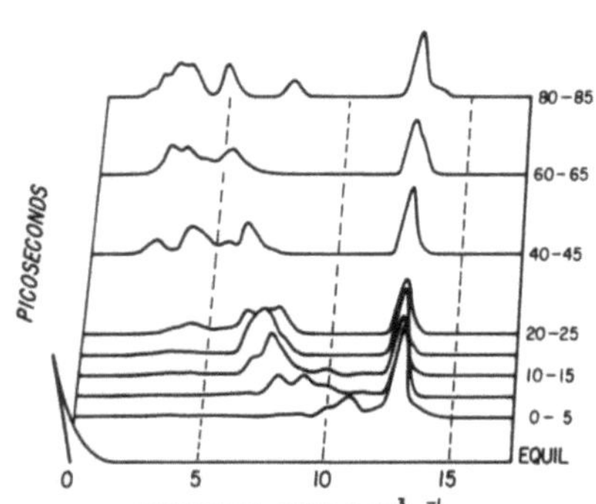

Fig. 1. Time evolution in liquid Xe solution of I_2 vibrational energy during reaction sequence of photodissociation (at time zero), solvent caging of some of the recoiling I atoms, radical recombination, and vibrational loss to solvent. Also shown is the equilibrium Boltzmann vibrational energy distribution before photodissociation. The vibrational energy distribution already bifurcates into two branches within the 0-5 ps period, the higher one corresponding to those I atom pairs which have escaped the cage and whose minimum energy is the I_2 dissociation energy, and the lower one corresponding to recombined I_2 progressively losing vibrational energy to the solvent.

In addition, theoretical calculations by NESBITT and HYNES[5, 6] for I_2 in rare gases and in CCl_4 and by our group[3, 4] in liquid Xe indicate that the decay to the solvent of the vibrational energy in the newly reformed I_2 molecule will require the order of a hundred to hundreds of picoseconds, as shown in Fig. 1, considerably slower than the time required for geminate recombination. Note that these calculations are for solvent atoms or molecules which are very weakly bound to one another, and that the vibrational relaxation might be quite different, for example, in a strongly hydrogen-bonded liquid.

From our calculated molecular dynamics, plus the potential curves and transition dipole moments, we can compute transient electronic absorption spectra[7], as shown in Fig. 2, which includes the $A \leftarrow X$, $B \leftarrow X$, and the $B'' 1u(^1\Pi) \leftarrow X$ transitions. A related nonmolecular dynamics spectral calculation has been carried out by NESBITT and HYNES[6]. A small quantum correction by temperature scaling, which would be exact for the coordinate distribution in the harmonic limit, is made to the equilibrium spectra which agree well with the known I_2 experimental spectral points[8]. Note that spectra measured at different wavelengths follow different time histories, which can in principle be used to follow the time evolution of the vibrational energy distribution of the relaxing I_2 molecules.

3. Experiment

The earliest picosecond experimental results were by the EISENTHAL group[9, 10], who measured the transient electronic absorption spectra after excitation at 530 nm. Decay times of ~70 ps for I_2 in hexadecane and ~140 ps in CCl_4 were observed. Subsequently these studies were extended by the EISENTHAL group to I_2 in aromatic solvents[11] which are believed to form complexes with I atoms, and by LANGHOFF[12] who observed I_2 photodissociation in several weakly associated liquids, finding decay times in the ~100-150 ps range. More recently, KELLEY and

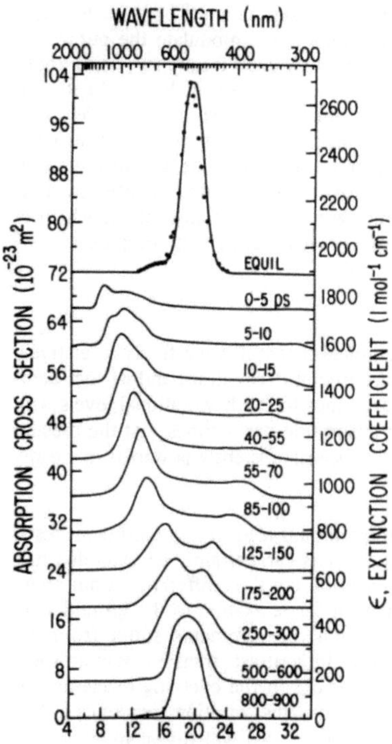

Fig. 2. Transient electronic absorption spectra computed from molecular dynamics for I_2 (in Xe solution) reaction sequence of photodissociation, solvent caging, radical recombination, and vibrational decay. Time zero is the photodissociation. The top curve is the computed equilibrium spectrum before photodissociation on which are superimposed the experimental points from Tellinghuisen[8].

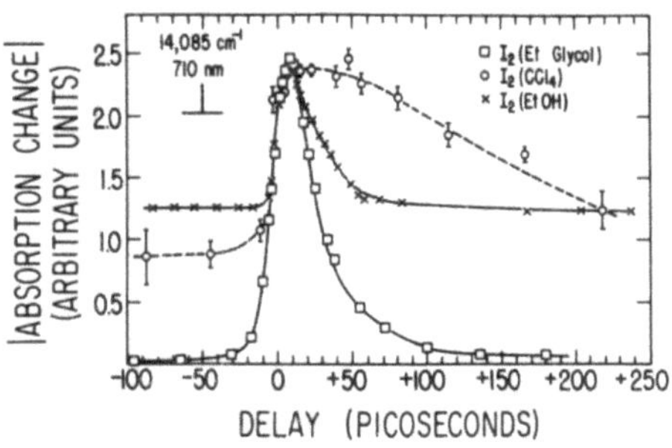

Fig. 3. Experimental transient electronic absorption spectra for I_2 in ethylene glycol (0.2 molar, 1.3×10^{-2} mole fraction), in ethyl alcohol (0.33 molar, 1.9×10^{-2} mole fraction), and in CCl_4 (0.096 molar, 9.3×10^{-3} mole fraction). All the spectra are for perpendicular orientation of linearly polarized pump and probe beams.

RENTZEPIS[13] have observed I_2 photodissociation in fluid and liquid Xe with a decay time of ~40 ps, as well as in CCl_4, and similar experiments have been carried out by the PETERS group[14]. All of the above studies used second harmonic Nd pump light at ~530 nm, which, at least in the gas phase, results in excitation largely to the bound B $0_u^+(^3\Pi)$ state which presumably predissociates, but may absorb another photon in the meantime[15, 16]. A delay of ~20 ps between excitation and the maximum in the absorption curve has been attributed to absorption from the B state[9, 13, 17]. Thus, the linking of these experimental transient spectra to the molecular dynamics of the I_2 photodissociation and recombination reaction is made difficult by the probable presence of the additional processes of B state predissociation and absorption. We have recently reported[3] transient absorption for I_2 excited using an Ar^+ – dye (DCM) synchronously pumped source at 710 nm into the dissociative A state, thus avoiding the problem of predissociation. In order to achieve the sensitivity needed to detect the very weak[8] I_2 A state absorption, we use a multiple modulation system[18] based on the discovery by HERITAGE[19] and LEVINE and BETHEA[20, 21] that the noise in Ar^+ synchronously pumped dye lasers falls off by several orders of magnitude in going from the audio to the radio frequency region. We modulate the pump and probe beams at two different radio frequencies and detect at the difference frequency, using inexpensive and readily available radio amateur equipment[18]. In addition, we audio modulate the pump beam and synchronously detect at that frequency. Decay times are shown in Fig. 3. We suggest the hypothesis that vibrational decay might be expected to be faster for the more strongly hydrogen-bonded solvents which are expected to have a greater spectral mode density in the range of I_2 vibrational frequencies. All the above experimental measurements suffer in interpretation from the rather weak connection between molecular dynamics and electronic absorption, which is further complicated by the possible presence of I_2 – solvent , I – solvent, and $I_2 - I_2$ complexes.

4. Discussion and Conclusion

In summary, two hypotheses have been advanced: i) that geminate recombination for I_2 is relatively slow and thus accounts for the observed range of transient absorption decay times and ii) that geminate recombination is relatively fast and that vibrational decay times to reach vibrational levels with higher absorption instead account for the observed transient absorption decay times. At the present time, the available theoretical and experimental tools have not been sufficiently powerful to cleanly disprove either or both of these hypotheses.

Improvements in both theoretical and experimental tools should lead to more stringent tests. On the theoretical side, molecular dynamics and spectral calculations can certainly be extended to a variety of different solvents, and to different pump photon energies and thus different I atom recoil energies. In addition, different assumptions as to the mechanism for relaxation to the ground state potential curve can be tried out. Transient electronic absorption measurements suffer from the intrinsic limitation of a weak connection to molecular dynamics. In contrast, transient Raman spectra can directly reveal, for example, rotational and vibrational periods in the evolving reactants and products. We have computed such transient spectra [3, 4] for this reaction sequence. The equivalent resonance Raman spectra can also be calculated and they also will reveal much of the underlying molecular dynamics, possibly including upper state recurrence times[22]. Thus Raman spectra could provide a definitive test of the above hypotheses.

It is surprising, but true, that for no chemical reaction in solution, not even for one as simple and as well studied as I_2 photodissociation and recombination, are the detailed atomic motions by which it occurs yet known. Even such basic aspects are not certain as the order of magnitude of the time required for caging and geminate recombination, and whether a direct deterministic or a stochastic diffusional approach to geminate recombination is most appropriate.

This situation may soon change, as there is now a four order of magnitude time range, ~100 fs to 1 ns, over which molecular dynamic calculations and short light pulse experiments can overlap. Transient infrared, Raman and electronic absorption spectra all reflect the underlying molecular dynamics of chemical reactions and can provide an interface at which theory and experiment may meet. By comparing transient spectra computed from molecular dynamics with the equivalent measured spectra, one can hope to discover the microscopic dynamics by which many chemical processes occur.

Thanks for the support which has made this work possible to NSF Chemistry, ONR Chemistry, NASA-Ames, NIH Division of Research Resources, and Fonds National Suisse for fellowship support to P. Bado.

References

1. D. L. Bunker and B. S. Jacobson, *J. Amer. Chem. Soc.* **94**, 1843 (1972).
2. J. N. Murrel, A. J. Stace, and R. Dammel, *J. C. S. Faraday Transactions II* **74**, 1532 (1978).
3. P. Bado, P. H. Berens, and K. R. Wilson, in *Picosecond Lasers and Applications*, edited by L. S. Goldberg (Proc. Soc. Photo-Optic. Engin., Bellingham, WA, 1982) Vol. 322, p. 230. (See ref. 16 below.)
4. P. H. Berens, J. P. Bergsma, and K. R. Wilson, *J. Chem. Phys.*, to be submitted (1982).
5. D. J. Nesbitt and J. T. Hynes, *Chem. Phys. Lett.* **82**, 252 (1981).
6. D. J. Nesbitt and J. T. Hynes, *J. Chem. Phys.*, in press.
7. P. H. Berens, J. P. Bergsma, K. R. Wilson, and E. J. Heller, to be submitted.
8. J. Tellinghuisen, *J. Chem. Phys.* **76**, 4736 (1982).
9. T. J. Chaung, G. W. Hoffman, and K. B. Eisenthal, *Chem. Phys. Lett.* **25**, 201 (1974).
10. K. B. Eisenthal, in *Ultrashort Light Pulses; Picosecond Techniques and Applications*, edited by S. L. Shapiro (Springer-Verlag, Berlin, 1977) p. 275.
11. C. A. Langhoff, K. Gnädig, and K. B. Eisenthal, *Chem. Phys.* **46**, 117 (1980).
12. C. A. Langhoff, B. Moore, and W. Nugent, in *Picosecond Phenomena II*, edited by R. Hochstrasser, W. Kaiser, and C. V. Shank (Springer-Verlag, Berlin, 1980) p. 249.
13. D. F. Kelley and P. M. Rentzepis, *Chem. Phys. Lett.* **85**, 85 (1982).
14. K. Peters, Harvard University, (private communication).
15. G. E. Busch, R. T. Mahoney, R. I. Morse, and K. R. Wilson, *J. Chem. Phys.* **51**, 837 (1969).
16. R. K. Sander and K. R. Wilson, *J. Chem. Phys.* **63**, 4242 (1974).
17. W. S. Struve, *Chem. Phys. Lett.* **51**, 603 (1977).
18. P. Bado, S. B. Wilson, and K. R. Wilson, *Rev. Sci. Instrum.*, in press, plus addendum noting that response of receiver is quadratic, as pointed out by D.B. McDonald and G.R. Fleming, which explains difference in decay times from ref. 3.
19. J. P. Heritage, in *Picosecond Phenomena II*, edited by R. Hochstrasser, W. Kaiser, and C. V. Shank (Springer-Verlag, Berlin, 1980) p. 343.
20. B. F. Levine and C. G. Bethea, *Appl. Phys. Lett.* **36**, 245 (1980).
21. B. F. Levine and C. C. Bethea, *IEEE J. Quantum Electron.* **QE-16**, 85 (1980).
22. E. J. Heller and K. R. Wilson, to be submitted.

Vibrational Predissociation of S-Tetrazine-Ar van der Waals-Molecules

J.J.F. Ramaekers, J. Langelaar, and R.P.H. Rettschnick

Laboratory for Physical Chemistry, University of Amsterdam,
Nieuwe Achtergracht 127, NL-1018 WS Amsterdam, The Netherlands

Introduction

The visible fluorescence spectroscopy of s-tetrazine in the vapour phase at room temperature is reasonably well understood. Vibrational relaxation data were obtained in the single collision limit for several collision partners [1]. The mode to mode energy flow turned out to be highly selective but not very efficient. With argon as a collision partner for instance the complete removal of one quantum of the 6a mode (703 cm^{-1}) in the electronically excited state was highly improbable, on the other hand with benzene as collision partner at least 20 hard sphere collisions are needed. S-tetrazine photodecomposes with high yield ($\geqslant$ 99%) after the absorption of visible light, a not yet understood process. Furthermore, the decay time of s-tetrazine in the electronically excited state ranges from $\leqslant$ 100 ps up to approximately 2 ns, depending on the vibrational mode excited [2]. Van der Waals complexes between s-tetrazine and atomic gases are easily formed in a supersonic beam, which presents a direct method to study photodissociation of the complex [3]. Vibrational predissociation of the Van der Waals complex can be seen as the half collision analog of the vibrational relaxation process.

Experimental

We have performed vibrational predissociation experiments, both time and spectrally resolved, on s-tetrazine-argon Van der Waals complexes. The complexes were formed in a supersonic expansion of argon gas seeded with ~ 0.03% s-tetrazine. The expansion was conducted through a nozzle of 50µm with an Ar stagnation pressure between 1 and 1.5 bar. The molecular species as well as the Van der Waals complexes could be selectively excited with a synchronously pumped cw dye laser [4]. The fluorescence emission was dispersed through a high resolution monochromator and detected by a photon counting system (spectra) or a single photon counting system (decay times).

Results and discussion

The dispersed emission spectra of the molecular species excited in different vibrational modes only reveal resonance emissions, indicating that vibrational relaxation as a result of collisions does not take place in the supersonic jet under the experimental conditions used.

Excitation to the zero point level (0^0) shows that in the emission spectra from the uncomplexed s-tetrazine as well as from both Van der Waals complexes (T-Ar and T-Ar_2) all emission comes from the originally prepared level with a decay time of 800 ps; the same value for all three species and identical to the collision free molecular lifetime at room temperature.

The $6a_0^1$ excitation bands of the molecular and complex species show a red shift of 22 and 44 cm^{-1} for respectively the T-Ar and T-Ar_2 complexes, indi-

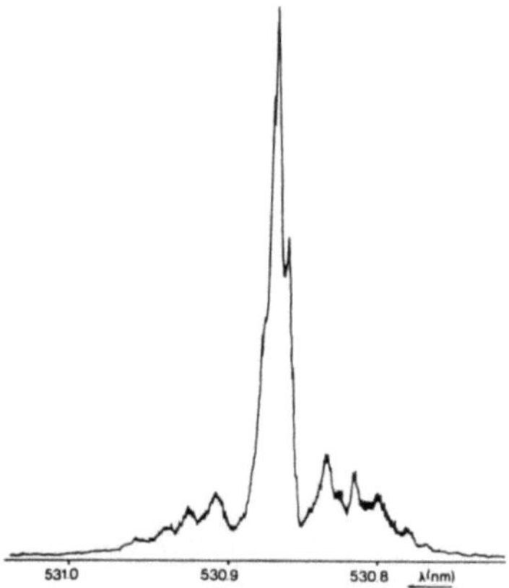

Fig 1 Fluorescence excitation spectrum of the $6a_0^1$ s-tetrazine molecular transition in a supersonic jet.

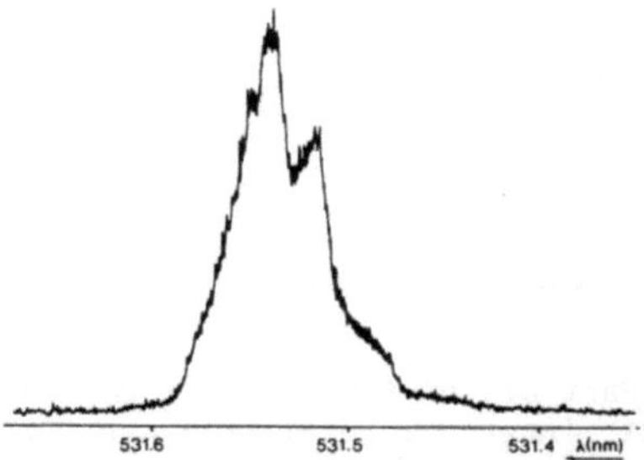

Fig 2 Fluorescence excitation spectrum of the $6a_0^1$ transition in the s-tetrazine-Ar Van der Waals complex.

cating a slight increase of the Van der Waals binding energy in going from the ground to the excited singlet state. The rotational population observed from the fluorescence excitation spectrum for the $6a_0^1$ transition of the molecular species (fig 1) can be characterized by a Boltzmann distribution with a temperature of 4 K. Two Van der Waals vibrational modes could be assigned in the $^1B_{3u}$ -state i.e. 35 and 43 cm^{-1}; most probably the Van der Waals bend and stretch vibration. This puts a lower limit to the Van der Waals binding energy of approximately 65 cm^{-1}.

If a Van der Waals complex is prepared in the excited state with one quantum of a s-tetrazine mode then the energy might flow to the Van der Waals stretch mode dissociating the molecule into an argon atom and an electronically excited s-tetrazine molecule in a lower vibrational state i.e. $T^*\text{-Ar} \rightarrow T^* + \text{Ar}$.

The emission spectra then might show us which product is formed in what quantum state. In addition, measurements of the decay times might indicate the dominant decay channels.

Unfortunately only two levels, i.e. $6a^1$ and $16a^2$, can be excited from the zero point level in the ground state to the excited electronic state ($^1B_{3u}$) with sufficient population to perform reliable experiments. Selective excitation can be performed for the molecular and complex species as shown in the figures 1 and 2 for the $6a^1$ level. Features observed in the emission spectra of the Van der Waals complex then have to be analysed in terms of complex bands or molecular product species.

We will now look in more detail into the T-Ar Van der Waals complex emission of the $6a^1$ vibrational storage mode (fig 3). The three bands indicated by an "X" are emissions from the prepared $6a^1$ level of the Van der Waals complex. They show an identical red shift of 22 cm^{-1} as compared to the well known $6a^1$ emission from the molecular species as observed under the same conditions in a separate experiment (fig 4).
The extra emission bands can be assigned to the $16a^1$, $16a^2$, 5^1 and (not shown in fig 3) $16a^1 16b^1$ vibrational levels of the complexed or to the uncomplexed tetrazine molecule. For these bands we are not able to make a discrimination for the assignments between the molecular and complex species, because the spectral resolution in emission was too low. In a separate experiment T-Ar complexes were prepared in the $16a^2$ vibrational level. These experiments indicate only a slight red shift (~7cm^{-1}) of the Van der Waals complex as compared to the molecular species. Thus a nearly identical Van der Waals binding

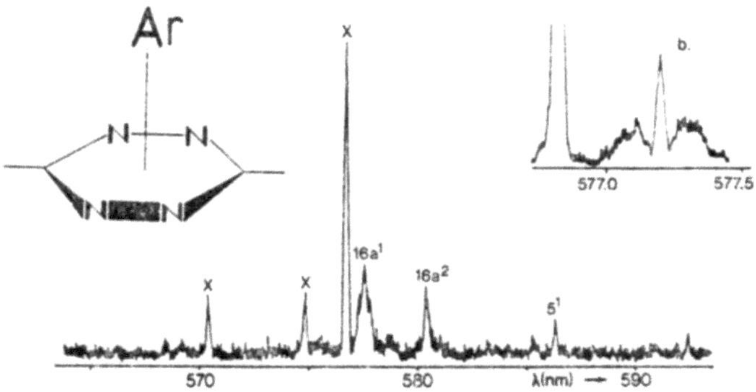

Fig 3 Part of the fluorescence emission spectrum of the $6a^1$ T-Ar Van der Waals complex. Insert b: rotational contour of the $16a^1$ emission around 577.2 nm.

energy in ground and excited state is concluded. The decay time of the $16a^2$ complex is much shorter than observed for the molecular $16a^2$ emission in contrast to the $6a^1$ complex and the 0^0 complex, where the decay times are observed to be identical to the lifetime of the molecular species (table 1).

This indicates that another decay channel is open for the $16a^2$ level of the Van der Waals complex in competition with the radiationless photochemical channel. This extra channel might be due to vibrational predissociation of the Van der Waals complex. The levels observed after excitation of the Van der Waals complex in the $6a^1$ vibrational storage mode all show a decaytime similar to that of the molecular species (table 1), except for the $16a^2$ and $16a^1 16b^1$ vibrational levels. Here again we might conclude that vibrational predissociation of the Van der Waals complex is responsable for the decaytime reduction. The $16a^2$ and $16a^1 16b^1$ vibrational levels must be formed as a result of collision induced vibrational relaxation within the Van der Waals complex T-Ar ($6a^1$). It is believed that soft collisions (or orbiting collisions) are

Table 1 Decaytime results of s-tetrazine (T) and s-tetrazine-Ar in a supersonic jet (in picoseconds).

A. For resonance emission from the prepared molecular level and compared with room temperature results [2].
B. For dispersed emission after selective excitation of the T-Ar Van der Waals complex in the $6a^1$ and $16a^2$ vibrational level. All emissions are assumed to originate from the complex, except $16a^1$ (see text).

A

($^1B_{3u}$) level	ΔE [cm^{-1}]	T (jet)	T (300 K p=0.5 Torr)
0^0	0	800	820
$16a^1$	256	1450	1470
$16a^2$	512	1540	1410
$16a^1 16b^1$	664	-	1420
5^1	681	-	500
$6a^1$	703	520	510

B

detected level / level excited	$16a^1$	$16a^2$	$16a^1 16b^1$	5^1	$6a^1$
$6a^1$ (T-Ar)	1500	560-680	900	520	520
$16a^2$ (T-Ar)	1400	800	-	-	-

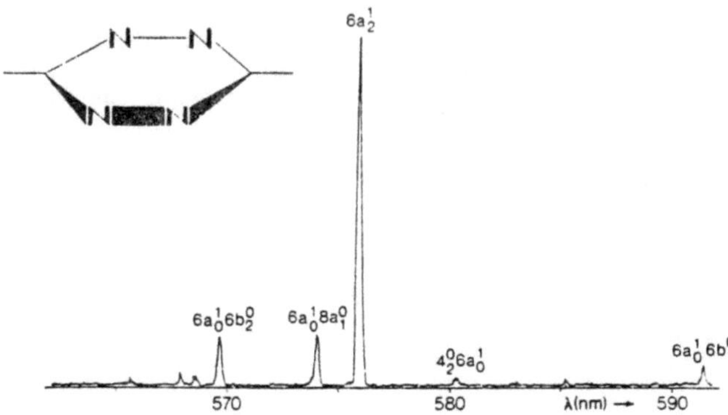

Fig 4 Part of the fluorescence emission spectrum of the $6a^1$ level of the s-tetrazine molecule. Note: all emission is from the $6a^1$ level.

sufficient to induce vibrational relaxation in the s-tetrazine-Ar complex assisted by the distortion of the molecular vibration by the Van der Waals bounded Ar atom. After vibrational relaxation to the $16a^2$ level, predissociation is favoured into a $16a^1$ molecular species and an argon atom because only one quantum of a 16a mode (256 cm^{-1}) has to be redistributed to the Van der Waals stretching mode for dissociation. This gives us an upper limit for the Van der Waals binding energy in s-tetrazine-Ar of 256 cm^{-1}. The broad rotational contour of the $16a^1$tetrazine fragment band indicates that part of the energy is redistributed into rotational energy indicating a binding energy < 256 cm^{-1}. A similar interpretation might hold for the $16a^1 16b^1$ tetrazine-Ar complex, where we also observed a reduction of the lifetime as compared to the molecular species.

For the tetrazine-Ar_2 complex several new bands were observed (fig 5). Unfortunately we do not have reliable decaytime-results due to the low intensity. The spectral positions indicate an assignment to the $16a^1$, $16a^2$, $16a^3$,

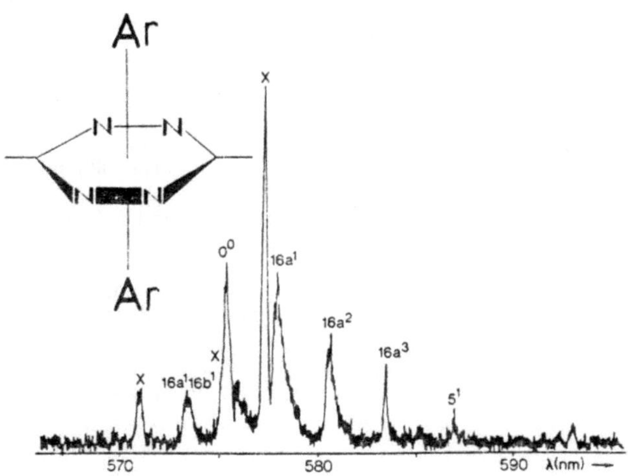

Fig 5 Part of the fluorescence emission spectrum of the T-Ar_2 ($6a^1$) Van der Waals complex.

0^0 and $16a^1 16b^1$ levels. From the known red shift of the complex the 0^0 band has to be assigned to the molecular species. The other bands might be interpretated as molecular as well as Van der Waals-Ar_n complex bands.
Following the interpretation of the T-Ar complex we might presume the following processes for the T-Ar_2 complex:

$$T^*\text{- } Ar_2\ (6a^1) \rightarrow T^*\text{- } Ar_2\ (16a^3) \qquad (1)$$
$$T^*\text{- } Ar_2\ (6a^1) \rightarrow T^*\text{- } Ar_2\ (16a^2) \qquad (2)$$

Consecutive vibrational predissociation from (1) gives

$$T^*\text{- } Ar_2\ (16a^3) \rightarrow T^*\text{- } Ar\ (16a^2) + Ar$$
$$T^*\text{- } Ar\ (16a^2) \rightarrow T^*(16a^1) + Ar$$

and from (2):

$$T^*\text{- } Ar_2\ (16a^2) \rightarrow T^*\text{- } Ar\ (16a^1) + Ar$$
$$T^*\text{- } Ar\ (16a^1) \rightarrow T\ (0^0) + Ar$$

From the results we conclude that vibrational predissociation in s-tetrazine-Ar_n (n=1,2) Van der Waals complexes is favoured through the 16a molecular storage mode lather. These results imply that coupling between the 16a out of plane tetrazine vibration and one of the Van der Waals vibrations is strong compared to the coupling of other molecular modes. Furthermore we conclude that in the T-Ar_2 complex the argon atoms are ejected one by one during predissociation. The observed vibrational predissociation rate is approximately $10^9 s^{-1}$ for the $16a^2$ vibrational level. This rate is a factor of two greater than the radiationless photochemical process observed in s-tetrazine for the same level. The results indicate that the Van der Waals binding energy lies between 65 and 250 cm^{-1}.

Acknowledgements

The authors are indepted to Mr. H.K. van Dijk for experimental assistance. The investigations were supported in part by the Netherlands Foundation for Chemical Research (SON) with financial aid from the Netherlands Organization for the Advancement of Pure Research (ZWO).

References

1. a. M.W. Leeuw, Thesis, University of Amsterdam (1981).
   b. M.W. Leeuw, J. Langelaar and R.P.H. Rettschnick. Proc.of the UPS 80, Reinhardsbrunn DDR, p. 220-224 (1980).
2. J. Langelaar, D. Bebelaar, M.W. Leeuw, J.J.F. Ramaekers and R.P.H. Rettschnick. Picosecond Phenomena II, Ed. R.M. Hochstrasser, W. Kaiser, C.V. Shank, p. 171-174. Springer Verlag (1980).
3. a. R.E. Smalley, L. Wharton, D.H. Levy. J.Chem.Phys. 68 2487 (1978)
   b. J.E. Kenny, D.V. Brumbaugh and D.H. Levy, J.Chem.Phys. 71 4757 (1979)
4. a. J. de Vries, D. Bebelaar, J. Langelaar. Opt.Comm. 10 24 (1976)
   b. D. Bebelaar, J.J.F. Ramaekers, M.W. Leeuw, R.P.H. Rettschnick and J. Langelaar. Proc.of the UPS 80, Reinhardsbrunn DDR, p. 212-217 (1980).

Picosecond Laser Induced Fluorescence Probing of NO_2 Photofragments

P.E. Schoen, M.J. Marrone, and L.S. Goldberg

Naval Research Laboratory, Washington, D.C. 20375, USA

Introduction

The UV photolysis of nitromethane has been studied extensively for many years [1-12]. It is now generally accepted that the primary photodissociation process leads to formation of the free radicals CH_3 and NO_2 in the reaction:

$$CH_3NO_2 \xrightarrow{h\nu} CH_3 + NO_2 \qquad (1)$$

We have performed 264 nm photolysis of nitromethane in the gas phase and have observed formation of ground-state NO_2 product within the 5 ps pulse resolution time of our laser. This represents the first direct observation of the initial fragments of nitromethane photodecomposition with picosecond time resolution. We have employed a newly developed Nd:phosphate glass laser system, mode-locked and repetitively pulsed at 0.2 Hz [13]. The presence of NO_2 radicals was probed by its fluorescence induced with a second harmonic pulse at 527 nm. Thus using picosecond-delayed laser induced fluorescence (LIF) in the gas phase (0.1 to 10 Torr), we have been able to observe ground-state product formation in the collision free regime. The technique of exciting fragment molecules to a fluorescent state has proved to be a particularly well adapted approach to detecting gas phase species at low concentrations.

Fig.1 shows the time integrated fluorescence intensity as a function of probe pulse delay. While the photolyzing UV pulse itself produces a fluorescing species, the subsequent arrival of the probe pulse causes the observed signal to increase by a factor of ~ 3. The dashed line in Fig.1 represents the system temporal response as determined by a photobleaching experiment in rhodamine 6G dye solution. The UV pulse initially depopulates the ground state of the dye molecule, allowing a subsequent 2nd harmonic pulse to be transmitted.

The fluorescence signal was weak so that its spectrum could only be examined by the use of a sequence of long-wavelength-pass color filters. These indicated that the emission was characteristic of NO_2, extending from the region near the probe wavelength to beyond 750 nm. This was true both for fluorescence induced by the UV pulse alone and also for the UV plus probe pulses. Signal at shorter wavelengths, on the antistokes side of the probe, was too weak to be detected in either case. Also characteristic of NO_2, the lifetime of the emitting species was long [14], as indicated by Stern-Volmer plots of decay rate vs. nitromethane pressure, for UV alone and UV plus probe induced fluorescence.

Fluorescence intensity for both processes was linear over 2 decades of UV pulse energy, indicating a single quantum process was occurring. An NO_2

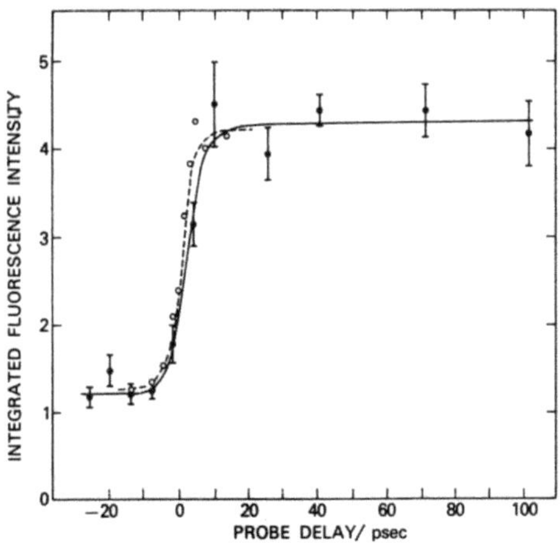

Fig.1 LIF signal as a function of probe time delay for photolysis of nitromethane at 0.8 Torr. Intensity is normalized to UV and probe pulse energies. Each data point represents an average of 10 laser shots. Dashed line: System temporal response function determined by photobleaching experiment in rhodamine dye solution.

fragment produced initially in an excited state by absorption of a single UV photon would have sufficient excess energy above that required for C-N bond scission [15] to fluoresce only at wavelengths beyond ~ 590 nm, which is consistent with our observations.

Therefore considerations of probe absorption, fluorescence spectrum and lifetime, and bond energy give strong evidence that we have observed photodecomposition of nitromethane in the collision free regime to yield CH_3 and NO_2 radicals.

We have measured the quantum yield for formation of NO_2 by comparing the signal intensity for LIF in photolyzed nitromethane with that obtained in neat NO_2 gas. Taking into account the relative quenching rates observed in these two experiments, and using an absorption cross-section of 3×10^{-20} cm^2 at 264 nm [16], we calculate the yield of ground-electronic-state NO_2 to be 70% ± 30%. In a representative experiment (0.63 Torr nitromethane, 185 μJ UV energy), 2.2×10^{11} ground-state NO_2 photofragments are produced in an observable volume element of 0.25 cm^3. A small fraction of these are excited by the LIF probe pulse: for a probe energy of 940 μJ, and an NO_2 absorption cross-section extrapolated to 527 nm of 1.4×10^{-19} cm^2 [17], we obtain 6.2×10^8 NO_2 molecules (0.3%) excited by the probe. Only about 4.6×10^8 molecules (0.2%) are produced initially as electronically-excited NO_2^* in the UV-only experiment.

The dependence of quantum yield on photolyzing wavelength was examined using stimulated Raman scattering in H_2 gas (5-20 Bar) to produce wavelength-shifted excitation pulses at 296 nm and 337 nm (first and second Stokes) and at 238 nm (first anti-Stokes). Fig.2 (upper half) shows the wavelength dependence of the quantum yields obtained for excited-state NO_2^* (UV only) and ground state NO_2 (UV + probe). The yield for ground-state NO_2 does not vary appreciably over these photolyzing wavelengths. However, the yield for NO_2^* increases sharply for 237 nm excitation and becomes essentially zero at 296 nm.

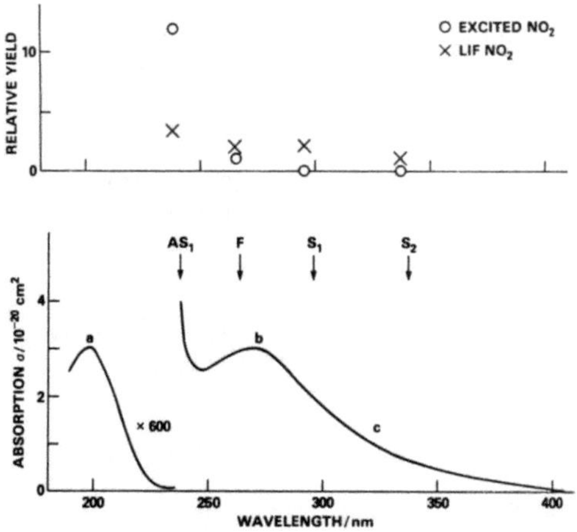

Fig.2 Upper half: Relative quantum yield of NO_2^* [0] and NO_2 [X] versus photolyzing wavelength. The yields are not shown to the same scale.

Lower half: UV absorption spectrum of nitromethane gas indicating known transition states as discussed in text below. The arrows indicate the UV photolyzing wavelengths obtained from stimulated Raman scattering in H_2.

Discussion

Rebbert and Slagg [2] suggested that more than one excited state of nitromethane was involved in its decomposition, and Honda et al. [6] and Flicker et al. [18] have supported this idea. Two states are observed in its optical absorption spectrum Fig.2 (lower half) [16, 18, 19]: a strong feature at ~ 198 nm (a) assigned to a $\pi \rightarrow \pi^*$ singlet-singlet transition, and a weak satellite at ~ 270 nm (b) suggested to be an $n \rightarrow \pi^*$ singlet-singlet transition. A third, still lower energy state has been found by electron energy-loss spectroscopy at ~ 326 nm (c) [18], which the authors suggest has $n \rightarrow \pi^*$ singlet-triplet character, but which may be a composite of overlapping transitions of different character. Most investigators have supported process (1) as the main primary photolysis channel for nitromethane [1-5]. Honda et al. [6] and Flicker et al. [18] suggest that the lower energy transition to the triplet state at ~ 326 nm addresses process (1) particularly while the singlet-singlet excitation near 270 nm induces a different reaction.

Our observations that ground-state NO_2 fragments are formed promptly and with high quantum yield from 264 nm photolysis of low pressure nitromethane support the conclusion that excitation of the $n \rightarrow \pi^*$ transition near 270 nm results in dissociation predominantly via reaction (1). Our Raman shifting experiments indicate that the yield of the primary product ground-state NO_2, as distinct from final products, does not vary appreciably over the excitation range studied. The lower energy transition near 326 nm does not appear to have a significant effect upon this reaction. These experiments do show, however, that the higher energy $\pi \rightarrow \pi^*$ transition near 198 nm is probably responsible for NO_2^* production.

Acknowledgments

The authors wish to thank R. G. Weiss and J.M. Schnur for many useful discussions. We also gratefully acknowledge the Office of Naval Research for partial support of this research.

References

1. E. Hirschlaff and R. G. W. Norrish, J. Chem. Soc. 1580 (1936).
2. R. E. Rebbert and N. Slagg, Bull. Socs. Chim. Belges. 71, 709 (1962).
3. B. H. J. Bielski and R. B. Timmons, J. Phys. Chem. 68, 347 (1964).

4. I. M. Napier and R. G. W. Norrish, Proc. Roy. Soc. A299, 317, (1967).
5. R. B. Cundall, A. W. Locke and G. C. Street, in "The Chemistry of Ionization and Excitation," G. R. A. Johnson and G. Scholes, eds. (Taylor Francis, Ltd., London, 1967) pp. 131-140.
6. K. Honda, H. Mikuni and M. Takahasi, Bull. Chem. Soc. Jpn. 45, 3534 (1972).
7. M. J. Colles, A. M. Angus, E. E. Marinaro, Nature, 262, 681 (1976).
8. K. G. Spears and S. P. Brugge, Chem. Phys. Lett. 54, 373 (1978).
9. W. L. Faust, L. S. Goldberg, T. R. Royt, J. N. Bradford, J. M. Schnur, P. G. Stone and R. G. Weiss, in "Picosecond Phenomena," C. V. Shank, E. P. Ippen and S. L. Shapiro, eds. (Springer, N.Y. 1978), p. 43.
10. Ph. Avouris, I. Y. Chan and M. M. T. Loy, J. Photochem. 13, 13 (1980).
11. B. H. Rockney and E. R. Grant, Chem. Phys. Lett. 79, 15 (1981).
12. H. S. Kwok, G. Z. He, R. K. Sparks and Y. T. Lee, Int. J. Chem. Kinetics, 13, 1125 (1981).
13. L. S. Goldberg, P. E. Schoen and M. J. Marrone, Applied Optics 21, 1474 (1982).
14. L. F. Keyser, S. Z. Levine and F. Kaufman, J. Chem. Phys. 54, 355 (1971).
15. R. Kandel, J. Chem. Phys. 23, 84 (1955).
16. W. D. Taylor, T. D. Allston, M. J. Moscato, G. B. Fazekos, R. Kozlowski, and G. A. Takacs, Int. J. of Chem. Kinetics, 12, 231 (1980).
17. T.C. Hall, Jr. and F.E. Blacet, J. Chem. Phys. 20, 1745 (1952).
18. W. M. Flicker, O. A. Mosher and A. Kuppermann, Chem. Phys. Lett. 60, 518 (1979).
19. S. Nagakura, Mol. Phys. 3, 152 (1960).

Excited State Proton Transfer in 2- (2-'Hydroxylphenyl)-Benzoxazole

G.J. Woolfe, M. Melzig, S. Schneider, and F. Dörr

Institut für Physikalische und Theoretische Chemie der Technischen Universität München, Lichtenbergstrasse 4, D-8046 Garching, Fed. Rep. of Germany

1. Introduction

The important roles of excited state proton transfer processes in such diverse fields as photostabilization, photochromism in certain compounds [1] and certain biochemical processes are increasingly being succesfully elucidated by photophysicists. Excited state intramolecular proton transfer usually results in the extremely rapid formation of an excited state tautomer of the originally excited species [2-5] . The possibility exists, therefore, of a dramatic modification of the expected photophysical or photochemical behaviour ie that which might be anticipated on the basis of naive considerations of the starting molecule.

As a consequence of the high rate of proton transfer and the subsequent rapid deactivation of the resulting tautomeric excited state, time resolved picosecond spectroscopy has the capacity to provide significant advances in the understanding of system exhibiting exciting state proton transfer [2-6] .

2. Experimental

Fluorescence decay curves were measured using a "synchroscan" streak camera system. The excitation source was a Rhodamine 6G dye laser (Spectra Physics 375) synchronously pumped by an acousto-optically mode-locked Ar^+ laser (Spectra Physics 171-18W). This combination delivers pulses with a temporal width of ca. 3 ps fwhm and tunable in wavelength from 580 nm to 610 nm, at a repetition rate of 81 MHz. These pulses are frequency doubled with an angle tuned $LiIO_3$ crystal and after passing through a filter to remove the remaining fundamental radiation, the second harmonic (300 nm) is focussed onto the sample. The resulting fluorescence is viewed by the streak camera (Imacon 675) at 90° to the direction of excitation.

The effects of the finite response time of the system were accounted for by numerical iterative convolution using an instrument response function obtained by recording the scatter of the fundamental laser pulses from chalk. Data fitting was accomplished using a non-linear least squares routine based on the algorithm of MARQUARDT [7] .

Steady state emission and excitation spectra were recorded with an Aminco-Bowman spectrofluorimeter and have not been corrected.

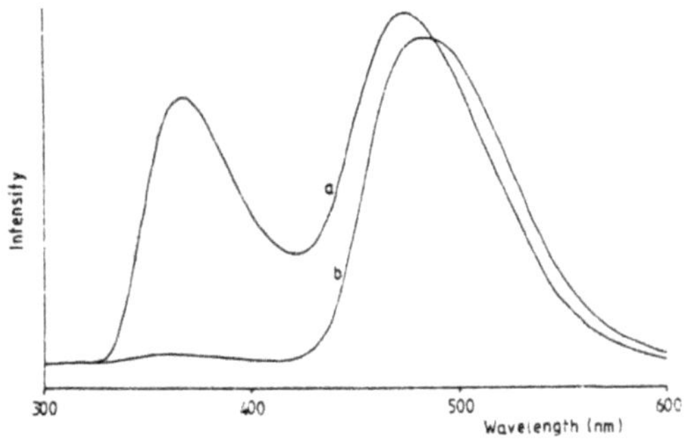

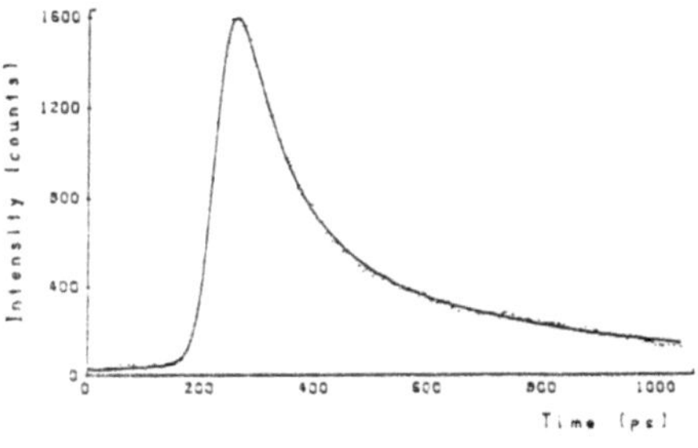

Fig. 1 Fluorescence spectra of HBO in (a) Ethanol (b) Cyclohexane

Fig. 2 Fluorescence decay of HBO in DMSO

3. Results and Discussion

3.1 The Proton Transfer

The occurrence of an excited state intramolecular proton transfer in 2-(2'hydroxyphenyl) benzoxazole (HBO) is clearly indicated by its fluorecence spectrum [8-10] . The spectrum consists of two emission bands, one of which has an anomalously large stokes shift (Fig.1). The relative intensities of these bands are both solvent and temperature dependent. The long wavelength emission (λ_{max} = 475) is attributable to the tautomeric species in which the proton has been transferred to the nitrogen heteroatom. The short wavelength emission (λ_{max} = 375 nm) is assigned to the normal form of the molecule, I.

Despite the very low transmission of our streak camera slit imaging optics at wavelengths below 400 nm, the short wavelength emission of HBO in dimethylsulphoxide (DMSO) is sufficiently intense that the lifetime of this emission can be determined by measurements on the tail (λ >400 nm) of the band. The fluorescence decay of HBO in DMSO is shown in Fig. 2. The decay is clearly nonexponential but has been well fitted by the sum of two decaying exponentials (1).

$$I(t) = A_1 \exp(-t/\tau_1) + A_2 \exp(-t/\tau_2) + B \quad (1)$$

The least-squares optimized lifetimes are τ_1 = 68.2 $\pm$ 7,5 ps, and τ_2 = 1080 $\pm$ 360 ps. A crude wavelength resolution has also been made by measuring the decay with various short-pass filters between the sample and the streak camera slit. In all cases the data are well fitted by the same two lifetimes (within experimental error), but with different values for the ratio A_1/A_2. The results are summarized in Table 1.

These results clearly indicate that the shorter lifetime is associated with the long wavelength emission band of HBO in DMSO. The fact that the short and long wavelength emission bands have different lifetimes and that the long wavelength emission shows no detectable risetime indicates that the proton transferred and normal excited states do not interconvert and, in particular, that an equilibrium between them is not established in the excited singlet state.

Table 1 Dependence of A_1/A_2 on wavelength for HBO in DMSO

Wavelength region [nm]	A_1/A_2
400 - 440	0
400 - 460	0.79 $\pm$ 0.27
400 - 480	1.89 $\pm$ 0.23
400 - 700	2.84 $\pm$ 0.20

The two obvious possiblities are that the normal and tautomeric excited states both arise from the same Franck-Condon excited state but become independent after their initial formation or that they result from the excitation of different ground state species (conformers) and remain independent after excitation. The excitation spectra of the short and long wavelength emission bands have been found to be significantly different from one another in all solvents we have measured, supporting the second of the above possibilities.

The most probable ground state precursor of the tautomeric excited state is a conformer exhibiting intramolecular hydrogen bonding between the hydroxy group and the nitrogen heteroatom There are two possible ground state precursors of the normal excited state - a "trans" structure with intramolecular hydrogen bonding between the hydroxy group and the oxygen heteroatom and a "strongly solvated" structure (II). The latter species is expected to exist in appreciable amounts only in solvents capable of intermolecular hydrogen bonding with HBO.

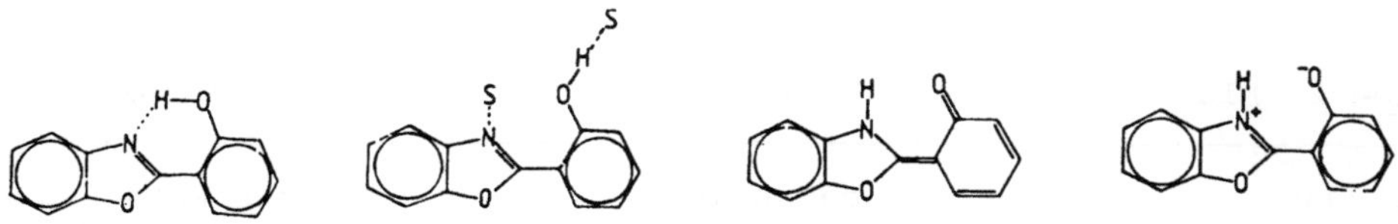

I II III IV

s=solvent

The foregoing discussion illustrates that the relative intensities of the normal and tautomeric emissions of HBO are determined not only by the rate constants for excited state deactivation mechanisms, but also by the relative populations of the various ground state conformers. The effects of solvent and temperature upon the steady-state emission spectrum is therefore quite complicated and a correct understanding of the photophysics of this molecule is possible only with the aid of direct measurements of the fluorescence decay.

The time dependence of the tautomeric emission has been measured in a large range of solvents. In all cases the risetime of the fluorescence was faster than the time resolution of the measurement system (5 ps), placing a lower limit on the rate constant for excited state intramolecular proton transfer in HBO of $2 \times 10^{11} s^{-1}$.

3.2 Deactivation of the Tautomeric Excited State

The fluorescence lifetimes and spectral maxima of the HBO tautomer in a range of solvents are listed in Table 2. As the dielectric constant of the solvent increases there is a general trend towards a shorter lifetime and a shift to shorter wavelengths of the tautomer fluorescence. This spectral shift is somewhat unusual and unexpected.

One possible explanation for this is that the tautomeric excited state can exist in two distinct forms - quinoidal (III) and zwitterionic (IV). Their excistence as distinct entities rather than simply as resonance forms of a single species is supported by studies of molecular models. These show that while IV is planar, III appears most stable in a bent conformation. Similar findings exist for a related benzothiazole [8] .

The existance of two distinct tautomeric species is also suggest by the shape of the long wavelength emission band of HBO. This band appears rather broad with a shoulder on the long wavelength side (ca. 520 nm). The shoulder is relatively more intense in solvents of lower dielectric constant. The more polar IV should be favoured in solvents of high dielectric constant while III exists preferentially in lower dielectric constant solvents. In all solvents the decay of the entire tautomer emission was well fitted by a single exponential decay law, suggesting that III and IV are in equilibrium in the excited state. The results can now be interpreted if IV emits at shorter wavelengths and has a shorter lifetime than III. The actual measured lifetime and spectral maximum then depends on the influence of the solvent on the position of the excited equilibrium between III and IV.
One difference between III and IV is the bond order of the bond linking the two rings of the molecule. The lower bond order in

Tabe 2 Fluorescence decay time and spectral maximum for HBO

Solvent	Dielectric constant at 25°C	Fluorescence lifetime [ps]	spectral Maximum [nm]
DMSO	46.68	69.0 ± 4.0	465.2
Acetonitrile	37.5	84.8 ± 4.0	476.8
N,N-Dimethyl-formamide	36.71	70.7 ± 4.0	466.9
Methanol	32.70	78.0 ± 4.0	470.7
Ethanol	24.58	112.1 ± 5.0	472.7
Propanol	20.33	147.3 ± 6.0	474.6
Pyridine	12.3	96.4 ± 5.0	485.5
Tetrahydrofuran	7.58	112.8 ± 5.0	482.4
Chloroform	4.806	344 ± 25	477.6
Carbon Tetrachloride	2.238	413 ± 30	482.7
Benzene	2.275	309 ± 15	482.0
Cyclohexane	2.023	257 ± 10	482.0
Octadecane Liquid (31°C)	⩽ 2	358 ± 25	482.8
Octadecane solid (25°C)	⩽ 2	516 ± 40	474.8

IV gives rise to the possibility of internal rotation of the two rings relative to one another. Such internal rotation results in rapid excited state deactivation in certain dye molecules and in 3-hydroxyflavone [6,11,12]. This is a possible justification for IV having a shorter lifetime than III. The large difference between the lifetimes in solid octadecane and the liquid hydrocarbon solvents suggests that such internal rotation may also play an important role in this molecule.

Another interesting trend emerging from the results in Table 2 is the effect of intermolecular hydrogen bonding between solute and solvent. The lifetimes in pyridine and tetrahydrofuran, are quite short despite the relatively low dielectric constants of these solvents. These solvents act as H-bond acceptors and the short lifetimes may be a consequence of the intermolecular H-bond vibrations acting as a strong accepting mode. Alternatively the hydrogen bonding may give rise to anharmonic contributions in the vibrational degrees of the HBO tautomers, leading to enhanced deactivation rates.

Acknowledgements

The financial assistance of the Deutsche Forschungsgemeinschaft is gratefully acknowledged.

References

1. W. Klöpffer, Adv. Photochem., 10, (1977), 311.
2. D. Ford, P.J. Thistlethwaite and G.J. Woolfe, Chem. Phys. Lett. 69, (1980), 246
3. P.J. Thistlethwaite and G.J. Woolfe, Chem. Phys. Lett., 63, (1979), 401
4,5,6. G.J. Woolfe and P.J. Thistlethwaite, J.Am. Chem. Soc., 102,(1980), 6917; 103 (1981), 3849; accepted (1981).
7. D.W. Marquardt, J. Soc. Indust. Appl. Math., 11, (1963),431
8. D.L. Williams and A. Heller, J. Phys. Chem., 74, (1970),4473
9. M.B. Strykov, A.E. Lyubarskaya and M.I. Kryazhanskii, Zh. Prikl. Spektrosk., 27, (1977), 1055
10. M.D. Cohen and S. Flavian, J.Chem. Soc. (B), (1967), 317
11. G. Oster and J. Nishijima, J.Am.Chem.Soc. 78,(1956), 1581
12. E.P. Ippen, A. Bergman and C.V. Shank, Chem. Phys. Lett. 38, (1976), 611.

Picosecond Dynamics of Unimolecular Ion Pair Formation

K.G. Spears, T.H. Gray, and D. Huang

Department of Chemistry, Northwestern University,
Evanston, IL 60201, USA

Introduction

The ultimate understanding of a complex process like electron transfer in solution depends on our knowledge of more elemental steps such as solvent stabilization of charge and diffusional motion of oppositely charged ions. We report additional work on a molecular system [1] that allows a kinetic analysis of these two effects. Malachite green leucocyanide (MGCN) can be photoexcited to create MG^+ and CN^- ions in some solvents. Fig.1 shows the MGCN tetrahedral structure and the MG^+ planar structure. The lowest excited singlet state of MGCN is localized in the dimethylaniline groups with absorption and fluorescence maxima at 270 and 350 nm, respectively. The MG^+ cation absorbs strongly at 620 nm and this absorption was used to follow the MG^+ concentration following photoexcitation of MGCN.

Experimental Description and Results

We have measured the steady state $MGCN^*$ fluorescence yields, the $MGCN^*$ fluorescence decay kinetics, and the MG^+ rise kinetics as a function of solvent.

The fluorescence yields were done with a commercial fluorimeter and the fluorescence decays were analyzed by time-correlated photon counting [2,3]. A mode-locked argon ion laser pumped a dye laser to provide $\sim$3 ps pulses at 600 nm which were then doubled by a temperature phase matched [4] ADA crystal to 300 nm. The fluorescence was polarization analyzed to remove rotation effects and the system response was $\sim$300 ps FWHM.

Time resolved absorption spectra were done with an amplified dye laser system similar to the design of IPPEN et al. [5]. A Quanta Ray Nd^{3+} YAG

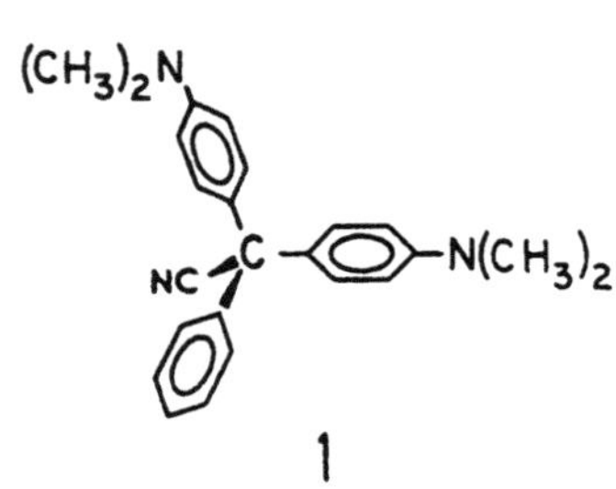

1

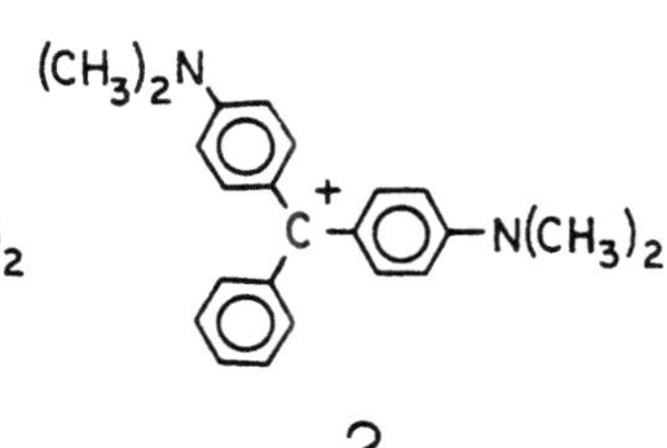

2

Fig.1 Structures of malachite green leucocyanide (1) and malachite green dye cation (2)

oscillator was doubled to 532 nm and the output beam of 80 mJ was used to amplify the ~3 ps pulses (0.5 nJ/pulse) from the input dye laser pulse train. The three stage amplifier delivered ~100 μJ/pulse at 10 Hz with good preservation of pulse shape. Our electronics and computer system allowed selection of acceptable energy limits, ratio calculations on each shot, and real time averaging. The transient absorbance of MG^+ at 600 nm was monitored as a function of time delay after the 300 nm second harmonic excitation pulse. For the data reported here we used a 15 shot average at each time delay point and measured I_0/I_T for the sample. Noise in I_0/I_T was 0.2% and spatial overlap and spatial noise with the 300 nm pump beam was the main noise source in our data.

We studied a variety of solvents having different dielectric constants and hydrogen bonding ability. The most systematic variation of dielectric constant (ε) at approximately constant viscosity was achieved by making mixtures of ethyl acetate (ε = 6) and acetonitrile (ε = 38). Low dielectric and inert solvents like cyclohexane and benzene (ε = 2.3) are reported to have zero ionization yield so that we expect $MGCN^*$ in benzene, for example, to have fluorescence decays and yields identical to dimethylaniline [6]. The fluorescence decays of $MGCN^*$ in the ethylacetate/acetonitrile mixtures showed a progressively shortened lifetime compared to benzene solutions and the short time (<100 ps) fluorescence had non-exponential behavior exhibiting a much faster decay rate than expected from the exponential decay occurring at times greater than 1 nanosecond after excitation. These latter times can be converted to ionization rates by assuming that the radiative and singlet-triplet lifetime of $MGCN^*$ is constant in all solvents. These rates are reported in Table 1 with fluorescence quantum yield results and lifetime results.

The early portion of the time resolved absorbance curves of MG^+ are shown in Fig.2 for a series of dielectric constants. There is a sudden initial rise in all solvents but the low dielectric constant solutions show a slow

Table 1 Fluorescence Decay Rates[b] and Yields of MGCN

Solvent[a]	Dielectric Constant	Fluorescence[b] Lifetime [ns]	Fluorescence[c] Yield	Ionization[d] Lifetime [ns]
Benzene	2.3	2.7	.11	-
EtOAc	6.0	1.9	.069	6.4
90.4% EtOAc/ACN	9.1	1.0	.035	1.6
62.4% EtOAc/ACN	18	0.4	.011	0.47
31.2% EtOAc/ACN	28	-	.0050	-
ACN	38	-	.0031	-
Ethanol	24.3	-	.0047	-
Methanol	37.2	-	.0023	-
Ethyl Ether	4.3	-	.11	-
Dioxane	2.2	-	.045	-

a. EtOAc is ethyl acetate and ACN is acetonitrile.
b. The fluorescence lifetime is exponential only for times greater than 1 nanosecond after excitation, the short time behavior is non-exponential.
c. The yield of benzene was assumed to be 0.11 as found for cyclohexane in [6]. Our relative yields (±5%) were scaled to this value.
d. The ionization lifetime is calculated by assuming a constant total decay of 2.7 ns for fluorescence and intersystem crossing in all solvents.

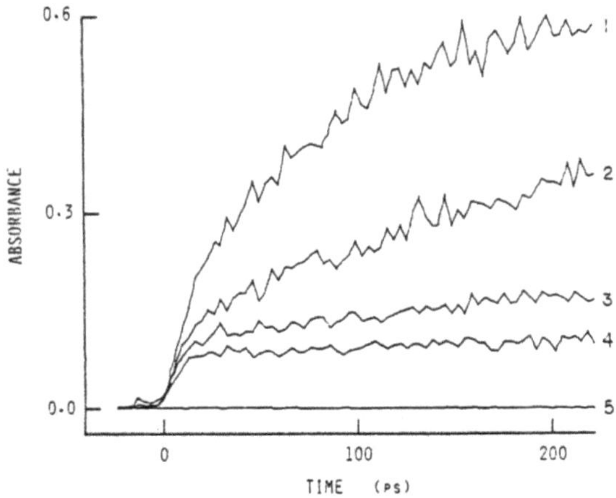

Fig.2 Time evolution of MG^+ after photo-excitation of MGCN in ACN (1, ε = 38), EtOAc/ACN (2, ε = 18), EtOAc/ACN (3, ε = 9.1), EtOAc (4, ε = 6.0) and benzene (5, ε = 2.3).

growth of MG^+ starting immediately after the fast rise. Absorbance measurements at longer time delays can yield relative yields of MG^+ and these will be reported for times longer than our current limit of 600 ps. The quantum yield of ionization was previously measured [7] in 95% ethanol as 0.91 and it has been postulated [6] that the excited singlet of MGCN ($MGCN^*$) directly decays to the ion pair. Our measurements address this unresolved question by comparing the fluorescence rate of $MGCN^*$ with the rate of rise of MG^+. The appearance of MG^+ qualitatively correlates with the rate of decay of $MGCN^*$ fluorescence.

Discussion

The data show a direct correlation of $MGCN^*$ singlet decay and MG^+ creation with no evidence for a long-lived intermediate state or species. The rate limiting step for $MGCN^*$ ionization is very dependent upon ion solvation energies as evidenced by the strong dependence of rate on the solvent dielectric constant. The non-exponential rise of MG^+ concentration suggests that the first 100 ps of the decay will eventually allow testing of realistic models. However, the exponential behavior at longer times for the lower dielectric constant solvents (ε from 6-28) appears to be consistent with an activated process having a barrier controlled by ion solvation energies. This simple theory has been used in bimolecular exciplex systems [8] and predicts our observed linear correlation between $1/\varepsilon$ and the ionization rate.

The fluorescence yields and approximate MG^+ ion yields can be used to evaluate the possibility of ion pair recombination. Our current estimate suggests that ~20% of the ion pairs never separate into MG^+ and CN^- in our solvent mixtures having a dielectric constant of $\varepsilon < 18$. Direct measurement of these rates is in progress.

This molecule and derivatives formed from other anions promises to create a data base for understanding solvent stabilization of incipient ionic charge and diffusion of ion pairs. We are actively pursuing this problem both experimentally and theoretically.

Acknowledgments

We gratefully acknowledge support of this research by Northwestern University and the NSF under Grant No. CHE-7714668. D. Huang gratefully acknowledges the People's Republic of China for financial support.

References

1. K.G.Spears, T.H.Gray, D.Huang: in Picosecond Lasers and Applications, Proc. SPIE, (1982) p. 75
2. K.G.Spears, L.E.Cramer, L.D.Hoffland: Rev. Sci. Instr. 49, 255 (1978)
3. K.G.Spears, K.M.Steinmetz-Bauer, T.H.Gray: in Picosecond Phenomena II, ed., R.M.Hochstrasser, W.Kaiser, and C.V.Shank, (Springer, Berlin, Heidelberg, New York 1980) p. 106
4. K.G.Spears, L.Hoffland, R.Loyd: Applied Optics, 16, 1172 (1977)
5. E.P.Ippen, C.V.Shank, J.M.Wiesenfeld, A.Migus: Phil. Trans. R. Soc. Lond. A 298, 225 (1980)
6. R.G.Brown, J.Cosa: Chem. Phys. Lett. 45, 429 (1977)
7. G.L.Fischer, J.C.LeBlanc, H.E.Johns: Photochem. Photobiol. 6, 767 (1967)
8. H.Mashuhara, T.Hino, N.Mataga: J. Phys. Chem. 79, 994 (1975)

Effect of Polymerization on the Fluorescence Lifetime of Eosin in Water

Wei-Zhu Lin, Yong-Lian Zhang, and Xin-Dong Fang

Laser Optics and Spectroscopy Laboratory, Physics Department, Zhongshan University, Guangzhou, China

1. Introduction

It is known that the fluorescence lifetime and the quantum yield of some organic compound solutions decrease as the solution concentration increases and this phenomenon has been attributed to the interaction of the excited molecules with the nonexcited molecules of the same kind and the formation of polymers.[1]-[4] We report further study of the effect of polymerization on the fluorescence lifetime of eosin in water using the ultrafast optical gate technique to measure the fluorescence lifetime of this solution in various concentrations and various temperatures.

2. Experimental

The experimental arrangement is shown in Fig.1. A passive mode-locked Nd^{3+}:YAG laser produces a pulse train of 1.06um, each pulse of 30 to 40ps duration, and separated by 10ns interval. The pulses are frequency doubled in a KDP crystal. The second harmonic pulses of 530nm are separated by a beam splitter and then sent through a variable optical delay to excite the sample con-

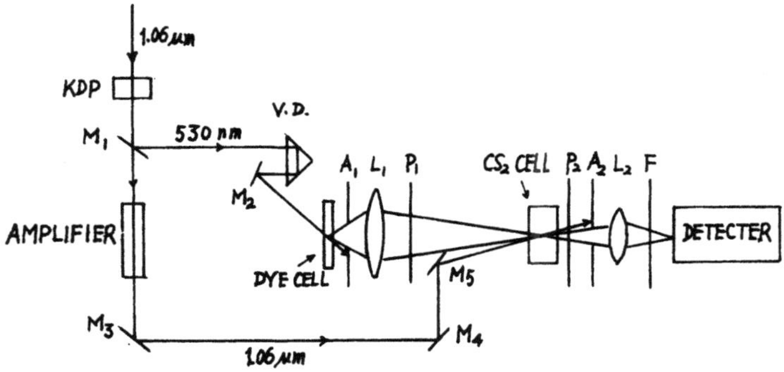

Fig.1 The schematic arrangement for measuring the fluorescence lifetime. P_1 and P_2: two crossed polarizers, A_1 and A_2: diaphragms for blocking 530nm and 1.06um respectively, F: filter.

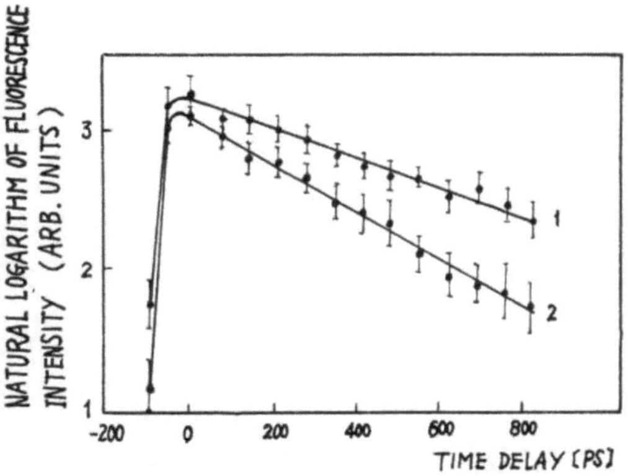

Fig.2 Eosin fluorescence
1: 5×10^{-4}M, 2: 1×10^{-2}M

tained in a cell of 1mm in thickness. The fluorescence from the sample is collected and focused to a Duguay shutter[5] and detected by a 1P28 photomultiplier. This shutter is opened by the 1.06um pulses from an amplifier. The signal to noise ratios at peak transmission in these experiments are 40:1

3. Results and discussion

The decay curves of the fluorescence intensity of eosin in water at two concentrations are shown in Fig.2. Each data point in the figure represents the mean of ten separate firings of the laser. The fluorescence lifetime has been obtained by a least squares fit of the experimental data. The measured fluorescence lifetime at concentration of 5×10^{-4} M is consistent with the value obtained by PORTER et al. [6].

Table 1 shows the experimental values of the fluorescence lifetime of eosin in water in the range of 5×10^{-4}M to 1×10^{-2}M.

Table 1 Fluorescence lifetime of eosin in water

Concentration (mole/liter)	lifetime (PS)
5×10^{-4}	962 ±135
1×10^{-3}	970 ±109
2.5×10^{-3}	900 ±82
4×10^{-3}	884 ±115
5×10^{-3}	827 ±114
6×10^{-3}	763 ±82
8×10^{-3}	661 ±79
9×10^{-3}	556 ±59
1×10^{-2}	500 ±83

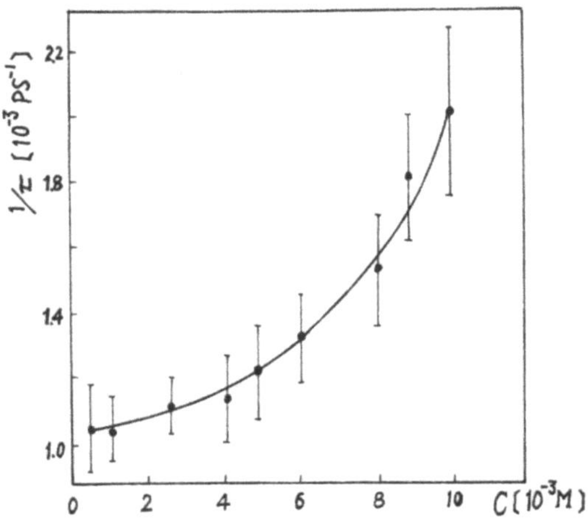

Fig.3 Fluorescence lifetime of eosin in water with increasing concentration C

It has been observed that the reciprocal of lifetime of the dye solution depends nonlinearly on the concentration (refer to Fig.3). This nonlinear dependence can be fit with the following equation

$$\frac{1}{\tau} = \frac{1}{\tau_M} + A_1C + A_2C^2 \qquad (1)$$

where τ is the lifetime of the solution, τ_M is the lifetime at very low concentration, C is the solution concentration, while A_1 and A_2 are constants. Supposing dimers as a kind of quencher besides nonexcited monomers, (1) can be derived from Stern-Volmer law.

Eq.(1) indicates that the effect of dimers, represented by the term A_2C^2, is chiefly responsible for the decrease of the fluorescence lifetime in concentrated solutions, while the influence of the nonexcited monomers, represented by the term A_1C, may be of less importance.

The absorption spectra of these solutions also show that the absorption band characterizing the dimers rises as the concentration increases. It means that the fraction of the dimers in the solution increases. This fact, togeter with the fact that the dimers might have a large specific quenching efficiency for the excited monomeric molecules, would lead to the strong fluorescence quenching in concentrated solution.

The effect of temperature on the fluorescence lifetime may be another evidence for the quenching by polymers. Previously, it was observed that heating a concentrated aqueous solution of eosin was accompanied by a corresponding increase in luminescence yield caused by the dissociation of the dimers. Our experiments have also shown that increasing the temperature of solution leads

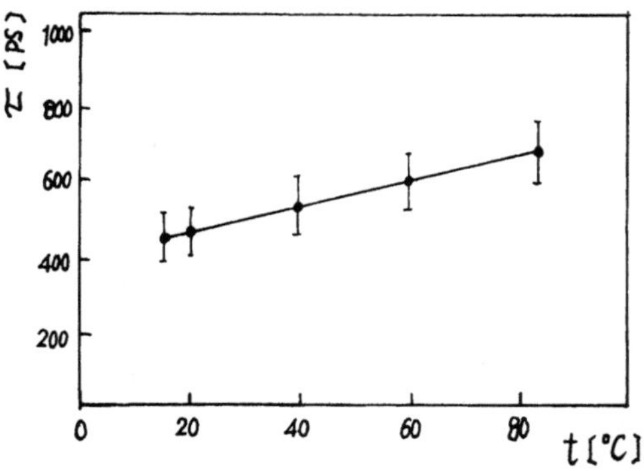

Fig.4 Influence of temperature on the fluorescence lifetime of eosin in water at concentration of 1×10^{-2}M

to an increase in fluorescence lifetime (refer to Fig.4). As the temperature of the solution at the concentration of 1×10^{-2}M rises from 20°C to 84°C, the fluorescence lifetime increases from 446±50ps to 661±72ps.

Reference

1. Peter Pringsheim, Fluorescence and Phosphorescence (Interscience, New York, 1949) P.349.

2. Michael D.Lumb, Luminescence Spectroscopy (Academic Press, London, 1978) P.130.

3. E.G.Baranova and V.L.Levshin, Optics and Spectroscopy, Vol. 10, 182 (1961).

4. R.R.Alfano, S.L.Shapiro and W.Yu, Opt. Commun., Vol. 7, 191 (1973).

5. M.A.Duguay and U.W.Hansen, Appl. Phys. Lett., Vol.15, 192 (1969).

6. G.Porter, E.S.Reid and C.J.Tredwell, Chem. Phys. Lett., Vol. 29, 469 (1974).

7. L.V.Levshin and V.G.Bocharov, Optics and Spectroscopy, Vol. 10, 330 (1961).

Part VI

Ultrashort Processes in Biology

Picosecond Processes Involving CO, O_2, and NO Derivatives of Hemeproteins

P.A. Cornelius and R.M. Hochstrasser

Department of Chemistry, University of Pennsylvania,
Philadelphia, PA 19104, USA

Introduction

Already it is well known that many ultrafast processes occur in biological systems. Picosecond pulses are used to initiate biological reactions in which light is either a natural constituent, such as in photosynthesis and vision, or a device to create nonequilibrium states of complex systems, such as with oxyhemoglobin. We have studied several heme proteins by picosecond transient absorption spectroscopy, a technique that is capable of revealing both ligand dynamics and changes in the protein structure.

For some time now, we have been concerned with the possible systematic errors associated with those picosecond spectroscopic methods that use the nonlinearly generated continuum. It is especially important to account for such errors in the case of hemeproteins, since most of the picosecond transient spectra reported have shown distinct differences from spectra known on longer timescales.

The resolution of this problem is hindered by the lack of any "standard" picosecond spectrum. In this article we discuss possible sources of systematic error in experiments of this type, in an attempt to begin a quantitative evaluation of picosecond spectral accuracy.

The present paper first presents a brief review of our recent studies of myoglobin photochemistry, then some new results on the nitric oxide derivative of hemoglobin. The interpretation of these results is critically dependent on the details of the spectral shapes which are then discussed. Finally, we present our assessment of all known picosecond timescale work on hemeproteins and possible relationships to the fundamental biological questions that first stimulated these studies.

Hemeprotein-Diatom Dynamics

A hemoglobin protein consists of four subunits, each possessing an iron porphyrin (heme) group to which a ligand such as O_2, CO, or NO can bind. Each of these derivatives (HbO_2, HbCO, HbNO, and Hb) has a characteristic visible absorption spectrum whose most prominent feature is an intense band (Soret) located between 415 and 445 nm. Because the Soret band is sensitive to ligation and to the heme environment, the time evolution of the spectrum can be used to study processes occurring at the heme site. For example, all of these liganded forms dissociate under visible or uv illumination. Three general questions arise concerning this photodissociation:

(1) What is the mechanism by which the energy of the photon is used to break the heme-ligand bond? (2) What is the subsequent behavior of the free ligand? (3) How does the Hb-like photoproduct respond to the loss of the

ligand? Because the primary steps in these mechanisms lie in the picosecond domain, these heme systems are unusually attractive subjects for picosecond transient kinetics and absorption work [1-11]. A series of experiments conducted in this laboratory [2-4,8,9] has studied the effect of varying the ligand, the protein structure, and the excitation wavelength.

Myoglobin (Mb) is a heme protein whose spectroscopy and electronic structure are virtually identical to hemoglobin, but it has only one subunit. In our experiments on MbCO and MbO_2 (using 5 ps pulses), less than ca. 15% geminate recombination was observed during the first 10 ns, in contrast to the Hb case. Thus myoglobin is a simpler system than hemoglobin for studying the photochemistry.

We have examined the transient spectra of MbO_2 and MbCO excited with second (527 nm) and third (351 nm) harmonic pulses [13]. In the case of 351 nm excitation, both molecules showed strong transients consisting of three features: (1) The bleaching of the Soret band representing the disappearance of ground-state population; (2) The appearance of new absorption representing the production of unliganded (Mb-like) product; and (3) a rapidly evolving difference feature which we associated with an excited state of Mb. In MbO_2 only, the appearance of the new absorption (2) was delayed in time, suggesting the presence of an intermediate step in the dissociation. In the MbO_2 case the rapidly evolving feature (3) was much more prominent than in MbCO. These spectra are presented in Fig. 1. The results for MbCO were unchanged when a second harmonic pulse was used as the excitation source, but MbO_2 exhibited only a weak signal under these conditions indicating a wavelength dependence for the conversion to photoproduct.

We have interpreted these results as follows: The optically pumped state of MbCO either promptly dissociates or first relaxes rapidly to the

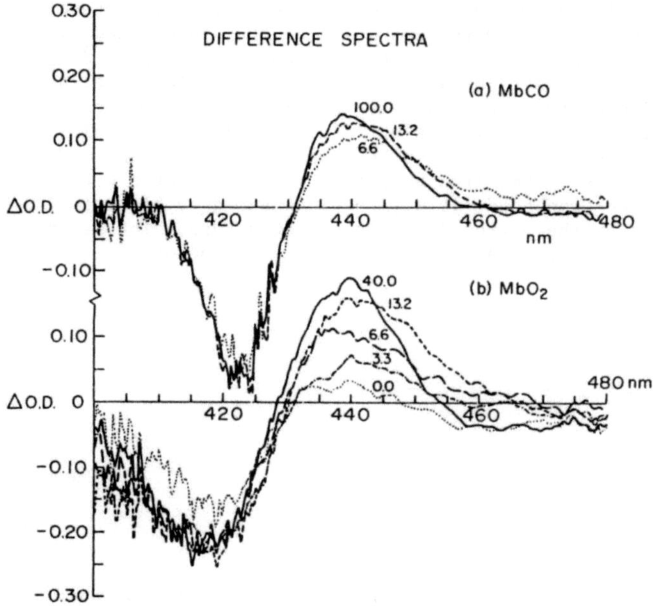

Fig. 1 Picosecond transient difference spectra of MbCO(a) and MbO_2(b). Time delay in ps is indicated on each curve; see [4].

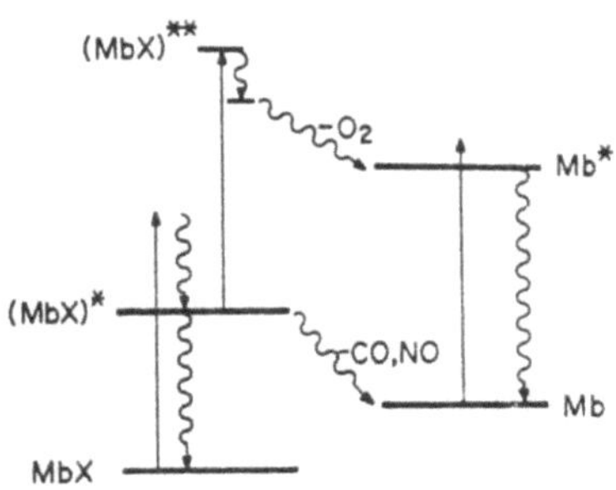

Fig. 2 Schematic energy level diagram for photolysis of heme proteins, showing the different pathways believed to operate for different ligands

lowest excited level and then dissociates. MbO_2 is more complicated (See Fig. 2). Optical pumping produces excited species that relax rapidly; during this relaxation process only a small probability exists for reaching those states through which dissociation occurs. Most of the population relaxes to the nondissociative excited level $(MbO_2)^*$. The large amplitude of the rapidly evolving feature (3) in MbO_2 suggests that, in this case, the Mb photoproduct is born in an excited state when a second photon excites $(MbO_2)^*$. In MbCO feature (3) is small, from which we conclude that this photoproduct is produced in its ground state.

We have now studied HbCO, HbO_2, and HbNO under the same experimental conditions: 353 nm excitation of pH7 solutions (phosphate buffer) with a heme concentration of $\sim 10^{-4}$ M. All three ligands are efficiently photolyzed under these conditions, producing transient spectra that show a strong bleaching of the Soret band. The behavior of the different ligands after photodissociation is strikingly dissimilar, despite the fact that all are diatomics of approximately equal size. For NO and to a lesser extent O_2 a diminution of the Soret bleaching amplitude was observed during the first nanosecond, suggesting the occurrence of geminate recombination of the dissociated ligands. For NO the bleaching signal decayed in two phases (τ_1 = 17 ps and τ_2 = 100 ps) and had completely recovered by 400 ps (See Fig. 3). In O_2 ca. 40% of the total bleaching signal recovered in a few hundred ps. In CO the difference signal was constant on this timescale. A simple picture explaining these results is that two barriers stand between the bound ligand and the protein-solvent interface. The first barrier (I) is associated with the formation of the ligand-heme bond; it will be influenced by electron spin effects and possibly by the surrounding protein structure. The second barrier(II) is an effective steric barrier opposing the escape of the

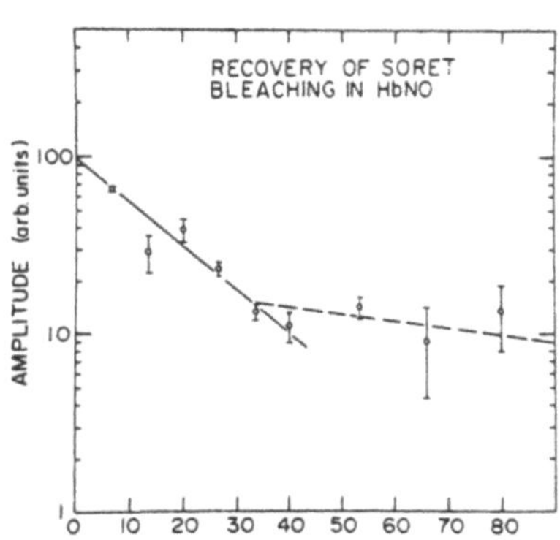

Fig. 3 Temporal behavior of HbNO difference spectrum. Decay of signal has two components and is associated with the occurrence of geminate recombination.

ligand. We have assumed that barrier II is approximately the same height for all three ligands. Barrier I is assumed to be strongly dependent on electronic factors such as spin-orbit coupling; we note that the formation of a bond between CO and heme requires that two spins flip, whereas NO and O_2 can bond with only one spin flip. For this reason we expect that barrier I will be higher for CO than for the other two ligands. The excitation pulse breaks the ligand-heme bond, in effect transferring population across barrier I. This population can either recombine with the heme by recrossing barrier I or dissociate by crossing barrier II. The partitioning of population between these two routes depends on the relative heights of the two barriers. Thus the lack of any recombination in CO indicates a a high barrier I and is consistent with the spin-orbit arguments above. Our data suggests that NO possesses a very small barrier I, while O_2 is intermediate between NO and CO.

The events that occur at the heme site are strongly influenced by the surrounding protein, which is in turn influenced by its aqueous environment. A complete understanding of the photokinetic behavior of heme systems must, therefore, include an understanding of aqueous solutions on the picosecond timescale. We have recently started experiments on simple ions in aqueous solutions aimed at understanding energy relaxation processes. Preliminary experiments on CN^- in H_2O showed that the v = 1 state of the ion decays in 5-10 ps in a process that is both solvent and counterion dependent [12]. This result illustrates that a wider range of fast phenomena might occur in aqueous solutions and emphasizes the need to understand better the full details as to how these systems recoverto thermal equilibrium after picosecond laser pulse excitation.

Picosecond Spectrophotometric Inaccuracies

There are several possible sources of systematic errors in our spectrometer: (1) a frequency chirp in the continuum pulse; (2) optical damage to the sample; (3) large (3 nm) spectral bandpass of the spectrometer; (4) imperfect pump and probe beam overlap. We have attempted to evaluate each of these effects and to minimize them whenever possible. We have found no evidence for chirping of the probe beam - the simultaneous appearance of both bleaching and absorption in MbCO precludes the presence of significant chirp. Optical damage to the hemeprotein samples seems unlikely, since this would be expected to produce ligand-independent spectral distortion. Computer calculations have demonstrated that our spectral bandpass of 3 nm does not seriously perturb the measured difference spectra. The effect of beam overlap is difficult to estimate, but it is clear that the probe beam traverses a sample volume that is not uniformly populated with photoproducts.

Under such circumstances the difference spectrum is distorted in shape, with the amount of distortion dependent upon the geometrical details of the pump and probe beams. We have previously carried out calculations that consider a probe continuum having a rectangular transverse profile and a pump beam having a Gaussian profile [13]. In this case we expect some spectral distortion but no significant spectral shifts. We also note that imperfect beam overlap cannot affect the position of an isosbestic point (a wavelength where the reactant and product have the same extinction coefficient), because such a point will always show $\Delta(OD) = 0$ in a difference spectrum. Isosbestic points are therefore the most reliable means of comparing results obtained by different investigators.

	530nm	353nm	OTHER
Mb		1. Soret bleached <5ps; DSP shows IP at 442nm; bleaching fully recovered in 11ps (4).	
MbCO	1. Like 353nm excitation (4) in all respects (9). 2. Soret bleached <25ps; DSP virtually identical to CW (+0.5nm shift?); t_D <25ps; no TDSC (25ps resolution); no bleaching recovery (10). 3. Like 353 excitation (5)(6) in all respects.	1. Soret bleached <5ps; DSP shows IP shifted by + 1.5nm from CW; Mb-like; t_D <5ps; weak TDSC; no bleaching recovery (4). 2. Soret bleached <6ps; no DSP; t_D=7-10ps; no bleaching recovery (6). 3. Soret bleached <6ps; no DSP; bleaching recovery t_R=125ps (5); (retracted in ref.6).	1. 265nm excitation - bleaching shows two component growth <5ps and ca. 100ps. Slow component represents ca. 10-20% of signal (energy transfer from tryptophan?) (9).
MbO_2	1. Much weaker DSP than for similar energy 353nm excitation (4); bleaching <8ps; no recovery of bleaching; other details not observable (9). 2. Soret bleached <25ps; DSP shows no shift from CW but slight shape difference; t_D<25ps; no TDSC; no bleaching recovery (7). 3. Soret bleached <6ps; no DSP; t_D=7-10ps; no bleaching recovery (6).	1. Soret bleached <5ps; DSP shows IP shifted +3nm from CW; Mb-like; formation of deoxy peak delayed by ∿ 5ps; no bleaching recovery (4). 2. Soret bleached <6ps; no DSP; t_D=7-10ps; no bleaching recovery (6).	
Hb	1. No transient observed (2).	1. Soret bleached <5ps; DSP similar to 353nm excitation of Mb; bleaching recovered by 20ps (4).	
HbCO	1. Soret bleached <5ps; DSP shows IP shifted + 3.5nm from CW; Hb-like; t_D<5ps; no TDSC; no bleaching recovery (3). 2. Soret bleached <8ps; DSP shows IP shifted + 3.5nm from CW: Hb-like; t_D longer than bleaching risetime; slight TDSC during first 24ps; no bleaching recovery (10). 3. Bleaching not studied; no DSP; t_D=12ps single wavelength kinetic study (11).	1. Soret bleached <5ps; DSP shows IP shifted by + 2.5nm from CW; t_D<5ps; no TDSC; no bleaching recovery (2)(3).	1. 615nm excitation: Induced absorption <0.5ps; no decay up to 20ps (1).
HbO_2	1. Soret bleaching <5ps; DSP shows IP shifted + 5nm from CW; Hb-like t_D<5ps; strong TDSC and ∿ 40% bleaching recovery in 500ps (3).	1. Soret bleaching <5ps; DSP shows IP shifted by + 5nm from CW; Hb-like; strong TDSC; no bleaching recovery; visible DSP shows IP at 559nm.	1. 615nm excitation: Induced absorption <0.5ps; decay of 2.5ps (1).
HbNO		1. Soret bleached <8ps; DSP. Shifted + 4nm from CW, Hb-like; t_D<8ps; no TDSC; t_R= 40ps (slower component ∿ 100ps), >95% recovery (3).	

Summary of Picosecond Studies of Heme Proteins

The table below presents a summary of all known picosecond absorption work on heme proteins. The table lists the following information about each experimental result: (1) the risetime of the bleaching signal (τ_B); (2) the general appearance of the difference spectrum (DSP), if any, and the positions of isosbestic points (IP); (3) the risetime of the deoxy peak (τ_D); (4) the presence or absence of any time-dependent spectral shape changes (TDSC); and (5) the rate at which the bleaching signal decayed (τ_R). The table clarifies the areas of agreement and disagreement between the various reported results, and also allows a quick comparison of the behavior of the different hemeprotein species studied.

Myoglobin has been the more extensively investigated system. It seems well established that no subnanosecond geminate recombination occurs in either MbO_2 or MbCO. Both species are efficiently photolyzed at 353 nm, but at 530 nm the situation is less clear. HbCO exhibits relatively straightforward behavior, dissociating efficiently under both 353 nm and 530 nm light showing little if any recombination up to 5 ns when pumped with 5 ps pulses. Only one set of results exists for HbO_2 and indicated significant bleaching recovery interpreted as geminate recombination in the 530 nm case. The most controversial aspect of this work concerns the details of the spectral shapes, where there is little agreement among the various workers. Only one study has reported spectra which closely match those obtained on longer timescales, so much further work is needed to resolve this question.

Acknowledgement This research was supported by grants from NSF (CHE 8000016) and NIH (GM12592). We thank W. A. Eaton and J. Hofrichter for suggestions concerning problems on spectrophotometric accuracy.

References

1. C. V. Shank, E. P. Ippen & Bersohn, R., Science 193, 50 (1976).
2. B. I. Greene, R. M. Hochstrasser, R. B. Weisman & W. A. Eaton, Proc. Natl. Acad. Sci. USA 75, 5255 (1978).
3. D. A. Chernoff, R. M. Hochstrasser & A. W. Steele, Proc. Natl. Acad. Sci. USA 77, 5606 (1980).
4. P. A. Cornelius, A. W. Steele, D. A. Chernoff & R. M. Hochstrasser, Proc. Natl. Acad. Sci. USA 78, 7526 (1981).
5. W. G. Eisert, E. O. Degenkolb, L. J. Noe & P. M. Rentzepis, Biophys. J. 25, 455 (1979).
6. A. H. Reynolds, S. D. Rand & P. M. Rentzepis, Proc. Natl. Acad. Sci. USA 78, 2292 (1981).
7. A. H. Reynolds & P. M. Rentzepis, Biophys. J. 38, 15 (1982).
8. P. A. Cornelius, A. W. Steele & R. M. Hochstrasser, "Ultrafast Relaxation in Picosecond Photolysis of Nitrosylhemoglobin" submitted to J. Mol. Biol.
9. P. A. Cornelius, unpublished data from this laboratory.
10. J. A. Hutchinson, T. G. Traylor & L. J. Noe, submitted to J.Am.Chem.Soc.
11. L. J. Noe, W. G. Eisert & P. M. Rentzepis, Proc. Natl. Acad. Sci. USA 75, 573 (1978).
12. E. J. Heilweil, F. E. Doany, R. Moore & R. M. Hochstrasser, J. Chem. Phys. 76, 5632 (1982).
13. B. I. Greene, R. M. Hochstrasser & R. B. Weisman, J. Chem. Phys. 70, 1247 (1979).

Femtosecond and Picosecond Transient Processes After Photolysis of Liganded Hemeproteins

J.L. Martin, C. Poyart[1], A. Migus, Y. Lecarpentier, R. Astier, and J.P. Chambaret

Laboratoire d'Optique Appliquée, Ecole Polytechnique - ENSTA, F-91120 Palaiseau, France

Dynamical analysis of photolysis products in the picosecond time scale have allowed new insights in the early processes of heme-ligand interactions in myoglobin and hemoglobin [1-4]. In a preliminary experiment SHANK et al [3] demonstrated that photolysis of HbCO occurs in less than 0.5 ps.

We report here informations upon the photolysis pathways of CO or oxygen liganded myoglobin and hemoglobin in the 100 fs to 200 ps time range. To reach this time resolution, we have developed a femtosecond transient electronic absorption spectrometer using an amplified ring laser. We observed that photodissociation of the liganded species occurs with a 250-300 fs time constant followed in the next 200 ps by different spectral evolutions of the photoproducts as the ligand is O_2 or CO. A two-step recovery process including the relaxation of an excited deliganded species and a geminate recombination, is only apparent in the O_2 liganded species. These data provide a plausible explanation for the large difference in the quantum yield of photodissociation between CO and O_2 liganded species.

Purified hemoglobin solutions were prepared from fresh human blood through DEAE sephadex column chromatography and diluted in 0.05 M bis Tris or 0.1 M phosphate buffer, pH7 or 8 at ambient temperature. Myoglobin (sperm whale type II (SW) or horse heart type III (HH) from Sigma) was prepared following the procedure of ROTHGEB and GURD [5]. Dilutions of the stock solutions were made so as to obtain an absorbance of 2.2 at the peak of the Soret band after full ligation with either O_2 or CO. These solutions were introduced in a 0.1 or 0.05 cm light path quartz cuvette which was moved horizontally so that each pulse at 10 Hz excites a new region of the sample.

Photolysis was performed by 100 fs pulses generated by a passively modelocked CW dye ring laser. Amplification of these pulses to the 10 GW regime is described in detail in this issue by MIGUS et al [6]. Briefly, the technique for generating optical pulses in the 100 fs range follows the principles described by R.L. FORK et al [7] with a slightly different configuration (fig. 1). Almost transform limited pulses are obtained by a spectral selection using coated mirrors and does not need intra cavity pellicle etalon. These pulses passed through a four stage amplifier pumped by a Q-switched frequency doubled Nd-Yag laser. A dispersive delay line is introduced at the output of the second stage to compensate the group velocity dispersion in the dye and optics at the level of the sample. This set up produces finally pulses at 10 Hz with energy above 1 mJ and 160 fs duration.

Half the energy is used to generate a broad band femtosecond continuum while the other half is focused into a 1.5 mm thick KDP crystal to produce

[1]INSERM U27, F-92150 Suresnes, France.

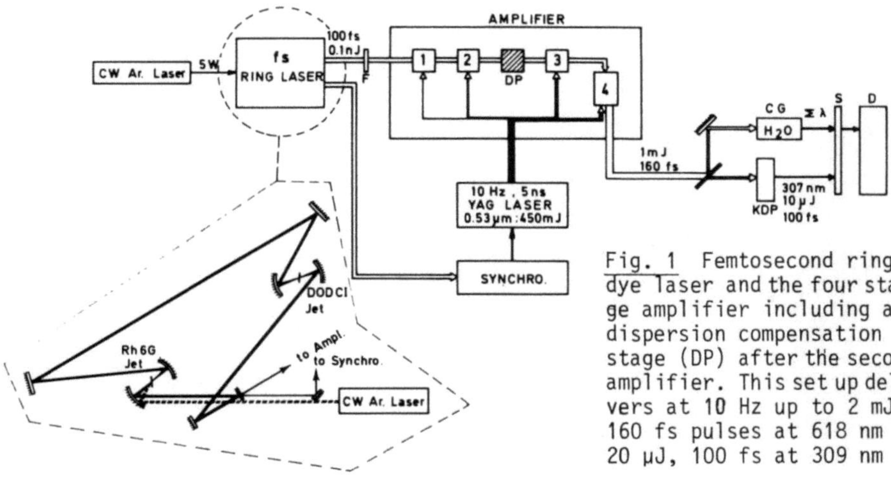

Fig. 1 Femtosecond ring dye laser and the four stage amplifier including a dispersion compensation stage (DP) after the second amplifier. This set up delivers at 10 Hz up to 2 mJ, 160 fs pulses at 618 nm or 20 µJ, 100 fs at 309 nm

(S) Sample
(D) Detector
(CG) Continuum generation

up to 20 µJ, 100 fs pulses at 309 nm. The UV beam is the excitation beam which can be delayed by an optical delay line driven by a step motor. The continuum beam is split into two parts and used in a double beam spectrophotometric technique. Transient spectra in the Soret region (380 nm-470 nm) were recorded with a 2D optical multichannel analyzer Vidicon detector. Kinetics at given wavelengths were obtained by two diodes placed at the exit slit of a 0.25 m spectrometer. The two signals corresponding to the probe and reference pulses are directed through an electronic chain to a HP1000 computer integrating at least, in each run, 6000 laser pulses of discriminated energy.

CO-Liganded Hemeproteins

In HbCO the rise of the bleaching at 419 nm (fig. 2), corresponding to the disappearance of the ground state liganded species, occurs within the pulse duration τ_p = 115 fs (instantaneous response). By contrast the induced absorption at 433 nm, taking into account the dispersion delay in the continuum beam, cannot be fitted to an instantaneous response. The best fit of the rate of appearance of the deoxy Hb species is obtained with τ_{Hb} = 300 fs (fig. 2). This value was found to be independent of pH, ionic strength and of CO concentration from 0.01 to 1 atm. Similar results were obtained with HH or SW MbCO solutions.

Fig. 3 illustrates the ΔA spectra recorded 5 and 50 psec after excitation of the MbCO sample. No other change in the ΔA spectra have been observed up to 200 ps after excitation. It may be noted also that the shapes of the transient spectra are similar to equilibrium ΔA spectra with no broadening at the longest wavelengths.

These results favor the proposal of a simple scheme of the photolytic pathways in CO liganded hemeproteins. These pathways imply the formation of a highly photodissociative excited species (XCO*) which relaxes for 50 % directly to the stable ground state deliganded species (τ = 300 fs) and to excited non dissociative liganded state(s). No recombination processes have been observed in the 200 ps time scale following excitation.

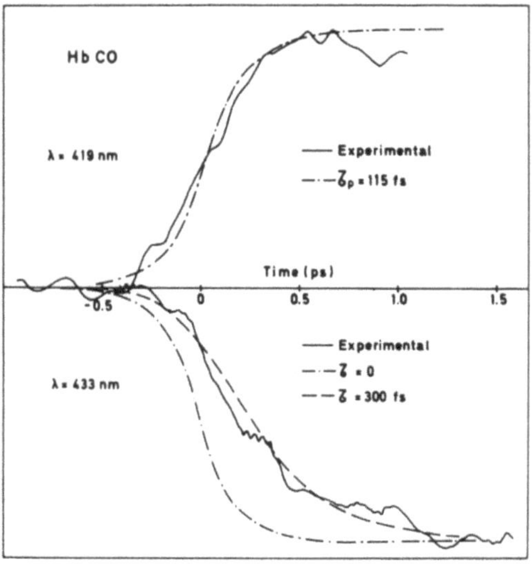

Fig. 2 HbCO : Kinetics of the bleaching at 419 nm and of appearance of the deoxy species at 433 nm. Centerline in the upper part corresponds to the instantaneous response computed with a pulse of duration τ_p = 115 fs. Centerline and dashed lines in the lower part correspond to exponential responses with τ indicated in the figure

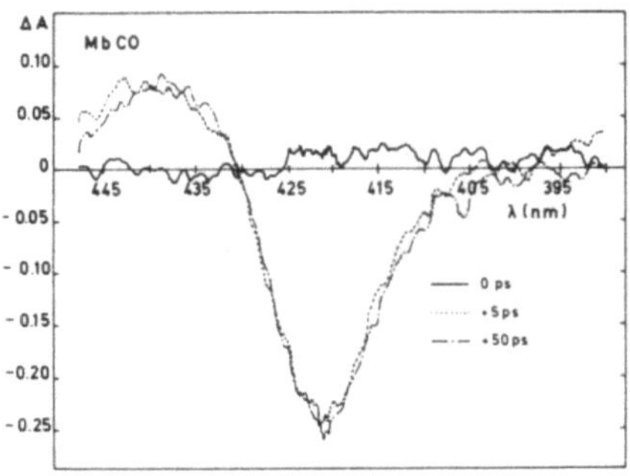

Fig. 3
MbCO : Transient difference spectra of deoxy Mb - MbCO at different delays after pulse excitation at 309 nm

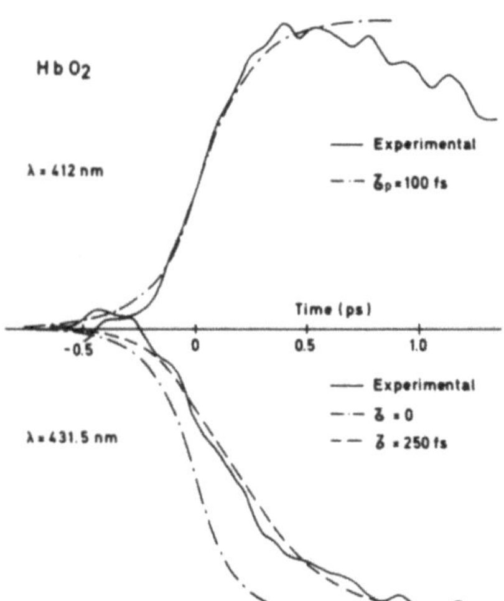

Fig. 4
HbO_2 : Kinetics of the bleaching at 412 nm and of appearance of the deoxy species at 431.5 nm. Normalized induced transmission (upper part) and induced absorption (lower part).

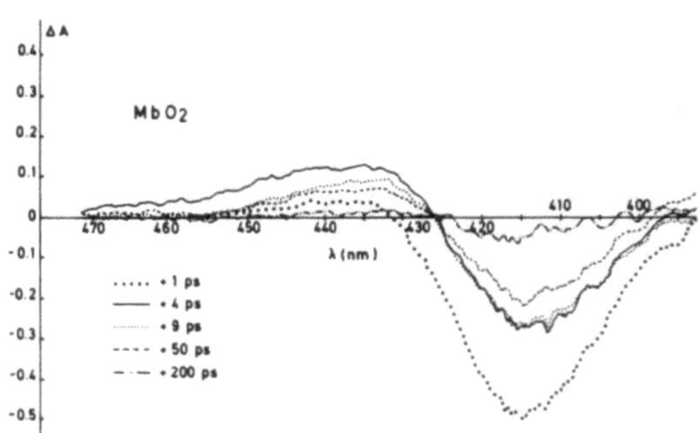

Fig. 5 Picosecond difference spectra following photodissociation of MbO_2. Note the broad red-shifted band occuring at + 4 ps which relaxes to stable Mb species (9ps) and the symmetrical changes occuring in the spectrum after this initial relaxation.

O_2 liganded hemeproteins

Fast kinetics of photodissociation in HbO_2 (fig. 4) are similar to those observed in HbCO or MbCO solutions. The time constant corresponding to the appearence of the deoxy species at 431.5 nm is 250 ± 20 fs. Identical values were measured in MbO_2 solutions. The fully developed initial ΔA spectrum corresponding to the 1 to 4 ps time range after excitation is shown in figure 5 (spectrum 4 ps). This reveals the formation of a deliganded Mb species with a broader absorption band extending up to 470 nm compared to a stable Mb spectrum. This transient excited species fully relaxes to the stable Mb product in the following 5 ps (+ 9 ps ΔA spectrum) (figure 5). The time constant of the decay of this excited species is 1.4 ps as measured from the kinetics of absorbance changes at, or near, the isosbestic point (424.5 nm). As shown in figure 5 no further change was noticed in the shape of the ΔA spectra following this early relaxation. However, we observed a simultaneous recovery of the bleaching at 414 nm and of the induced absorbance at 438 nm during the following 200 ps.

Conclusion

The present results demonstrate that photodissociation of CO or O_2 liganded hemeproteins proceeds through a very short-lived excited state which relaxes (τ=250 fs-300 fs) into a deliganded species [8,9]. The presence of O_2 as ligand induces an *excited* deliganded state as indicated by the red-shifted band in the ΔA spectra. Following these early events our results demonstrate also the existence of a geminate recombination phase of O_2 molecules with a time constant τ = 100 ps. These data give new informations in the pathways of photodissociation and imply that the low quantum yield of O_2 liganded hemeproteins is primarily due to relaxation of excited states to nondissociative states *and* also to a geminate recombination of O_2 molecules [9].

References

1 B.I. Greene, R.M. Hochstrasser, R.B. Weisman, W.A. Eaton, Proc. Natl. Acad. Sci. USA, 75, 5255-59 (1978).

2 L.J. Noe, W.G. Eisert, P.M. Rentzepis, Proc. Natl. Acad. Sci. USA, 75 573-577 (1978).

3 C.V. Shank, E.P. Ippen, R. Berson, Science, 193, 50-51 (1976).

4 P.A. Cornelius, A.W. Steele, D.A. Chernoff, R.M. Hochstrasser, Proc. Natl. Acad. Sci. USA 78, 7526-7529 (1981).

5 T.M. Rothgeb, F.R.N. Gurd, Methods in Enzymol, 52, 473-482 (1978).

6 A. Migus, J.L. Martin, R. Astier, A. Antonetti, A. Orszag, This issue (1982).

7 R.L. Fork, B.I. Greene, C.V. Shank, Appl. Phys. Lett. 38, 671 (1981).

8 J.L. Martin, A. Migus, C. Poyart, Y. Lecarpentier, R. Astier, A. Antonetti, Proc. Natl. Acad. Sci. USA, to be published.

9 J.L. Martin, A. Migus, C. Poyart, Y. Lecarpentier, A. Antonetti, A. Orszag, Biochem. Biophys. Res. Comm. In press.

Picosecond Fluorescence Spectroscopy of Hematoporphyrin Derivative and Related Porphyrins

M. Yamashita and T. Sato
Laser Research Section, Radio- & Opto-Electronics Division, Electrotechnical Laboratory, 1-1-4 Umezono, Sakura-mura, Niihari-gun, Ibaraki-ken 305, Japan

K. Aizawa and H. Kato
Tokyo Medical College, 6-7-1 Nishishinjuku,
Shinjuku-ku, Tokyo 160, Japan

Recently it has been demonstrated that tunable laser irradiation permits the treatment of malignant tumors with a light-activated dye [1]. The success of this technique is due to the ability of the dye, hematoporphyrin derivative (HpD), to accumulate to a higher degree (~ ten times) in malignant tumors than in normal tissues. If HpD in a cancerous cell is exposed to red light from a laser, it absorbs enough energy to destroy the cells but allows the full recovery of neighboring normal cells containing a lower concentration of the dye. However, little is yet known about the dynamic mechanisms of photochemical reactions occurring in the treatment, including the reason that HpD selectively accumulates in the tumors. In order to make clear the dynamics it is important to investigate quantitatively relaxation processes from excited electronic states of HpD and related molecules. A few spectroscopic studies of metal-free porphyrins which have low-intensity fluorescences have been performed [2-4], but no direct measurements of the fluorescence decay have been reported [5]. In this paper, we report the direct measurement of picosecond fluorescence lifetimes τ_M of HpD, hematoporphyrin IX (HpIX), hematoporphyrin IX diacetate (HpIXDA), protoporphyrin IX disodium salt (PpIXDS), and protoporphyrin IX dimethyl ester (PpIXDE), and the effects of oxygen molecules and acid solutions on the lifetimes by a use of a synchroscan streak camera [6]. In addition, fluorescence quantum yields of the metal-free porphyrins are measured, and natural radiative lifetimes and radiationless transition rates from the first excited singlet state are determined.

HpD was directly offered from DOUGHERTY's group [1]. Their recent analysis has shown that HpD is a mixture of ten metal-free porphyrins including HpIX, HpIXDA, and PpIX. HpIX and PpIXDE, and PpIXDS and HpIXDA were obtained from the Sigma Chemical Company and the Nakarai Chemical LTD., respectively. Each porphyrin was dissolved in a phosphate buffer saline solution (PBS, pH=7.2) which is the same condition as an aqueous solution in living cell, or in an ethanolic solution (EtOH), at the concentration of 0.5 mg/ml. The porphyrin solution in a 1x1x5 cm quartz cell was circulated at the flow speed of 20 cc/min to prevent any photodegradation of its solution.

A schematic of the experimental arrangements is shown in Fig.1. A synchronous-passive hybrid mode-locked CW dye (Rhodamine 6G) laser provides the source of excitation pulses of ~0.8 ps duration at 572 nm [7]. Average powers of ~50 mW are obtained at a repetition rate of 12.2 ns. The continuous train of excitation pulses is focussed by a lens (f=3 cm) into the sample cell containing the circulating porphyrin solution. The line of focus is directed just inside the output face of the cell to minimize any reabsorption of the fluorescence. The fluorescence band in the region from 600 to 700 nm from the porphyrins is efficiently selected by using a interference filter which eliminates the excitation radiation at 572 nm. The fluorescence at 90° with respect to the input direction is focussed on the input slit (slit-width 20 μm) of the streak camera by a lens after passing through the filter and a polarizer set at 55°.

The fast fluorescence lifetime τ_M of each porphyrin solution having the low fluorescence quantum yield Φ_{FM} (~10^{-3}) was measured by using a Hamamatsu TV Co. Ltd. Model C1587X synchroscan streak camera [8]. The streak-camera tube with multi-channel plates (S20 UV photocathode) in the case of the single shot mode has a time resolution limit of 9.0 ps at 532 nm. A synchronous output (41.2 MHz) from a frequency synthesizer (1×10^8 of frequency stability) driving a mode-locking modulator of a pumping Ar ion laser is used for the source of synchronization of the fluorescent event with the continuous sinusoidal ramp voltage to the streak-camera deflection plates, as shown in Fig.1. This means that ~10^8 fluorescent events are precisely superimposed on the streak camera phosphor. In order to examine the overall resolution of the system at the present experimental condition of excitation average powers of ~50 mW at 572 nm, the pulse duration of the dye laser beam was measured on the TV monitor and shown to be ~26 ps. Fluorescence quantum yields were determined by measuring fluorescence spectra by means of a calibrated fluorometer (Hitachi Perkin-Elmer Model 650) and using a rhodamine B ethanolic solution (2×10^{-7}M) as a standard [9]. The excitation wavelength of the solution was 360 nm, and a value of 0.69 was used as its quantum yield.

The porphyrin solution in the cell is excited from the ground state to the lowest excited singlet state by continuous pulse trains of the dye laser beam at 570 nm to avoid photodegradation due to irradiation around 400 nm. Some parts of the porphyrin molecules in the excited state decay to the ground state with emitting the fluorescence from 600 to 700 nm. From the decaying fluorescence profile displayed on the TV monitor the fluorescence lifetime is determined. In order to examine the effect of oxygen quenching of the fluorescence, the fluorescence decays of HpD in aero-, oxygen-bubbled, and oxygen-outgassed PBS solutions were measured, and those fluorescence lifetimes were determined to be 239, 176, and 333 ps, as shown in Fig.2-(1), (2), and (3), respectively. This indicates that oxygen molecules quench some excited singlet state porphyrins by the enhanced $S_1 \rightarrow T_1$ intersystem crossing due to diffusion controlled collisions [9, 10] or by photooxidation of porphyrins due to surrounding singlet oxygen molecules produced [9, 11]. The fluorescence decay of HpD in ethanol (EtOH) shown in Fig. 2-(4) indicates that the fluorescence lifetime in the case of EtOH is almost similar to that in the case of PBS.

Furthermore, fluorescence lifetimes of HpIX in PBS, PpIXDS in PBS + EtOH (1:1), HpIXDA in PBS, PpIXDE in EtOH + HCl (pH=0.9), and PpIXDS in EtOH + HCl

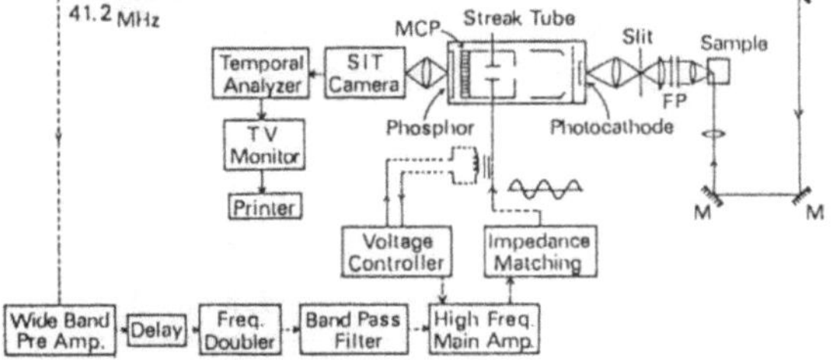

Fig.1 Schematic of experimental arrangement

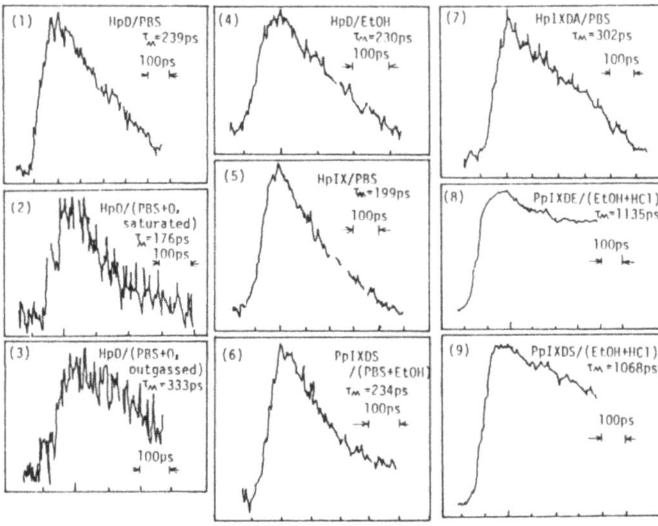

Fig.2. Recorded fluorescence decay profiles for HpD and related metal-free porphyrin solutions

(pH=1.0) were determined to be 199, 234, 302, 1135, and 1068 ps as shown in Fig. 2-(5), (6), (7), (8), and (9), respectively. Since PpIXDE hardly dissolved in PBS or EtOH, HCl was added to the PpIXDE ethanolic solution. The measurement for the PpIXDS ethanolic solution with HCl addition was made in order to examine the effect of the acid solution on the lifetime. From Fig. 2-(5), (6), and (7) it is found out that the average value of fluorescence lifetimes of the three porphyrin solutions except for the HCl additive solutions is approximately equal to that of the aero-PBS solution of HpD including those porphyrins. It seems that the lifetime of the former three porphyrin solutions increases as the size of the molecular groups attaching to the main ring instead of the pyrrole exo-hydrogens 6 and 7 increases. The remarkable lengthening of lifetimes of the latter two porphyrins in the acid solutions is presumably due to photoreduction of the porphyrins in protonic solvents to the phlorins [3, 11]. For all the porphyrin solutions, the depolarization effect on the fluorescence lifetime was also examined by setting the angle of the polarizer in Fig.1 at 0 and 90°. However, any remarkable change in the lifetime was not detected.

Table 1. Obtained fluorescence lifetimes τ_M, fluorescence quantum yields Φ_{FM}, natural radiative lifetimes τ_{FM}, and radiationless transition rates k_{NR} for metal-free porphyrins

metal-free porphyrins	τ_M [ps]	Φ_{FM}	τ_{FM} [ns]	k_{NR} [sec^{-1}]
HpD in PBS	239	2.9×10^{-3}	82	4.2×10^{9}
HpIX in PBS	199	5.3×10^{-3}	38	5.0×10^{9}
HpIXDA in PBS	302	3.5×10^{-3}	86	3.3×10^{9}
PpIXDS in PBS+EtOH	234			
PpIXDE in EtOH+HCl	1135	1.8×10^{-2}	63	8.6×10^{8}
PpIXDS in EtOH+HCl	1068	1.6×10^{-2}	68	9.2×10^{8}

Fluorescence quantum yields measured for each porphyrin solution are summarized in Table 1 with the fluorescence lifetime. By using the quantum yield and lifetime, the natural radiative lifetime τ_{FM} and radiationless transition rate k_{NR} from the first excited singlet state are obtained [9], and also given in Table 1. The values of those natural radiative lifetimes accord with the order of the value of that estimated from absorption spectra of tetraphenyl porphyrin in benzene [2]. From the comparison between the radiationless transition rates in Table 1 it is found out that their rates of porphyrins in PBS solvents are relatively larger than those in protonic solvents. This may suggest that energy of the first singlet excited state porphyrins in the PBS solution is spent more effectively than that in the protonic solution, for chemical reactions with target-biomolecules important for the treatment of malignant tumors.

In conclusion, we have directly measured picosecond fluorescence lifetimes of HpD and related metal-free porphyrin molecules by the technique of the synchroscan streak camera. In addition, the effects of oxygen molecules and acid solutions on the fluorescence lifetime have been discussed. Furthermore, the natural radiative lifetimes and radiationless transition rates from the first singlet excited state have been determined.

References

1. T.J.Dougherty and R.E.Thoma, in *Lasers in Photomedicine and Photobiology*, Springer Series in Optical Sciences Vol.22, p.67, eds. by R.Pratesi and C.A.Sacchi, (Springer-Verlag, Berlin, 1980).
2. P.G.Seijbold and M.Gouterman, J.Mol. Spectry. 31 (1969) 1.
3. M.Gouterman and G.E.Khalil, J.Mol. Spectry. 53 (1974) 88.
4. R.C.Srivastava, V.D.Anand, and W.R.Carper, Appl. Spect. 27 (1973) 444.
5. M.Gouterman, in *The Porphyrins*, Vol.III, Part A, p.24-47, ed. by D.Dolphin (Academic Press., New York, 1978).
6. W.Sibbett, J.R.Taylor, and D.Welford, IEEE J.Quantam Electron. QE-17 (1981) 500.
7. M.Yamashita and T.Sato to be published in Applied Optics.
8. M.Yamashita et al., to be presented at 1982 Conference on Precision Electromagnetic Measurements.
9. J.B.Birks, *Photophysics of Aromatic Molecules,* p.84-141 and 496-504 (John Wiley & Sons Ltd, London, 1970).
10. D.R.Keans, Chem. Rev. 71 (1971) 395.
11. F.R.Hopf and D.G.Whitten, in *The Porphyrins*, Vol.II, Part B, p.171-174 and 179-180, and D.Mauzerall, Vol.V Part C, p.36-45 and 50, ed. by D.Dolphin (Academic Press., New York, 1978).

Resonance Raman Spectra of Picosecond Transients: Application to Bacteriorhodopsin

M.A. El-Sayed[+]

Physikalische Chemie der Technischen Universität München, D-8046 Garching, Fed. Rep. of Germany

Chung-Lu Hsieh and M. Nicol

Chemistry Department, UCLA, Los Angeles, CA 90024, USA

Resonance Raman Spectra of Picosecond Transients

Picosecond pulses have been used extensively in determining the optical (mostly the visible) spectra of picosecond transients. These spectra, unfortunately, did not prove useful in most cases in imparting structural information about these picosecond transients. They gave the rise and, if not too long, the decay times of these transients. These results form the fact that most of the systems studied with the available lasers are too large with broad absorption spectra. Most small molecules with well resolved spectra absorb in the deep UV for which continuous monitoring picosecond lasers are not yet conveniently available.

Vibration spectroscopy has proven very useful over the years for the synthetic organic chemist in determining the structure of chemicals he prepares. Both, infrared and Raman spectroscopy, have been used for this purpose. For this reason, time resolved resonance Raman techniques are very essential to develop /1/. In spite of this need and the fact that picosecond lasers have been used to determine the spontaneous spectra of stable liquids /2, only a handful of papers appeared in the literature which report on the Raman spectra of picosecond transients /3/. The reason for this undoubtedly is the low signal expected to be observed. Raman scattering, even after enhancement, has a probability which is far smaller than that for optical absorption or fluorescence (if it has non-zero quantum yield). This, together with doing the monitoring on the picosecond time scale, makes the number of collected Raman scattered photons by picosecond transients very small. Of course, one may think of increasing the laser intensity. If the transients are formed photochemically by the same laser pulse, as in the case of all our systems studied so far, the Raman signal should increase quadratically with the laser intensity barring saturation effects. However, the other nonlinear processes taking place in solution, e.g. filamentation, stimulated Brillouin and Raman processes, multiphoton ionization and dissociation, etc., would compete (and probably win) if one would try to strongly increase the intensity of the pulsed laser system. In our work we have used as a source a mode-locked cavitydumped synchronously pumped dye system /4/. Each pulse has only few nanojoules in energy but it produces a million pulses per second. The scattering radiation is collected with a spectrometer fitted with vidicon detection at such a high duty cycle.

[+]permanent address: Chemistry Department, UCLA, Los Angeles, CA 90024, USA

This system, however, introduces another problem if the picosecond transient changes chemically or photochemically to another species in a longer time scale. Using a large volume of the sample with continuous stirring (a method used with high power pulsed lasers) might not assure a new fresh sample in the laser focal volume between pulses (which is μsec in synchronuous lasers instead of fraction of a second for the YAG lasers or even minutes for the glass laser). Electronic gating is of no use for solving this problem.

Sample flow, when combined with laser microbeam (tight focusing) techniques, can be used to secure that a fresh sample is present for each picosecond laser pulse when using the synch-pump system. Using available pumps, the solution sample can be flowed through a syringe needle with speeds of up to 40 meters per second. If the laser is focused on the sample as it comes out of the syringe needle to a spot of a few microns (using a microscope lense) /5/, the sample residence time in the laser beam is 4×10^{-4}cm/40×10^{2}cm/sec = 0.10μsec which is ten times shorter than the time between the pulses. This insures that, by the time a new pulse exposes the flowing sample, a complete new fresh supply of the sample would indeed have replaced the sample exposed by the previous pulse. The details of this technique are given elsewhere/6/.

Resonance Raman Spectra of Nanosecond - Microsecond Transients /7/

Just on passing, one should mention that the above technique has been used previously to obtain the resonance Raman spectra of microsecond and down to 80 nanosecond transients without the need of a pulsed laser. A CW laser initiates the photochemistry as well as acts as a source for the Raman scattering. The time resolution in this case is determined by the residence time of the photolabile sample in the laser beam. Under tight focus and rapid flow a resolution of 50 - 10 nsec can be obtained. For slow flow and more diffuse focus, scattering from transients appearing at a longer time scale can be detected and studied.

The System

The above techniques can be used for any photolabile /6/ systems such as:

$$A \xrightarrow{< \text{ psec}} B \overset{> \text{ psec}}{\leadsto} C$$

where both, A and B have strong overlapping absorption bands as is the case for many photobiological systems. The Raman spectrum is obtained at low laser powers (low photolysis efficiencies) as well as high laser powers (or tight focus). Computer subtraction techniques are used to determine the characteristic bands of the picosecond transient B. It greatly helps if the photochemical yield of B is high, its resonance enhancement is large, and neither A not B is fluorescent. To avoid Raman photon losses from absorption by A and B, the concentration of A, the laser wave-length as well as the optical arrangement have to be carefully adjusted to give a maximum resonance Raman signal of B.

The system discussed here is bacteriorhodopsin (bR) /8/, the second photosynthetic system in nature (the first being chlorophyl).

The primary processes resemble /8/ those of rhodopsin. Using optical flash techniques /9/, a scheme is derived showing the kinetic behaviour of the different intermediates from the changes of the retinal absorption in the visible region. It was believed for a long time that 11-cis to all-trans isomerization is the first step in vision occurring on the picosecond time scale. Some kind of isomerization was also believed to occur for bR on a similar time scale /8/. Recently, it was observed that the rate of the first step in these two systems was slower /9/ in D_2O than in H_2O. The authors of Ref. 9 questioned whether cis-trans isomerization can occur on this time *cale. Instead they suggested that the first step involves a proton translocation, e.g. the schiff base proton on the nitrogen connecting the retinal system with the protein.

In the present work we present the resonance Raman spectra of bacteriorhodopsin transients in the 30 - 50 psec time scale of two vibrational regions of retinal, one (1100-1400 cm^{-1}, the fingerprint region) is sensitive to retinal conformational changes and the other (1646 cm^{-1}) is of protonated C=N stretching of the schiff base. The results in the fingerprint region are shown in Fig. 1. A comparison of the lower two spectra shows that the spectrum of the parent (containing all-trans retinal) and that for the picosecond transient are indeed different. This fact strongly suggests a change in the retinal conformation //10/ during the 30 - 50 psec pulse width of the laser used.

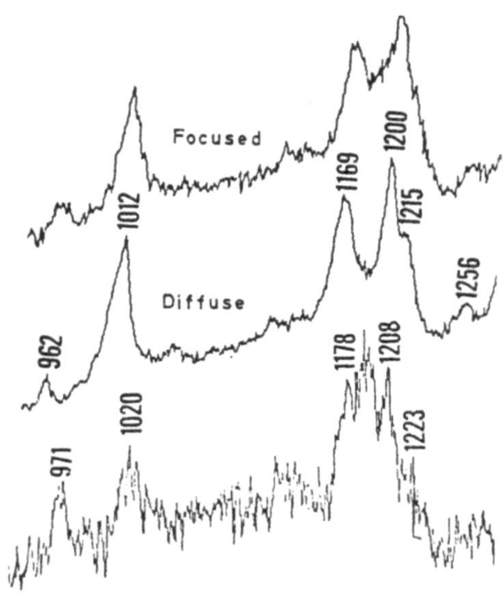

Fig.1 The effect of photolysis of 50 psec, 15 nanojoules, MHz-modelocked synchronously pumped dye laser at 587 nm on the resonance Raman spectrum of bacteriorhodopsin in the fingerprint region of retinal (1000-1400 cm^{-1}). The third spectrum shows the best computer difference betrween the "unphotolyzed" spectrum (using diffuse laser focus) and the photolyzed spectrum (focused laser). The result suggests that retinal isomerization indeed takes place in less than 50 psec within the protein pocket.

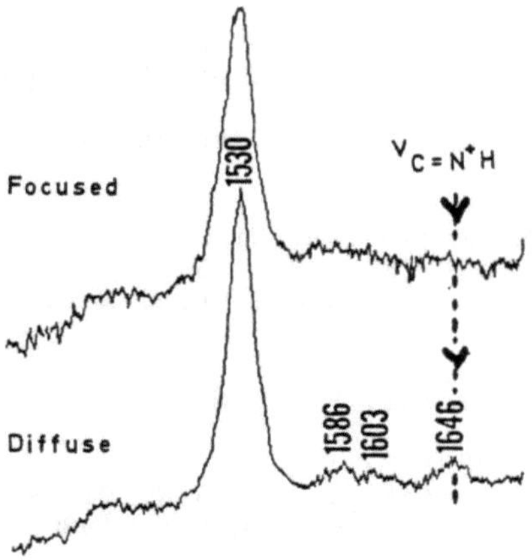

Fig.2 The effect of photolysis of the same laser described in Fig.1 on the $C=N^+H$-stretching region of the schiff base of retinal observed at 1646 cm^{-1}. Increasing the laser intensity (by focusing the laser) is found to decrease the relative scattering in this region. This could be a result of the inhomogenous broadening of this vibrational transition for the picosecond transient or due to a shift in the frequency of this vibration for the picosecond transient. In any case, this might suggest a change in the electronic structure of this group upon picosecond photolysis.

Fig. 2 shows the effect of the photolysis (focused) on the $C=N^+H$ region (at 1646 cm^{-1}) of the parent (diffuse). While the signal to noise ratio in this region is not very good, the results suggest a change in the C=N-stretching frequency upon transformation into the picosecond transient(s). Could this be due to a change of the degree or the strength of protonation of the schiff base nitrogen or a result of isomerization or perhaps both?

Acknowledgement

M.A. El-Sayed would like to thank the Alexander von Humboldt-Foundation for the award granted. He also thanks Professor E.W. Schlag for his hospitality and for stimulating scientific interactions with him and his group. The financial support of the Department of Energy for this research is also acknowledged.

References

1 For a modest review: M.A. El-Sayed, Time Resolved Resonance Raman Spectroscopy in Photochemistry and Photobiology, in Multichannel Image Detectors in Chemistry, ACS Symposium Series Bk 102, Chap. 10, 1979, p. 215-227.
2 see references in 1; e.g., M. Bridoux and M. Delhaye, in Advances in Infrared and Raman Spectroscopy", Vol. 2, eds. R.J.H. Clark and R.E. Hester, Heyden 1976, p. 140.
3 J. Terner, T. Spiro, M. Nagumo, M.F. Nicol, and M.A. El-Sayed, J. Am. Chem. Soc. 102 (1980) 3238;

M. Coppey, H.Tourbez, P. Valat, B. Alpert, Nature (London) 284 (1980)568; G. Howard, W. Carlsen, A. Siegman, and L. Stryer, Science 211 (1981) 942.
4 M. Nicol, Y. Hara, J. Wiget, M.F. Anton, J. Mol. Struct. 47 (1977) 371.
5 M.W. Berns, Exp. Cell Res. 65 (1971) 470.
6 J. Terner, J.D. Strong, T. Spiro, M. Nagumo, M. Nicol, and M.A. El-Sayed, Proc. Nat. Acad. Sci.USA 78 (1981) 1313.
7 J. Terner, C.L. Hsieh, A.R. Burns, M.A. El-Sayed, Proc. Nat. Acad. Sci. USA 76 (1979) 3046.
8 R. Henderson, Ann. Rev. Biophys. Bioeng. 6, (1977) 87;
W. Stockenius, R.H. Lozier, R.A. Bogomolni, Biochim.Biophys. Acta 505 (1979) 215;
W. Stockenius, Acc. Chem. Res. 13 (1980) 337;
M. Ottolenghi, in Advances in Photochemistry, eds. J.N. Pitts, G.S. Hammond, K. Gollnick, D. Grosjean, Wiley, New York 1980, p. 97.
9 R.H. Lozier, R.A. Bogomolni, W. Stockenius, Biophys. J. 15 (1975) 955.
10 For more details of the conclusions: see C.L. Hsieh, M. Nagumo, M. Nicol, and M.A. El-Sayed, J. Phys. Chem. 85 (1981) 2714.

Picosecond Studies of Bacteriorhodopsin Intermediates from 11-cis Rhodopsin and 9-cis Rhodopsin

J.-D. Spalink
Bell Laboratories, 1600 Osgood St., North Andover, MA 01845, USA

M.L. Applebury
Department of Biochemical Sciences, Princeton University, Princeton, M.J. 08544, USA

W. Sperling
Institut für Neurobiologie, KFA Jülich, D-5170 Jülich, Fed. Rep. of Germany

A.H. Reynolds and P.M. Rentzepis
Bell Laboratories, 600 Mountain Ave., Murray Hill, N.J. 07974, USA

An improved double beam laser spectrophotometer is described, developed to perform a comparative study of the bathorhodopsin photoproducts of native 11-cis rhodopsin and regenerated 9-cis rhodopsin. This instrument measures bathorhodopsin-rhodopsin difference spectra of these two visual pigments under identical experimental conditions, 25ps to 8ns after excitation of 532 nm. The spectra are taken over the entire wavelength range between 400 nm and 650 nm using a single monitoring pulse. Particular attention was given to optimizing the signal to noise ratio of measured absorbance changes. This allows data collection with excitation pulse energies low enough to avoid multiphoton events. In this way, we sought data which would allow a careful investigation of the first metastable products resulting from initial photochemical events. Our goal was to establish, as best possible, characteristic absorption spectra of the transient photoproducts arising from 11-cis or 9-cis rhodopsin at room temperature and to compare these with the published spectra obtained by photostationary studies carried out with aqueous-glycerol rhodopsin glasses at low temperature.

The rhodopsin samples were prepared from frozen bovine retinae (G. Hormel) solubilized in Ammonyx detergent and purified by hydroxyapatite chromatography. This procedure allows us to obtain highly purified material which is free of any excess retinal chromophore and has excellent optical clarity.

The samples were photoexcited by a single, 25 ps light pulse (FWHM) at 532 nm. This second harmonic was generated by a KDP crystal from the 1064 nm pulse emitted by Nd^{3+}/YAG laser (Quantel). Absorbance changes were monitored by a broadband picosecond continuum, generated in a 10 cm quartz cuvette containing water. Chirp was found to be $\leq$15 ps over the 400 nm to 650 nm range and therefore did not influence our results obtained for delays of 85 ps or longer between excitation and monitoring pulses.

Scattered light from the excitation pulse was blocked from entering the monitoring pathway by a narrow band rejection filter (OMEGA OPTICAL, Brattleboro, VT) which transmits 80% of the light between 440 nm and 650 nm, but has a transmission of about 10^{-5} in the 525 nm primary and 10^{-3} in the 440-450 nm secondary rejection band. Application of this filter allowed us to monitor absorbance over the entire wavelength range for each interrogation pulse without the problems related to the use of cross-polarizers in excitation and monitoring pathways.

The interrogation pulses were spectrally resolved by a PAR polychromator, detected by a ISIT/OMA2 detector system, and analyzed on an Eclipse S/130 minicomputer. A wavelength calibration of the detection system was carried out at the beginning of each experiment, resulting in a relative error between the experiments of less than 0.5 nm. Calibrating the whole system against different emission lines resulted in an absolute wavelength error of less than 2 nm. The detection system was checked for a linear response in absorbance changes; with 10 prescans and 20 read out scans (140 μsec/channel) the error was less than 10% with a signal of $\Delta OD = 0.2$.

To obtain reproducible absorption amplitudes, a quartz flow cell of 2 mm pathlength, holding 50 μl sample, was used (Hellma 138 QS). Samples were stirred between each shot and discarded after two excitation shots. At the excitation energies used here the signal amplitudes of the first and second shots differed by less than 5%. No change in shape of difference spectra was observable between first and second shots. Four to six pairs of data (excitation/no excitation) which corresponded to similar excitation energy pulses were averaged to smooth the data. Experiments were performed under as nearly identical conditions as possible for 11-cis and 9-cis rhodopsin samples.

The energy of each excitation pulse was measured with a new energy meter (RJ5200, Laser Precision Corp.) calibrated against the NBS standard.

The difference spectra of bathorhodopsin produced minus rhodopsin bleached, taken in the manner described above, show that the isosbestic point (≈520 nm) of the 11-cis rhodopsin spectra is red shifted about 5 nm compared to that (515 nm) of 9-cis rhodopsin. There is no indication of positive absorption change in the region of 400 nm to 450 nm. The spectra shows that the maximum ratio of absorbance decrease due to rhodopsin bleached is 2.3 for 11-cis and 1.5 for 9-cis. This ratio for 11-cis is more than twofold higher than previously measured at room temperature but is nearly the same as that measured following photolysis of rhodopsin at low temperatures. Similar observations were made at 600 ps and 8 ns.

These results do not support the classical interpretation given to early photochemical events in the process of vertebrate visual transduction. Photostationary studies of rhodopsin-bathorhodopsin in aqueous glasses at low temperature have been accounted for by a model which suggests the following photoequilibrium is established:

$$\text{11-cis rhodopsin} \xrightarrow{0.67} \text{bathorhodopsin} \xleftarrow[0.1]{} \text{9-cis rhodopsin.}$$

This mechanism has been one of the key paradigms dictating that the primary event in vision is cis-trans isomerization. The model indicates that a common batho photoproduct arises upon absorption of a photon by either the 11-cis or 9-cis retinal chromophore bound to opsin - hence the common intermediate must be all-trans.

Our study of these two rhodopsins indicates that the respective bathorhodopsins differ in their extinction coefficients and that their absorption maxima are shifted in wavelength by about 10 nm when observed as transients at room temperature between 85 ps and 8 ns. This suggests that there is not a common bathorhodopsin state, whether it be all-trans or distorted all-trans, when starting from two different isomers of the visual pigment chromophore.

This work was supported in part by a grant from the Deutsche Forschungsgemeinschaft to J.D.S. and a grant from the National Eye Institute to M.L.A. The authors wish to thank Wolfgang Baehr and Kathy Savoie-Luisi for their expert help in preparation of rhodopsin samples.

Multiple Photon Processes in Molecules Induced by Picosecond UV Laser Pulses

V.S. Antonov, E.V. Khoroshilova, N.P. Kuzmina, V.S. Letokhov, Yu.A. Matveetz, A.N. Shibanov, and S.E. Yegorov

Institute of Spectroscopy, USSR Academy of Sciences, Troitzk, 142092 Moscow Region, USSR

1. Introduction

Multistep and multiple photon excitation of atoms and molecules by laser radiation makes a base for nonlinear laser photochemistry and their applications (see, e.g. [1,2]). Multiple photon processes dealing with vibrational and electronic excitation of molecules have been studied mainly in the gas phase when a comparatively long time of excitation and relaxation allows the use of nanosecond pulses. In condensed media at room temperature, due to a fast energy relaxation efficient multiple photon excitation is possible only when one applies laser pulses of picosecond duration. This approach has already resulted in the first successful experiments on multiquantum photochemistry of molecules in condensed media (aqueous solutions of nucleic acid bases [3,4], porphyrine [5]).

The report briefly presents the recent results of studying photochemical transformations of polyatomic molecules in a condensed medium under multiple photon excitation of molecular electronic states induced by UV picosecond laser radiation. Two examples are considered:

a) Excitation of high-lying electronic states of molecules in solutions, provoking new photochemical reactions, which were not previously observed in linear photochemistry. The yield of the new photochemical reactions is nearly 100% [6].

b) The high rate excitation of a chromophore group in a large molecule at the surface of a crystal by a picosecond UV laser pulse resulting in a predominant detachment of the chromophore ions. When nanosecond laser pulses are used a detachment of molecular ions predominates.

2. Picosecond UV Nonlinear Photochemistry of Unsaturated Organic Acids and Their Ammonium Salts

Photochemical transformations of some unsaturated acids and their salts in acqueous solutions induced by powerful picosecond pulses of 4th harmonic YAG-ND^{3+} laser (λ = 266 nm, τ = 30 psec) were studied.

It was found that the irradiation of ammonium salts in aqueous solutions of maleic, fumaric and citraconic acids brings about the formation of corresponding α-amino acids [6]. The reaction is depicted thus:

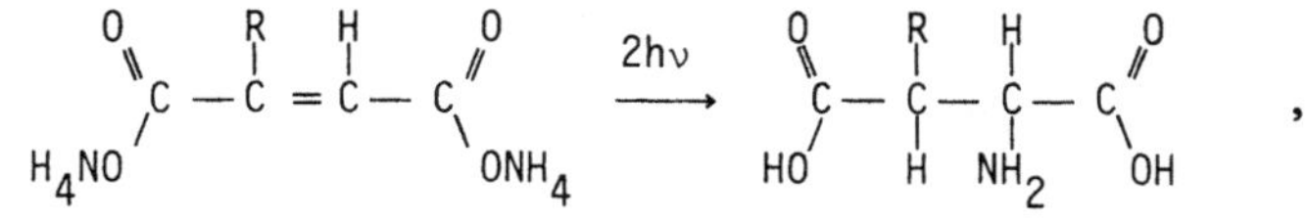

where R = H for maleic (cis-isomer) and fumaric (trans-isomer) and R = CH_3 for citric acids.

The quantum yield of aspartic acid formed as a result of the irradiation of an aqueous solution of maleic acid ammonic salt is enhanced by the intensity increase and reaches 0.4 when the intensity is more than 1 GW/cm^2. The yield dependence upon the intensity indicates the nonlinear two-step character of the photochemical process. The reaction product yield saturation is caused by that of the intermediate, in particular, first singlet state.

The dependence of photoproduct yield for a wide range of intensities starting with the values which provoke one-quantum linear photochemical processes up to intensities which activate multiquantum photochemical processes was studied using an aqueous solution of maleic acid [9]. It is shown that the process, well-known in linear photochemistry, of cis-trans isomerization takes place when the laser radiation pulse intensity does not exceed $5 \cdot 10^7$ W/cm^2. With further intensity rise the efficiency of the above-mentioned process starts to decrease with a simultaneous yield enhancement of a new photoproduct (Fig.1). The radiation intensity being 10^8 W/cm^2, the quantum yield for this photoproduct exceeds unity which reveals the free-radical character of its formation. The analysis of the photoproduct by the gas chromatography method and the interpretation of its infrared spectrum

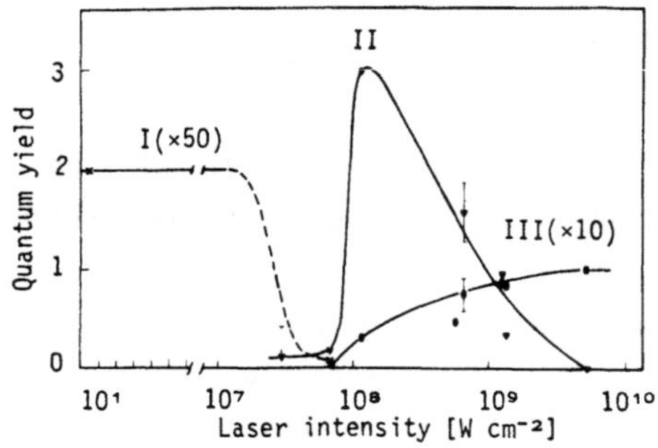

Fig.1. Dependence of quantum yield of photoproducts resulting from irradiation of maleic acid aqueous solution on the radiation intensity.
I - fumaric acid (cis-trans isomerization of maleic acid);
II - maleic acid dimer (radical dimerization);
III - malic acid (water addition to -c=c-bond of maleic acid)

indicates the formation of a maleic acid cyclic dimer having the following structure:

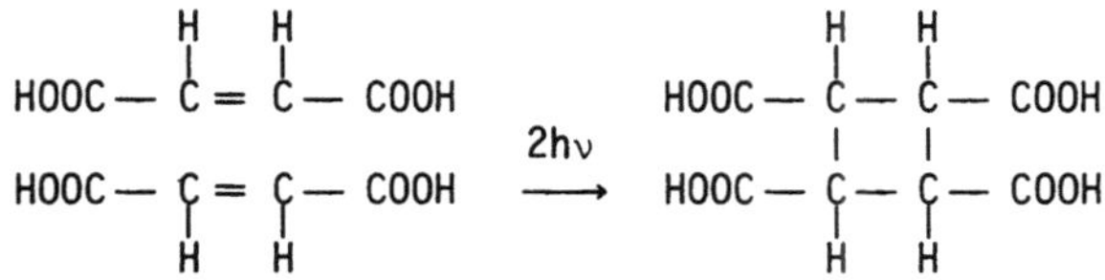

With a further increase in the radiation intensity the quantum yield of dimer formation becomes lower, while there is a simultaneous gain of the next product of the nonlinear photoreaction. This product is a malic acid formed as a result of water photoaddition to -C=C-bond of maleic acid:

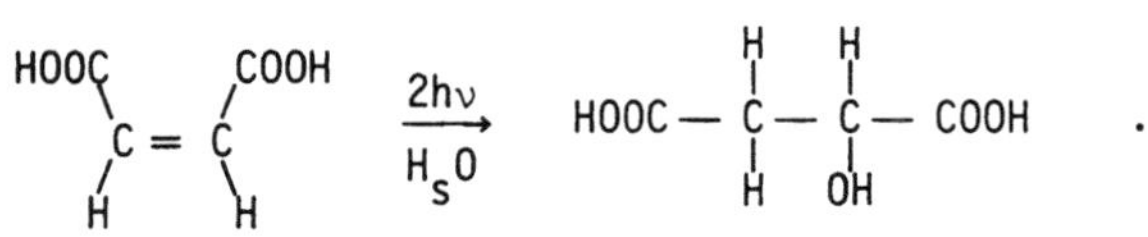

Under a laser pulse intensity higher than 10^9 W/cm^2 the malic acid yield dependence on the intensity becomes saturated.

Thus, by changing only one photoprocess parameter, namely the laser pulse intensity, we managed to selectively produce three different photoprocesses having an essential quantum photoproduct yield which results from the following photoreactions violating the -C=C-bond of unsaturated organic acids (Fig.2):

1. cis-trans isomerization;
2. the radial break of a double bond with dimer formation;
3. addition of ammonium and water to double bond with the subsequent formation of oxi- and amino acids.

This experiment confirms a principal possibility of new methods of laser photochemistry making use of highly excited electronic states.

3. Chromophore Detachment off a Large Molecule Induced by UV Picosecond Pulses

The problem of detachment of chromophore group ions off large molecules on surfaces is of interest due to a project which uses a laser ionic microscope to visualize the structure of large molecules [10]. A number of our works demonstrated that when irradiating DNA bases [7,8] and other polyatomic molecules [11,12] in the solid state phase by UV laser radiation pulses there occurred a photodetachment of molecular and quasimolecular ions. The

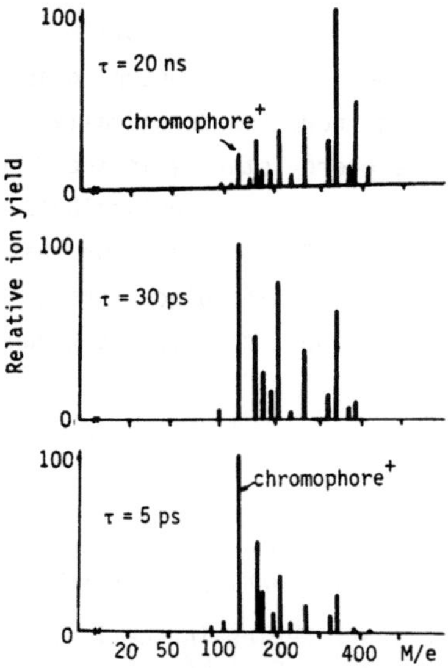

Fig.2. Mass-spectra of photoions of peptide molecules desorbed off the surface with UV laser pulses (λ = 266 nm)

heating of the crystal's surface is about 20-70°C. Hence the ion formation has a nonthermal origin.

To study the role of chromophores in a complex molecule during the process of photoionization and photodesorption of photoions the experiments [13] were performed with peptide molecules consisting of a chain of amino acids - triptophane, alanine and glycine with acetate and ether groups attached.

Fine-dispersive crystal peptide powder was positioned in the ionization chamber of a time-of-flight mass spectrometer. The sample was irradiated by laser pulses with λ = 266 nm and durations τ = 20 nsec, 30 psec, 5 psec. The absorption of the peptide molecule at the 266 nm wavelength is caused by the presence of aromatic chromophore-tryptophane amino acid. Desorbed ions were analyzed by the mass-spectrometer.

In Fig.2 the mass-spectra of laser-desorbed ions off the sample surface are presented. The mass-spectra consist of several sharp peaks corresponding to molecule fragmentation mainly by peptide bonds and charge localization on the aromatic chromophore. All three mass-spectra were obtained under one and the same energy fluence. Comparison of these mass-spectra reveals the fact that reducing the pulse duration results in a large increase in the fraction of chromophore desorbed ions so that under τ = 5 psec the value of a related peak reaches its maximum.

The result obtained can be explained in the following way. Deactivation of the first excited electronic state occurs within the time 10^{-9} sec [14] and that of highly-excited states much faster. Therefore, when a peptide molecule is irradiated by picosecond pulses the probability of absorbing additionally one or more photons by excited molecules increases drastically in comparison to nanosecond pulses. Due to a large energy storage in the chromophore the latter detaches before the absorbed energy can redistribute over the whole molecule and further along the crystal. Thus it is shown that a detachment of the chromophore from a peptide molecule takes place during the radiation by a UV laser pulse. The relative yield of a chromophore group drastically increases for picosecond laser pulses.

References

1 A. Ben-Shaoul, Y. Haas, K.L. Kompa, R.D. Levine: *Laser and Chemical Change*, Springer Series in Chemical Physics, Vol.10 (Springer, Berlin, Heidelberg, New York 1981)
2 V.S. Letokhov: *Nonlinear Laser Chemistry*, Springer Series in Chemical Physics, Vol.22 (Springer, Berlin, Heidelberg, New York 1982)
3 P.G. Kryukov, V.S. Letokhov, D.N. Nikogosyan, A.V. Borodavkin, E.I. Budovsky, N.A. Simukova: Chem. Phys. Lett. *61*, 375 (1979)
4 D.N. Angelov, P.G. Kryukov, V.S. Letokhov, D.N. Nikogosyan, A.A. Orayevsky: Kvantovaya Elektron. (Russian) *8*, 595 (1981)
5 T.I. Karu, V.S. Letokhov, P.G. Kryukov, Yu.A. Matveetz, V.A. Semchishen: Appl. Phys. *24*, 245 (1981)
6 V.S. Letokhov, Yu.A. Matveetz, V.A. Semchishen, E.V. Khoroshilova: Appl. Phys. B*26*, 243 (1981)
7 V.S. Antonov, V.S. Letokhov, A.N. Shibanov: Pis'ma Zh. Exp. Teor. Fiz. *31*, 471 (1981)
8 V.S. Antonov, V.S. Letokhov, A.N. Shibanov: Appl. Phys. *25*, 71 (1981)
9 E.V. Khoroshilova, N.P. Kuzmina, V.S. Letokhov, Yu.A. Matveetz: Appl. Phys. B (in press)
10 V.S. Letokhov: Kvantovaya Elektron. *2*, 930 (1975)
11 V.S. Letokhov, V.G. Movshev, S.V. Chekalin: Zh. Exp. Teor. Fiz. *81*, 480 (1981)
12 V.S. Antonov, V.S. Letokhov, A.N. Shibanov: Laser Chem. (in press)
13 V.S. Antonov, S.E. Yegorov, V.S. Letokhov, A.N. Shibanov: Pis'ma Zh. Exp. Teor. Fiz. (in press)
14 T.S. Werner, L.S. Foster: Photochem. and Photobiology *29*, 905 (1979)

P-BR and Its Role in the Photocycle of Bacteriorhodopsin

T. Gillbro and V. Sundström

Division of Physical Chemistry, University of Umeå,
S-901 87 Umeå, Sweden

Introduction

The initial photochemical events occuring in the chromophore (bacteriorhodopsin) of the purple membrane of Halobacterium halobium belongs to one of the most studied subjects in the literature of picosecond spectroscopy. Both fast fluorescence [1-3] and absorption recovery [4-7] has been studied. However, there is rather poor agreement between experimental data reported. In some recent flash photolysis [8] and fluorescence studies [9-10] it has been shown that there is a long-lived photoproduct of bacteriorhodopsin, called P-BR, which is formed even at 77 K. We have undertaken this study in order to investigate if some of the results given in the literature might be better understood assuming that P-BR contributes to the data reported.

Experimental

The picosecond laser system consists of a mode-locked Ar^+-laser which pumps a dye laser equipped with a cavity dumper (Spectra-Physics) [11]. The pulsewidths used in these experiments were typically 5-8 ps and all the measurements reported here were done with the pump and probe technique. The pulse frequency was 82 kHz and the intensity of the excitation beam about 0.5 mW at the sample. For some of the experiments the sample was pumped through the 0.5 mm sample cell with a peristaltic pump (Cole-Palmer). At subambient temperatures we use a cryostat (Oxford Instruments) cooled with liquid nitrogen. Purple membrane preparations were made from Halobacterium halobium S-9 according to the procedure given by Oesterhelt and Stoeckenius [12].

Results

In order to check the possible influence of a long-lived photoproduct on the absorption recovery kinetics of purple membrane in water at room temperature we performed measurement on the same sample at different flow rates. At steady state (no flow) a bleaching signal, which relaxed exponentially with $\tau = 16 \pm 2$ ps to a constant value was observed (Fig. 1) at all wavelengths. At $\lambda > 600$ nm the constant value was negative, which indicates that a photoproduct was formed, which absorbs more than the species that are excited at these wavelengths. When we increased the flow rate of the sample through the cell the shape of the signal drastically changed until a critical flow rate was reached at which no further change of the signal intensity and shape could be observed. The signal recorded at high flow rate (at which a sample stayed less than 1.6 ms in the excitation beam) is shown in Fig. 2. In this case only a pulse width limited absorbance increase is observed at 630 nm. Our interpretation of these results is as follows:

At steady state a photoproduct which absorbs in the same wavelength region as light adapted bacteriorhodopsin (BR) is formed. The high concen-

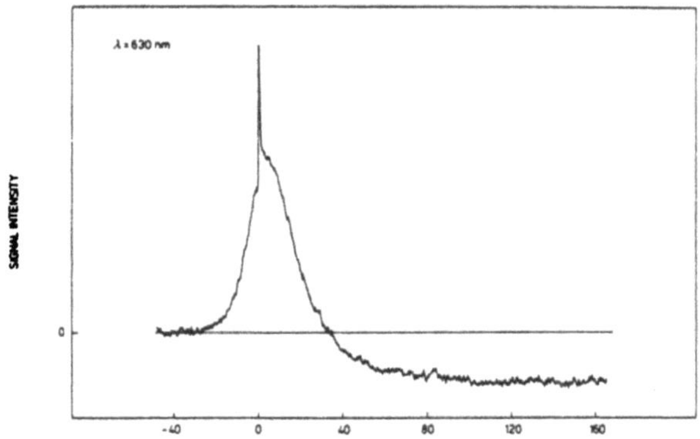

Fig.1 The kinetics of the absorbance change of a bacteriorhodopsin suspension (298 K) at steady state. Excitation and analyzing light of 630 nm.

tration of this photoproduct can be explained if it is an intermediate of the photocycle of bacteriorhodopsin and if it has a long lifetime relative to other intermediates. A good candidate for this intermediate is P-BR, which has an absorption spectrum similar to bacteriorhodopsin and a lifetime of about 100 ms at room temperature [8]. Other considerations show that the K,L,M and O-forms are poor candidates for the photoproduct [13]. At high flow rates the concentration of P-BR and other intermediates is low and the fast signal observed ($\tau \lesssim 2$ ps) is due to the reaction BR $\xrightarrow{h\nu}$ K-BR. This number compares favourably with the lifetime of 1.0 ± 0.5 ps obtained by IPPEN et al. [7] for this reaction. At other wavelengths, e.g. 576 nm, the bleaching due to the depopulation of the BR groundstate was observed. Also the repopulation of BR occured with $\tau \lesssim 2$ ps.

The absorption recovery kinetics of light adapted purple membrane in glycerol/water was studied at temperatures from room temperature to 77 K (Fig. 3). At 77 K and 100 K the recovery lifetimes were 62 ± 5 and 48 ± 5 ps, respectively. The lifetime at 77 K is in good agreement with the fluorescence lifetime measured by SHAPIRO et al. [3] of 60 ± 15 ps. These authors assigned the fluorescence to BR although it has been shown that P-BR is the strong emitter at 77 K [9,14]. The fluorescence quantum yield measurement performed by SHAPIRO et al. showed an unusual temperature dependance. Furthermore, the ratio between the quantum yields at 77 K and 100 K was about 3.5. This deviates significantly from the ratio of the lifetimes at these temperatures, which is only 1.2. The large temperature effect on the emission quantum yield is easily explained if one considers that P-BR is unstable already at T = 86 K [10]. From Fig. 3 it also follows that log k vs. 1/T for the excited state relaxation obeys an Arrhenius equation rather nicely. It is thus likely that the long-lived intermediate at room temperature is identical to the strongly emitting primary photoproduct trapped at 77 K [8].

P-BR in the Photocycle

In the following we will make a tentative assignment of P-BR and discuss its possible role in the photocycle of bacteriorhodopsin. It seems likely from this and other studies [8-10] that P-BR is a primary photoproduct of BR as well as the last intermediate of the photocycle. This circumstance is

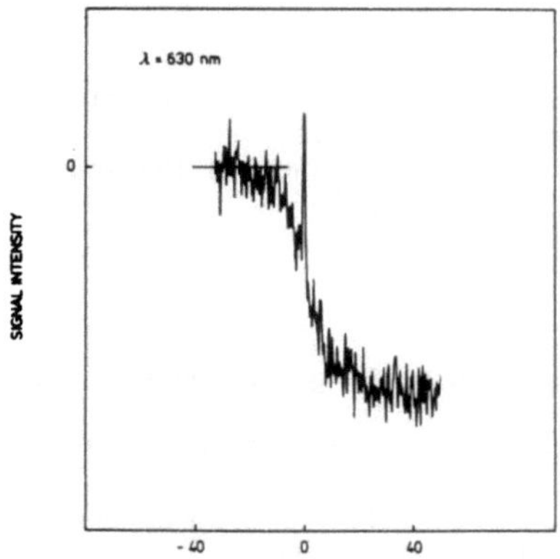

Fig.2 The kinetics of the absorbance change (λ = 630 nm) of bacteriorhodopsin at high flow rate

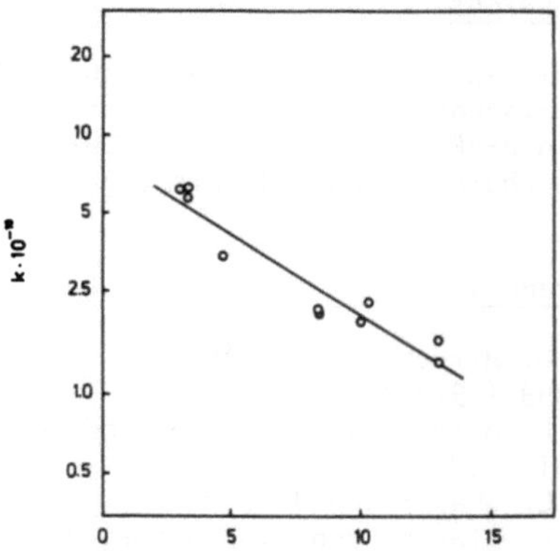

Fig.3 The relaxation rate vs. 1/T for bacteriorhodopsin at steady state

important for our interpretation. P-BR is also formed in competition with K-BR at 77 K. This suggests that P-BR is formed by relaxation from the same excited state potential surface as K-BR. We think that there are two parallel relaxation processes that bring BR to K-BR. First, the chromophore undergoes a partial trans-cis isomerization in the excited state. Second, while the chromophore is still in the excited state the surrounding protein structure relaxes into a new equilibrium conformation. According to our model P-BR is formed only if the chromophore loses its excitation energy before the protein relaxation takes place. Thus the partial trans-cis isomerization has occured but the protein surrounding is unchanged. This means that P-BR is not stabilized by the matrix, which explains for the thermal relaxation back to BR already at 86 K [10] and its low activation energy. A further implication of this model is that the bathochromic shift (∿ 20 nm) of K-BR as compared to P-BR is caused by the protein matrix. According to the external point-charge model of NAKANISHI et al. [15] this would mean that some negative charges in the protein (opsin) moves closer to the β-ionone ring double bond or away from the protonated Schiff base of the retinal chromophore during the excited state process P-BR → K-BR and in the opposite direction during the ground state relaxation P-BR → BR.

It is also interesting to see how our model can explain for the last steps of the photocycle. In this case we consider O-BR as an intermediate with a protonated Schiff base chromophore, which is still not in the relaxed all-trans conformation, and with an unrelaxed opsin structure around the chromophore. The next step in the photocycle would be a relaxation of the opsin matrix to form P-BR, which thus still has an unrelaxed chromophore. Finally the chromophore relaxes to its most stable conformation, which is BR.

We propose that a similar mechanism as for bacteriorhodopsin can explain for the occurrence of hypsorhodopsin in competition with bathorhodopsin in for instance octopus rhodopsin at 4 K [16]. In this case, however, hypsorhodopsin seems to be able to relax thermally to rhodopsin on the ground state energy surface.

Conclusions

The picosecond lifetime data reported in this work lend further support to the existence of a long-lived primary photoproduct of bacteriorhodopsin, called P-BR, which has been observed in previous fluorescence [9,10] and flash photophysics [8] investigations on bacteriorhodopsin.

References

1. M.D. Hirsch, M.A. Marcus, A. Lewis, H. Mahr and N. Frigo: Biophys. J. *16*, 1399 (1976)
2. R.R. Alfano, W. Yu, R. Govindjee, B. Becher and T. Ebrey: Biophys. J. *16*, 541 (1976)
3. S.L. Shapiro, A.J. Campillo, A. Lewis, G.J. Perreault, J.P. Spoonhower, R.K. Clayton and W. Stoeckenius: Biophys. J. *23*, 383 (1978)
4. K.J. Kaufmann, P.M. Rentzepis, W. Stoeckenius and A. Lewis: Biochem. Biophys. Res. Com. *68*, 1109 (1976)
5. K.J. Kaufmann, V. Sundström, T. Yamane and P.M. Rentzepis: Biophys. J. *22*, 121 (1978)
6. M.L. Applebury, K.S. Peters and P.M. Rentzepis: Biophys. J. *23*, 375 (1978)
7. E.P. Ippen, C.V. Shank, A. Lewis and M.A. Marcus: Science *200*, 1281 (1978)
8. T. Gillbro: Biochim. Biophys. Acta *504*, 175 (1978)
9. T. Gillbro, A.N. Kriebel and U.P. Wild: FEBS Letters *74*, 57 (1977)
10. A.N. Kriebel, T. Gillbro and U.P. Wild: Biochim. Biophys. Acta *546*, 106 (1979)
11. V.Sundström and T. Gillbro: Appl. Phys. *24*, 233 (1981)
12. D. Oesterhelt and W. Stoeckenius: Methods Enzymol *31 A*, 667 (1974)
13. T. Gillbro and V. Sundström, to be published.
14. T. Gillbro, A. Holzwarth and V. Sundström, to be published.
15. K. Nakanishi, V. Balogh-Nair, M. Arnaboldi, K. Tsujimoto and B. Honig: J. Am. Chem. Soc. *102*, 7945 (1980)
16. M. Tsuda, F. Tokunaga, T.G. Ebrey, T.K. Yue, J. Marque and L. Eisenstein: Nature *287*, 461 (1980)

Picosecond Linear Dichroism Spectroscopy of Retinal

M.E. Lippitsch, M. Riegler, and F.R. Aussenegg

Institut für Experimentalphysik, Universität Graz, A-8010 Graz, Austria

L. Margulies[1] and Y. Mazur[2]

(1) Isotope and (2) Organic Chemistry Departments, The Weizmann Institute of Science, Rehovot, Israel

1. Introduction

The molecular mechanism of vision was the first biological problem to be tackled by picosecond spectroscopy. BUSCH et al. [1] showed that the first intermediate in the photoreaction of rhodopsin, the visual pigment in vertebrates, is formed in at least 6 ps after absorbing a photon of proper wavelength. Rhodopsin consists of opsin (a large protein) and a chromophore called retinal. The absorption properties of retinal are strongly influenced by its surrounding. So in solution it has its main absorption band at 375 nm, while in rhodopsin it absorbs at 498 nm.

Despite the different spectroscopic properties, the photochemical behaviour of retinal in rhodopsin and in solution is not very different. The all-trans configuration is the energetically most stable, but can be converted by illumination into various cis-configurations (9-cis, 11-cis, 13-cis) and also back to the all-trans form. In rhodopsin retinal is found to be 11-cis, and the major (if not the only) action of light is supposed to be the photoisomerization into all-trans. There has been a controversy over the last years, whether in rhodopsin isomerization is the very first step following absorption or not. This controversy has not been settled so far.

Surprisingly, most picosecond studies start with rhodopsin without prior examination of the properties of retinal itself. In our opinion it would be helpful to know the behaviour of the simple chromophore before passing over to the complex molecule. In this context it seems desirable to conduct a study of the primary photoprocesses in retinal using picosecond spectroscopic techniques capable of detecting conformational changes following excitation.

2. Linear Dichroism Spectroscopy

Usually non-spherical molecules in solution are randomly distributed with respect to the directions of their axes. So absorption is independent of polarization direction, irrespective of the shape of the molecule and the direction of its transition dipole moment. If the molecules are directionally ordered, however, measurement of absorption parallel and perpendicular to the preferential direction can yield information on the direction of the transition dipole and hence on the conformation of the molecule. A method to measure linear dichroism by incorporating the molecules in stretched polyethylene films has been developed and studied extensively over the past years ([2] and references herein). One of the aims of the present work is to find out if this technique is also useful in picosecond experiments. Furthermore with respect to retinal, to study ordered molecules is of advantage, since most likely also in the disk membrane of the eye the rhodopsin molecules are ordered, the retinal chromophores being oriented parallel to the membrane surface.

3. Experimental

Experiments were performed using single pulses of 6 ps duration extracted from a passively mode-locked Nd^{+}-glass laser. The pulses were amplified in a double-pass amplifier, coupled out by a combination of waveplate and polarizer, frequency doubled and tripled and then split up spectrally. The green pulse was used to excite a picosecond continuum in a CCl_4 cuvette providing a source for absorption spectroscopy up to 340 nm. The polarization of the continuum was rotated by rotating the green pulse using a waveplate. The continuum was split into a probe and a reference beam. Observation was performed using optical multi-channel detection. The pulse duration was recorded by a two-photon-fluorescence device with TV-readout for every shot. Pulses were carefully selected regarding duration, energy , and spectral shape of the continuum.

The technique for incorporating the retinal into the polyethylene films was previously described [3]. Conventional dichroic spectra were measured on a Cary 118 spectrophotometer using the PNP technique [3].

4. Results and Discussion

The dichroic spectrum of all-trans retinal in a stretched polyethylene film is shown in fig.1. The dichroic ratio is $d_o = 1.8$ in the band maximum. This is a surprisingly low value compared with other polyenes of comparable length (e.g. 1,8-diphenyl-1,3,5,7-octatetraene: $d_o = 14,5$). By replacing the methyl groups in position 1 and 5 by hydrogens, the expected increase in the dichroic ratio value could be obtained ($d_o = 6,3$). This fact is interpreted as due to the nonplanarity of all-trans retinal. Because of steric hindrance by the methyl

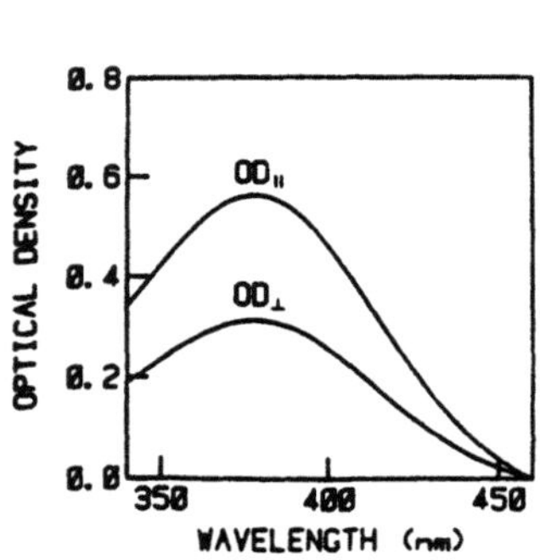

Fig.1: Dichroic spectrum of all-trans retinal in stretched polyethylene film. $OD_{\parallel}$ and $OD_{\perp}$: optical densities for light polarized parallel and perpendicular to the stretching direction, respectively

groups the plane of the ring is tilted by about 59° with respect to the symmetry plane of the chain, as revealed by X-ray studies [4]. As is known from other measurements [2], planar molecules are oriented in a stretched film with their plane parallel to the stretching direction. In the case of all-trans retinal this means that the directions of the chains (and hence of the transition dipoles of the S_1-band) form relatively large angles with the stretching direction, yielding a small value of the dichroitic ratio.

In the time-resolved experiment excitation of the sample was performed by pulses of 354 nm. The excitation was always polarized with its electric vector parallel to the stretching direction. The absorption was probed in both directions of polarization varying the delay between excitation and probing from -30 to +50 ps in steps of 5 ps. Bleaching of the ground state is observed immediately after excitation. The bleaching recovers with a time constant of approximately 35 ps (fig. 2). The experimental values are consistent with the assumption of a single exponential decay. This time behaviour was more or less the same for parallel and perpendicular polarization. However, the extent and spectral shape of the bleaching is significantly different for the two polarization directions. In the perpendicular polarization case, the whole absorption band bleaches in a fairly uniform way. In the parallel case, efficient bleaching is found only in the short-wavelength region of the absorption band, while the long-wavelength part bleaches only slightly. Accordingly, the dichroic ratio changes remarkably over the absorption band (fig. 3).

All-trans retinal in unstretched polyethylene film as well as in n-hexane solution showed a behaviour resembling that

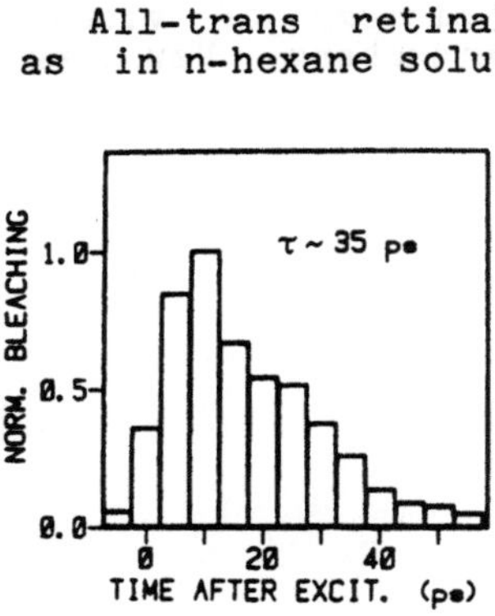

Fig.2: Time dependence of ground-state bleaching of all-trans retinal in stretched polyethylene film after excitation with 354 nm

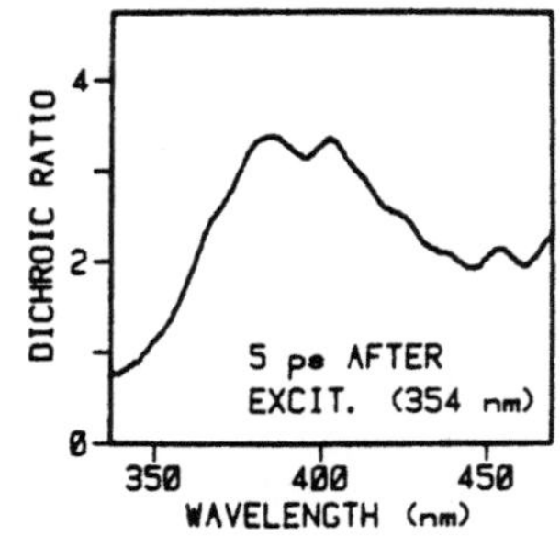

Fig.3: Dichroic ratio measured 5 ps after excitation

observed for perpendicular polarization in the stretched films, irrespective of the relative polarization of the excitation and probe beam. Also the relaxation time was approximately the same in solution and in the polyethylene film.

The results show that incorporation of retinal in polyethylene films does not alter its ultrashort time response to photoexcitation significantly. Yet, there is a significant difference when probing the excited molecule with polarizations parallel and perpendicular to the stretching direction. A tentative interpretation could be that there exists a $S_1 - S_n$ absorption with transition dipole moment along the stretching direction and partially overlapping with the $S_o - S_1$ absorption.

Acknowledgement:

Financial support of this work by the Jubiläumsfonds der Österreichischen Nationalbank, grant no. 1814, is gratefully acknowledged.

References :

1. Busch, G.E., M.L. Applebury, A.A. Lamola and R.M. Rentzepis, Proc.Nat.Acad.Sci.USA 69, 2802 (1972)
2. Margulies, L. and A. Yogev, Chem.Phys. 27, 89 (1978)
3. Yogev, A., L. Margulies, J. Sagiv and Y. Mazur, Rev.Sci. Instr. 45, 386 (1974)
4. Sperling, W. in: Biochemistry and physiology of visual pigments (H. Langer, Ed), Berlin Heidelberg New York Springer 1973

Picosecond Absorption Spectroscopy of Biliverdin

M.E. Lippitsch, M. Riegler, A. Leitner, and F.R. Aussenegg

Institut für Experimentalphysik, Universität Graz,
A-8010 Graz, Austria

1. Introduction

Among the various pigments found in biological systems biliverdin is of special interest because of the important role it plays in different photobiological processes. The most thoroughly investigated of these processes is plant morphogenesis governed by the chromoprotein phytochrome, containing a biliverdin chromophore. Recently, a number of important spectroscopical observations concerning the photochemical actions of phytochrome have been published [1,2,3] . It became apparent, that the primary processes in phytochrome phototransformation are going on in picoseconds [2,4].

There is a strong need for spectroscopic investigations on the picosecond kinetics in simple biliverdins, to yield a sound basis for interpreting the data known from phytochrome. In this work, picosecond absorption spectroscopy of etiobiliverdin-IV- γ in chloroform (10^{-4} mol/l) was performed using a passively modelocked Nd^{+}- glass laser system and optical multichannel detection. This molecule, which has two strong absorption bands at 630 nm and 375 nm, has been studied extensively by FALK and coworkers [5,6,7]. From that work it is known, that torsions around single bonds severly affect the absorption properties, while E-Z-isomerizations around double bonds give only minor changes. From measurements of the fluorescence quantum yield it is apparent, that fast radiationless processes compete successfully with radiation in this molecule. These radiationless processes proceed essentially via a singlet, as was proven for bile pigments in general by the lack of phosphorescence, missing heavy-atom effects, and missing influence of triplet quenchers [8]. Results of force-field as well as PPP-LCAO calculations [6,7] give a good basis for understanding transient spectra and interpreting intramolecular motions following excitation.

2. Results and Discussion

A representative result for the time-evolution of the transient absorption following excitation with a pulse of wavelength λ= 628 nm (generated by stimulated Raman scattering in propanol) is shown in fig.1a. A new absorption band is found to emerge immediately after excitation. A thorough analysis of the spectra shows that this absorption consists of two bands, one centred at 460 nm and decaying with a time constant of $\tau = 7 \pm 2$ ps, and a second, broader one with a decay time of $\tau = 23 \pm 4$ ps (fig.1c), which is centred at 480 nm in the beginning and shifts to longer wavelengths while decaying. The original absorption is bleached with a small delay ($\tau = 12 \pm 4$ ps) and recovers within 22 ± 4 ps (fig.1d). An interpretation of this transient spectra can be given using the theoretical results of FALK et al.[6,7]: The 460 nm absorption cannot originate from a S_o species, because not any ground-state configuration of the molecule shows an absorption in that region. So it is assigned to a S_1-S_n absorption, which assignement fits well to the prompt appearance. The fast relaxation of this excited state is attributed to internal conversion, possibly assisted by excited-state proton tunneling between N_{22} and N_{23} [9]. As can be seen from the energy level diagram (fig.3a), the relaxation can end up in an open configuration of the molecule with $\beta = 212^o$. We assume this path of relaxation to be strongly favoured. From a Fokker-Planck analysis [10] it is found, that this state should revert back to the original groundstate conformation within about 20 ps. In addition, a number of molecules should undergo a small twist in α (fig.3b), leading to an absorption around 650 nm. The spectra calculated theoretically under these assumptions (fig.1b) are in excellent agreement with the experimental ones, thus confirming our interpretation.

When the sample is excited with a pulse of $\lambda = 354$ nm, a transient absorption in the red part of the spectrum is found (fig.2a). The absorption relaxes with $\tau = 11 \pm 3$ ps (fig.2c).

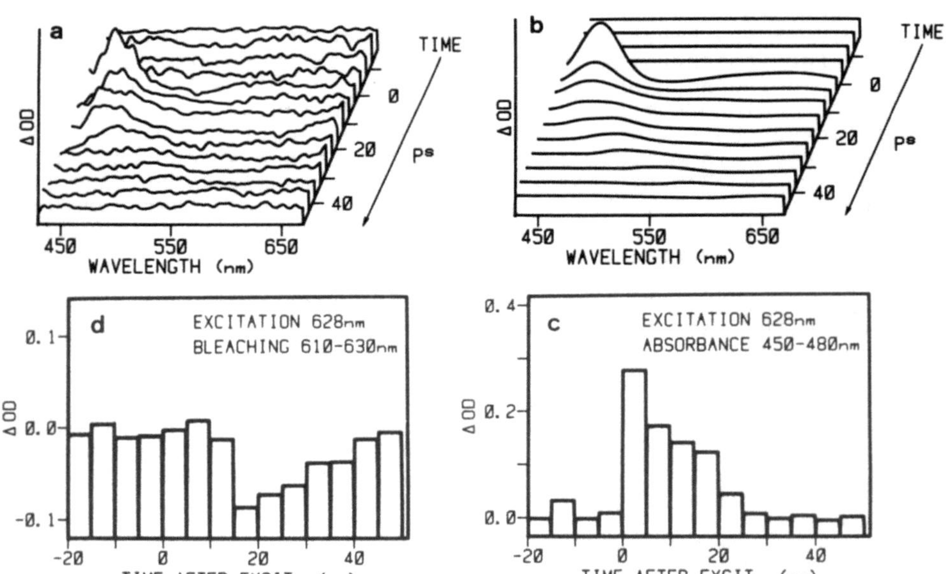

__Fig.1:__ a) Transient absorbance changes after excitation with $\lambda = 628$ nm
b) theoretical curves, calculated using results from [6,7]
c) time-histogram for the band 460/480 nm
d) time-histogram for bleaching at 630 nm

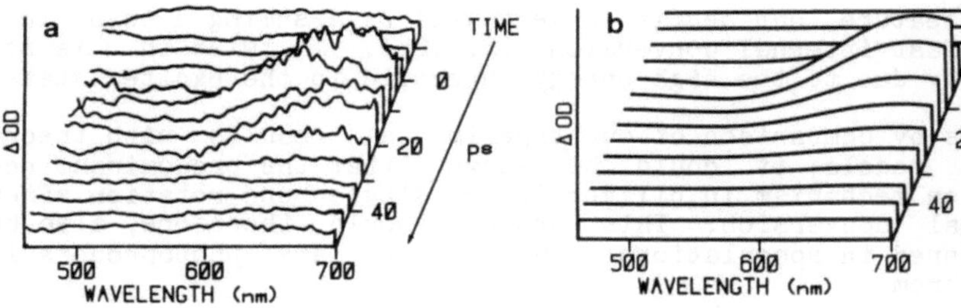

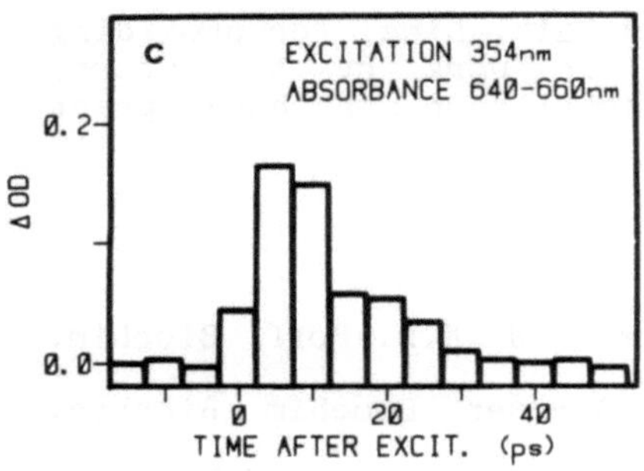

<u>Fig.2:</u> a) Transient absorbance changes after excitation with λ = 354 nm b) theoretical curves, calculated, using results from [6,7] c) time histogramm for the band λ = 650 nm

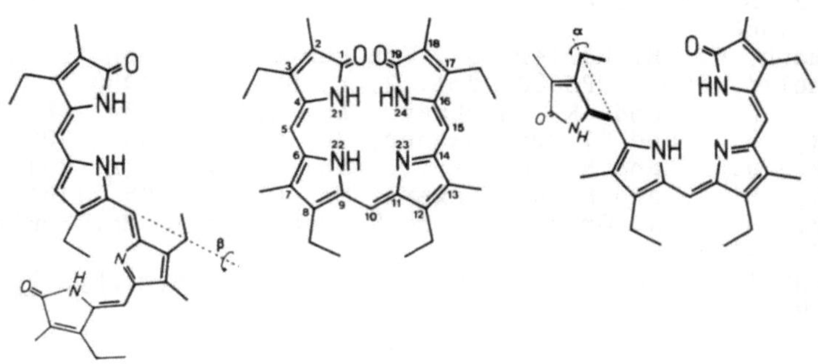

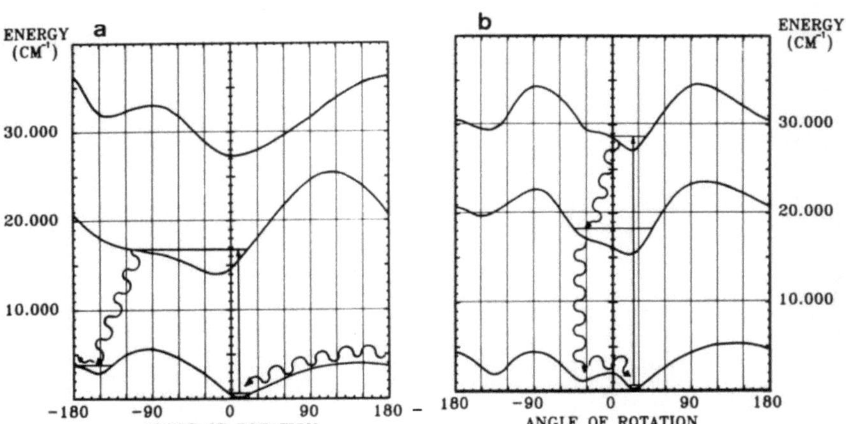

<u>Fig.3:</u> Energy of ground and excited states a) vs. β b) vs. α

This feature can be fitted very well by assuming a twist in α after fast internal conversion (fig.2b). A twist in β is now unlikely due to the high energy threshold in the excited state.

Thus by comparison of our experimental results with theoretical models it could be shown, that the predominant relaxation mechanism in biliverdin is single-bond rotation after internal conversion. This process should, therefore, also be considered in speculations about the primary photoprocess in phytochrome.

Acknowledgement: We thank Prof.H.Falk, Linz, for providing the samples. Financial support of this work by the Österr. Fonds zur Förderung der wissenschaftlichen Forschung, grant no.4031, is gratefully acknowledged.

References :

1. Song, P.S., H.K. Sarkar, I.S. Kim and K.L. Poff, Biochim. Biophys. Acta 635, 369 (1981)
2. Song, P.S., Q. Chae, and J.D. Gardner, Biochim. Biophys. Acta 576, 479 (1979)
3. Braslavsky, S.E., J.I. Matthews, H.J. Herbert, J.De Kok, C.J.P. Spruit and K. Schaffner, Photochem.Photobiol . 31, 417 (1980)
4. Hermann, G., E. Müller, D. Schubert, H. Wabnitz and B. Wilhelmi, Proc. UPS 80, Vol. 2, p.386, Jena 1980
5. Falk, H. and K. Grubmayr, Angew.Chemie 89, 487 (1977)
6. Falk, H. and G. Hoellbacher, Mh. Chemie 109, 1429 (1978)
7. Falk, H. and N. Müller, Mh. Chemie, 112, 791 (1981)
8. Falk, H. and F. Neufingerl, Mh. Chemie 108, 1181 (1977)
9. Falk, H. , K. Grubmayr, and F. Neufingerl, Mh.Chemie 110, 1127 (1979)
10. McCaskill, J.S., and R.G. Gilbert, Chem.Phys.44, 389 (1979)

Picosecond Time-Resolved Resonance Raman Spectroscopy of the Photolysis Product of Oxy-Hemoglobin

J. Terner* and T.G. Spiro

Department of Chemistry, Princeton University,
Princeton, NJ 08544, USA

D.F. Voss, C. Paddock, and R.B. Miles

Department of Mechanical and Aerospace Engineering, Princeton University,
Princeton, NJ 08544, USA

The ligand binding processes of hemoglobin and myoglobin are among the most extensively studied kinetic mechanisms in biology. In recent years, the photodeligation mechanisms of hemoglobin and myoglobin have been investigated by a number of groups using picosecond absorption spectroscopy [1-4]. In this paper, we present some of our recent results, using a newly developed technique, picosecond time-resolved resonance Raman spectroscopy [5-7], which produces vibrational spectra of transient rather than stable molecules. Flash photolysis techniques, such as picosecond absorption spectroscopy, have been powerful methods for identifying chemical intermediates and providing kinetic rate information. The broad absorptions though, contain only limited structural information. Resonance Raman spectroscopy, on the other hand, is capable of providing detailed vibrational information [8] and can be time-resolved using pulsed laser techniques [9].

Excitation of carbonmonoxy-hemoglobin with 10 picosecond laser pulses, has given resonance Raman photoproduct spectra resembling that of deoxy-hemoglobin, but with small downshifts in frequency [5,7]. Similar resonance Raman spectra had been observed with 30 nanosecond laser pulses [10]. The downshifts were interpreted as reflecting a high-spin heme with the Fe atom closer to the heme plane than in deoxy-hemoglobin, presumably due to unrelaxed protein forces [7]. The relaxation time for the disappearance of these downshifts, reflecting the movement of the Fe atom to its out-of-plane position in deoxy-hemoglobin [11] has since been determined to be in 30 nanosecond range [12]. This is consistent with a large scale tertiary structure change, but it is shorter than the known relaxation time for the R-T quaternary structure change in deoxy-hemoglobin [13]. The picosecond time-resolved resonance Raman results on CO-hemoglobin are consistent with the picosecond absorption data of GREENE et al. [3]. Their results showed an absorption spectrum of the product of CO-hemoglobin photolysis that was similar to that of deoxy-hemoglobin, but broader. It was developed within 10 picoseconds and persisted for at least 680 picoseconds.

Transient absorption studies of oxy-hemoglobin photolysis [4] gave somewhat different results than what had been seen for CO-hemoglobin [3] in the picosecond timescale. Two distinct absorptions were detected, one within 6 picoseconds, that evolved into a second that was stable within 93 and 1206 picoseconds. These transient absorptions were similar to the deoxy-hemoglobin absorption except weaker, broader, and shifted in frequency. We have followed up our previous picosecond time-resolved resonance Raman work on CO-hemoglobin [5,7] with the present study of oxy-hemoglobin photolysis in order to provide a more complete picture of this interesting mechanism.

---

*Current Address: Dept. of Chemistry, Virginia Commonwealth University, Richmond, Virginia 23284, USA.

1. Experimental

Hemoglobin was isolated from human red blood cells, and deoxygenated and stored under N_2, at 4°C. Experimental samples were diluted to ~1 mM in 0.5 M phosphate buffer at pH 7.0.

The detection system was an Optical Multichannel Analyzer (Princeton Applied Research model 1205A) with extended delay accessory (model 1207), and a silicon intensified vidicon (model 1205D) in a dry-ice cooled housing (model 1212). The vidicon was attached to a Spex model 1401 0.75 meter double spectrometer equipped with a 0.85 meter extended focal length exit mirror and a 10 mm central slit. Data was output onto paper tape and processed on a MINC (Digital Equipment Corp.) computer.

The passively mode-locked Nd:YAG laser system was the design of MILES *et al.* [14]. Single pulses of 30 picosecond duration were switched out and amplified with a double pass Nd:YAG amplifier. The pulse energies were 0.2 mJ, at 10 Hz, giving an average power of 2 mW. The frequency doubled pulses (5320 Å) were used to partially photolyse the oxy-hemoglobin, and to simultaneously Raman scatter from whatever was present during the 30 picosecond pulsewidth [15].

Oxy-hemoglobin was pumped through a 30 gauge syringe needle at a velocity of 5 m/sec and recirculated. Since the laser pulse frequency was 10 Hz, the sample volume was excited by only one pulse and had several seconds to relax and recombine with oxygen before being pumped again through the syringe needle. The method of obtaining the picosecond resonance Raman spectrum of photolyzed oxy-hemoglobin was similar to the method we have previously described for the picosecond resonance Raman studies of CO-hemoglobin [7]. With high oxy-hemoglobin sample concentrations and a diffusely focused laser beam, a negligible proportion of the sample was photolyzed. To increase this proportion, the sample concentration was lowered by dilution with buffer, and the laser beam was focused to approximately 20 μm. In this manner, we were able to photolyze 20-30% of the irradiated volume. We had to be careful that the photon flux was not too high, however, or stimulated processes would be observed, which were characterized by a broad emission and burning of the sample. The resonance Raman spectrum of deoxy-hemoglobin was obtained by recirculating through a 1.0 mm glass capillary with the reservoir stirring under N_2.

2. Results

Figures 1a and b show picosecond Nd:YAG laser excited resonance Raman spectra with maximal and minimal attainable photolysis of the oxy-hemoglobin sample. While Fig. 1b is the resonance Raman spectrum of essentially unphotolyzed oxy-hemoglobin, Fig. 1a is a superposition of the transient upon the spectrum of unphotolyzed oxy-hemoglobin. The resonance Raman spectrum of the picosecond transient of oxy-hemoglobin, shown in Fig. 1c, was obtained by computer subtracting [16, 17] spectrum 1b from spectrum 1a with a weighting factor adjusted to blank out the well isolated oxy-hemoglobin band at 1640 cm^{-1}. This spectrum is quite different from that of previously reported picosecond resonance Raman spectra of oxy- and CO-hemoglobin [5-7, 18] which were similar to the resonance Raman spectrum of deoxy-hemoglobin (Fig. 1d) except for slight downshifts in frequency (2-3 cm^{-1}). These earlier results were ascribed to a deoxy-hemoglobin like species with a high-spin heme but with an expanded porphyrin core and the Fe atom closer to the heme plane than in stable deoxy-hemoglobin [5, 7].

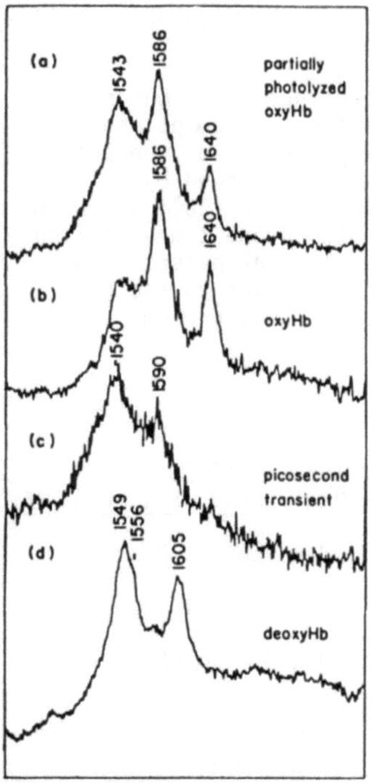

Figure 1

Our present results show resonance Raman 10-15 cm^{-1} lower than those observed for deoxy-hemoglobin. From polarization studies, we have determined that the 1590 cm^{-1} band in Fig. 1c is depolarized (dp), and that the 1540 cm^{-1} band has two components, at 1538 (dp) and 1550 cm^{-1} (anomalously polarized, ap). This is similar to the polarization pattern of the main bands in the deoxy-hemoglobin spectrum, Fig. 1d, (the bands at 1549 and 1605 are dp; the band at 1556 cm^{-1} is ap).

3. Discussion

The polarization pattern of the photoproduct indicates that the 1538, 1550, and 1590 cm^{-1} bands may be the ν_{11}, ν_{19} and ν_{10} porphyrin modes [19, 20]. The large downshifts of these frequencies compared with those of deoxy-hemoglobin cannot however be ascribed to furthur core expansion over the CO-hemoglobin picosecond transient [7]. In $(THF)_2$-FeTPP [21] whose protoporphyrin analog gives the same ν_{11} and ν_{19} frequencies as the CO-hemoglobin photoproduct [7], the Fe atom is exactly in the porphyrin plane, and the porphyrin is expanded as far as in any known Fe porphyrin complex. There is no obvious mechanism for producing a larger expansion. However, ν_{10} and ν_{11} are also sensitive to the effects of electron donation into the lowest unoccupied π-orbital of the porphyrin, e_g^* [22, 23]. Thus, both ν_{10} and ν_{11} are strongly downshifted in $(ImH)_2Fe(II)MP$ relative to $[(ImH)_2Fe(III)MP]^+$, reflecting the $d_\pi \rightarrow e_g^*$ backbonding in the former, and the downshifts are progressively diminished as ImH is replaced by π-acid ligands, which compete with the porphyrin for the d_π electrons [24]. Moreover bands of ZnEP which are assignable to ν_{10} and ν_{11}, at 1613 and 1562 cm^{-1}, have been observed to shift down, by 32 and 25 cm-1, upon formation of the porphyrin dianion, $ZnEP^{-2}$ [25].

These considerations lead us to attribute the 1590 and 1538 cm^{-1} oxyhemoglobin photoproduct bands to ν_{10} and ν_{11} arising from an electronically excited high-spin heme. If this state has appreciable π-π^* character, the 12 and 5 cm^{-1} downshifts, relative to the CO-hemoglobin photoproduct, would be consistent with the sensitivity of ν_{10} and ν_{11} to e_g^* orbital occupancy.

It is plausible that in the present experiment we are observing the rapidly relaxing spectral intermediate of oxy-hemoglobin described by CHERNOFF et al. [4], and that this intermediate is electronically excited deoxy-hemoglobin, with substantial π-π^* character. This might be a triplet π-π^* state, formed directly by the dissociation of triplet O_2 from photoexcited oxy-hemoglobin (and therefore not formed via dissociation of CO from CO-hemoglobin). It should also be noted that CORNELIUS et al. [26] have just recently reported a prompt spectral intermediate upon excitation of oxy-, CO- and deoxy-myoglobin; the extent of formation decreasing in the order oxy-deoxy- > CO-. The intermediate was suggested to be an excited state of deoxy-myoglobin, which could be formed directly from oxy-myoglobin but not from

CO-myoglobin (the small amount of intermediate formed from CO-myoglobin could have been generated from photo-produced deoxy-myoglobin).

The envelope of our oxy-hemoglobin photoproduct resonance Raman bands is broad, and it seems likely that we are observing a mixture of prompt and delayed photoproducts. This is consistent with the disappearance of the prompt oxy-hemoglobin transient absorption within 90 picoseconds followed by a longer lived (delayed) intermediate [4], the recent myoglobin data [26]; and with the observation by NAGUMO et al. [18], in a parallel picosecond time-resolved resonance Raman study of oxy-hemoglobin photolysis, using weak pulses at a different excitation frequency (5750 Å), of a transient similar to the previously observed picosecond CO-hemoglobin photoproduct [7, 9], presumably the longer lived intermediate of ref. 4.

References

1. C.V. Shank, E.P. Ippen, and R. Bersohn, Science, 193, 50 (1976).
2. L.J. Noe, W.G. Eisert, and P.M. Rentzepis, Proc. Natl. Acad. Sci. USA, 76, 573 (1978).
3. B.J. Greene, R.M. Hochstrasser, R.B. Weisman, and W.A. Eaton, Proc. Natl. Acad. Sci. USA, 75, 5255 (1978).
4. D.A. Chernoff, R.M. Hochstrasser, and A.W. Steele, Proc. Natl. Acad. Sci. USA, 77, 5606 (1980).
5. J. Terner, T.G. Spiro, M. Nagumo, M.F. Nicol, and M.A. El-Sayed, J. Amer. Chem. Soc., 102, 3238 (1980).
6. M. Coppey, H. Tourbez, P. Valat, and B. Alpert, Nature (London) 284, 568 (1980).
7. J. Terner, J. Stong, T.G. Spiro, M. Nagumo, M.F. Nicol, M.A. El-Sayed, Proc. Natl. Acad. Sci. USA, 78, 1313 (1981).
8. T.G. Spiro, Acc. Chem. Res., 7, 339 (1974).
9. J. Terner, A. Campion, and M.A. El-Sayed, Proc. Natl. Acad. Sci. USA, 74, 5121 (1977).
10. K.B. Lyons, J.M. Friedman, and P.A. Fleury, Nature (London) 275, 565 (1978).
11. M.F. Perutz, Brit. Med. Bull., 32, 195 (1976).
12. P. Stein, J. Terner, and T.G. Spiro, J. Phys. Chem., in press (1982).
13. C.A. Sawicki and Q. H. Gibson, J. Biol. Chem., 251, 1533 (1975).
14. R.B. Miles, G. Laufer, C. Paddock, and G. Faris, Applied Optics, 19, 3593 (1980).
15. A. Campion, J. Terner, and M.A. El-Sayed, Nature (London) 265, 659 (1977).
16. J. Terner, C.-L. Hsieh, and M.A. El-Sayed, Biophys. J., 26, 527 (1979).
17. J. Terner, C.-L. Hsieh, A.R. Burns, and M.A. El-Sayed, Proc. Natl. Acad. Sci. USA, 76, 3046 (1979).
18. M. Nagumo, M. Nichol, and M.A. El-Sayed, J. Phys. Chem., 85, 2435 (1981).
19. M. Abe, T. Kitagawa, and Y. Kyogoku, J. Chem. Phys., 69, 4526 (1978).
20. S. Choi, T.G. Spiro, K.C. Langry, K.M. Smith, D.L. Budd and G.N. LaMar (1982) submitted for publication.
21. C.A. Reed, T. Mashiko, W.R. Scheidt, K. Spartalian and G. Land, J. Amer. Chem. Soc., 102, 2302 (1980).
22. T.G. Spiro, and T.C. Strekas, J. Amer. Chem. Soc., 96, 338 (1974).
23. T. Kitagawa, T. Iizuka, M. Saito, and Y. Kyogoku, Chem. Lett. (1975) 849.
24. T.G. Spiro and J.M. Burke, J. Amer. Chem. Soc., 98, 5482 (1976).
25. N.M. Ksenoforitova, W.G. Maslov, A.N. Sidorov, and Ya. S. Bobovich, Opt. Spectrosc., 40, 462 (1976).
26. P.A. Cornelius, A.W. Steele, D.A. Chernoff, and R.M. Hochstrasser, Proc. Natl. Acad. Sci. USA, 78, 7526 (1981).

Part VII

Applications in Solid-State Physics

Picosecond Time-Resolved Detection of Plasma Formation and Phase Transition in Silicon

J.M. Lui, H. Kurz and N. Bloembergen

Division of Applied Sciences, Harvard University
Cambridge, MA 02138, USA

1. Introduction

The interaction of laser pulses with strongly absorbing media has received a great deal of attention during the past five years, especially in silicon [1-3]. An important issue has been the time scale of energy transfer between a dense electron-hole plasma and the lattice. If this energy transfer were insignificant during a picosecond pulse, it is clear that the plasma temperature would attain much higher values. We have presented ample evidence [4,5] that melting of a silicon surface layer can take place during a 20 ps laser pulse. There is a sharply defined fluence threshold (0.2 J/cm^2 for a 20 ps pulse at λ = 532 nm on silicon) above which the reflectivity changes to a value characteristic of liquid phase, the evaporation of silicon atoms becomes appreciable, and an amorphous phase may be created by rapid resolidification, with cooling rates exceeding 10^{13} °C/sec, of a thin (10 nm) molten layer on top of a cool crystalline substrate.

In this paper we present new data where the changes in the complex index of refraction of silicon are determined with a time resolution of 30 ps both below and above the melting point. The classical technique is used, where a picosecond excitation or heating pulse is followed by a probe pulse with variable time delay.

2. Experimental Technique

A single 30 ps pulse with Gaussian spatial and temporal profile is switched out of a mode-locked Nd-YAG laser pulse train. Doubling crystals provide a Gaussian pulse of 20 ps duration at the second harmonic λ = 532 nm, or a 14 ps pulse at the fourth harmonic λ = 266 nm. The 532 nm pulse is focused to a spot of 240 µm diameter at the surface of a silicon sample. The latter is either a single crystal Si-wafer or a (100) silicon film of 0.5 µm thickness on sapphire. Since the absorption depth of crystalline silicon at λ = 532 nm is 1.25 µm, a nearly uniform electron-hole plasma and heating profile is created throughout the film thickness of the SOS sample.

The probe pulse at λ = 1.06 µm is used to determine the reflectivity of the silicon wafer and the reflectance and transmittance of the SOS sample. The probe pulse is focused to a spot of only 30 µm diameter. Its alignment with the center of the heating pulse is verified by recrystallization by the probe pulse of the center of the amorphous spot produced by a heating pulse of suitable fluence [4]. The probe pulse was then attenuated to less than three percent of the pump pulse, so that it would not alter the induced heating profile significantly. Temporal overlap of the pulses was ascertained by sum frequency mixing in a KDP crystal. The probe pulse could be given a variable time delay. The combined spatial and temporal definitions are essential to obtain reliable information on the index changes. Inadequate definition in earlier experiments has led to erroneous conclusions [6].

3. Experimental Results

Figure 1 shows the results of the reflectivity at λ = 1.06 µm of the silicon wafer as the function of the energy fluence of the heating pulse at λ = 532 nm for zero time delay and for a delay of 100 ps. Note the abrupt rise of the reflectivity to a value of 0.76 characteristic of molten silicon. This is prima facie evidence for the occurrence of a first-order phase transition. The decrease at higher fluences is due to overheating of the liquid. For zero time delay one has a time convolution of effective pump pulse fluence with the probe pulse. At 0.1 J/cm^2 no melting occurs. The reflectivity in this case shows a minimum as a function of time, due to the formation of an electron-hole plasma. More precise information about the variation of the complex index of refraction can be extracted from the observed reflectance and transmittance of the silicon film, shown in Fig. 2, by the use of the equations of thin film optics [7].

The data at 0.1 J/cm^2 pump fluence are consistent with the following events: 1) More than 80 percent of the absorbed energy is transferred to the lattice during the 20 ps pulse. 2) The lattice is heated to about 900 K.

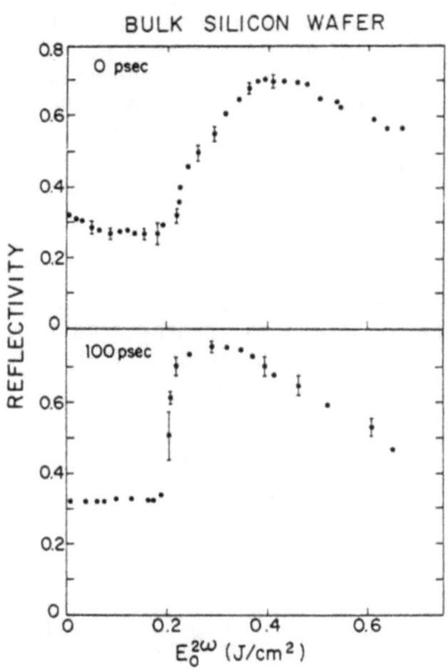

Fig.1 Reflectivity changes at λ = 1.064 µm induced by pump pulse at 0.532 µm for pump-probe delay times τ = 0 and τ = 100 ps, as a function of pump fluence on a bulk silicon crystal

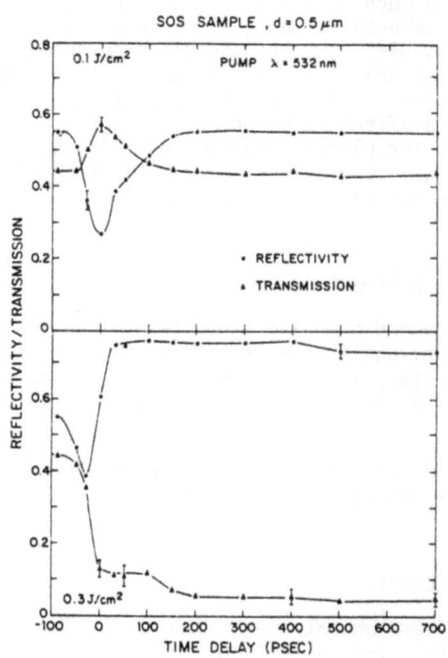

Fig.2 Reflectance and transmittance changes at λ = 1.064 µm for a silicon on sapphire (SOS) film of 0.5 µm thickness, induced by pump fluences at λ = 0.532 µm of 0.1 J/cm^2 and 0.3 J/cm^2, respectively

This causes band shrinking and increases the lattice absorption at λ = 1.06 μm. 3) About 20 percent of the energy of the pump pulse is stored in the electron-hole plasma with a carrier density between 2 and 5×10^{20} cm^{-3}. 4) The free carrier absorption contributes about one third of total absorptivity change at λ = 1.064 μm. For a probe wavelength λ = 532 nm, the plasma absorption was found to be negligible. 5) The observed recovery time of 100 ps for changes in the reflectivity and transmission is consistent with an Auger recombination constant c = 2.8×10^{-31} cm^6/sec [7], and an electron-hole density of about 2×10^{20} cm^{-3}. 6) Auger recombination limits the maximum electron-hole plasma density to less than 10^{21} cm^{-3} before melting occurs on a time scale of 10 ps.

At 0.3 J/cm^2 the reflectivity first drops because of suppression of multiple reflections as the film becomes more absorbing; then melting occurs. At the same time the film becomes opaque. The residual background is due to light scattering from vapor formation. It is reduced by placing a pinhole, smaller than the pump spot but larger than the probe spot, just behind the sample. This background cannot be used to derive a thickness of the molten layer.

Additional data, taken with an ultraviolet pump wavelength λ = 266 nm where the absorption depth in crystalline silicon is only 5 nm, confirm the same general picture. The heated layer in this case is determined by thermal diffusion during the pulse, $(2 D t_p)^{1/2} \sim 45$ nm. The absorption depth does not change much on melting. Overheating of the liquid at high fluences, reducing the reflectivity at 1.06 μm below 0.76, is much less pronounced. At low ultraviolet fluences the electron-hole plasma effects are not appreciable, because the absorption depth is smaller and the fluence threshold for melting is lower. Data have also been taken with a probe pulse λ = 532 nm. No plasma effects are noticeable at this wavelength. Clearly a more sensitive probe for plasma densities of 10^{20} to 10^{21} cm^{-3} would be an infrared pulse with wavelengths between 3-5 μm. Such picosecond pulses could be produced by an optical parametric downconverter (OPO).

4. Conclusion

Picosecond pump and probe techniques permit the detailed study of the kinetics of phase transitions and of energetic dense plasmas in condensed matter. Characteristic interaction times are shorter than 10^{-11} sec [8]. Lattice heating, melting and vaporization can occur during a 20 ps pulse.

This research was supported by the Joint Services Electronics Program under contract N00014-75-C-0648.

References

1. *Laser and Electron Beam Processing of Materials*, ed. by C.W. White and P.S. Peercy (Academic Press, New York 1980)
2. *Laser and Electron Beam Solid Interactions and Material Processing*, ed. by J.F. Gibbons, L.D. Hess and T.W. Sigmon (North-Holland, Amsterdam (1981)
3. *Laser and Electron Beam Interactions with Solids*, ed. by B.R. Appleton and G.K. Celler (North-Holland, Amsterdam 1982)
4. J.M. Liu, R. Yen, H. Kurz and N. Bloembergen: Appl. Phys. Lett. 39, 755 (1981)

5. R. Yen, J.M. Liu, H. Kurz and N. Bloembergen: Appl. Phys. A 27, 153 (1982)
6. A. Aydinli, H.W. Lo, M.C. Lee and A. Compaan: Phys. Rev. Lett. 46, 1640 (1981); see also, A. Compaan: In Ref. 3
7. O.S. Heavens: *Optical Properties of Thin Solid Films* (Dover Publications, Inc., New York 1965), p. 46
8. A. Lietoila and J.F. Gibbons: Appl. Phys. Lett. 40, 624 (1982)

Spectroscopy of Picosecond Relaxation Processes in Semiconductors

D. von der Linde and N. Fabricius
Universität Essen, Fachbereich Physik, D-4300 Essen, Fed. Rep. of Germany

J. Kuhl and E. Rosengart
Max-Planck-Institut für Festkörperforschung,
D-7000 Stuttgart, Fed. Rep. of Germany

1. Photoluminescence Correlation Technique for Measuring Carrier Dynamics

Time-resolved photoluminescence is a very useful probe of transient physical phenomena. Much progress has recently been made in the generation of extremely short, femtosecond laser pulses which can be used for photoexcitation. However, there are still severe experimental limitations of direct detection of luminescence with time resolution matching the available pulses. We have recently introduced a versatile correlation technique permitting femtosecond resolution measurements of weak light scattering[1] and spontaneous emission[2]. Here, the application of this method to ultrafast hot carrier phenomena is discussed, and some experimental results for III-V semiconductors are presented.

In the physical situation of interest an equal number of photoexcited electrons and holes is distributed over the conduction and the valence band according to distribution functions f_e and f_h. Consider direct transitions corresponding to the emission of photons of energy $\hbar\omega$. The instantaneous intensity of the photoluminescence $I_{LU}(t)$ is proportional to the product $f_e f_h$, the joint probability that the states coupled by the transition are occupied. Changes in the electron or hole distribution will be reflected in changes in the luminescence. For example, when a "hot" distribution cools down to some lower temperature, states at high energy are depleted while states near the band gap are filled up. This redistribution of carriers is accompanied by a decrease in the emission at short wavelengths, and an increase at long wavelengths. We will now show that carrier redistribution gives also rise to characteristic correlation effects, which reveal many details of the carrier dynamics.

Let us consider an experiment in which two consecutive short light pulses photoexcite carriers and populate electron and hole states in the bands. We assume that these states are subsequently depleted by relaxation processes. In Fig.1 two situations are compared: a) the pulses arrive at the same time; b) the pulses are separated by a time Δt. In the latter case the first pulse generates about half the electron and hole population of the two coincident pulses. However, if Δt is shorter than the lifetime of the electrons or holes, a larger population exists after the second pulse due to the surplus of carriers surviving from the first pulse. The total luminescence is determined by the time integral of the product $f_e f_h$ (hatched area). Therefore, two coincident pulses produce four times as much emission as one pulse alone, and twice as much as two pulses with very large Δt. The total luminescence is thus a function of Δt, having a maximum at Δt=0, and decaying to a background equal to one half of the maximum when Δt is increased. The decay of the correlation with Δt indicates the

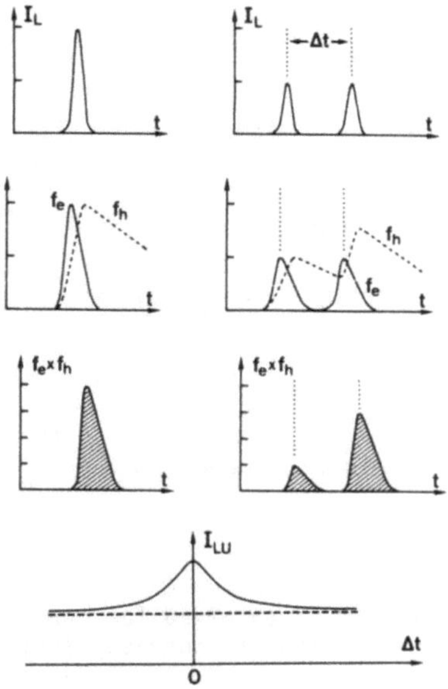

Fig. 1

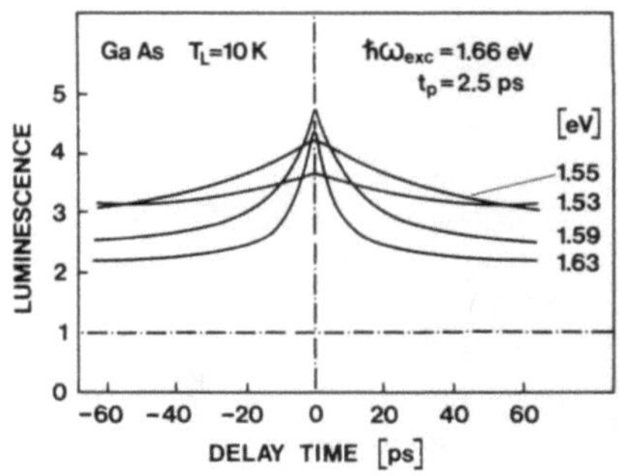

Fig. 2

diminishing temporal overlap of the carrier populations and the changes of the carrier distribution with time.

Photoluminescence correlation experiments were carried out in GaAs. A continuous pulse train from a synchronously mode-locked oxazine dye laser (t_p=2.5 ps; λ=745 nm) is used for photoexcitation. The output pulse train is divided into two beams of equal intensity, and orthogonal polarization. The beams are recombined on the sample surface with an adjustable delay time Δt, and the luminescence is measured by a photon counting spectrometer. In Fig.2 the total time-integrated photoluminescence is plotted versus Δt for various photon energies $\hbar\omega > E_{gap}$=1.52 eV. Unity of the vertical scale represents the intensity of the luminescence generated by one excitation beam alone (dashed-dotted line).

The following features should be noticed in Fig.2: (i) a signal maximum is observed at Δt=0 for all photon energies; (ii) the luminescence signal decays with increasing Δt; the higher the photon energy, the faster the decay. We find that at still higher photon energies the correlation curves approach the autocorrelation function of the laser pulses; (iii) for photon energies close to the gap energy only weak, slow correlations are observed; the background level is not reached on the time scale of Fig.2.

Figure 3 shows results of correlation experiments performed in a GaAs-$Ga_{0.65}Al_{0.35}As$ sandwich structure. The excitation wavelength of the laser is chosen such that the light is absorbed only in the 0.6 µm GaAs layer. The photoexcited carriers remain confined to the GaAs layer, because diffusion is hindered by the potential barriers formed by the larger band gap of the GaAlAs layers. On the left of Fig.4 luminescence versus Δt is shown for constant laser excitation (0.75 MW/cm^2 peak) and three different photon energies. It is interesting to note that at the lowest photon energy ($\hbar\omega \simeq E_g$) an anti-correlation is now observed, in the sense that the signal has a minimum near Δt=0 and increases for larger Δt. The anti-correlation disappears at higher photon energies, e.g., at $\hbar\omega$=1.61 eV we find again a maximum at Δt=0.

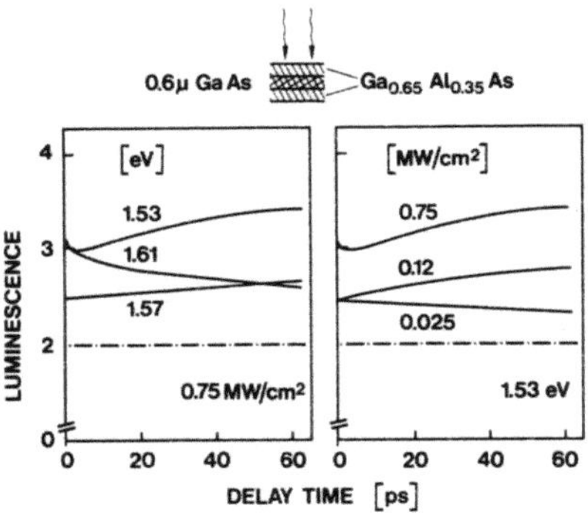

Fig. 3

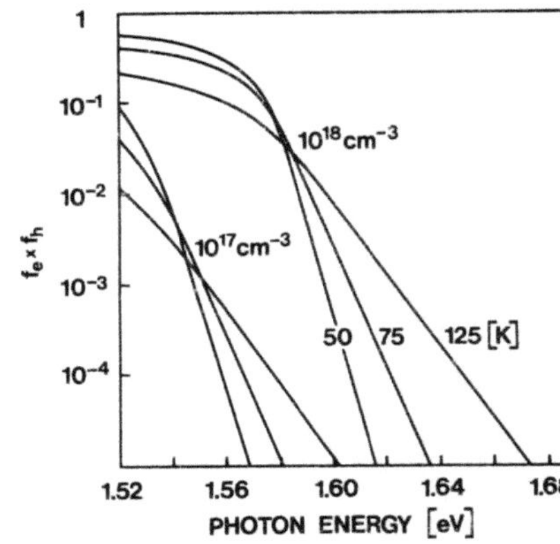

Fig. 4

A similar transition to an anti-correlation effect is observed for constant photon energy $\hbar\omega$=1.53 eV, when the laser excitation is increased from 25kW/cm^2 to 0.75 MW/cm^2 (right side of Fig.3).

For a qualitative interpretation of these correlation effects let us consider the curves of Fig.4 which show products $f_e f_h$ of Fermi distribution functions for different carrier densities and temperatures. First we note that for constant density the decrease in the carrier temperature leads to an increase of the product function at low energies, and a decrease at high energies. This situation is typical of the GaAs-GaAlAs structure in which diffusion and surface or interface recombination is strongly reduced; the carrier density can be regarded as constant over the range of delay times of the experiment. The observed anti-correlation effect indicates the filling of states in the vicinity of the band gap, while the decay of the correlation for higher photon energies directly reflects the depletion of high energy states due to energy relaxation. Fig.4 also explains the observed transition to anti-correlation behavior when the laser power is increased: a photon energy corresponding to states being depleted during cooling in a low density situation will correspond to states with growing population when the carrier density is higher.

In bulk GaAs, filling of low energy states due to cooling has to compete with depletion by surface recombination and diffusion. The product $f_e f_h$ will decrease for all states if the latter processes dominate, and an anti-correlation effect is not observed in this case. The absence of anti-correlations in bulk GaAs therefore indicates that the carrier concentration in the surface layer from which photoluminescence is observed is rapidly depleted by diffusion and surface recombination.

It should be emphasized that the correlation effects described here are an immediate consequence of the fact that the radiative transition rate is given by the product $f_e f_h$, which is a nonlinear function of the excitation pulse energy. Similar correlations are expected to occur in many other physical situations and systems, possibly also with different material probes, e.g., photoconductive, photoelectric or photoacoustic effects, if there is a suitable nonlinearity of the photoexcitation process.

2. Reflectivity and Transmission of Si During Picosecond Laser Annealing

In the last few years there has been a controversial discussion about the question of whether pulsed laser annealing involves ordinary thermal melting[3] or some other mechanism[4] in which the extremely hot and dense plasma of photoexcited carriers plays a dominant role. Optical reflectivity measurements during and after ns laser illumination of silicon showed an abrupt increase of the reflectivity[5] lasting for several ns and up to ~100 ns. The high reflectivity phase (HRP) has been attributed to a surface layer of molten Si. On the other hand, it has been proposed[4] that the HRP is due to a dense, long-lived plasma which must be postulated for the non-thermal model to be valid.

We have performed careful subnanosecond time scale measurements of reflectivity(R) and transmission(T) of crystalline Si during laser annealing with light pulses at 532 nm of 25 ps duration. A [100] surface is illuminated by a laser beam with a Gaussian profile 0.5 mm in diameter (FHWM). Reflectivity and transmission of a 100 µm central section of the irradiated surface area are probed with weak interrogation pulses at 1.064 µm.

Figure 5 shows R and T as a function of time for an exposure of 0.35 J/cm^2. Before the arrival of the pump pulse the reflectivity is found to be 32%, in agreement with the optical constants of c-Si at room temperature. About 20 ps before the maximum of the pulse, R goes through a minimum of 27%; during the pump pulse the reflectivity jumps to a plateau of R=76%. It was observed that R remains constant all the way from the picosecond time scale up to tens of ns. In the same experiment the transmission of the sample falls monotonically from 60% down to a few percent. We note that the detector of the transmitted light is also exposed to some photoluminescence around 1 µm, which should be taken into account if the sample transmission proper is to be evaluated.

When the pump energy is lowered to 0.11 J/cm^2 we obtain the result shown in Fig.6. Now the reflectivity drops to a minimum of 25% after the pump pulse maximum and recovers to a value of 30% at 50 ps. The transmission falls to a minimum of 28% and then increases slowly over several hundred picoseconds.

We find that there is a very distinct energy threshold of 0.21 J/cm^2 separating the two different types of behavior of R and T. In fact, if the pump pulse energy is allowed to vary by only 10% around 0.21 J/cm^2, the data clearly separate into two different subsets which correspond to the two types of curves, Fig.5 and Fig.6.

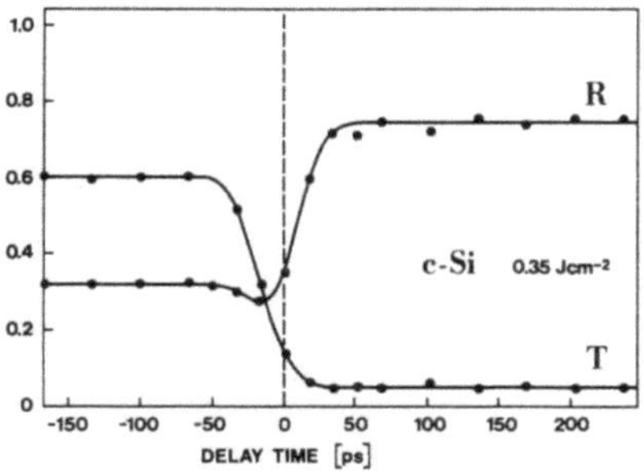

Fig. 5

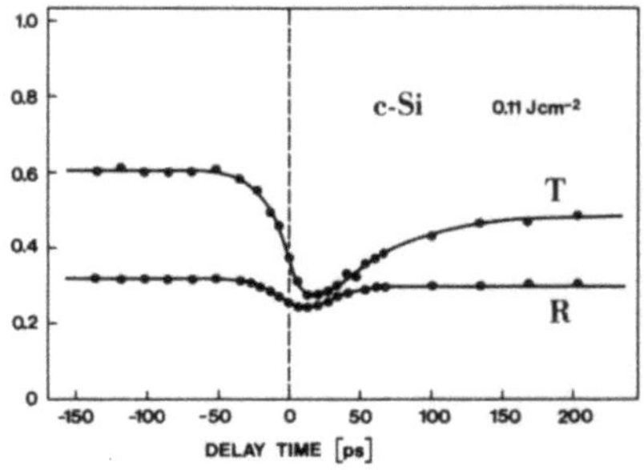

Fig. 6

These observations suggest the following conclusions:

(i) The reflectivity minimum - which has never been seen in ns experiments[5,6]-clearly indicates the presence of an extremely dense electronic plasma. The plasma frequency must be close to the frequency of the 1.064 µm probe pulse, which corresponds to the stunning density of about $10^{22}/cm^3$, approximately 10% of the total number of p-valence electrons of c-Si.

(ii) The lifetime of the plasma is of the order of 25 ps or shorter, because the induced change in the reflectivity follows the laser pulse in the low energy experiment (Fig.5).

(iii) The initial minimum of the reflectivity in the high energy case (Fig.6) also indicates the presence of a dense plasma. However, the subsequent reflectivity jump can hardly be attributed to the plasma, unless there is a mechanism for an abrupt increase by three orders of ten in the plasma lifetime caused by only a ~10% increase of the excitation energy. On the other hand, the observed absolute reflectivity value of 76% is identical to that of liquid silicon, suggesting that the reflectivity jump is rather due to the solid-liquid phase transition.

References

1. D. von der Linde, J. Kuhl and H. Klingenberg, Phys. Rev. Lett. 44, 1505 (1980)

2. D. von der Linde, J. Kuhl, E. Rosengart, J. Luminescence 24, 675 (1981)

3. R. F. Wood and G. E. Giles, Phys. Rev. B23, 2923 (1981)

4. J. A. Van Vechten, R. Tsu and F. W. Saris, Phys. Lett. 74A, 422 (1979)

5. D. H. Auston, J. A. Golovchenko, A. L. Simons and C. M. Surko, Appl. Phys. Lett. 34, 777 (1979)

6. A. Lietoila and J. F. Gibbons, J. Appl. Phys. 53, 32o7 (1982)

Picosecond Spectroscopy of Excitonic Molecules and High Density Electron-Hole Plasma in Direct-Gap Semiconductors

S. Shionoya

The Institute for Solid State Physics, The University of Tokyo,
Roppongi, Minato-ku, Tokyo 106, Japan

In the physics of highly excited direct-gap semiconductors, there are two important subjects to be investigated; one is the formation of excitonic molecules due to the attractive covalent interaction between two single excitons, and the other is the generation of high density electron-hole plasma (EHP), which is caused by the screening of the Coulomb interaction between the electron and hole in an exciton and should take place at higher electron-hole concentration than that for the formation of excitonic molecules. It is needless to say that picosecond spectroscopy is very important to investigate dynamical behavior of high density electron-hole systems in direct-gap semiconductors, in which carrier lifetime is very short, the order of nanosecond or less. This paper reviews recent results of such studies performed by us* for direct-gap materials such as CuBr, CdS and CdSe.

Excitonic Molecules

Previously we observed time-resolved luminescence spectra of excitonic molecules in CuCl, and obtained the lifetime of the molecule(1,2). Recently we have performed similar experiments for excitonic molecules in CuBr(3). The energy level structure of the excitonic molecule in CuBr is more complicated than in CuCl; the ground state of the molecule is split into three states, Γ_1, Γ_5 and Γ_3 states from the low energy side. This is due to the reflection of the difference of the valence band structure between the two materials. These three molecule states in CuBr are all allowed for two-photon transition, and three two-photon absorption lines are observed experimentally. The transition from each of these three molecule states to the single exciton state Γ_5 is allowed, but only the luminescence emitted from the lowest molecule state Γ_1 is observed. This luminescence is composed of two lines, M_L and M_T, leaving longitudinal and transverse single excitons, repectively.

We have measured in detail time-resolved spectra and time-dependence of the excitonic molecule luminescence in CuBr at 4.2 K under the one-photon band-to-band excitation and two-photon resonant excitation of excitonic molecules. A tunable picosecond laser system consisting of a mode-locked $YAG:Nd^{3+}$ laser and a $LiNbO_3$ parametric oscillator was used. Time-resolved spectroscopy was made by using a CS_2 optical Kerr shutter. Figure 1 shows four kinds of time-dependence of the intensity of the M_T luminescence line (422 nm) for the one-photon band-to-band excitation (B, 355 nm excitation) and for the two-photon resonant excitation of the three molecule states, Γ_1

---

* Coworkers are Yasuaki Masumoto, Yutaka Unuma, Hiroshi Saito and Hidemi Yoshida.

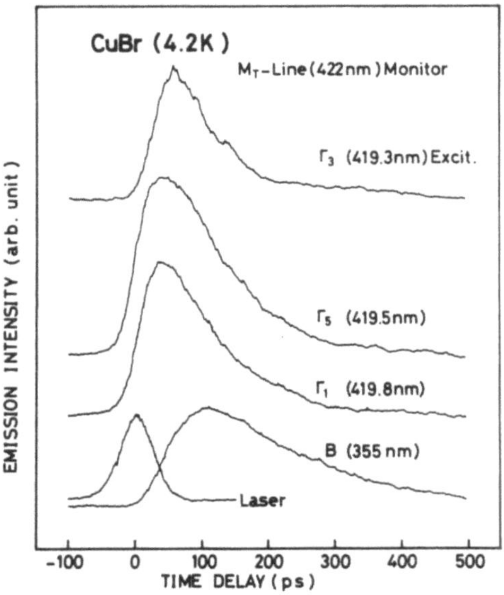

Fig. 1. Time-dependence of the intensity of excitonic molecule luminescence M_T under the band-to-band (B) and two-photon resonant excitation (Γ_1, Γ_5 and Γ_3) at 4.2 K. The shape of laser pulse is also shown.

(419.8 nm), Γ_5 (419.5 nm) and Γ_3 (419.3 nm). Under the excitation of two lower states Γ_1 and Γ_5, the intensity was found to rise almost in proportion to the time integration of the shape of exciting pulse, while in the case of the excitation of the highest state Γ_3 the intensity rises slightly delayed from the case of Γ_1 and Γ_5 excitation. This delay is so small compared with the present experimental accuracy that we cannot discuss the relaxation process from the Γ_3 to Γ_1 state, but it seems that the relaxation time may be a few picoseconds. Under the one-photon band-to-band excitation the peak of the luminescence intensity is delayed about 100 ps from the peak of exciting pulse.

The observed time-dependence of luminescence is convoluted with the duration of 1064 nm laser pulse used to open the Kerr shutter. The time resolution of the Kerr shutter of about 40 ps is not negligible compared with the observed decay time of luminescence of ~ 100 ps. Therefore, the observed data were deconvoluted by the calculation assuming that the opening laser pulse has a Gaussian shape. These deconvoluted data were used in the analysis,

The population dynamics of excitonic molecules and single excitons is expressed by the following rate equations.

$$dn_m/dt = -\alpha n_m + \beta n_{ex}^2, \qquad (1a)$$

$$dn_{ex}/dt = \alpha n - 2\beta n_{ex}^2 - \gamma n_{ex}. \qquad (1b)$$

Here, n_m and n_{ex} are the densities of molecules and single excitons, respectively. $\alpha = \tau_m^{-1}$ and $\gamma = \tau_{ex}^{-1}$, where τ_m and τ_{ex} are the lifetime of molecules and that of excitons, respectively. $\beta = \tau_f^{-1} n_{ex}^{-1}$, where τ_f is the formation time of molecules from single excitons. Using these equations we analyzed the observed time-dependence of the excitonic molecule luminescence, changing the values of three adjustable parameters α, β and γ, and obtained the lifetime of the excitonic molecule. The time duration of exciting laser pulse is not short enough to neglect compared to the decay

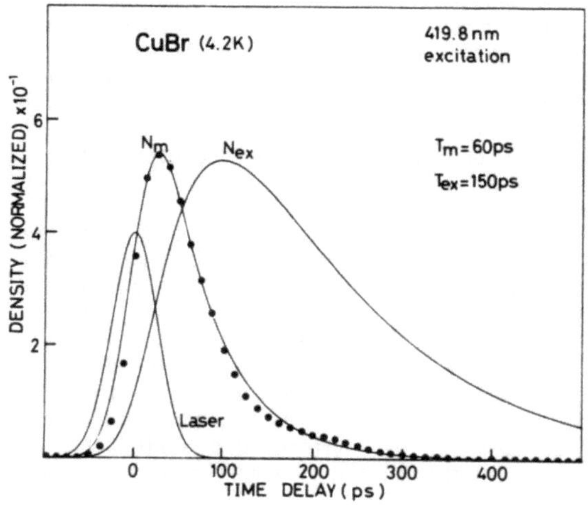

Fig. 2. Time-dependence of the molecule luminescence intensity (••••) (deconvoluted data of Fig.1) under the two-photon resonant excitation of the lowest state Γ_1. The solid curves show calculated dependence of the densities of molecules and excitons.

time of luminescence. Then, we have to add the term expressing the generation process of excitonic molecules or single excitons to eq. (1a) or (1b), respectively, for the case of the resonant two-photon excitation or band-to-band excitation.

Figure 2 shows results of the calculation for normalized densities N_m and N_{ex} in the case of the resonant two-photon excitation. N_m and N_{ex} are given by $N_m = n_m/n_m^0$ and $N_{ex} = n_{ex}/n_m^0$, respectively, where n_m^0 is the total density of excitonic molecules directly created by the laser pulse. The set of the parameters used is α_m = 60 ps, τ_{ex}= 150 ps and β_m^{-1}= $(\beta n_m^0)^{-1}$= 1 ns. It is seen that the observed time-dependence of luminescence agrees well with the calculated curve for N_m. Also for the case of the band-to-band excitation good agreement was obtained using the same values of τ_m and τ_{ex}. It was found that in this case of excitation the rise time of luminescence is mostly determined by τ_m, while the decay time by τ_{ex}.

The lifetime of excitonic molecules obtained 60 ps almost agrees with the radiative lifetime calculated by using existing theories. Further from results of the above-mentioned analysis, it was found that the probability of the formations of excitonic molecules per one collision of two single excitons is quite low, and that in the case of the resonant excitation the re-formation process of molecules is negligible.

High Density Electron-Hole Plasma

As to the high density EHP, time-resolved spectra of spontaneous luminescence were observed in detail for CdS and CdSe at 4.2 K under the band-to-band excitation(4,5,6). General features common to the two materials are that for the first 100-200 ps after pulse excitation the spectral shape changes drastically, the spectral width becoming much narrower. This indicates that hot carriers generated are rapidly cooled down during this period. Thereafter up to 500-800 ps delay times the spectral shape hardly changes in spite of the successive decrease of luminescence intensity. The spectral width in the late stage depends a little on excitation intensity. The spectral shape was analyzed, and the e-h concentration and effective carrier temperature were obtained. It was indicated that carriers are kept hot until they are radiatively annihilated.

In CdSe time-resolved spectra were measured with changing excitation photon energy in the vicinity of the bandgap energy. It was found that in the whole range of excitation photon energy the effective carrier temperature at the stage of 800 ps delay time does not fall below ∿20 K. This means that even if the excess energy is not given to carriers at the instance of pulse excitation, carriers are made hot, so that there exists a process which makes the EHP hot. Considering the phonon-assisted Auger process as such a process, it is shown that the EHP can be kept hot at ∿20 K.

As mentioned above the spectral shape of the EHP luminescence hardly changes until the very late stage. Although carriers are kept hot, the carrier temperature obtained from the spectral analysis is lower than the calculated critical temperature for the formation of e-h liquid. From these facts, it seems as if gaseous plasma of high density electrons and holes is transformed to a condensed e-h liquid state. However, as mentioned above, the spectral width, which corresponds to the e-h concentration, in the late stage depends a little on excitation intensity. Therefore, it is sure that the condensed e-h drop state, as observed in indirect-gap semiconductors such as Ge and Si, which is in equilibrium with the surrounding exciton gas, is not realized in the present case.

To obtain further information on this argument, time-resolved spectra were measured in a wide range of temperatures which are surely higher than the calculated critical temperature(7,8). It was found that in the case of CdS at 130 K aspects of changes of the spectral shape with time are a little different from the case of 4.2 K. Namely, the spectral width decreases slightly with time. Further the manner of shifts of the high energy edge and low energy edge of spectra with changing excitation intensity is different. Although we can not give reasonable interpretation for the observed dynamical behavior of the EHP for the moment, it would be sure that the EHP is in a non-equilibrium state until it is annihilated.

References

1) M. Ojima, T. Kushida, Y. Tanaka and S. Shionoya, J. Phys. Soc.Jpn. 44 (1978) 1294.
2) M. Ojima, T. Kushida, S. Shionoya, Y. Tanaka and Y. Oka, J. Phys.Soc.Jpn. 45 (1978) 884.
3) Y. Unuma, Y. Masumoto and S. Shionoya, J. Phys. Soc. Jpn. 51 (1982) 1200.
4) H. Yoshida, H. Saito, S. Shionoya and V. B. Timofeev, Solid State Commun. 33 (1980) 161.
5) H. Yoshida, H. Saito and S. Shionoya, J. Phys. Soc. Jpn. 50 (1981) 881.
6) H. Yoshida, H. Saito and S. Shionoya, Phys. Stat. Sol. (b) 104 (1981) 331.
7) H. Yoshida and S. Shionoya, J. Luminescence 24/25 (1981) 601.
8) H. Yoshida and S. Shionoya, in preparation.

Picosecond Time-Resolved Study of Highly Exited CuCl

D. Hulin[1], A. Antonetti, L.L. Chase[2], G. Hamoniaux, A. Migus and A. Mysyrowicz[1]

Laboratoire d'Optique Appliquée, Ecole Polytechnique - ENSTA,
F-911210 Palaiseau, France

[1]Groupe de Physique des Solides de l'Ecole Normale Supérieure,
2, place Jussieu, F-75005 Paris, France

[2]Physics Department, Indiana University, Bloomington,
Indiana 47405, USA

We present experimental results of a time-resolved investigation of pure CuCl crystals, in which extremely high free carrier densities ($n > 10^{20}$ cm^{-3}) are optically injected. Highly excited CuCl crystals have been extensively studied before [1], using nanosecond or subnanosecond optical pulses. However, the density of generated electron-hole pairs was limited so far by the onset of irreversible sample damage to values $n < 10^{20}$cm^{-3}. By using optical pulses of subpicosecond duration, it is possible to extend the range of carrier densities beyond the Mott dissociation limit [2]. Accordingly, a transition in the excited state of the system occurs, from an insulating phase of excitonic particles to a conducting phase consisting of a superdense plasma. Although the formation of an electron-hole plasma has been observed in several direct gap semiconductors, most notably GaAs [3], a dynamical study of the Mott transition in these materials has proved difficult, since the effective temperature of the plasma is comparable with the exciton binding energy. In CuCl, $E_X = 0.2$ eV so that the criterion $E_X/kT \gg 1$ should be more easily satisfied. Interest in the dynamics of superdense plasmas in direct gap semiconductors has been recently revived, as it may lead to picosecond laser action with broadband tunability [4], and also for its implication in the behavior of fast bistable optical devices [5].

To generate the carriers, a UV pulse is used ($\lambda = 309$ nm, pulse duration $\Delta t = 120$ femtoseconds, maximum energy per pulse = 20 µJ). It is obtained by frequency doubling in a KDP crystal the amplified red output from a passively mode-locked Rh 6 G dye ring laser [6]. When focused on the sample, it creates up to 10^{21} cm^{-3} electron-hole pairs, initially with an excess energy $\Delta E = 622$ meV with respect to the minimum energy band gap. In order to obtain nearly homogeneous volume excitation, very thin crystalline films of thickness $d < 1$ µm are used. The samples are obtained by sublimation of high purity anhydrous CuCl.

In fig. 1, time-integrated luminescence spectra are shown for different input light intensities for a sample temperature T = 15 K. The line at 392 nm is well-known and results from the radiative decay of excitonic molecules [1]. The hitherto unreported broad emission appearing on the lower energy side of the spectrum above a threshold input intensity is characteristic of the recombination of electron-hole pairs inside a variable density plasma : it appears above a well-defined excitation threshold and its width broadens, with the low energy side of the emission band shifting to longer wavelengths, as the excitation further increases.

An estimate of the plasma carrier density is obtained from the position of the renormalized energy gap, as inferred from the low energy edge of the plasma emission. A value $n_c \sim 3.10^{20}$ cm^{-3} is extracted near plasma appearance threshold. This is consistent with the experimental parameters and is

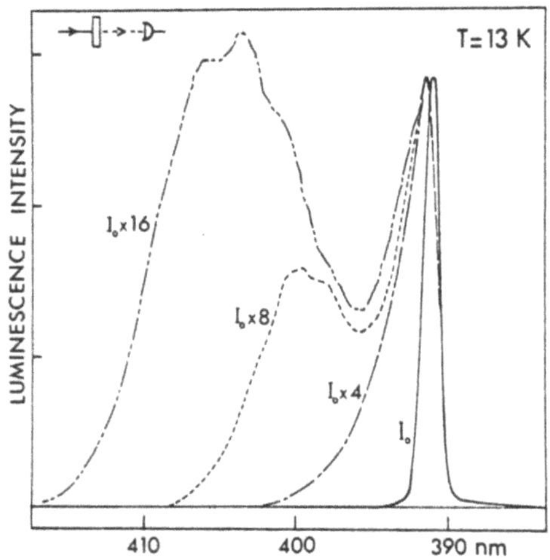

Fig. 1 Time-integrated emission spectra of a CuCl film (thickness = 50 nm) at T = 15 K obtained with different input intensities I_0 from a UV subpicosecond laser pulse. All curves are normalized to the same height for the biexciton emission line at 391.5 nm. The line at 391.5 grows superlinearly below I_c, and slightly sublinearly above I_c (I_c is the threshold intensity for plasma appearance).

close to the Mott criterion $n_c r_s^3 \leq 1$, where $r_s \cong 2a_X = 1.4$ nm is the average distance between particles (a_X = exciton radius).

The decay of the plasma emission has been measured with fast streak cameras (Thomson-CSF, and Imacon). The duration of the signal was found to be less than 10 ps, slightly above our highest camera resolution (see fig. 2). The emission at 391.5 nm, from the biexciton gas, was also detected, both below and above plasma threshold. In the latter case, it was found to succeed the signal from the plasma with a delay ($\sim$ 10 ps) more important than the 2 ps computed effect of group velocity dispersion. In the former case, the signal duration was dependent upon input intensity, ranging between $\tau \sim 10$ ps and $\tau \sim 45$ ps at lower intensities.

Information concerning the dynamics of the excited system may also be obtained in transmission experiments, using a weak subpicosecond continuum probe which traverses the excited region of the sample at different delays following the UV pump pulse [6,7]. Results taken at 15 K are shown in fig.3. The continuous curves is obtained when the probe pulse precedes the pump on

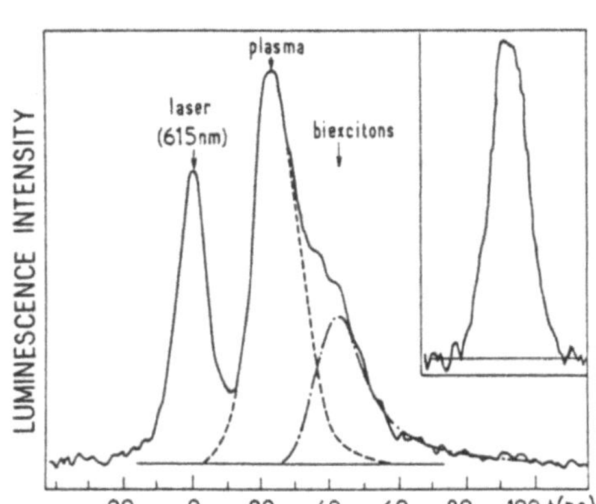

Fig. 2 Time evolution of the total luminescence of CuCl at 15 K, recorded in single shots by a fast streak camera (UV pulse excitation). The zero of the time scale is arbitrarily set at the position of residual red light from the primary laser pulse. The time delay between the red laser pulse and the luminescence is well accounted for by the dispersion of optical elements between the sample and the detector. The inset shows the signal from the plasma emission alone (same time scale).

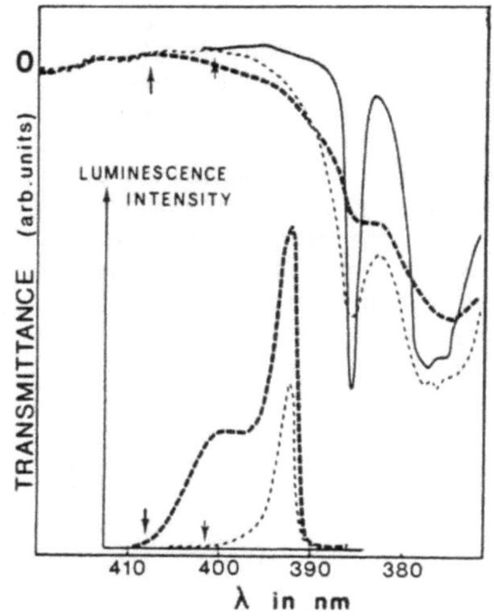

Fig. 3 Absorption spectra of CuCl at 15 K in the exciton region, recorded with a broad band subpicosecond pulse at the same delay ($\Delta t = 1$ ps) from the intense UV pump pulse but with two different excitation intensities. The corresponding time integrated luminescence spectra are also displayed. Note the correspondance between the low energy edge of the plasma emission and the band gap renormalization.

the sample and is therefore representative of the unexcited sample. The two lines at $\lambda = 387$ nm and 378 nm are the n = 1 terms of the excitonic series from the two upper valence bands (Γ_7,Γ_8) and the lowest conduction band (Γ_6).

For a time delay $\Delta t = 1$ ps, a reduction of the exciton oscillator strength occurs, as expected in the presence of the plasma phase. This effect is particularly well evident in very thin samples where a nearly complete disappearance of the exciton structures is observed. At the same time, a new absorption tail, extending below the excitonic structures is apparent with an edge shifting to lower energies under increasing excitation.

The excitonic oscillator strength recovers in accordance with the plasma decay time reported above, although a broadening and high energy shift of the lines persist for a longer period of the order of 100 ps. Time-resolved transmission spectra also show some evidence of a transition from optical loss to gain in the plasma spectral region below the excitons, occuring within a few ps from t = 0 (see fig. 4). This long life gain in the biexciton emission region may be due to collision processes between the excitonic molecules.

We now briefly comment on the measured plasma decay. The results of fig. 2 indicate that the lifetime of the free carriers is limited by excitonic par-

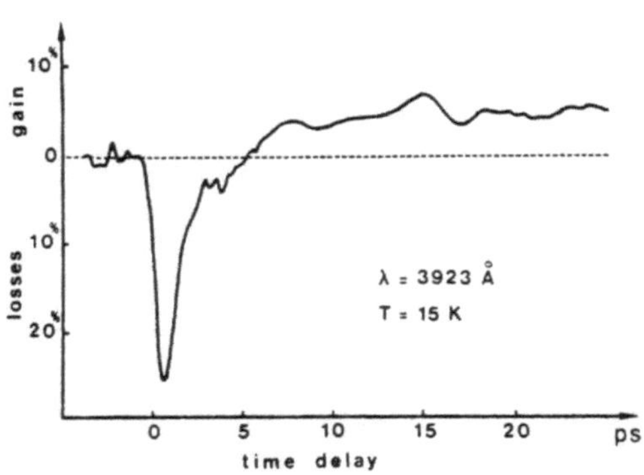

Fig. 4 Time-resolved transmission of the probe pulse at $\lambda = 3923$ Å in the plasma spectral region. The gain established at $\Delta t \cong 8$ ps remains stable for the longer time delay.

ticle formation when their density reaches n_c. The initial decrease of the carrier density down to n_c may have several origins : volume expansion of the excited region of the sample [8], non-radiative Auger decay involving three (or more) carriers [9], and direct annihilation of electron-hole pairs. The first process should not play a significant role here, in view of the nearly homogeneous volume excitation, large lateral dimensions and short times considered. We believe that direct electron-hole pair recombination dominates over Auger non-radiative decay in the density range of our experiments because the overall radiative efficiency from the plasma increases sharply with higher input intensities well above n_c (see fig. 1). To our knowledge, there are no known values for the Auger and bimolecular decay constants in CuCl. Assuming only radiative decay, we obtain for the electron-hole (bimolecular) recombination constant B a value $B \sim 5.10^{-10}$ cm^3/sec. However, neglect of Auger decay and stimulated emission may not be justified.

This work has been supported by the Direction des Recherches, Etudes et Techniques.

References

1 See for instance, E. Hanamura, H. Haug, Physics 33c, 209 (1977).
2 N.F. Mott, Proc. Phys. Soc. London Sec. 1 62, 416 (1949).
3 S. Tanaka, H. Kobayashi, H. Saito, S. Shionoya, J. Phys. Soc. Jap. 49, 1051 (1980).
E.O. Gobel, P.H. Liang, D. Von der Linde, Solid State Comm. 37, 609 (1981).
4 T.C. Damen, M.A. Dugay, J. Shah, J. Stone, J.M. Wiesenfeld, R.A. Logan, Appl. Phys. Lett. 39, 142 (1981).
5 H.M. Gibbs, S.L. Mc Call, T.N.C. Venkatesan, A. Passner, A.C. Gossard, W. Wiegmann, CLEA 1979, IEEE J. Quantum Electron. QE-15 (1979) 108 D.
6 E.P. Ippen, C.V. Shank, Topics in Applied Physics, Vol. 18, Springer Verlag (1977) S.L. Shapiro editor.
A. Migus, C.V. Shank, E.P. Ippen, and R.L. Fork, IEEE J. Quantum Elect. QE-18, 101 (1982).
R.L. Fork, C.V. Shank and R.T. Yen, to be published.
A. Migus, J.L. Martin, R. Astier, A. Antonetti and A. Orszag, this issue.
7 C.V. Shank, R.L. Fork, R.F. Leheny, J. Shah, Phys. Rev. Lett. 42, 112 (1979).
D. Von der Linde, R. Lambrich, Phys. Rev. Lett. 42, 1090 (1979).
8 D.H. Auston, C.V. Shank, Phys. Rev. Lett. 32, 1120 (1974).
A. Cornet, M. Pugnet, J. Collet, T. Amand, M. Brousseau, J. de Phys. C7, 471 (1981).
9 D.H. Auston, C.V. Shank, P. Lefur, Phys. Rev. Lett. 35, 1022 (1977).

Picosecond Dynamics of Excitonic Polariton in CuCl

Y. Aoyagi, Y. Segawa, and S. Namba

The Institute of Physical and Chemical Research, Wako-shi, Saitama, 351, Japan

Abstract

Dynamics of the excitonic polariton in CuCl was clarified by using a tunable picosecond induced absorption spectroscopy and a picosecond transient grating spectroscopy. The energy dependent lifetime of the excitonic polariton and the scattering mechanism were discussed.

1. Introduction

An excitonic polariton (polariton) - the mixed state of an exciton and a photon in a crystal - gives rise to a nonlinear dispersion in E-k (energy-momentum) relation (Fig.2-a), when the photon dispersion crosses the exciton band. Many workers have recently interested in the characteristics of the excitonic polariton [1]. However, the dynamics and the scattering mechanism of the polariton have not been clarified. We observed the energy dependent lifetime of the polariton and the temperature dependent diffusion coefficient by using a picosecond induced absorption spectroscopy and a picosecond transient grating spectroscopy, respectively.

2. Experimental Procedure

Figure 1 shows experimental block diagram for the tunable picosecond induced absorption spectroscopy [2] and transient grating spectroscopy [3]. In this configuration the polariton was resonantly excited by a narrow band tunable picosecond laser (band width .01 nm, tunable range 385-390 nm). The band width of the probe light was about 2nm and the arrival time of the probe light on the sample was delayed by an optical delay line. For the induced absorption spectroscopy one of the excitation beams was used for the excitation of the polariton and the other was cut in front of the sample. The time dependence of the induced absorption of the probe light (from a level of the excited polariton to that of the excitonic molecule) was observed as a function of the

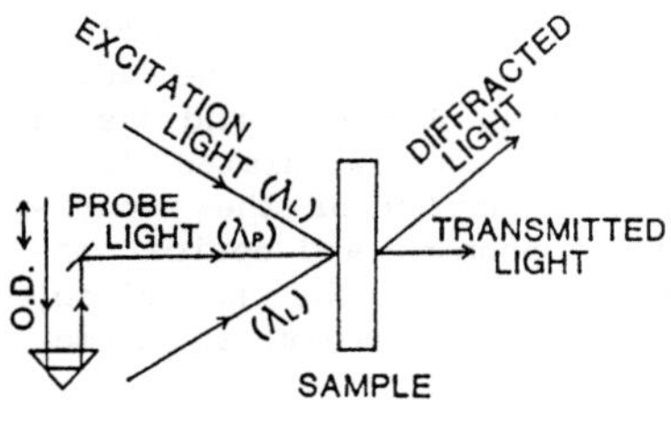

Fig.1
The Schematic diagram of the transient grating and induced absorption spectroscopy

excitation energy. The excited polariton moves to the direction of the incident laser with the group velocity of the polariton (dw/dk). The number of the polariton at the initial energy and momemtum decreases, if this polariton is scattered. This decrease is observable by the decrease of the induced absorption. For the transient grating spectroscopy two excitation beams in Fig. 1 were simultaneously used to make a transient grating of the excitonic molecule. Since the wavelengths of the peaks of the diffraction due to the excitonic molecule and the polariton differ each other, the spatial and energy dynamics of the polariton was examined independently from that of the excitonic molecule.

3. Results and Discussion

Figure 2 shows the intensity of the induced absorption for various delay time at two different excitation energy. These experiments show that the induced absorption increases rapidly and decay exponentially.

Figure 3 shows a lifetime of the polariton of CuCl at 1.8 K at various energies. As shown in this figure the life time of the polariton strongly depends on the energy of the polariton. The life time is given by

$$\tau=1/v\sigma n \qquad (1)$$

where v is the velocity of the polariton, n is the number of particles of the scattering source and σ is the cross section of the scattering source. The velocity of the polariton is given by the differential calculus of the dispersion curve shown in Fig. 1 (group velocity dw/dk). The solid line (a), dotted line (b) and dashed line (c) are calculated

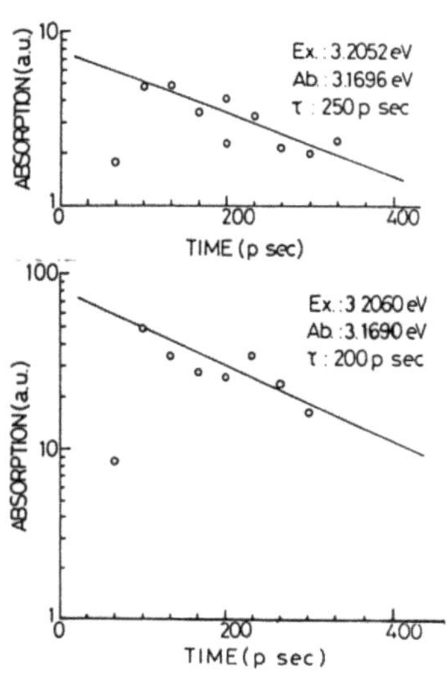

Fig.2
The intensity of the induced absorption for various delay times. Ex: the photon energy of the excitation, Ab: The photon energy of the induced absorption.

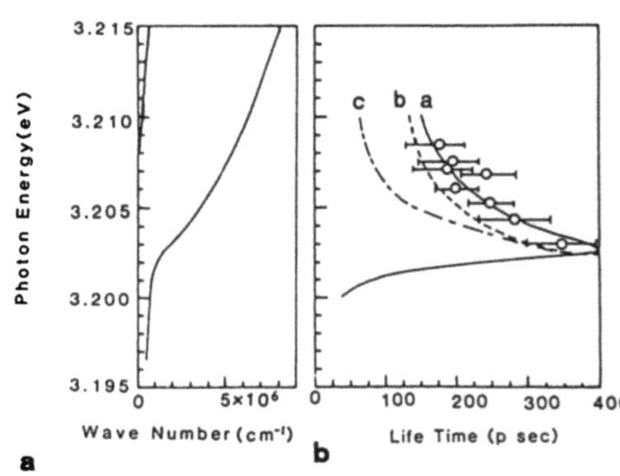

Fig.3
(A) The dispersion relation of the excitonic polariton in CuCl. (B) Observed life time of the polariton for various energies. The solid line (a), dotted line (b) and dashed line (c) show the calculated lines under the assumption of the impurity, acoustic phonon, and optical phonon scattering of the polariton, respectively. The maximum is normalized to be 400 ps.

from eq.(1) considering the dispersion of the group velocity of the polariton under the assumption of an impurity scattering, acoustic phonon sacttering and optical phonon scattering, respectively. The characteristics that the lifetime decreases with increasing the polariton energy were mainly explained by the dispersive characteristics of the group velocity of the polariton. However, the best fit of eq. (1) with experiment is obtained under the assumption that the polariton has the constant scattering cross section for the various momentum of the polariton. This suggests that the main scattering mechanism of the excitonic polariton at 1.8 K in CuCl is the impurity scattering.

Figure 4 shows the spectra of the diffracted light at different delay times. The narrow and intense diffraction spectra A and A' are the parametric coupling signals between the weak probe light and the strong excitation light in the sample due to large nonlinear susceptibility at the wavelength of the two-photon absorption. The broad spectrum B is due to the diffraction of the probe light by the index grating of the periodically induced density modulation of the excitonic molecule in the crystal. The diffracted light C builds up after the excitation of the excitonic molecule with some delay time and decays with some decay constant. This band is the diffraction from the grating of the excitonic polariton generated after the radiative relaxation of the excitonic molecule.

From the time behavior of these bands B and C for several grating periods we can estimate the diffusion coefficient and the lifetime of the excitonic polariton. The decay time of the diffracted light T observed under the various grating periods gives the diffusion coefficient D and the lifetime τ independently as an intercept of y-axis and the slope in 1/T vs. $4\pi^2/\Lambda^2$ plot according to the following equation

$$1/T=1/\tau+4\pi^2/\Lambda^2. \quad (2)$$

Figure 5 shows 1/T vs. $4\pi^2/\Lambda^2$ plot at two different temperatures. This figure shows that increasing the temperature, the diffusion coeffi-

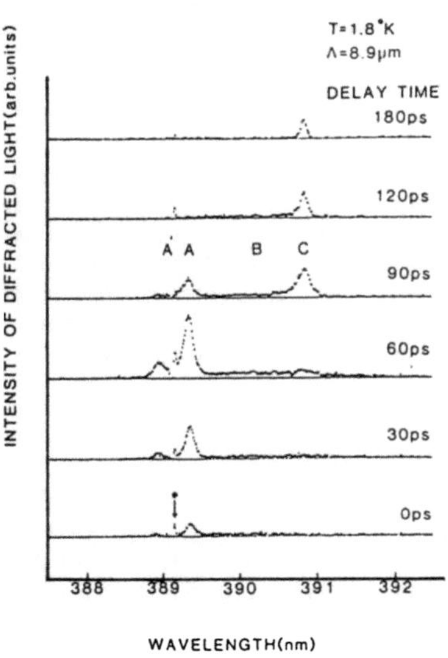

Fig. 4
The spectra of the diffracted light at different delay times. The asterisk indicates the excitation light scattered at the sample. As for the nomenclature A, A', B and C, see the text.

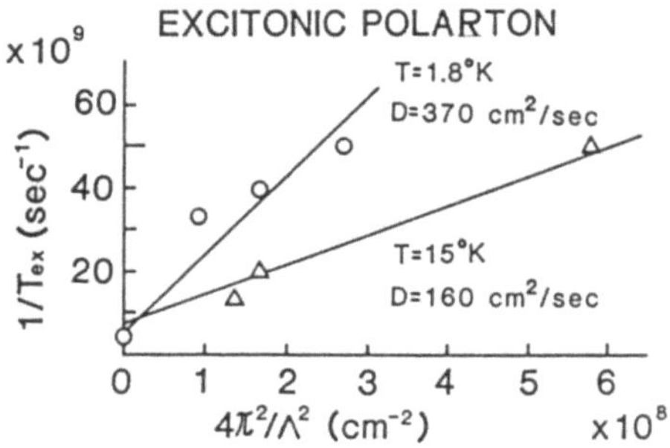

Fig.5
The inverse decay time (1/T) vs. the inverse sqaure of the grating period ($4\pi^2/\Lambda^2$) for the excitonic polariton at 1.8 K and 15 K

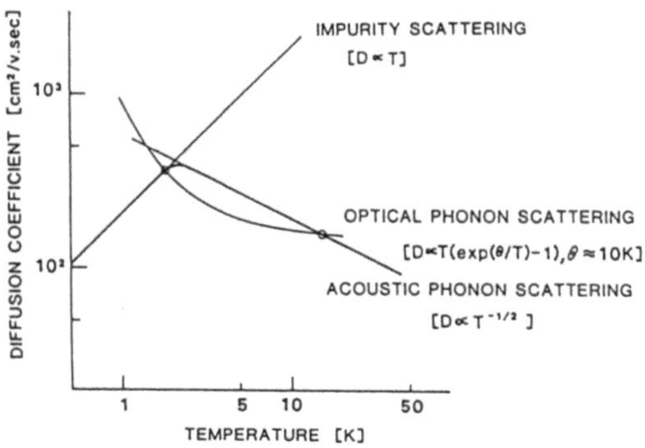

Fig. 6
Temperature dependence of the diffusion coefficient of the excitonic polariton

cient decreases. The temperature dependence of the diffusion coefficient observed suggests that the phonon (optical or acoustic) scattering is dominant for the polariton scattering mechanism in CuCl as shown in Fig.6.

Since the induced absorption experiment suggest that the impurity scattering mechanism is dominant at 1.8 K, the scattering mechanism should change from the impurity scattering to the phonon scattering in the temperature region of 1.8 K to 15 K to explain consistently the results obtained by the both experiments. To confirm our model detailed experiments are now in progress.

4. Conclusion

The simultaneous use of induced absorption spectroscopy and transient grating spectroscopy is a powerful technique to examine the scattering mechanism of the polariton. We can emphasize that the energy dependent life time is explained by the dispersive characteristics of group velocity of polariton and the impurity scattering and the temperature dependent diffusion coefficient is explained by the phonon scattering mechanism. These results suggest that the scattering mechanisms in 1.8 k and 15 k differ each other.

References

1) eg. T. Itoh, T. Suzuki and M. Ueta:J. Phys. Soc. Japan 44, 345 (1978).
2) Y. Segawa, Y. Aoyagi and S. Namba: Solid. State. Commun. 39, 535 (1981).
3) Y. Aoyagi, Y. Segawa and S. Namba: Phys. Rev. (B), 25, 1453 (1982).

Picosecond Spectroscopy of Highly Excited GaAs and CdS

H. Saito[+], W. Graudszus, and E.O. Göbel

Max-Planck-Institut für Festkörperforschung, Heisenbergstraße 1,
D-7000 Stuttgart 80, Fed. Rep. of Germany

1. Introduction

The realtime observation of relaxation, thermalization, and recombination of nonequilibrium carriers in direct gap semiconductors like GaAs or CdS requires picosecond time resolution. Various problems like the dynamics of e.g. hot carrier-phonon interaction, exciton screening or electron-hole-liquid (EHL) formation have been investigated by picosecond experiments [1-4]. In this paper we report application of picosecond spectroscopy for the study of a) free carrier relaxation within extended and localized continuum states in GaAs and b) the dynamics of the free carrier-exciton system in highly excited CdS.

2. Experimental

We have used picosecond absorption ("excite and probe") as well as luminescence spectroscopy. Light pulses of 25 ps duration are generated at 1064 nm by a passively mode locked Nd:YAG laser. Frequency tunable pulses of about 20 ps width are obtained in the entire visible and near infrared spectral range by synchronous pumping of a dye laser by the second or third harmonic of the Nd:YAG laser emission. The GaAs and CdS single crystals are excited by two photon absorption of 1064nm and 532nm pulses, respectively. Transmission spectra of CdS are obtained at different delay times using the synchronously pumped dye laser emission as the probe light. Time resolved luminescence spectra are measured with 25 ps time resolution using a CS_2 optical Kerr shutter and an intensified Si-diode array camera.

3. Experimental Results and Discussion

3.1 Free Carrier Relaxation Within Extended and Localized Continuum States in GaAs

The time constants associated with the relaxation of energy and momentum of nonequilibrium carriers within extended continuum states of pure GaAs single crystals are in the picosecond or subpicosecond range [1,2,4]. Relaxation within localized continuum states, however, can be slower by orders of magnitude as it has been shown for amorphous semiconductors [5]. We present data on free carrier relaxation in GaAs heavily doped with Si ($p=5\times10^{18}cm^{-3}$). In heavily doped semiconductors extended (high energetic) and localized (low energetic) states are present, very similar to amorphous materials [6]. The unique advantage of highly doped semiconductors, however, is the possibility of varying the density of localized states and the degree of locali-

---

[+] permanent adress: Dept. Appl. Phys., Okayama Coll., Okayama 700, Japan zation in a wide range by simply changing the doping. The experimental results directly reveal the slowing of the relaxation as localized states are involved.

The time resolved luminescence spectra obtained from heavily doped and pure GaAs exhibit characteristically different time behavior as shown in Fig.1. The emission of the doped sample shifts towards lower energies with increasing time, in contrast to the results for the pure material. This pronounced but opposite shift disappears for both samples at higher temperatures. Fig.2 depicts emission spectra for the heavily doped and pure GaAs at 200 K and 150 K, respectively. The low energy onset of the luminescence remains constant in both cases. The shift of the maximum to lower energies and the de-

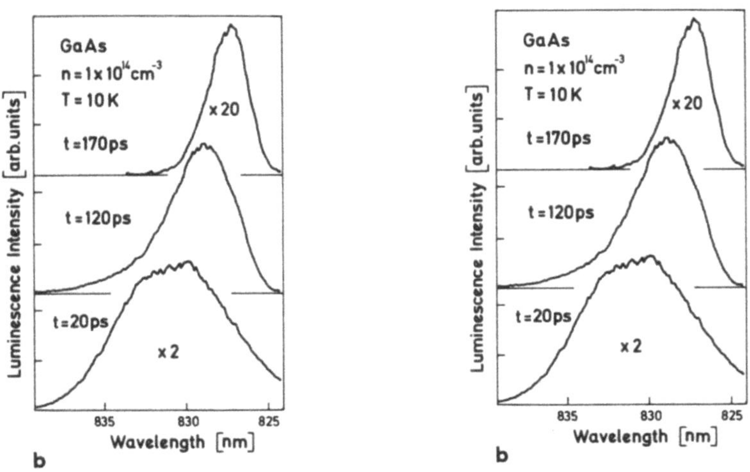

Fig.1 Time resolved luminescence spectra of heavily doped (a) and pure (b) GaAs at a temperature of 10 K

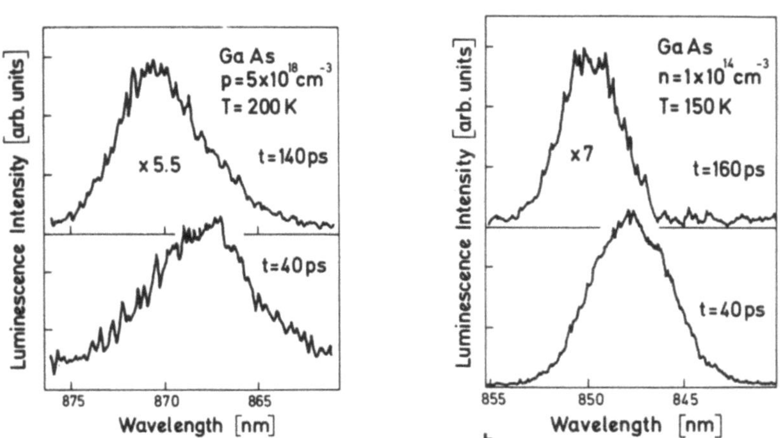

Fig.2 Time resolved luminescence spectra for heavily doped (a) and pure (b) GaAs at temperatures of 200 K and 150 K, respectively

crease in halfwidth with time arises from the decrease of the carrier density due to recombination.

A detailed analysis of the temperature dependence of the time resolved luminescence spectra shows that the relaxation within the localized states of the heavily doped material occurs via multiple trapping of the carriers [7]. The slow energy relaxation at low temperatures ($\partial E/\partial t \sim 10^8$ eV/s) is due to the thermally activated nature of this process. The data for the pure GaAs are consistent with an extremely fast relaxation. The high energy shift of the spectra with time is explained by a reduction of the band gap shrinkage because of free carrier recombination. This density dependent gap shrinkage, however, seems to be less important at higher temperatures. This indeed is expected, because exchange and correlation interaction are effective only as long as the thermal energy is smaller than Fermi and plasmon energy, respectively.

3.2 Dynamics of the Free Carrier-Exciton System in Highly Excited CdS

A stable EHL is expected in CdS at low temperatures on the base of energetic arguments [8]. It is questionable, however, to which extent separation of the excited carrier system into a low density (gaseous) and a high density (liquid) phase occurs within the short lifetime. We report time resolved transmission and luminescence experiments which provide first insight into this problem.

Figure 3 depicts transmission spectra of CdS at 10 K and 70 K for different excitation intensities. Optical gain corresponding to negative values for the optical density is observed for delay times larger than 80 ps. In the earlier time regime only absorption occurs. The low temperature trans-

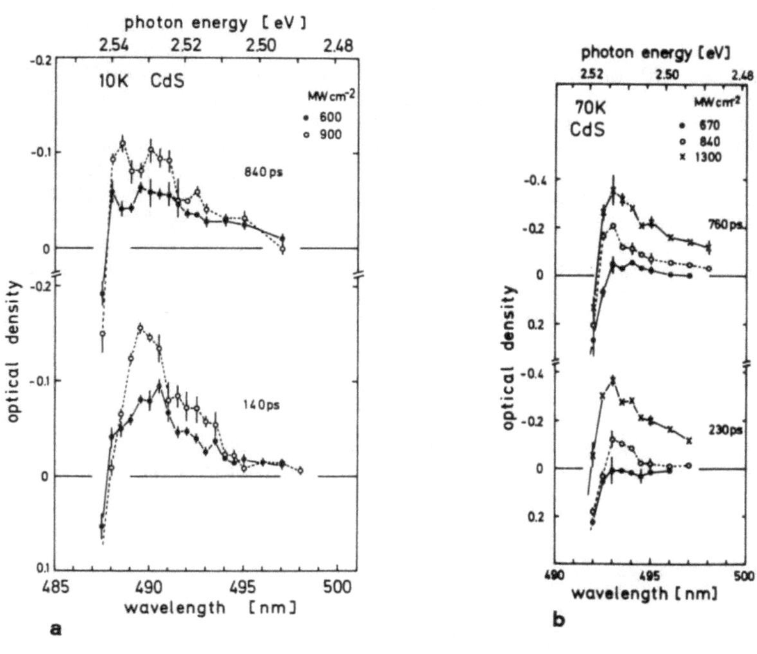

Fig.3 Time resolved transmission spectra of CdS at 10 K (a) and 70 K (b) for different excitation intensities. Optical density (αd) is plotted vs. wavelength for different delay times.

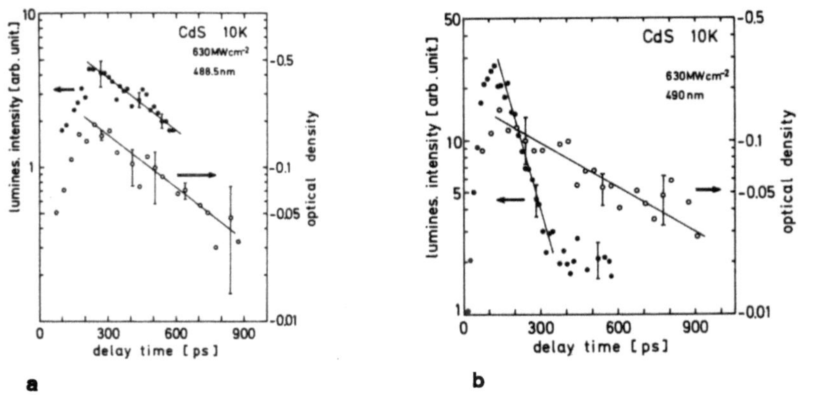

Fig.4 Luminescence intensity and optical density at 488.5 nm (a) and 490 nm (b) vs. time at a temperature of 10 K

mission spectra consist of several components due to the coexistence of the various excited species. The high energy part of the spectra at about 2.54eV can be attributed to exciton molecule recombination, whereas the low energy part is due to free carrier and excitonic recombination [9]. At 70 K the exciton molecule is thermally dissociated and furthermore no EHL exists. The spectra at 70 K thus are dominated by exciton recombination (exciton scattering).

The results of the luminescence and transmission experiments are summarized in Fig.4, where luminescence intensity and optical density at 488.5nm and 490nm is plotted vs. time. At 488.5 nm (exciton molecule recombination) the same time behavior is found for luminescence and optical gain (time constant ~ 400 ps). At 490 nm, however, a fast initial decay of the luminescence is observed, whereas the gain decreases with almost the same time constant as seen for the molecule recombination (Fig.4a). We therefore conclude that the gain at 490 nm is dominated by excitonic processes, too, in accordance with the observed maximum gain values of about 30 cm^{-1}. The initial fast component observed in luminescence is attributed to free carrier recombination. Within this time regime the initially hot free carrier system cools and transforms into the various coexisting components, namely excitons, molecules, and possibly small EHL clusters. The luminescence decay thus is determined by the characteristic times related to the cooling of the free carrier gas and the formation of the various coexisting elementary excitations.

References

1. D.v.d.Linde, J.Kuhl, R.Lambrich, in Picosecond Phenomena II, ed. R.M. Hochstrasser, W.Kaiser, C.V.Shank (Springer Verlag, Berlin, N.Y., 1980)
2. D.H.Auston, S.McAffe, C.V.Shank, E.P.Ippen, O.Teschke, Solid State Electr. 21, 147 (1978)
3. S.Shionoya, J.Luminesc. 18/19, 917 (1979)
4. J.Shah, Journ. de Phys. 42, C7-445 (1981)
5. see e.g. R.A.Street, Adv. in Phys. 30, 593 (1981)
6. D.Redfield, Adv.in Phys. 24, 463 (1975)
7. E.O.Göbel, W.Graudszus, Phys.Rev.Lett. 48, 1277 (1982)
8. M.Rösler, R.Zimmermann, Phys.Stat.Sol.(b) 83, 85 (1977); G.Beni, T.M. Rice, Phys.Rev. B18, 768 (1978)
9. H.Saito, S.Shionoya, J.Phys.Soc.Japan 37, 423 (1974)

Non-Linear Attenuation of Excitonic Polariton Pulses in CdSe

P. Lavallard and Pham Hong Duong*

Groupe de Physique des Solides de l'E.N.S., Université Paris VII,
Tour 23, 2 Place Jussieu, F-75221 Paris Cedex 05, France

1. Introduction

Optical transmission and polariton time of flight measurements are done with a very thin CdSe sample ($e = 0.93\mu \pm 0.02\mu$) immersed in pumped liquid helium. The source is a synchronously pumped dye laser (Rhodamine 640 or DCM). The pulse width is 6 ps ; the spectral width of the pulse is 2 Å. The mean wavelength of the pulse can be easily tuned in the region of interest. The rate of repetition is 80 Mhz. The propagation time of the light through the sample is measured by the usual autocorrelation technics [1]. Our experimental arrangment allows us to measure also the reflection and the luminescence spectra.

2. Experimental results

2.1 The light beam was focused on the sample down to a 100μ spot size. Even with a rather low average power ($P \lesssim 5$ mW), a large non linear effect occurs in the frequency range between ω_T and ω_L.

By putting neutral filters on the incident beam, we observed that the total transmitted light intensity saturated when the incident light intensity was increased. With a lens, we made an enlarged image of the spot on a screen : the center of the beam was black and surrounded by a bright circle. At higher power a small bright spot appeared in the center. With a diode array, we analyzed the spatial repartition of the intensity in the beam. The figure 1

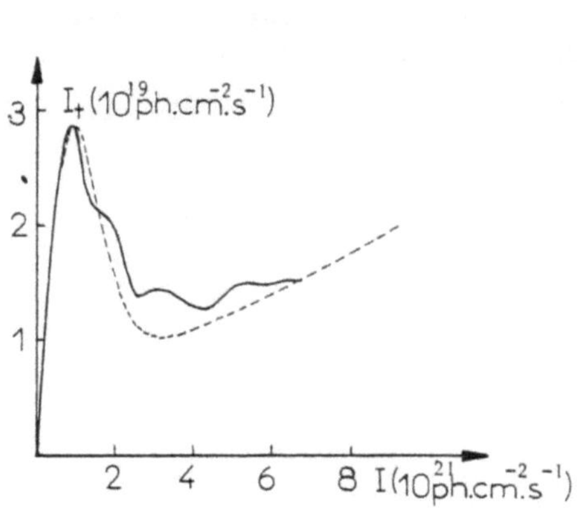

Fig.1 The transmitted intensity as a function of the incident intensity

* On leave of absence from Institut of Physics, Center for Scientific Research of Viet Nam, Hanoi, Viet Nam

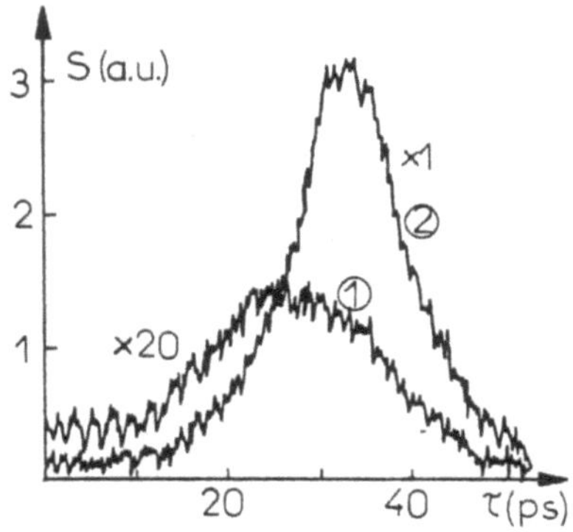

Fig.2 Cross correlation signal as a function of the delay time (curve 1 is obtained with an incident intensity 20 times larger than for curve 2)

shows the transmitted intensity as a function of the incident intensity ; the curve was obtained by comparing the signal given by each diode at low and high intensity. The critical incident intensity which corresponds to the maximum of transmission was minimum for the light frequencies which propagate with the slowest group velocity. Non linear transmission was also observed with a CW dye laser for an incident power ten times larger than the average power of the pulses.

2.2 The cross-correlation measurements show that at high intensity, the pulse shape is deformed and the peak of the pulse is shifted to earlier time. This is well explained by assuming that the early part of the pulse creates diffusing (or absorbing) centers which are responsible for an increased absorption of the remaining later part. (Figure 2)

2.3 In a high intensity transmission experiment, the beam behaves as a pump and a probe, at the same time. In order to distinguish between the two roles of the beam, we studied the transmission of a weak intensity laser beam as a function of the wavelength of a second strong intensity laser beam. Both lasers are CW. Figure 3 shows the results we obtained when the wavelength of the first laser beam is fixed near the polariton resonant frequency (6791 Å). The absorption is very much increased when one excites bound excitons (I_2) or free excitons. We conclude that bound excitons are responsible of the induced absorption of polaritons (the creation of free excitons is only an indirect way to produce the bound excitons).

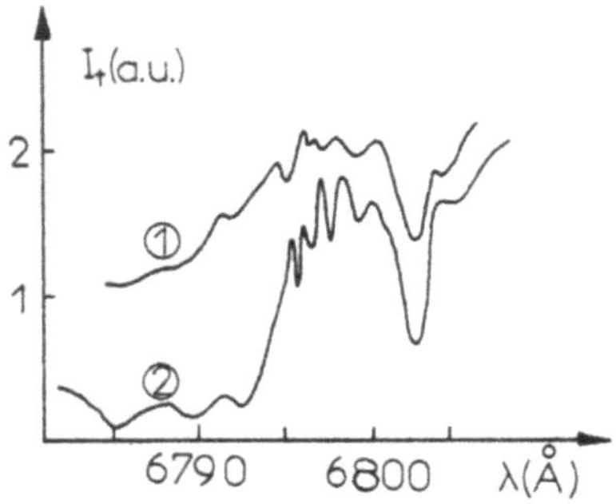

Fig.3 Transmitted light intensity at 6791 Å as a function of the second laser beam wavelength (curve 2 is obtained with a pump intensity 4 times larger than for curve 1)

3. Interpretation

We developed a model which explains the experimental results:

At first, polaritons are diffused on bound excitons :

$$\mathrm{pol}_k + \oplus\overset{-}{-}+ \rightarrow \mathrm{pol}_{k'} + \oplus\overset{-}{-}+$$

Diffused polaritons do not contribute to the transmitted light ; they do other collisions, emit acoustical phonons and lose a part of their energy in the band [2]. When they have an energy close of the bound exciton resonance, they are easily captured by the donors :

$$\mathrm{pol}_{k'}\ \text{(relaxed)} + \oplus^{-} \rightarrow \oplus\overset{-}{-}+$$

The net result of the initial collision of the polariton is then, the creation of a new diffusing center. The process amplifies by itself and stops when all the impurities have captured an exciton. On figure 2, we have plotted the results of the calculation for realistic values of the parameters. The fitting is rather good. The diffusion section is found to be of the same order of magnitude as the orbit surface of bound excitons.

References

1. E.P. Ippen, C.V. Shank : "Ultrashort light pulses" edited by S.L. Shapiro (Springer-Verlag 1977) p. 91
2. Y. Segawa, Y. Aoyagi, S. Namba, Sol. State Com. 39, 535 (1981)

Time Resolved Photoluminescence Study of n Type CdS and CdSe Photoelectrode

D. Huppert, Z. Harzion and S. Gottesfeld

Department of Chemistry, University of Tel Aviv, 69978 Ramat Aviv, Israel

N. Croitoru

Department of Electronic Devicex, Tel Aviv University, 69978 Ramat Aviv, Israel

Introduction

The importance of kinetic parameters in the determination of the efficiency and stability of semiconductor photoelectrodes has been stressed by many authors [1]. Transient techniques have served as an important tool for the evaluation of kinetic parameters in both electrochemical [2] and solid state solar cells [3]. The use of such techniques in photoelectrochemical cells has been introduced recently [4,5].

We have previously studied the photocurrent transients in $CdSe/S^{=}$, S° cell [5]. It was found that photocurrent decay is sensitive to the state of the photo-electrode (CdSe crystal) surface. Chemical etching enhanced the amplitude and reduced the decay time of the transients and increased the photocharge obtained (per flash). Nelson et al [6] found earlier that the surface recombination of carriers at the interface n type GaAs/aq. alkaline Se^{2} $/Se_x^{2-}$ solution, which served as part of a liquid junction solar cell, severely limited its performance. They also found that adsorption of Ru ions on the surface of n-type GaAs decreased the surface recombination velocity, and enhanced the energy conversion efficiency of the liquid junction solar cell significantly. Utilizing time resolved emission techniques, the same authors [6] measured the photoluminescence decay time for the GaAs photoelectrode, and deduced from it the surface recombination velocity before and after Ru ion adsorption.

In this study we report the results of time resolved photoluminescence measurements on n-type CdSe and CdS single crystals at three different states of the surface. 1) an untreated crystal, 2) mechanically polished crystal, 3) chemically etched crystal.

Experimental

The schematics of the optical arrangement was described elsewhere [7]. CdSe single crystals were irradiated by 532nm 6 psec or 20 psec pulses and CdS single crystals by 353nm 6 psec or 20 psec pulses, (these are the second and third harmonics of a mode locked Nd^{3+}/glass and YAG laser respectively). The photoluminescence was collected from the sample front surface, at an angle of 45° relative to the incident laser light. Colored glass filters as well as interference filters were placed before the streak camera, (Hamamatsu Model C939), entrance slit. The output of the streak camera was imaged onto a silicon intensified Vidicon connected to an optical multichannel analyzer (PAR 1205 D). The streak records were averaged and analyzed by a Delta Microcomputer (Z-80 microprocessor).

The hexagonal n-type CdSe and CdS single crystals (Cleveland Crystals Incorporated) were 1-2 mm thick, had a resistivity of 12.2 Ωcm and 3.3 Ωcm for CdSe and CdS respectively, and a surface perpendicular to the C-axis. Polishing was performed by 0.3 μm alumina. Etching was carried out in 4:1 $HCl:HNO_3$ for CdSe while CdS was etched by concentrated HCl. The duration of the etching was 30 seconds, followed by rinsing in distilled water.

Results

CdS and CdSe edge luminescence (band to band transition) is observed at 500nm and 705nm at room temperature respectively. For both crystals, the luminescence intensities exhibit a quadratic dependence on the excitation intensities. The decay times of the edge luminescence for both crystals are very sensitive to the crystal surface condition. CdSe and CdS luminescence decay times, prior to any surface treatment are shorter than the laser excitation pulse width, i.e. <15 ps. The emission quantum yields are very low, estimated to be $<10^{-5}$. After mechanical polishing by 0.3μ alumina, the luminescence quantum yields are increased by a factor of 10, and the emission decays are longer, ≈ 50ps. Further polishing by finer alumina (0.05μ) causes some additional increase in the luminescence quantum yield and the decay times become somewhat longer. The quantum yields and the decay times of the edge luminescence are further increased by chemical etching: The emission decay for both crystals is then biphasic and the experimental decay curves are analyzed by computer fitting to biexponential decay. Following chemical etching, the short component of CdSe decay corresponds to τ_S = 400±100 ps, and the long one to τ_ℓ = 8±2 ns, while the quantum yield is estimated to be 10^{-2}. For etched CdS the relevant values are: τ_S = 100±20 ps, τ_ℓ = 700±100 ps, QY≈10^{-3}. The emission curves are shown in Fig.1. The optical spectra of the edge emission is very similar for polished crystals and chemical etched crystals.

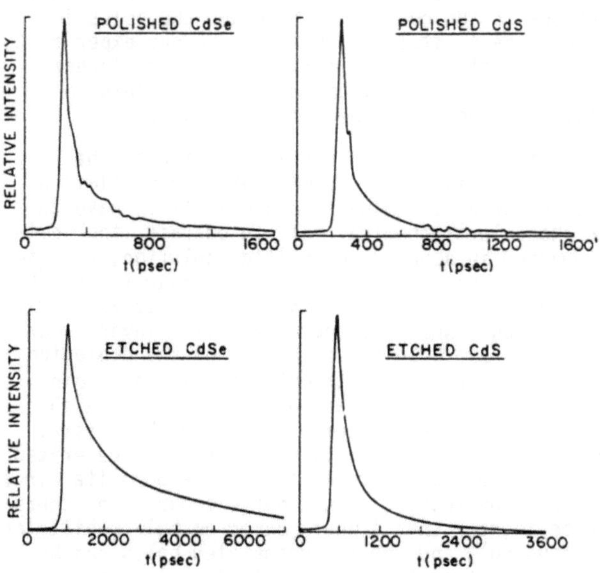

Fig.1 Streak camera records of CdS and CdSe time resolved luminescence. CdSe was excited by a 532nm (second harmonic) 15 ps pulse, and CdS by 353nm (third harmonic). Appropriate filters were used to eliminate penetration of extraneous light into the streak camera. Upper curves: CdSe and CdS luminescence after mechanical polishing with 0.05μ alumina. Lower curves: Chemically etched crystals.

Low Temperature Studies

At 77°K, both CdS and CdSe crystals exhibit an additional emission band at energies lower than the edge emission. The maximum of the sub-band gap emission at 77°K is located at 590 nm (~2.1 eV) and 850 nm (~1.45 eV) for CdS and CdSe respectively. These emissions are structureless and broad. The full width half maximum is ~0.3 eV for both crystals compared to 0.04 eV for the edge emission at 77°K. The decay rate of the sub band-gap emission of a chemically etched CdS crystal at room temperature is 16ns, 10 times longer than the edge emission decay rate (~1 ns). The sub band-gap emission decay rate is very sensitive to temperature. It slows down with the lowering of the temperature and reaches a value of τ=330 nsec at ~170°K. The sub band-gap emission at 77°K from a mechanically polished CdS crystal exhibits a similar spectrum to the one obtained from the etched crystal. However, the sub band-gap emission intensity from the mechanically polished crystal is estimated to be lower by more than 2 orders of magnitude. The edge emission band itself at 77°K is blue shifted with respect to its room temperature location and its band-width is narrower. The temperature dependence of the edge emission decay rates for CdS and CdSe is small. The emission quantum yield exhibits small sensitivity to temperature.

Discussion

The edge luminescence intensity was found to be quadratically dependent on the exciting laser intensity. This result is expected at high excitation intensities since the emission intensity is then proportional to the product of the densities of the photogenerated carriers: $I_{lum}(t)=\gamma_r n(t)p(t)$ where γ_r is the radiative recombination coefficient and n(t) and p(t) are the nonequilibrium densities of the electrons in the conduction band and the holes in valence band respectively. Under the excitation conditions in our experiment, the generated densities immediately after the short laser pulse are larger than the equilibrium densities for both minority and majority carriers, and hence the luminescence quadratic dependence on the laser intensity.

The edge luminescence decay rate and quantum yield are related to the photoelectrochemical cell conversion efficiency: To elucidate this relationship, the main component processes at an illuminated photoelectrode have to be examined: In order for the photoelectrochemical cell to operate, the semiconductor crystal has to be immersed in aqueous polysulfide solution, resulting in band bending of about 700mV. This potential variation near the crystal surface is forcing the majority carriers (electrons in n-type crystals), away from the semiconductor surface region, thus creating a depletion layer. For direct band-gap crystals such as CdS and CdSe the thickness of the depletion layer (calculated to be several thousands of Angstroms) approximately coincides with the crystal optical excitation depth, as calculated from the absorption coefficient (~$10^5 cm^{-1}$). Immediately after irradiating the crystal by an ultrashort laser pulse, the generated minority carriers are accelerated towards the surface, while the electrons are accelerated in the opposite direction towards the edge of the depletion region. Thus charge separation occurs on a time scale calculated to be shorter than 1 nsec (based on bulk mobilities and the field known from band bending), and the photogenerated holes can be utilized in an electrochemical charge transfer reaction taking place between the photoelectrode and sulfide ions. The photogenerated electrons are transferred via the external circuit to the counterelectrode, and react there with the polysulfide ions.

As found in our previous study [5], the amount of charge that is transferred from the excited crystal in a pulse experiment is about 5 times smaller in the case of mechanically polished crystals as compared with chemically etched crystals. The luminescence decay rate and its quantum yield respond qualitatively in the same way, i.e., the luminescence decay time for the etched crystals is 10 times longer than in mechanically polished crystals. The quantum yields show the same effect, and increase by more than an order of magnitude following chemical etching. In the mechanical polishing process, the crystal layer near the surface is damaged to some extent. The thickness of this layer is estimated to be of the order of the polishing material particle size, i.e., for alumina 0.05 it is ca. 1000 A, which is of the order of the optical excitation depth as well as the thickness of the depletion layer. In this damaged layer ultra fast recombination processes are taking place. Therefore, the luminescence decay rate, the quantum yield and the conversion efficiency are more than an order of magnitude lower than in chemically etched crystals, in which the nonradiative recombination rate is much slower.

Another important conclusion can be obtained from the intensity and decay rate of the sub-band emission: The number of photo carriers which recombine via these levels is much higher than by the direct band to band recombination mechanism. At room temperature it is expected that non radiative recombination through these levels might limit the efficiency of these cells. Further clarification of the nature of the midgap states, which cause this fast form of radiative and nonradiative decay, is required to show how they are related to material properties, and whether they may act also as mediating states in the heterogeneous charge transfer process at the semiconductor-electrolyte interface.

References

1. "Semiconductor Liquid-Junction Solar Cells," A. Heller, Editor, The Electrochemical Society Soft-bound Proceedings Series, Princeton, N.J. (1977).
2. D.D. MacDonald, "Transient Techniques in Electrochemistry," Plenum Press, New York (1977).
3. S.M. Ryvkin, "Photoelectric Effects in Semiconductors," Consultants Bureau, New York (1969) and Ref. therein.
4. S.P. Perone, J.H. Richardson, S.B. Deutscher, J. Rosenthal and J.N. Ziemer, J. Electrochem. Soc., 127, 2580 (1980).
5. Z. Harzion, N. Croitoru and S. Gottesfeld, J. Electrochem. Soc., 128, 551 (1981).
6. R.J. Nelson, J.S. Williams, H.J. Leamy, B. Miller, H.C. Casey, Jr., B.A. Parkinson, and A. Heller, Appl. Phys. Lett., 36, 76 (1980).
7. D. Huppert and E. Kolodney, Chem. Phys., 63, 401 (1981).

Time Resolved Spatial Expansion of the Electron-Hole Plasma in Polar Semiconductors

A. Cornet, T. Amand, M. Pugnet, M. Brousseau

Laboratoire de Physique des Solides, associé au C.N.R.S.
I.N.S.A., Avenue de Rangueil, 31077 Toulouse-Cedex, France

Introduction

We have studied the spatial expansion of hot electron-hole plasma generated in polar semiconductors (CdSe, CdS) by two photons absorption at λ = 1.06µm or λ = 0.532 µm, respectively, using a mode locked YAG laser pulse of duration 30 ps.

Two different experimental techniques, including space and time resolution, are used :

. A one dimensional expansion geometry (previously used in [1]) ;

. Spot size analysis. In that case the kinetics are compared for different spot size of excitation but with a constant observed region [2].

These two methods give convergent results which are consistent with an expansion of the electron hole plasma, at velocities in the range (5.10^7 cm/s - 10^8 cm/s) depending on the plasma and lattice temperatures.

Experimental set-up (Fig.1)

The experimental set-up has been described elsewhere [3, 4, 5] ; its characteristics are :

- Two photon excitation of the sample in order to obtain a homogeneous excitation. (For 30 µm thick samples, the inhomogeneities in the excitation are less than 10 % between the front and back faces) ;
- A resolution of 30 picoseconds in kinetics obtained by using a Kerr cell as an optical gate.

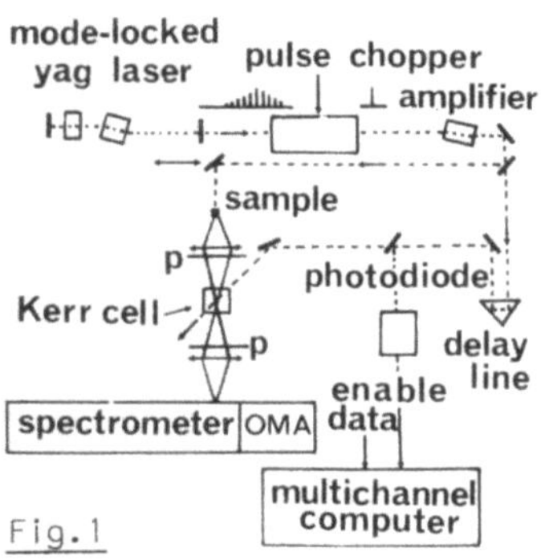

Fig.1

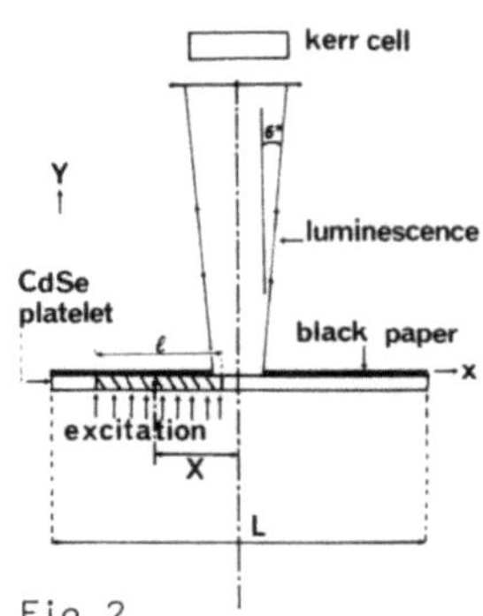

Fig.2

The sample faces were roughened to avoid the light stimulation due to the cavity effects, and were covered with black paper masks to reject the stimultated luminescence of the sample edges.

In the first experimental set-up, the excited region is a parallelipipedic volume (1 x 0.3 x 0.03 mm^3) (see Fig.2). The luminescence emerges through a slit (width 100 μm) made of black paper which covers all the back surface. The thinness of the sample minimize the stimulation in the detector direction (Y).

The principle of the experiment is to get a time and wavelength resolution of the luminescence which springs from the back side slit for various positions X of the excited region (Fig.2) so that we obtain the space and time resolved luminescence spectrum of the sample.

The experiments are performed, on CdS and CdSe, at $T_L \sim$ (10 - 20 K), $T_L \sim$ (80 - 90 K) and $T_L \sim$ 300 K.

In the second experimental set-up, the laser spot diameter on the sample was varied between 80 μm and 1 mm, and was defined by forming on the sample, the image of a slit with well defined dimensions. The luminescence is detected in standard backward configuration. The sample thickness was 30 μm.

Experimental results and discussion

A. One dimensional expansion

The Fig.3 shows, for CdSe in Fig.2 configuration, two time resolved luminescence spectra emerging from a non excited region, and previously identified as due to recombination of electron-hole pairs in a plasma assisted by a LO-phonon emission (E.H.P.-LO) [3, 5, 6].

We have shown that this luminescence band is an evidence for the presence (at the point X) of high density electron-hole plasma. The time resolved spatial distribution of the detected luminescence is strongly dependent on

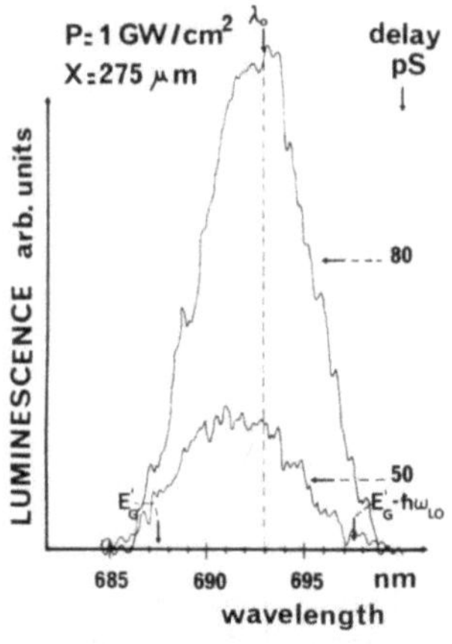

Fig.3 Time resolved LO-phonon assisted plasma luminescence originating from a non-excited region. (X = 275 μm). $T_L \sim$ 10 K.

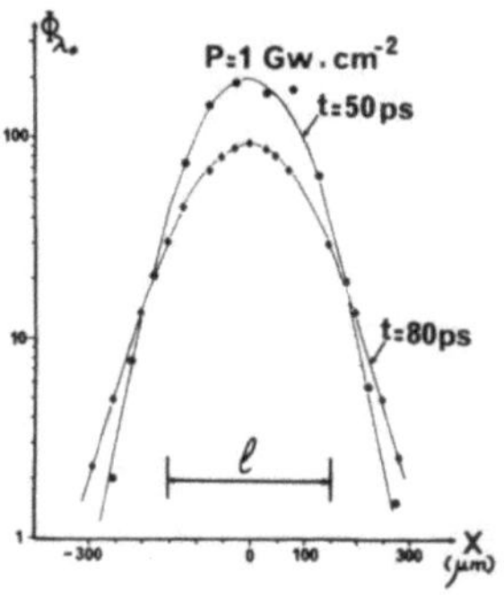

Fig.4 Two post-excitation distributions of time resolved luminescence. Theoretical fit with a classical diffusion model [1].

the excitation power : under a threshold P_e, the light springs only from the initially excited region, while above this critical value P_e, the light emerges from the whole sample.

Fig.4 shows the luminescence at a given wavelength in the plasma band versus X for an incident-time averaged power P = 1 GW/cm^2 at two time delays t_1 = 50 ps ant t_2 = 80 ps. The time origin is defined as the luminescence threshold. When P is greater than a critical value P_e (which is dependent on the relative direction x with respect to the C-axis) the light emerges from regions outside the excited zone. We find, for CdSe, $P_e (x \parallel c) \sim 0.9$ GW/cm^2 and $P_e (x \perp c) \sim 0.8$ GW/cm^2.

All these results are similar when the experiment is performed at $T_L \sim 10$ K and $T_L \sim 80$ K ; on the contrary, we cannot detect any luminescence outside the excited region when $T_L \sim 300$ K.

The fit of the luminescence data reported in Fig.4, using a classical diffusion model, would give a diffusivity value D $\sim 10^6$ cm^2/s which is more than four order of magnitude greater than the low excitation one : the electron-hole plasma blows up with an expansion velocity of about 10^8 cm/s. This very fast expansion is achieved few tens picoseconds after the end of the exciting pulse.

B. Influence of the excitation spot size on luminescence kinetics

We have excited CdS, using standard backward luminescence configuration at fixed power (P $\sim$ 250 MW/cm^2) with different spot diameter (d). We display the luminescence spectra obtained with d = 80 µm (Fig.5a) and d = 1 mm (Fig.5b).

Both spectra present a similar kinetic behaviour, i.e. the disappearance, with time, of the E.H.P.-LO band on the low energy side of the A-LO band (attributed to LO-phonon assisted recombination of free exciton) which do-

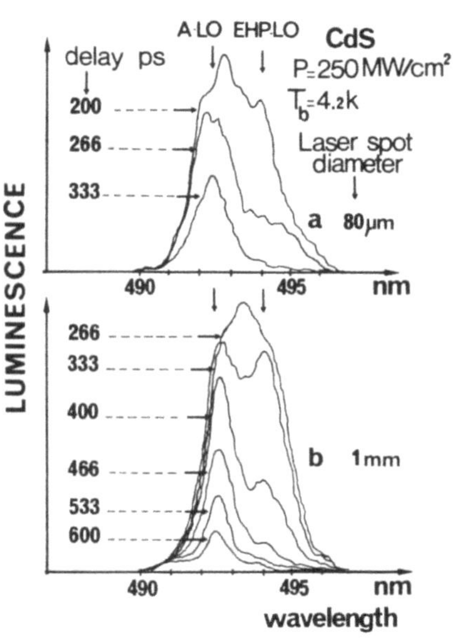

Fig.5 Time resolved luminescence spectra of CdS under intense two photon excitation (P = 250 MW/cm^{-2}, λ_{exc} = 530 nm, pulse duration : 30ps) for different laser spot diameter d : (a) d = 80 µm ; (b) d = 1 mm.

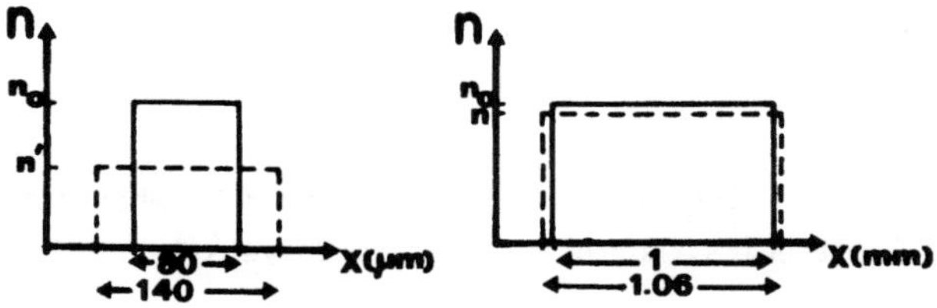

Fig.6 Lowering of (e-h) density by explosion effects. Profile: ___ initial ; - - - 30 ps later.

minates the spectrum for longer time delays [6, 7]. The acceleration of the recombination process observed in Fig.5a as compared to Fig.5b can be explained in terms of plasma explosion (Fig.6) : starting with the same excitation power per surface unit and, consequently, with the same electron-hole pair density n_o in the plasma, the initial velocity is identical in 5a and 5b and so the lowering of e - h pairs density is more efficient with a smaller excited region (Fig.6).

Then the Mott density is reached earlier and the A-LO band appear for shorter time delays in Fig.5a than in Fig.5b.

References

1. A. Cornet, M. Pugnet, J. Collet, T. Amand, M. Brousseau
3rd Int. Conf. on hot carriers, Montpellier (France), July 1981 ; Journal de Physique, C7, Suppl. 10, 42, 471 (1981).
2. A. Cornet, T. Amand, M. Pugnet, M. Brousseau
Sol. Stat. Comm. (In press).
3. M. Pugnet, A. Cornet, J. Collet, M. Brousseau, B.S. Razbirin, G.V. Michailov
Sol. Stat. Comm., 36, 85 (1980).
4. M. Pugnet
Thèse, Univ. Paul Sabatier, Toulouse (1981).
5. J. Collet, M. Pugnet, A. Cornet, M. Brousseau, B.S. Razbirin, G.V. Michailov
Phys. Stat. Solidi, 103, 367 (1981).
6. A. Cornet, J. Collet, T. Amand, M. Pugnet, M. Brousseau, B.S. Razbirin
Int. Conf. on Luminescence, Berlin (1981) ; J. of Luminescence, 24, 609 (1981).
7. A. Cornet, J. Collet, T. Amand, M. Pugnet, B.S. Razbirin, G.V. Michailov
J. of Physics and Chem. of Solids (In press).

Weak-Wave Retardation and Phase-Conjugate Self-Defousing in Si

E.W. Van Stryland, A.L. Smirl, T.F. Boggess, M.J. Soileau and B.S. Wherrett*

Center for Applied Quantum Electronics, North Texas State University, Denton, TX 76203, USA

F.A. Hopf

Optical Sciences Center, University of Arizona, Tucson, AZ 85721, USA

We describe and measure the effects of self-defocusing on the various coupling effects produced when two coherent, noncollinear, picosecond optical pulses (strong excitation and weak probe) are both spatially and temporally coincident in a thin silicon wafer. Specifically, we observe that the weak probe beam experiences considerably more defocusing than the excitation beam. We believe that this is the first direct confirmation of weak-wave retardation in light-by-light scattering experiments. We also observe the effects of this defocusing on the quality of the forward-traveling conjugate wave.

In our experiments, two 65-psec (FWHM) pulses at 1.06 μm (excitation and probe), separated by an angle θ = 1.2°, were focused to a 600 μm (FWHM) diameter spot on the surface of a thin (~270 μm-thick) Si wafer. The excitation and probe could be delayed by an amount τ with respect to one another by means of an optical delay line. When both excitation and probe are spatially and temporally coincident (τ = 0), the interference between these two parallel-polarized pulses modulates the intensity across the face of the sample. The indirect absorption of the two pulses produces a spatially-modulated optically-created carrier density that results in a spatial modulation of the refractive index.

The coherent interaction between the excitation pulse and the probe can be viewed as the self-diffraction of the excitation pulse from an optically-produced grating. That is, the excitation (E_e) and probe (E_p) pulses interfere to modify spatially the optical properties of the sample, as described above. The excitation pulse (E_e) is then self-diffracted by the grating produced by E_e and E_p to produce two first-order scattered beams. One first-order diffracted excitation beam is collinear with the transmitted probe beam; the other first-order beam E_c travels in the background-free direction $-\theta$. An alternate point of view is to consider the coherent interaction between the two pulses as a transient, degenerate, four-wave mixing process. In this case, the second self-diffracted beam E_c, discussed above, is easily recognized as the forward-traveling phase-conjugate of the probe beam. The self-diffracted excitation pulse that travels in the direction of the probe is responsible for the so-called coherent coupling "artifacts" (e.g., correlation spikes) that are observed in traditional picosecond excitation-and-probe experiments. These interactions have also been called real-time holography (e.g., Ref. 1) and light-by-light scattering (e.g., Ref. 2). Assuming that the sample is optically-thin and that E_c and $E_p \ll E_e$, the general form for the coupled equations for the excitation, probe, and conjugate polarizations in the transient regime are:

---

*Permanent Address: Heriot-Watt University, Edinburgh, Scotland.

$$P_e(z,t) = -iKE_e(z,t)\int_{-\infty}^{t} |E_e(z,t')|^2 \exp[-(t-t')/\tau_\ell]dt' \tag{1}$$

$$P_p(z,t) = -iKE_p(z,t)\int_{-\infty}^{t} |E_e(z,t')|^2 \exp[-(t-t')/\tau_\ell]dt' - iKE_e(z,t) \times \int_{-\infty}^{t} [E_e^*(z,t')\,E_p(z,t') + E_c^*(z,t')\,E_e(z,t')] \exp[-(t-t')/\tau_G]dt' \tag{2}$$

$$P_c(z,t) = -iKE_c(z,t)\int_{-\infty}^{t} |E_e(z,t')|^2 \exp[-(t-t')/\tau_\ell]\,dt' - iKE_e(z,t) \times \int_{-\infty}^{t} [E_p^*(z,t')E_e(z,t') + E_c(z,t')E_e^*(z,t')] \exp[-(t-t')/\tau_G]dt', \tag{3}$$

where K is a constant, τ_ℓ is the free-carrier lifetime and τ_G is the grating lifetime, $\tau_G^{-1} = \tau_D^{-1} + \tau_\ell^{-1}$. The grating in the nonlinear refractive index that is introduced by interference between probe and excitation (or conjugate and excitation) decays by recombination τ_ℓ^{-1} or diffusion τ_D^{-1}. For our geometry, the optical pulsewidth (65 psec) is much less than the grating lifetime (~47 nsec).

The various terms in (1) and (2) correspond to changes in the phase of the excitation (strong) and probe (weak) pulses, respectively. Notice that, for picosecond pulses, there is an additional phase delay for the probe wave. This additional increase in refractive index was named weak-wave retardation by Chiao and coworkers [2], who first predicted this effect. These workers later observed light-by-light scattering, but they did not verify weak-wave retardation [3].

If beams with Gaussian spatial profiles are used in these self-diffraction experiments, the changes in phase velocities predicted by (1) - (3) should result in differing degrees of self-defocusing for the various transmitted pulses. We measure the degree of self-defocusing by observing the transmitted beam profiles with a vidicon detector. Before summarizing our results, we remark that the self-defocusing of the transmitted probe, excitation and conjugate in Si have been studied recently by Hopf et al. [4] using various nonlinear interferometers. They observed a substantial self-defocusing of all beams, but they were unable to detect weak wave retardation. For their work, the pulsewidth was comparable to the diffusion-dominated grating lifetime. If this were the case, then the second terms in (2) and (3) would be small with respect to the first.

Figure 1 illustrates the distortion of the excitation and probe beam profiles during these self-diffraction studies. The fluence of the excitation pulse was 46 mJ/cm^2, and the fluence of the probe was a factor of 500 smaller. Figure 1a shows scans of the probe profile (in the far field) when the excitation was blocked - the profile is reasonably Gaussian. Figures 1b and 1c show profiles of the transmitted probe and excitation, respectively, when both were simultaneously present. The broadening of the excitation pulse caused by the optically-created free carriers in the Si is evident, and the additional self-defocusing of the probe (weak-wave retardation) is clear. We believe this to be the first direct observation of this effect.

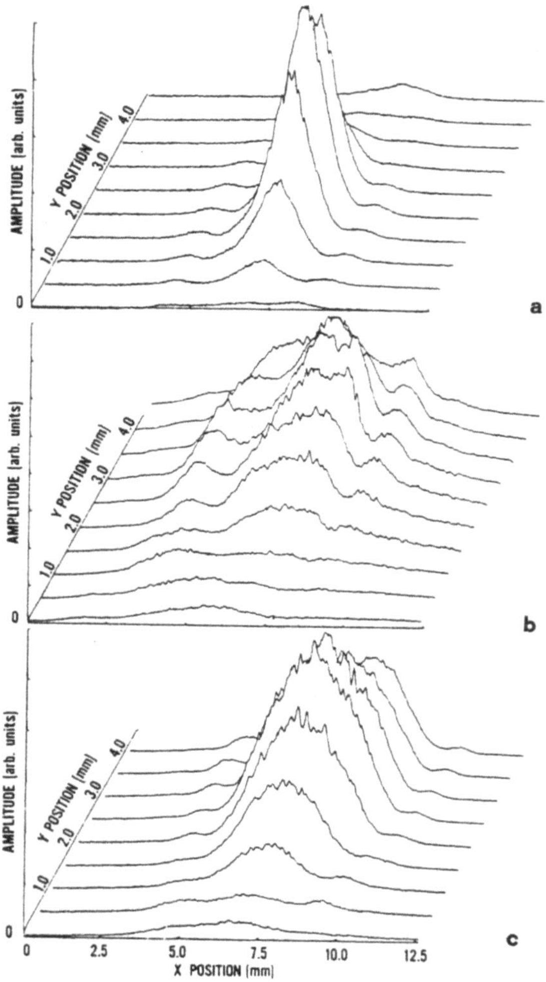

Fig.1 Vidicon scans of the spatial beam profiles of the (a) probe with pump blocked, (b) probe with pump, and (c) pump

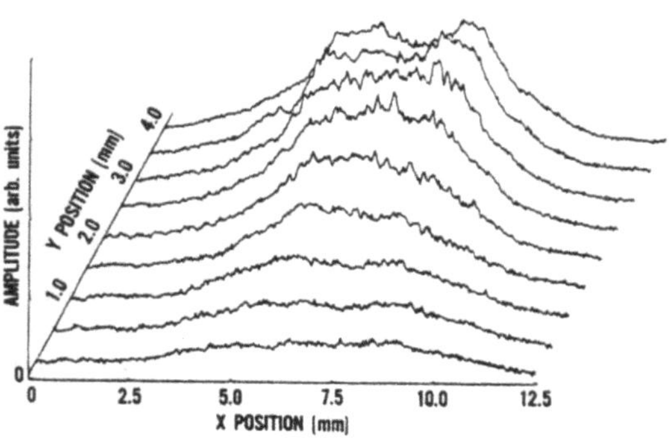

Fig.2 Vidicon scan of the spatial beam profile of the conjugate beam

In addition, we have measured the transmitted beam profile of the forward-traveling conjugate wave under experimental conditions identical to those of Figs.1b and 1c. The observed distortion of the conjugate (Fig.2) is different from the defocusing of either the probe or the excitation, contrary to the disparate conclusions of Refs. 2 and 4.

This work was supported by the Office of Naval Research, the National Science Foundation, and The Robert A. Welch Foundation.

References

1. J. P. Woerdman, Philips Res. Rep. Suppl. 7 (1971).
2. R. Y. Chiao, P. L. Kelley, and E. Garmire, Phys. Rev. Lett. 17, 1158 (1966).
3. R. L. Carmen, R. Y. Chiao, and P. L. Kelley, Phys. Rev. Lett. 17, 1281 (1966).
4. F. A. Hopf, A. Tomita, and T. Liepmann, Opt. Commun. 37, 72 (1981).

Ultrafast Relaxations of Photoinduced Carriers in Amorphous Semiconductors

Z. Vardeny, J. Strait and J. Tauc*

Division of Engineering and Department of Physics, Brown University, Providence, RI 02912, USA

Introduction

A passively modelocked dye laser producing subpicosecond pulses at 2 eV was used to study ultrafast dynamics of photogenerated carriers, by the pump and probe time resolved photoinduced absorption ($\Delta\alpha$), in doped and undoped amorphous hydrogenated silicon and chalcogenide glasses. We found that when the exciting photon energy $\hbar\omega_p$ is larger than the band gap E_g, we could follow thermalization of hot carriers ($t < 2$ ps) and the consecutive trapping and recombination processes (we follow it up to 200 ps). When $\hbar\omega_p < E_g$, the photogenerated electrons and holes are bound together forming pairs whose geminate recombination by tunneling was observed.

Experimental

The dye laser and experimental set-up have been described elsewhere 1,2 . The laser produces linearly polarized transform-limited light pulses at $\hbar\omega_p = 2$ eV with $t_p = 0.6 - 0.8$ ps duration, 1 - 2 nJ energy and repetition rate $10^4 - 10^6$ s^{-1}.
The pump and probe beams were focused non-colinearly on the sample. When the probe polarization was perpendicular ($\perp$) to the pump polarization, we found that $\Delta\alpha_\perp$ was free from the coherence artifact.[2] Measuring both $\Delta\alpha_{\parallel}$ and $\Delta\alpha_\perp$ enable us to get the exact shape of the coherent artifact and the delay zero. This is shown in Fig. 1 for a-Si:H where curve c is $\Delta\alpha_{\parallel} - \Delta\alpha_\perp$.

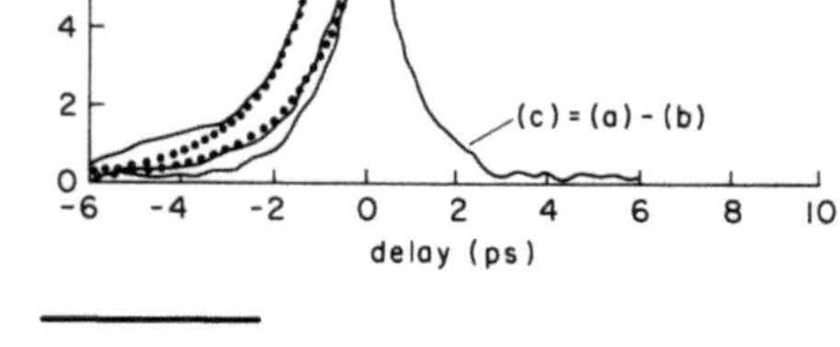

Fig. 1 - Photoinduced absorption decay $\Delta\alpha(t)$ in a-Si:H ($C_H = 10\%$) for and $\perp$ polarizations. Solid curves: experimental, dotted curves: calculated.

*On leave at Max Planck Institute Fur Festkorperforschung, 7000 Stuttgart 80, Heisenbergstrasse 1, W. GERMANY.

Transient induced absorption model

All samples showed an instantaneously induced absorption ($\Delta\alpha > 0$) which subsequently decays in time. $\Delta\alpha$ is positive because the excited carriers have higher optical absorption cross section σ_1 than the cross-section σ_o for band to band transitions. The decay of $\Delta\alpha(t)$ can be explained in two ways (i) σ_1 may decrease with time; this is the case for hot carrier thermalization and for trapping in deep defect-states (ii) carriers disappear from the excited state by recombination.

Hot carrier thermalization

For $\hbar\omega_p > E_g$, hot carriers with excess energy $\overline{\Delta E}(o) = (\hbar\omega_p - E_g)/2$ are excited across the band gap. These carriers thermalize to the bottom of the band losing their excess energy due to electron phonon interaction. Since σ of hot carriers increases with ΔE^3, it is possible to observe the fast thermalization process optically (Fig.1) and determine the excess energy dissipation rate $R = d\overline{\Delta E}/dt$. In a-Si:H $R \simeq 0.1$ eV/ps and in a-As_2Se_3 $R \simeq 0.25$ eV/ps. These dissipation rates may be explained by Frölich coupling to polar phonons.[4] In non-hydrogenated a-Si R is larger ($\simeq 0.5$ eV/ps); apparently the increase in disorder compared to a-Si:H, opens more channels for electron-phonon interaction.

Geminate recombination by tunneling

If the electron hole distance after thermalization r_o, is smaller than the Coulomb capture radius r_c (the "Onsager radius"), most of the electrons and holes form bound pairs that recombine geminately. For r_o smaller than the recombination radius r_p ($\simeq 10$ Å), carriers do not have to diffuse towards each other first to reach recombination conditions. In this case geminate recombination by tunneling occurs. This is observed if the excitation takes place in the Urbach-tail where $\hbar\omega_p < E_g$. This is the case for a-Se and a-$AsS_{2.4}Se_{0.6}$ as shown in Fig. 2. The observed decays are exponentials $\exp(-t/t_r)$ where t_r was found to depend exponentially on r_o, in accordance to a tunneling process.[5] In this model, the temperature dependence of t_r (shorter at low T) is produced by the temperature dependence of the gap E_g (which is larger at low T causing a smaller r_o).

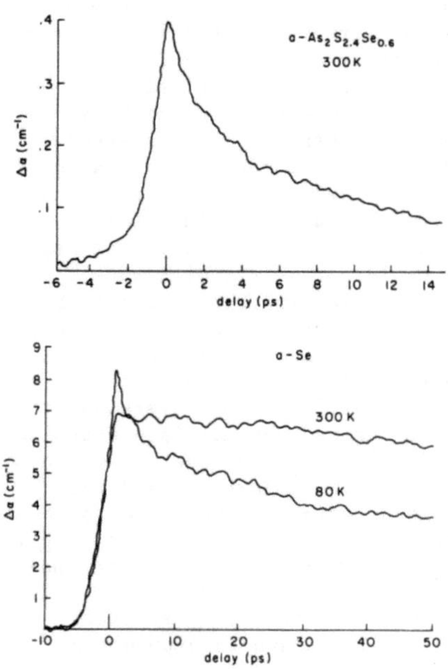

Fig. 2 - $\Delta\alpha_{\parallel}(t)$ in a-$As_2S_{2.4}Se_{0.6}$ and a-Se

Geminate recombination involving transport

For $r_p < r_o < r_c$, a large fraction of the carriers may still recombine geminately,[6] but the

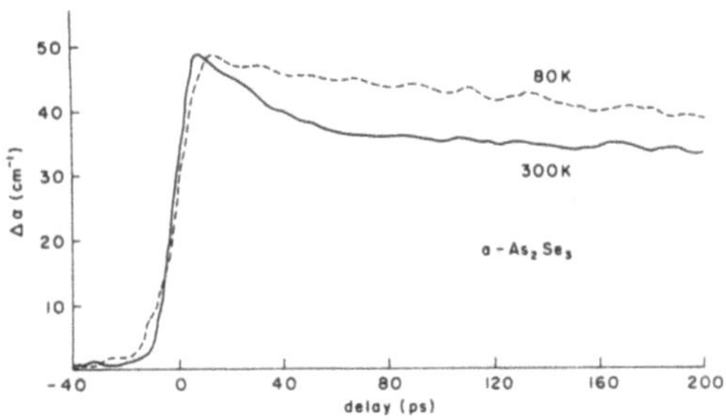

Fig. 3 - $\Delta\alpha_{\perp}(t)$ in a-As_2Se_3

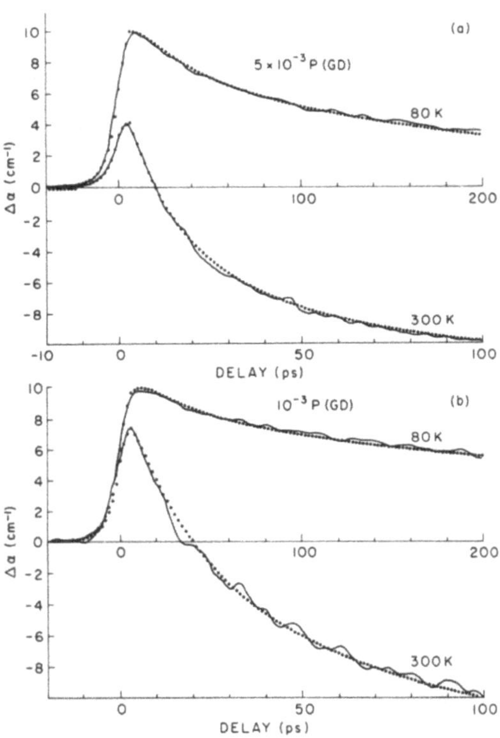

Fig. 4 - $\Delta\alpha_{\perp}(t)$ in 5 x 10^{-3} and 10^{-3} P-doped GD in a-Si:H. Solid lines: experimental; dotted: calculated

carriers have to drift towards each other to reach the distance r_p. This appears to be the case for a-As_2Se_3 (Fig. 3), in which $\hbar\omega_p > E_g$. The decay is slower at low T because the transport is slower at low temperatures.

Trapping by deep defect states

In contrast to undoped a-Si:H where $\Delta\alpha > 0$ was observed, the introduction of a sufficient level of phosphorous boron or both dopants accelerates the decay and the response eventually changes its sign, becoming induced transmission[7] ($\Delta\alpha < 0$). This is shown in Fig. 4 for glow discharge a-Si:H film with nominal P-doping level of 5 x 10^{-3} in the gas. At t = 0 $\Delta\alpha > 0$ is again induced instantaneously but at 300K it decays quickly and passes through zero becoming induced transmission. The decay is slower at lower doping levels and when the temperature is decreased. We associate the induced transmission with carriers trapped in the deep defects produced by P-doping. In these states the optical absorption cross-section σ_2 is smaller than σ_o. The observed decay corresponds to the transport of the thermalized carriers n(t) from the band edge into these states. The decay curves in Fig. 4 could not be fit with an exponential function for n(t) but with expressions following from the dispersive transport theory where the mobility is time dependent.[8] A simple version of this theory, the multiple trapping model,[9] gives the observed dependencies on temperature and trap concentrations. The motion of a photocarrier towards a deep trap is dominated by trapping in the shallow traps close to the band edge, whose distribution is assumed to be exponential. From our observation that the dispersive transport starts below 5ps it follows that the distribution of shallow traps is exponential very close to the band ($\leq$ 0.01 eV).

This work was supported in part by NSF grant DMR79-09819 and the NSF Materials Research Laboratory at Brown University. One of us (J.T.) would like to thank the Max Planck Institute for kind hospitality during his stay in Germany.

References

1. E.P. Ippen and C.V. Shank, Ultrashort Light Pulses, edited by S.L. Shapiro (Springer, New York 1977) p. 83.

2. Z. Vardeny and J. Tauc, Opt. Commun. 39. 396 (1981).

3. Z. Vardeny and J. Tauc, Phys. Rev. Lett. 46, 1223 (1981).

4. E.M. Conwell, Solid State Physics, Suppl. 9, Academic Press, New York (1967).

5. D. Ackley, J. Tauc and W. Paul, Phys. Rev. Lett. 43, 715 (1979).

6. J. Mort, I. Chen, M. Morgan and S. Gramatica, Sol. State Commun. 39, 1329 (1981).

7. Z. Vardeny, J. Strait, D. Pfost, J. Tauc and B. Abeles, Phys. Rev. Lett. 48, 1132 (1982).

8. H. Scher and E.W. Montroll, Phys. Rev. B12, 2445 (1975).

9. T. Tiedge and A. Rose, Sol. State Commun. 37, 49 (1980); J. Orenstein and M. Kastner, Phys. Rev. Lett. 46, 1421 (1981).

Periodic Ripple Structures on Semiconductors Under Picosecond Pulse Illumination

P.M. Fauchet, Zhou Guosheng*, and A.E. Siegman

Edward L. Ginzton Laboratory, Stanford University,
Stanford, CA 94305, USA

1. Introduction

We can routinely generate spontaneous periodic surface structures or ripples on both semiconductors and metals, using either single or multiple TEM_{00} Nd:YAG pulses with durations of $\leq$ 100 ps at 1.06 μm or 80 ps at 532 nm [1]. We believe these ripples evolve through stimulated growth of surface corrugations on molten surfaces, and offer theoretical evidence to this end. Results of two-pulse experiments which test another recently suggested model are also presented.

2. Experimental Facts

When we illuminate for example ion-implanted silicon surfaces with single picosecond laser pulses, we commonly observe the production of surface ripples occurring in an intensity range between the thresholds for amorphous to polycrystalline and amorphous to crystalline transitions (Fig. 1). We obtain similar periodic surface structures with picosecond pulses at λ = 1.06 μm and 532 nm on Si, GaAs, and Cu samples in various forms including amorphous layers, single crystals, and thin polycrystalline films. The

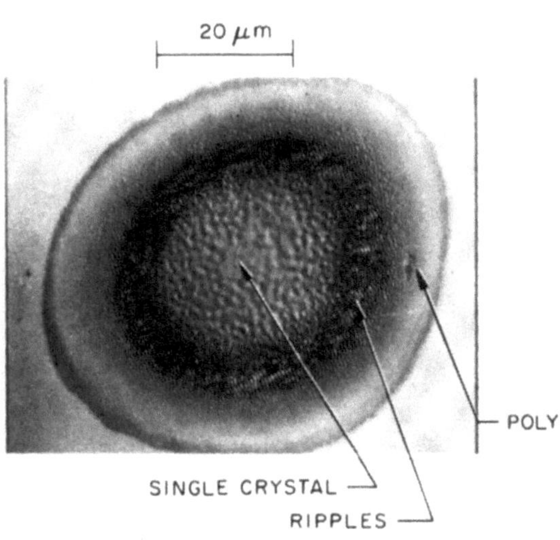

Fig.1 Annealing of As-implanted Si by a single 80 ps pulse at 532 nm. The peak intensity is 4 GW/cm^2

*Zhou Guosheng is currently a visiting scholar at Stanford University from Shanxi University, People's Republic of China.

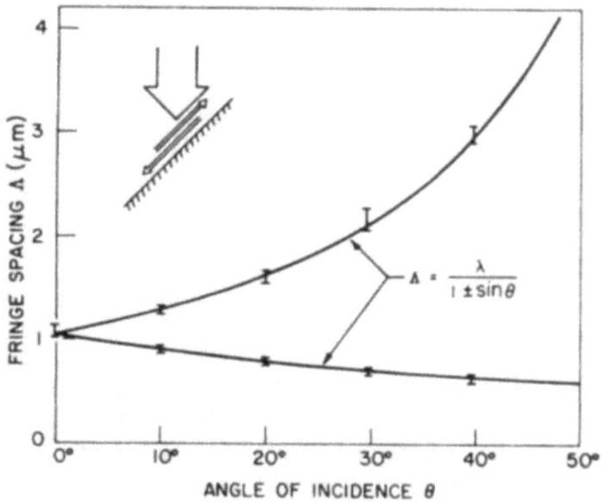

Fig.2 Dependence of the ripple spacing on the angle of incidence for p-polarized, λ = 1.06 μm pulses on Si and GaAs

peaks of these ripples always run perpendicular to the incident optical electric field vector and are not observed with circularly polarized light. For TM waves arriving at angle θ to the normal, the ripples occur randomly in two branches with spatial periods given by $\Lambda \approx \lambda/(1 \pm \sin\theta)$ (Fig. 2), while for TE polarization $\Lambda \approx \lambda$. Large and small spacing ripples may coexist in the same region, with varying relative strength. Covering the surface with a liquid of index n causes Λ to scale as $1/n$.

Illumination with hundreds of shots at intensities below the threshold for amorphous to polycrystal transition in ion-implanted silicon can also create ripples even when single shot illumination produces no observable effect. Under single shot conditions, if the laser beam is moved between spots in the direction parallel to the ripples, the ripples in one spot generally have no connection with those in an adjoining spot. If the laser spot is moved perpendicular to the ripples such that two successive shots overlap, the ripples can extend coherently *ad libitum* (Fig. 3).

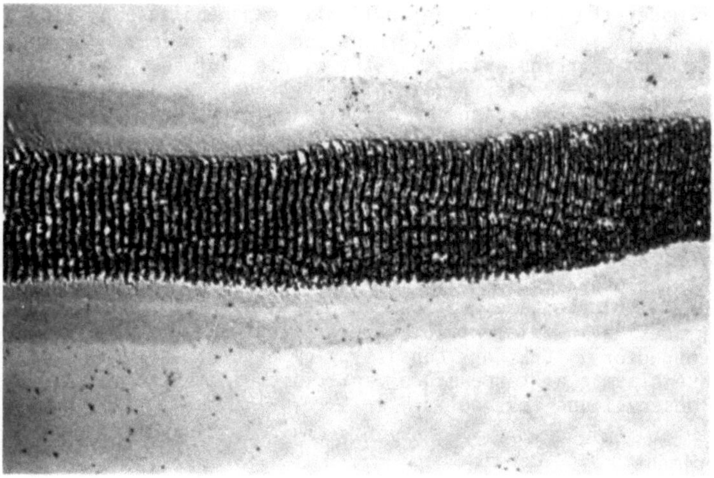

Fig.3 Coherent extension of ripples in raster scanning mode. The incident optical electric field is horizontal

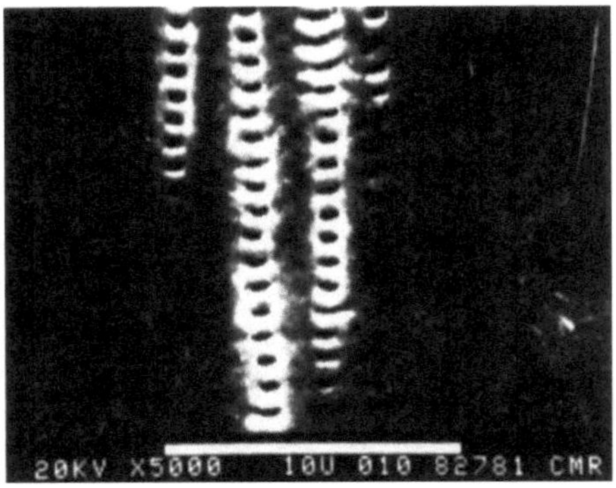

Fig.4 Ripples broken into narrow arrays of pits in GaAs. The incident optical electric field is vertical

The ripples are also sometimes organized into narrow strips whose width can be as small as a few microns. This effect can be quite spectacular in gallium arsenide: in Fig. 4 each strip is an array of pits spaced by Λ. Finally we note that the ripples seem most often to appear at pulse intensities close to the melting threshold for the material involved.

3. Theory

Similar surface ripples have been produced by others under widely different conditions, with illumination times varying from picoseconds to milliseconds, wavelengths from 10.6 µm to approximately 0.5 µm, and materials including dielectrics, semiconductors and metals. Qualitative models proposed to explain these ripples have included scattering from dust or scratches [2]; coherent emission from small metallic droplets [3]; effects of surface plasmons [4]; and surface polaritons [5]. We propose a new theory [6] which predicts most of our experimental findings and is well-suited to the case of absorbing surfaces.

According to our model the ripples develop from an exponentially growing interaction between the incident laser wave front and selected Fourier components of initially random irregularities of the sample surface. These Fourier components coherently scatter radiation travelling just along the surface, either upward (smaller spacing) or downward (large spacing), as shown by the inset of Fig.2. Interference between these scattered waves and the incident laser beam then produces optical intensity fringes with the same spacing, which may in turn cause the surface gratings to grow.

When we carry out electromagnetic theory calculations assuming periodic surface gratings of either temperature, electron density, or surface corrugation, we find that the observed polarization dependence, periodicity, and angular variation of the ripples are most compatible with a model that assumes corrugated surfaces having dielectric properties characteristic of metals or molten semiconductors. Gratings consisting of electron-hole density fluctuations in dense plasmas or temperature variations, in metals or semiconductors fit the observations less well.

4. Two-Pulse Experiment

We are also presently performing experiments in which two picosecond pulses with different wavelengths and linear polarizations and variable time delay overlap spatially on the surface of the sample. Van Vechten has suggested

[4] such an experiment as a test for the plasma model of pulsed laser annealing. In that model the high reflectivity phase that persists for nanoseconds after laser illumination is interpreted as the signature of a boson condensation undergone by the electron-hole plasma. A single plasmon mode would thus be present and cause the ripples as discussed above. If, during the high reflectivity phase, a second laser pulse of different polarization is applied, Van Vechten suggests that the fringe pattern would rotate by continuously variable amounts depending on pulse intensity and time delay.

We have performed experiments on Si and GaAs samples in which the polarization of the second pulse at 1.06 μm is perpendicular to that of the first pulse at 532 nm while the time delay varies from 0 to 4 ns. In all cases, except one experiment in silicon, the resulting ripple direction and spacing were determined solely by one or the other of the pulses, depending upon relative intensities and delay. In one experiment, which we were unable to duplicate, the resulting ripples occurred at 45° to each polarization with a spacing of 1 μm. We note that ripples at 45° have also been observed [7] following normal Q-switched ruby laser annealing of silicon.

Acknowledgement

This work was supported by a grant from the Air Force Office of Scientific Research.

References

1. P.M. Fauchet and A.E. Siegman, Appl. Phys. Lett. 40, 824 (1982).

2. D.C. Emmony, R.P. Howson, and L.J. Willis, Appl. Phys. Lett. 23, 598 (1973).

3. M.F. Becker, R.M. Walser, Y.K. Jhee, and D.Y. Sheng, Proc. SPIE 322 93 (1982).

4. J.A. Van Vechten, Solid State Commun. 39, 1285 (1981).

5. F. Keilmann and Y.H. Bai, Appl. Phys. A, to be published.

6. P.M. Fauchet, Zhou Guosheng and A.E. Siegman, to be published.

7. D. Haneman and R.J. Nemanich, Solid State Commun., to be published.

Transmission of Picosecond Laser-Excited Germanium at Various Wave-lengths

C.Y. Leung and T.W. Nee

Department of Physics, National Central University,
Chung-Li, Taiwan, R.O.C.

Interaction of germanium with mode-locked Nd:glass laser has been studied, both experimentally [1-3] and theoretically [4], in the past. In one experiment, an intense picosecond pulse creates a dense electron-hole plasma in a thin Ge slab and the interaction region is probed by measuring the transmission of a weak pulse delayed with respect to the excitation pulse [5]. Both excitation and probe have the same wavelength : 1.06μm, corresponding to a photon energy of 1.17eV which is larger than the bandgap energy of Ge. Measured probe pulse transmission is seen to rise, peak and then drop with increase in delay time. At a sample temperature of liquid nitrogen, peaking of probe transmission occurs at a delay time of about 100 psec.

The excite-and-probe experiment described above provides information on the ultrafast evolution of laser-generated electron-hole plasma in a semiconductor. Theoretically, it has been proposed that the rise in probe pulse transmission with delay time is due to a decrease in direct interband absorption when the optically-coupled states become clogged as the laser-generated hot carriers cool via emissions of phonons [4]. However, using known electron-phonon coupling coefficients, probe transmission rise time less than 20 psec instead of ∿100 psec observed is calculated with this theory [2]. An alternative theory which involves the decrease of electron-hole density by Auger recombination has also been suggested [6]. In order to explain the peaking of probe transmission, according to this theory, an absorption minimum must be encountered when carrier concentration is reduced. However, this minimum in absorption was not found experimentally [7].

Still another suggestion is that diffusion of laser-induced free carriers, which originally concentrate within a micron of the irradiated surface, toward bulk of the sample may play an important role in probe pulse transmission [8]. Preliminary calculations show that diffusion in the pulse propagation direction may lead to the rise, peak and fall characteristic of probe transmission. There is not enough experimental data to access the validity of this proposal. We have thus performed the following experiment where the sample is excited by an intense picosecond pulse at 1.054μm and the excited spot is probed by pulses at another wavelength.

Our experimental configuration is depicted in Fig.1. A neodymium-doped phosphate glass (Kigre Q-98) laser mode-locked by a saturable dye solution (Nippon Kankok Shikiso Kenkyusho NDL 112 dye in 1,2,-dichloroethane) produces 1.054μm pulses 11±2 psec in duration. It is forced to lase in the lowest transverse modes by an intracavity aperture. Typical amplified single pulse energy is 5-10mJ. An angular phase-matched temperature-stabilized KDP crystal is used to generate, at a conversion efficiency >30%,second harmonic wave at 0.527μm, which in turn pumps a 1.5 cm long temperature-tuned CDA crystal optical parametric amplifier (OPA). The CDA crystal is cut in

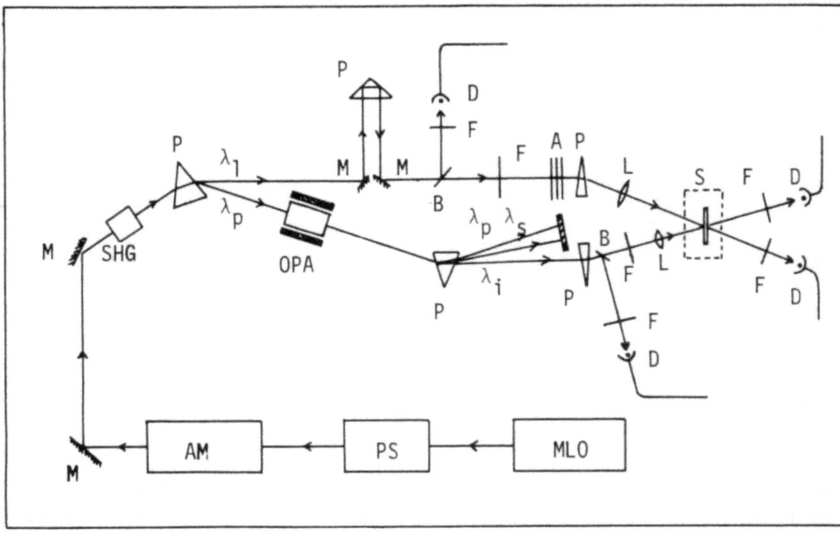

Fig.1 Experimental setup where MLO denotes the mode-locked oscillator, PS the pulse selector, AM the amplifier, SHG the second harmonic generator, OPA the optical parametric amplifier, M a mirror, P a prism, L a lens, D a detector, B a beam splitter, A an attenuator, F a narrow band filter and S the sample in a dewar. Laser, pump, signal and idler wavelengths are indicated by λ_l, λ_p, λ_s and λ_i, respectively.

the 45 degrees Z-cut orientation, with the pumping beam, polarized along the optic axis, propagates along the [110] direction. To avoid crystal damage, irradiance on crystal is kept below 500 MW/cm^2 by careful control of beam diameter. Output from the OPA is tunable over the range 0.85-1.4μm by varying crystal temperature between 40 and 70 ^{0}C [9]. Strongly dependent on pump power and output wavelengths, conversion efficiency of the OPA varies from 0.1% to a few percent in our experiment. The unconverted 1.054μm pulse, going through a variable delay path and a set of attenuators, is used to probe the interaction volume. To ensure good spatial overlapping of the two beams, the probe pulse is focused down to an area estimated to be at least 4 times smaller than the excitation spot. Before data taking, a pinhole at the sample position is used to check that the two beams overlap. Throughout our experiment, sample damage is checked from time to time by measuring linear transmission at low irradiance levels. A drastic change in linear transmission will be found whenever damage occurs. The Ge sample is an undoped high purity (min. ρ=40 Ω-cm) single crystal cut with the [111] plane as the face. It is fine polished and etched on both surfaces and is mounted on a glass substrate.

Probe pulse transmission vs delay at 77 ^{0}K sample temperature for four different probing wavelengths : 0.95μm, 1.054μm, 1.15μm and 1.25μm are depicted in Fig.2. The data are plotted as normalized ratio of probe pulse transmission to linear transmission of the sample at the wavelength under consideration, in arbitrary units. The arbitrary units are chosen such that the highest transmission ratio at 1.25μm is unity. The 1.054μm data shown is obtained by simply splitting the laser pulse. Each point plotted is an average of at least five data points. The error bars indicate typical spreading of data. For all data plotted, the excitation pulse contains approxi-

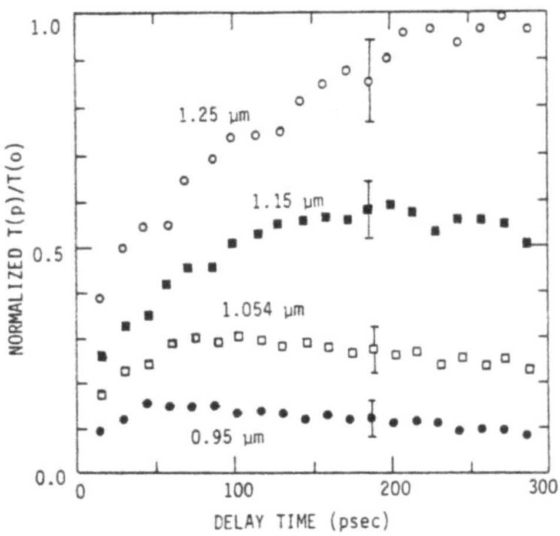

Fig.2 Probe pulse transmission vs delay at four wavelengths. The data are plotted as normalized ratio of probe transmission to linear transmission at the wavelength under consideration, T(p)/T(o), in arbitrary units. Excitation pulse is 11 psec long at 1.054μm, containing 3.8×10^{14} photons incident on a spot of 1 mm diameter. Sample temperature is 77 ^{0}K.

mately 3.8×10^{14} quanta, equivalent to an irradiance of approximately 0.85 GW/cm^2, which is far below the damage threshold of 7 GW/cm^2. We have omitted data for delay times less than 20 psec, where coherent coupling between the two pulses occur [4]. Also, we must note that actual probe transmission differs in different runs of the experiment. We attribute this to a strong dependence on the degree of spatial overlapping of the two pulses. In Fig.2 we have plotted the highest transmission curves obtained. We believe these have been obtained with the best overlapping of the pulses. We notice the following from experimental data. First, probe pulse transmissions are higher for longer wavelengths. Second, peaking of probe transmission occurs at a longer delay time for the longer wavelength.

We have performed numerical calculations [10] to investigate the effect of diffusion under the above experimental conditions. It is found that diffusion in the pulse propagation direction may lead to a rise, a drop, or both in probe transmission. When the number of free carriers created by excitation is large enough, when they migrate from the front surface, they can fill the direct transition-coupled states in the rear without emptying those states in the front. Probe pulse transmission rises monotonously in this case. On the other hand, when the number of carriers is small, diffusion will deplete the optically-coupled states in the front without significantly filling the states in the back, thus transmission decreases. When the number of carriers is intermediate between the two cases above, diffusion may lead to an intial increase in transmission which peaks and fall afterwards. The time delay that transmission reaches its maximum depends on free carrier concentration. For a higher concentration, this time is longer.

At all our probe wavelengths, photon energy is higher than direct bandgap energy, thus direct interband absorption is always the dominant absorption process at our excitation levels. When we excite with one wavelength and probe with a longer wavelength, the states coupled by probe photons are located closer to the band minima. The carrier density needed to saturate this transition is lower. In this case, calculations indicate that probe transmission will be higher and it takes a longer delay for the peaking to occur, similar to the higher concentration case when probing with excitation wavelength.

We have theoretically estimated [10] the carrier density averaged over interaction volume to be about 4×10^{19} cm^{-3} at our excitation level. At this carrier concentration, Auger recombination is a slow process on the picosecond time scale. Also, phonon-assisted relaxation of hot carriers is more or less a completed process for delay times longer than 20 psec. With these considerations, we thus conclude that our experimental results seem to support that diffusion plays the major role in probe pulse transmission observed. In fact, we have recently performed an excite-and-probe experiment in which the excitation level is varied. Our data clearly show that probe transmission is higher and peaking occurs at a longer delay time when excitation level is increased.

References

1. C. J. Kennedy, J. C. Matter, A. L. Smirl, H. Weichel, F. A. Hopf, S. V. Pappu, M. O. Scully, Phys. Rev. Lett. 32, 419 (1974)
2. W. P. Latham, Jr., A. L. Smirl, A. Elci, Solid-State Electron. 21, 159 (1978)
3. C. Y. Leung, Appl. Phys. B27, 201 (1982)
4. A. Elci, M. O. Scully, A. L. Smirl, J. C. Matter, Phys. Rev. B16, 191 (1977)
5. A. L. Smirl, J. C. Matter, A. Elci, M. O. Scully, Opt. Commun. 16, 118 (1976)
6. D. H. Auston, S. McAfee (unpublished)
7. A. L. Smirl, J. R. Lindle, S. C. Moss, Phys. Rev. B18, 5489 (1978)
8. A. Elci, A. L. Smirl, C. Y. Leung, M. O. Scully, Solid-State Electron. 21, 151 (1978)
9. G. A. Massey, R. A. Elliott, IEEE J. Quant. Electr. QE-10, 899 (1974)
10. Using theory of Ref.3, extended to include direct intra-valence band transition and spatial dependence of parameters.

Nonlinear Interactions in Indium Antimonide

M. Hasselbeck and H.S. Kwok

Department of Electrical and Computer Engineering, State University of New York at Buffalo, Amherst, NY 14226, USA

1. Introduction

The investigation of the phenomenon of laser-induced breakdown in a semiconductor is important for high power laser optics. Traditionally, the emphasis was on the measurement of the damage thresholds and its scaling with respect to pulse area, duration and other laser parameters. We have performed some experimental investigations on the prebreakdown stage. The interaction of a strong laser field with a semiconducting material below the breakdown threshold is important not only to the breakdown process, but is also interesting fundamentally because of the high nonlinearities involved. Incidentally, of more practical interest, this is the intensity region where most of the laser processing of semiconducting materials are performed.

InSb was chosen to be studied because of its well-known large two photon absorption (TPA) coefficient and its importance as a nonlinear optical material. The Burstein shift of InSb is such that TPA is allowed in intrinsic InSb, but not allowed in n-InSb when the doping concentration exceeds 0.8 x $10^{18}cm^{-1}$. This enables us to investigate different types of nonlinear processes in the same material. In our studies, we employed intrinsic InSb with n = 1.2 x $10^{-16}cm^{-3}$ and highly doped n-InSb with a concentration of 1.3 x $10^{18}cm^{-3}$.

2. Experimental

The laser system used has been documented elsewhere [1]. Basically, it is an Optical Free Induction Decay system capable of producing continuously adjustable 30-300 ps duration laser pulses. The laser system can also produce smooth single longitudinal mode 100 ns TEA laser pulses using the hybrid high pressure and low pressure double discharge arrangement.

To measure the electron-hole plasma generation in the semiconductor during the intense laser pulse interaction process, a transient reflectivity arrangement was employed. The laser beam was focussed onto the sample at a Brewster angle and the integrated reflectivity was measured as a function of the laser intensity. Optical microscopy and SEM were used to examine the surface of the crystal after the experiments. For 100 ns pulses, the time-resolved reflectivities were also observed. The change in reflectivity was based on the modification of the dielectric constant of the material by the electron-hole plasma. For our particular experiment, the critical plasma density where the dielectric constant undergoes drastic change was 2.4 x $10^{18}cm^{-3}$. Therefore, changes in the plasma concentration below this value will not be observed.

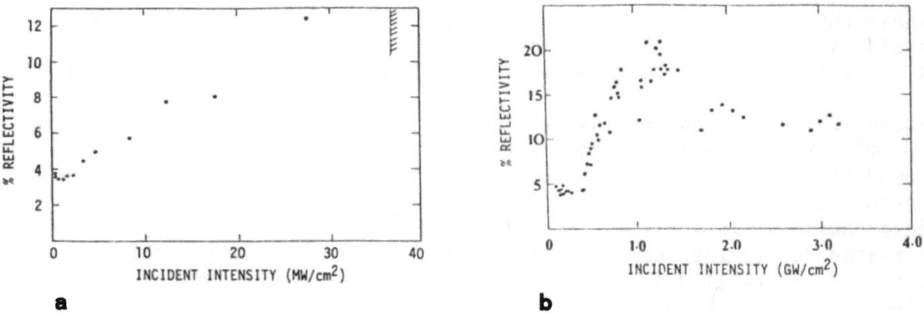

Fig. 1 Reflectivity of i-InSb with 100 ns and 75 ps pulses

3. Experimental Results

3.1 Intrinsic InSb

Figures 1(a) and 1(b) depict the measured integrated reflectivities of the intrinsic sample as a function of the incident laser intensity with 100 ns and 75 ps pulses respectively. The damage thresholds for the sample have been carefully measured by observing a small surface spark in pitch darkness. For 100 ns pulses, the breakdown threshold is 37 MW/cm^2 while for 75 ps pulses, it is 4.3 GW/cm^2.

The increase in reflectivity for the case of 100 ns pulses can be explained by TPA. Using the established value of 0.23 cm/MW for the TPA coefficient β, the intensity required to create a plasma with $2.4 \times 10^{18} cm^{-3}$ density is estimated to be 2 MW/cm^2. This value corresponds to the observed nonlinear reflection onset of 1.5 ± 0.5 MW/cm^2 quite satisfactorily. We also examined the time-resolved increase in reflectivities for these 100 ns pulses. The results are shown in Fig. 2. In Fig. 2(a), the smooth 100 ns pulse is shown. The nonlinearly reflected pulse at an intensity of $\sim$15 MW/cm^2 is depicted in Fig. 2(b). It can be seen that the reflectivity increases continuously as a function of time. Fig. 2(c) corresponds to an intensity of 37 MW/cm^2 just at the breakdown threshold. Visible tiny surface sparks were observed for these pulses. The reflected pulse shows sudden changes in reflectivity corresponding to surface plasma formation.

The picosecond reflectivity curve in Fig. 1(b) is more difficult to interpret. Using the same TPA coefficient as in the ns case, one would expect a

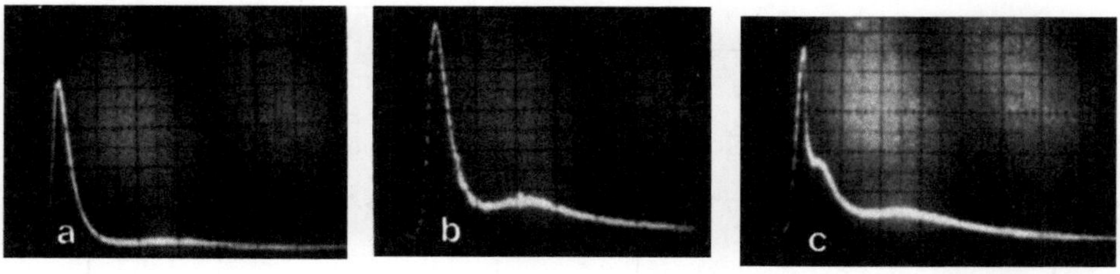

Fig. 2 Time-resolved reflectivities of the i-InSb sample with 100 nsec TEA pulses. (a) incident pulse, (b) 15 MW/cm^2, (c) 37 MW/cm^2. Horizontal scale, 0.2μs/div.

nonlinear threshold of 0.06 GW/cm^2. However, the experimentally observed threshold is 0.4 GW/cm^2. Since β was calculated from perturbation theory, it is not unlikely that its value will be different at high laser intensities.

Figure 1(b) also shows an interesting decrease in reflectivity above 1.2 GW/cm^2. We have examined the surface of the crystal at these intensities and found that physical changes were induced in the sample by the laser pulse (comparable to Fig. 4). It is fairly certain that the sample was melted by the laser pulse at intensities above 1.2 GW/cm^2 and hence the reflectivity is decreased due to the increased absorption. To our knowledge, this is the first direct observation of nonlinear melting in a semiconductor without the simultaneous occurrence of a breakdown damage. It should be emphasized that the surface remains smooth and without topography when this nonlinear melting occurs as opposed to a crater formation usually observed in laser damage experiments.

It is interesting to compare this nonlinear melting process with ordinary melting by linear interband transition. Using the example of 0.532 nm laser-induced phase transformation on Si [2], the ratio of the damage to melting threshold is 10 for this linear system, while in our case, the ratio is 3. There is indeed a "comfortable" region of laser intensities where nonlinear melting occurs without the complications of laser damage.

3.2 Highly Doped InSb

The 100 ns and 65 ps reflectivity curves for n-InSb are shown in Figs. 3(a) and (b). For this material, the bandgap is 0.25 eV which is larger than $2h\nu$. Therefore, TPA is not allowed and a different type of nonlinearity is expected to generate the dense electron-hole plasma. The measured curves reflect this important difference. Fig. 3(a) shows that the 100 ns pulse reflectivity remains quite constant and then changes abruptly near the damage threshold intensity. This observation is consistent with the avalanche ionization picture of laser-induced breakdown. However, on closer examination of the reflectivity below the breakdown, it was found that an enhancement in the reflectivity could be observed without the occurrence of a breakdown. This effect can be measured more precisely with the time-resolved reflectivity in the same manner as Fig. 2. By translating the focussing lens to adjust the laser intensity more precisely, a smoothly increasing nonlinearly reflected pulse can be observed. By increasing the intensity further, one would suddenly observe sparks on the crystal surface and the corresponding reflected pulse would show ragged changes in the same manner as Fig. 2(c). It is believed that the nonlinearity is due to melting of the surface. Experimentally, the ratio of the damage threshold to the melting threshold is 1.20.

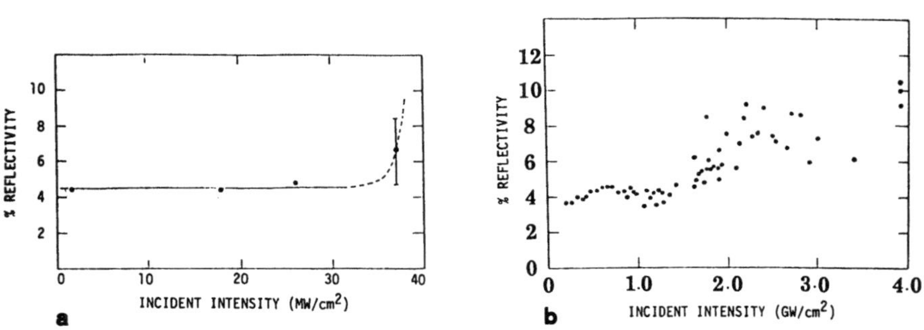

Fig. 3 Reflectivity of n-InSb with 100 ns and 65 ps pulses

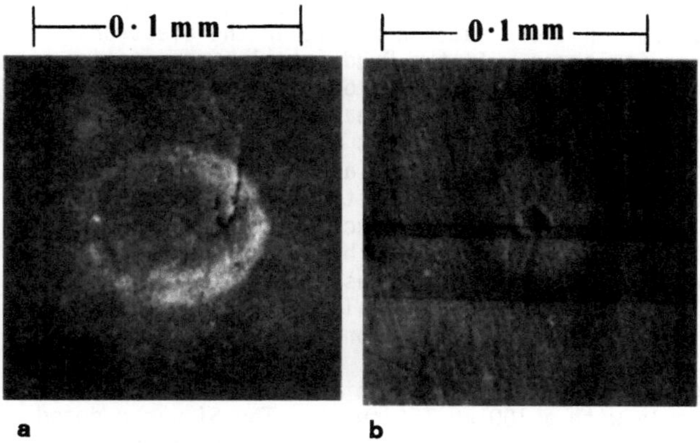

Fig. 4 SEM photographs of the n-InSb crystal surface after irradiation by 65 ps pulses. (a) slightly below damage at 4 GW/cm^2, (b) slightly above damage at 5 GW/cm^2

The 65 ps pulse curve in Fig. 3(b) appears to be similar to Fig. 1(b). However, the threshold for enhanced reflectivity occurs at higher intensities. Moreover, the enhanced threshold corresponds to the melting threshold of the intrinsic sample in Fig. 1(b). Upon examination of the surface of the crystal using an SEM, it was found that indeed the surface has been modified by the 65 ps laser pulses above 1.6 GW/cm^2. Fig. 4(a) shows an SEM scan of the surface at 4 GW/cm^2 and 4(b) shows the surface at 5.1 GW/cm^2 which is just at the breakdown threshold. It should be emphasized that the surface of the crystal corresponding to Fig. 4(a) was very smooth and without topography. As a matter of fact, the ring shown in 4(a) was not observable using optical microscopy, but was readily found using SEM with secondary emission. We are not certain what type of modification has been impounded on the crystal. Possibilities are shallow evaporation of the surface, change in relative composition of In and Sb, or changes in the crystal lattice due to crystallization and surface oxidation.

Figure 4(b) is the familiar picture of a damaged surface. The central crater corresponds to the plasma plume formation. Notice that beyond the central crater, there is an outer ring structure which looks the same as Fig. 4(a), confirming the speculation that the color change on the smooth surface is due to recrystallization of a molten surface layer. The depth of the molten layer can be estimated to be 4.8μm, using a hole absorption cross section of 8.6 x $10^{-16}cm^{-1}$.

The question of the initial generation of the dense electron-hole plasma still remains. We propose that avalanche electron multiplication, similar to the breakdown process in transparent dielectrics, is responsible for the above observations. This multiplication has been analyzed by CANTRELL et al [3] and NURMIKKO et al [4]. Assuming that the avalanche ionization coefficient $\eta(E)$ is known, the temperature rise in the sample can be calculated

$$N(t) = N_o \exp\left[\int_0^t \eta(E)\, dt'\right] \tag{1}$$

$$\Delta T(t) = 1/C \int_0^t N(t')\, \sigma I(t')\, dt' \tag{2}$$

where N_0 = initial electron (or hole) density, C = volumic specific heat of the crystal and σ is the absorption cross section of either the electrons σ_e or the holes σ_h, whichever is larger. In the Drude model approximation, σ_e and σ_h are inversely proportional to the relaxation times τ_e and τ_h respectively. In this manner, the analysis of nonlinear melting in a semiconductor is very similar to the case of transparent dielectrics. However, there is one important distinction, namely the initial concentration N_0 is much larger ($\gtrsim 10^{16} cm^{-3}$ as opposed to 1-10 for dielectrics). This much larger N_0 will result in a weaker dependence of ΔT as a function of the laser intensity, enabling melting to be observable prior to breakdown. It is believed that with careful pulse shaping, the different stages in the development of melting and vaporization/plasma formation can be realized in sequence and melting without damage can be achieved in any semiconductor.

Along the same reasoning, we have performed an initial investigation on the melting of a silicon sample with a 100 ns TEA pulse. The Si was n doped to $1.5 \times 10^{18} cm^{-3}$. Indeed, it was found that the behavior of n-Si is very similar to n-InSb. The ratio of the damage threshold to the onset of the nonlinear reflection was also 1.2.

In summary, we have demonstrated that the picosecond CO_2 laser can be used to melt semiconductors without damage by a nonlinear mechanism. The mechanism responsible for the nonlinear generation of a dense plasma is TPA for i-InSb and avalanche ionization for larger bandgap materials. The induced melting should have potential applications in the laser processing of semiconductors.

Support by the NSF under Grant No. ECS 8106007 is gratefully acknowledged.

References

1. H.S. Kwok, E. Yablonovitch and N. Bloembergen, Phys. Rev. A23, 3094 (1981).
2. J.M. Liu, R. Yen, H. Kurz and N. Bloembergen, Appl. Phys. Lett, 39, 755 (1981).
3. T.W. Nee, C.D. Cantrell, J.F. Scott and M.O. Scully, Phys. Rev. B17, 3936 (1978).
4. S.A. Jamison and A.V. Nurmikko, Phys. Rev. B19, 5185 (1979).

Picosecond Relaxation Kinetics of Highly Photogenerated Carriers in Semiconductors

S.S. Yao, M.R. Junnarker, and R.R. Alfano

Ultrafast Spectroscopy and Laser Laboratory, Physics Department,
The City College of New York, New York, NY 10031, USA

1. Introduction

High-density nonequilibrium electrons and holes can be produced in semiconductors through interaction with picosecond laser pulses. Picosecond photoluminescence and absorption spectroscopy is an extremely useful experimental tool for probing the time evolution of the elementary excitations in semiconductors.

The picosecond excite and probe absorption technique [1] was used to study the relaxation processes of photogenerated hot carriers in ε-GaSe at room temperature. The dominant relaxation mechanism of the photogenerated hot carriers in GaSe is attributed to the emission of nonpolar optical phonons $A_1'^{(1)}$. A smaller value for the deformation potential extracted from the measurements is attributed to the screening of hole-phonon interaction by the photogenerated carrier density.

The time-resolved photoluminescence kinetics of GaSe were measured at room temperature [2]. The rise time of the spontaneous emission band increases with increasing carrier density, and it is attributed to the screening on nonpolar optical phonon $A_1'^{(1)}$ emission from hot photogenerated carriers due to high photogenerated carrier density. On the other hand, the rise time of the stimulated emission band cannot be time resolved.

The hot photoluminescence kinetics of GaAs under intense picosecond excitations ($10^{28}/cm^2s$) were also measured [3]. A slow rise time of near band-edge luminescence was observed to arise from a slowed cooling of the electron distribution. The slowed electron kinetics of over 50 fold are attributed to the screening of the electron-phonon interaction.

2. Experimental Method, Results, and Theoretical Explanation

2.1 Time-Resolved Absorption Spectroscopy of GaSe

A schematic diagram of the picosecond absorption apparatus has been described elsewhere [4]. The absorption curves measured at different time delays are displayed in Fig.1. The absorption change at different time delays was fitted (dots) using:

$$\alpha = \alpha_0(1 - f_e)(1 - f_h) \quad , \tag{1}$$

where α and α_0 are the absorption of the sample with or without excitation, and f_e and f_h are the Fermi distributions of electrons and holes. The carrier

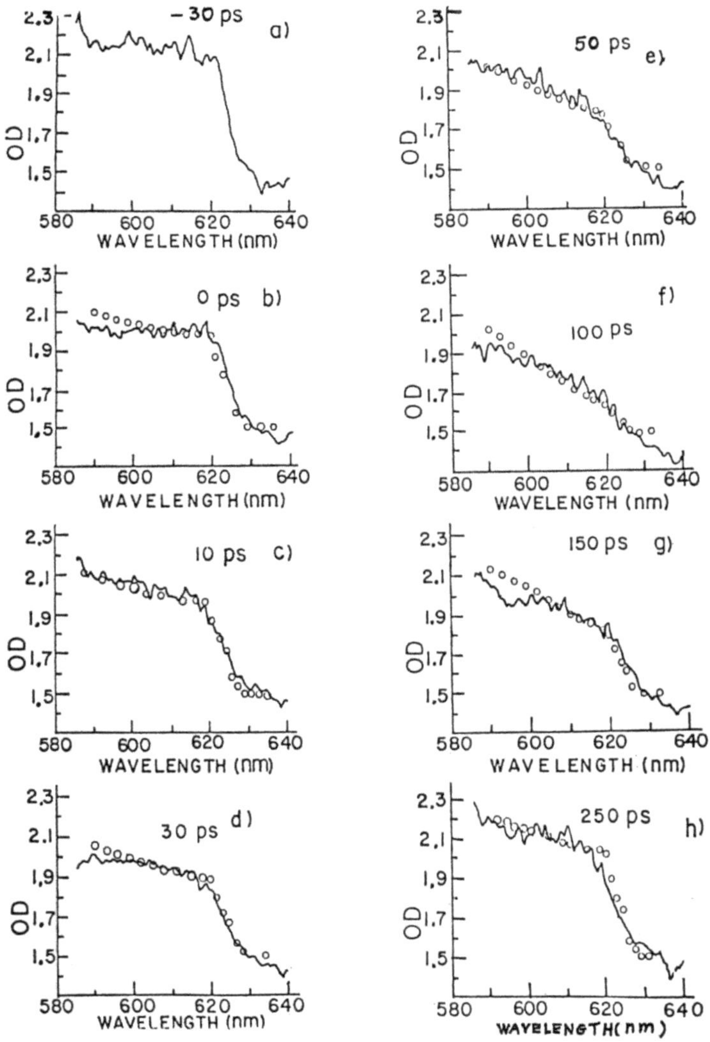

Fig.1. The absorption curves measured at different time delays. b) $T_e=T_h=2000$ K, $n=10^{19}$ cm^{-3}, $\tilde{E}_g=1.950$ eV; c) $T_e=1700$ K, $n=9\times10^{18}$ cm^{-3}, $\tilde{E}_g=1.950$ eV; d) $T_e=1000$ K, $n=8\times10^{18}$ cm^{-3}, $\tilde{E}_g=1.955$ eV; e) $T_e=600$ K, $n=7\times10^{18}$ cm^{-3}, $\tilde{E}_g=1.955$ eV; f) $T_e=320$ K, $n=6.5\times10^{18}$ cm^{-3}, $\tilde{E}_g=1.955$ eV; g) $T_e=300$ K, $n=3\times10^{18}$ cm^{-3}, $\tilde{E}_g=1.960$ eV; h) $T_e=300$ K, $n=7\times10^{17}$ cm^{-3}, $\tilde{E}_g=1.965$ eV

density for a degenerate distribution is $n = 2(2\pi m\ kT_i/h^2)^{3/2}F_{1/2}(\mu_i/kT_i)$, $i = e,h$, where $F_{1/2}(\mu_i/kT_i)$ is a Fermi function. Both carrier temperature and density were varied to fit the experimental results. The reduced band gap is obtained [2] for a carrier density below 10^{18} cm^{-3} from luminescence peak minus 1/2kT.

In Fig.2, the carrier temperature T_e is drawn as a function of the time delay. The carrier temperature cools at a rate of 30 K/ps. The temperature decay time is 42 ps.

The average rate of change of carrier energy due to nonpolar optical phonon interaction [5] is described by:

$$\langle d\varepsilon/dt\rangle = -(2/\pi)\cdot(D^2 m_\perp m_{||}^{1/2})/(\pi\hbar^2 MN)\cdot(kT_e)^{1/2}[e^{x_0-x_e}-1]/(e^{x_0}-1)$$
$$\cdot\left[\frac{x_e}{2}\right]K_1[x_e/2]e^{x_e/2} \quad , \tag{2}$$

where K_1 is a Bessel function of the second kind, $x = \hbar w/kT$, $x = \hbar w/kT_e$, T and T_e are the temperatures of lattice and carriers, $\hbar w$ is the nonpolar optical phonon energy, D is the deformation potential, M is the reduced ionic mass per unit cell, and N is the number of cells per unit volume. Following SCHMID [6], we have replaced the crystal density in (2) by MN, because the interaction matrix of the carrier optical phonon is different for layer compounds. Equation (2) is used to fit the experimental data in Fig.2 by the solid line. We found the coefficient $1/3(2/\pi)^{1/2}(D^2 m_\perp m_{||}^{1/2}/\pi\hbar^2 MN)$ $k^{-1/2} = 1.11\times10^{12}$, and $\hbar w = 16.7$ meV fits the experimental data very well. Therefore the dominant relaxation mechanism of the photogenerated hot carriers is the emission of nonpolar optical phonons $A_1'^{(1)}$ with an energy of 16.7 meV. The deformation potential between the holes and nonpolar optical phonons is calculated to be 1.3 eV/Å, which is about 5 times smaller than the result [6] measured by SCHMID and VOITCHOVSKY (6.6 eV/Å). We attributed

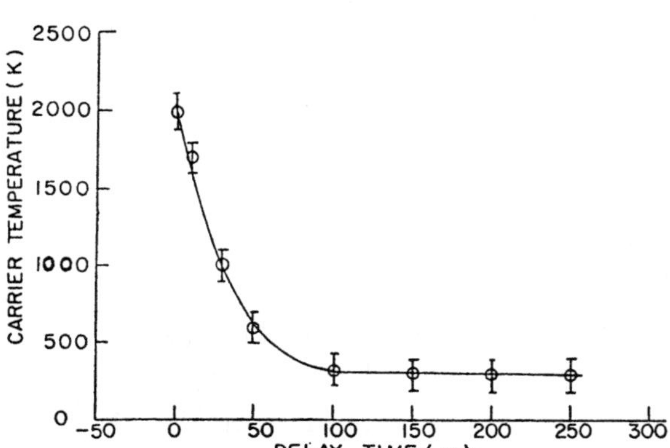

Fig.2. The carrier temperature vs the delay time between the pump and probe pulses. The dots with error bars are experimental results, and the solid line is a theoretical fit

this to the screening of the interaction between holes and nonpolar optical phonons $A_1'^{(1)}$. The interaction between electrons and $A_1'^{(1)}$ phonons can be neglected as compared with that between holes and $A_1'^{(1)}$ phonons [6].

2.2 Time-Resolved Emission Spectroscopy of GaSe

The photoluminescence from GaSe excited by intense picosecond laser pulse consists of two components [2]. The luminescence was collected by a Hamamatsu streak camera with a 100-μ slit. Time-resolved photoluminescence spectra for a spontaneous emission component with a wavelength of 610 to 630 nm from GaSe at room temperature is displayed in Fig.3. We used $I(t) = A(\exp(-t/\tau_d) - \exp(-t/\tau_r))$ to fit the profile of the time-resolved emission kinetics, where τ_d and τ_r are the decay and rise times of the emission, respectively. The rise time increases with increasing carrier density. Time-resolved photoluminescence for stimulated components was also measured. The rise time remains within the resolution of the measuring system ($\leq$ 20 ps). The wavelength is above 640 nm.

From the theory developed by YOFFA [7], the emission rate ν of nonpolar optical phonons from hot carriers in semiconductors is proportional to $(1+(n/n_c)^2)^{-1}$, where n is the carrier density and n is equal to $\varepsilon_0\hbar/8\pi e^2 (zm_i^*\omega/\hbar)^{3/2}(1/\beta m_i^*)^{1/2}$. In GaSe, ε_0 = 10.6 when the incident light is normal to C axis, $\hbar w$ = 16.7 meV, m_i^* = 0.5 m_0 for holes, and $\beta = 1/kT_e$. At the temperature decay rate of 30 K/ps from time-resolved absorption kinetics measurement, the temperature will drop from 2000 K to 400 K during the rise time of spontaneous photoluminescence of 60 ps for a photogenerated carrier density of 10^{19} cm^{-3}. The inverse rise time should be proportional to the emission rate ν of phonons from the relaxation of hot holes in GaSe. In Fig.4, the inverse of rise time at various carrier densities is plotted. The values of $\nu \propto 1/(1 + (n/n_c)^2)$ are also plotted for $n_c = 4.8\times10^{18}$ cm^3 at T_e = 2000 K. These values fit the experimental data well when the carrier density is above 5×10^{18} cm^{-3}.

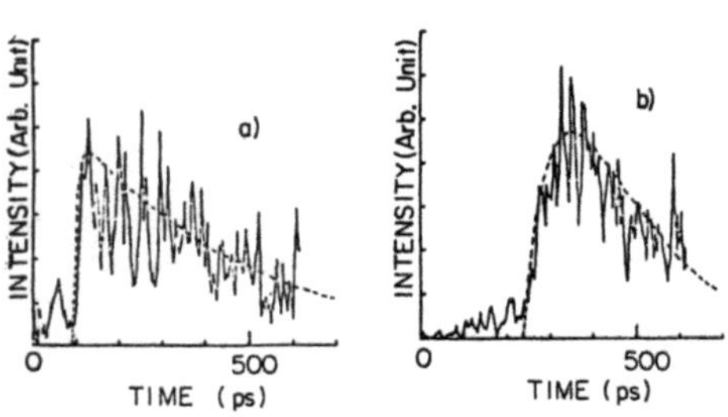

Fig.3. Spontaneous emission from GaSe at room temperature.
a) $n=4.8\times10^{18}$ cm^{-3}, τ_r=10 ps;
b) $n=1.4\times10^{19}$ cm^{-3}, τ_r=70 ps

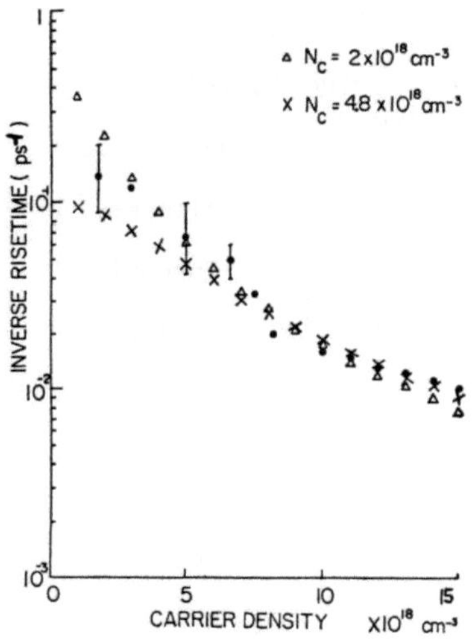

Fig.4. The inverse of the rise time at various photogenerated carrier densities (dots). The crosses represent $\nu=A/(1+(n/n_C)^2)$, where A=0.1, and $n_C=4.8\times10^{18}$ cm^{-3}. The triangles represent ν, with A=0.45, and $n_C=2\times10^{18}$ cm^{-3}

The emission rate at 10^{19} cm^{-3} and T_e = 2000 K is about $1/(1 + (10^{19}/4.8\times10^{18})^2) \cong 0.2$, or 5 times smaller than the unscreened value. From the absorption measurement, we found that the deformation potential is reduced 5 times when the carrier density is 10^{19} cm^{-3}; the energy decay rate would be 25 times smaller than the unscreened value. Since the rise time is below 20 ps, which is the resolution of our measuring system when the carrier density is below 5×10^{18} cm^{-3}, it is impossible to know experimentally whether the emission rate at 10^{19} cm^{-3} will be 25 times smaller than the unscreened value. In Fig.4, the values of ν are also plotted with $n = 2\times10^{18}$ cm^{-3}. In this case, the value of ν at 10^{19} would be 25 times smaller than the unscreened value. These values do not fit the experimental results as well, compared with that of $n_c = 4.8\times10^{18}$ cm^{-3} above 1.5×10^{19} cm^{-3}. There may be three reasons to account for this difference of emission rate ν at 10^{19} cm^{-3}. The first is that the carrier density of 10^{19} cm^{-3} and the carrier temperature of 2000 K from absorption measurements could be off from exact values. The second reason is that this sample is a layered semiconductor; therefore the theory of YOFFA probably should be modified to account for this change to obtain a different value of n_c by a factor 2. The third reason may be that the unscreened deformation potential is smaller than 6.6 eV/Å by a factor 2; therefore the emission rate ν is about 5 times smaller at a photogenerated carrier density of 10^{19} cm^{-3}.

Acknowledgment. We are grateful for the AFOSR 80-0079 grant in support of this research.

References

1 S.S. Yao, J. Buchert, R.R. Alfano: Phys. Rev. B, May 15, 1982
2 S.S. Yao, R.R. Alfano: Submitted for publication
3 R.J. Seymour, M. Junnarker, R.R. Alfano: Solid State Commun. *41*, 657 (1982)
4 A.G. Doukas, V. Stefancic, J. Buchert, R.R. Alfano, B.A. Zalinskas: Photochem. Photobiol. *34*, 505 (1981)
5 E.M. Conwell: In *Solid State Physics*, ed. by F. Seitz, D. Turnbull, H. Ehreneich (Academic, New York 1967), Suppl.9, p.155
6 Ph. Schmid, J.P. Voitchovsky: Phys. Status Solidi b *65*, 249 (1974)
7 E.J. Yoffa: Phys. Rev. B*23*, 1909 (1981)

Picosecond Radiative and Nonradiative Recombination in Amorphous As_2S_3

T.E. Orlowski and B.A. Weinstein

Xerox Research Center, Webster, NY 14580, USA

W.H. Knox, T.M. Nordlund, and G. Mourou

University of Rochester, Laboratory for Laser Energetics, Rochester, NY 14627, USA

Current understanding of localized states in amorphous (a-) semiconductors has been strongly influenced by studies of sub-bandgap photoluminescence (PL). In particular, the PL kinetics and temperature dependence bear directly on the mechanisms for recombination and separation of localized carriers. In the prototype chalcogenide glass As_2S_3 previous experiments have found that the PL is characterized by a large distribution of monomolecular decay times extending from less than ten nanoseconds to several milliseconds [1,2]. It is thought that different recombination processes acount for slow PL decaying in $10^{-6}-10^{-3}$ s and fast PL decaying in $< 10^{-6}$ s [3]. The principle time resolved PL spectrum is a broad sub-bandgap peak (~0.5eV FWHM) whose center shifts from ~1.6→1.1 eV with increasing time following excitation [1-3]. Band-tail or higher energies (>2 eV) are required for photoexcitation. A substantial part of the energy difference between emission and excitation has been attributed to a Stokes shift accompanying localization at defect sites [1,2,4] or small polaron formation [5].

The maximum radiative rate ν_1 has not been measured directly for any amorphous semiconductor. This is because the largest decay rates (even at low T) are considerably beyond the temporal resolution ($\geq$10ns) of the fastest experiments reported [2,3]. In the present work picosecond laser and jitterfree streak camera techniques have been exploited to shorten this limit to ~7ps. Consequently, for the first time we were able to isolate the most rapid radiative processes and thereby determine ν_1. We also determined the temperature dependence (8-204K) of the effective nonradiative rate $\bar{\nu}_{nr}$ in competition with ν_1 and found that, even at T = 0K, $\bar{\nu}_{nr}$ is substantial. In addition we have investigated the spectral dependence of both the build-up and initial decay kinetics for emission energies between 2.1 and 1.5 eV.

The experiments were performed using the apparatus shown in Fig. 1. An active-passive mode-locked Nd^{3+}:YAG oscillator [6] provided very reproducible pulse trains. Single pulses (30ps FWHM, $\lambda = 1064$nm) were extracted using a double Pockels cell scheme, amplified, and split into two beams. The first beam was frequency doubled (2.33 eV) in KD*P, filtered and weakly focused (0.5 mm) at an intensity of 0.2GW/cm^2 upon the sample housed in a variable temperature helium refrigerator. The second beam was directed upon a GaAs photoconductive high voltage switch used to provide the deflection voltage ramp for a streak tube (S-20 response). This arrangement has been shown [7] to synchronize the streak tube sweep to within 2ps with the excitation pulse, allowing accurate averaging of ps optical information from successive laser shots. Front surface PL was measured. Data from up to 300 laser shots were averaged to obtain the results. The As_2S_3 glass was high quality optical window material from Servo. The visible and IR absorption and CW luminescence spectra agreed with previous work [1,3].

The measured instantaneous PL intensity I(t) from 0 to 300 ps at 8 and 154K is shown in Fig. 2; I(t) corresponds to the total intensity in the bandwidth 2.1-1.5 eV. By virtue of these time and energy constraints our experiments probe only the most rapid PL processes [3]. Using different high energy cutoff filters (2.1, 1.9 and 1.7 eV) we studied the dependence on PL energy of both the build-up and decay rates at 8 and 154K. Within experimental accuracy <u>no</u> variation in rates was found, although the integrated intensity did decrease for the lower energy cutoffs. These results are consistent with a single process contributing to PL in this energy regime. Whether picosecond processes extend to still lower energy with any intensity awaits future measurements with an S-1 response streak tube.

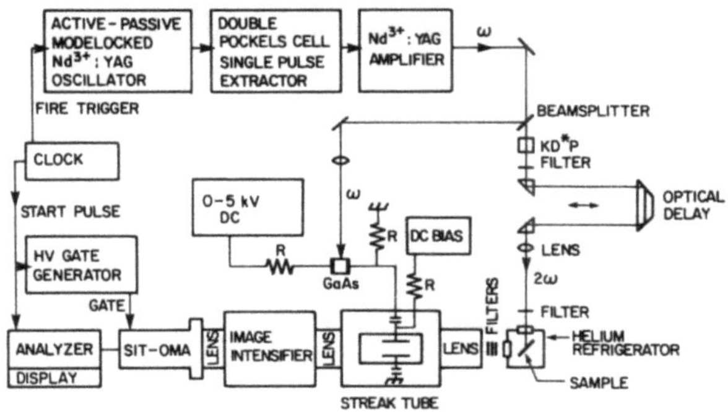

Fig. 1. Schematic diagram of the pico-second laser and jitter-free streak camera system

In all our measurements the PL build-up was observed to follow an instantaneous response to the laser pulse. Considering our signal/noise ratios we find that the deconvoluted build-up time is ≲10 ps except for the lowest temperature data illustrated by the 8K results shown in Fig. 2. In that case a slightly longer build-up time (≲20ps) is found which we are presently investigating in greater detail. Consequently we can set an upper limit of ~20 ps on the time for photoexcited carriers to relax into emitting states, regardless of the energy of these states (within 2.1-1.5 eV) or the temperature. Since several tenths of an eV are lost during relaxation, our PL observations support a rapid strong coupling channel. The Stokes shift mechanisms which have been proposed, involving bond switching or breaking at defect sites [1,2,4], or small polaron formation [5], fit this criterion. A fraction of the energy lost could also be accounted for by tunneling to lower energy band-tail states; however this multi-step process should be slower because of the decreasing density and increasing localization of successive states. We can not rule out an important role for band-tail tunneling in the relaxation accompanying longer time and CW emission.

In Fig. 2 the inserts show the best fit of an exponential decay to our data for t > 30ps (i.e., after the laser pulse). Typical uncertainties in the decay times are 15-20%. At 8K the observed decay time (1/e) is 1150ps but by 204K it has decreased to 45ps. Above this temperature we were unable to measure the PL decay time due to unacceptable signal/noise ratios. The detailed temperature dependence of the corresponding decay rate $\nu(T)$ is shown in Fig. 3.

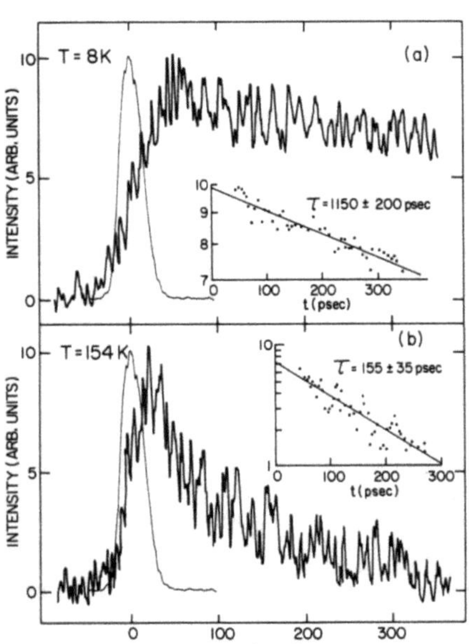

Fig. 2. Measured build-up and decay of PL (1.5 - 2.1 eV) in a-As_2S_3. Inserts show least-squares fits assuming an exponential decay. The laser excitation pulse is also displayed. 200 shots were averaged in (a), 300 in (b).

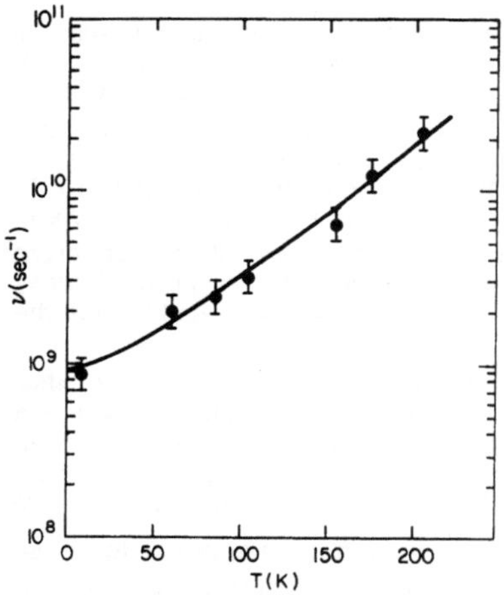

Fig 3. Temperature dependence of the observed PL decay rate in a-As_2S_3. At high-T the decay rate is determined primarily by nonradiative processes; the dependence is similar to that for η_{cw}^{-1} (see text). The solid line representing $\nu_1 + \bar{\nu}_{nr}(o)$ exp (T/To) fits all the data quite closely.

We also attempted to fit the PL decay to a power law $t^{-m(T)}$ and the product t^{-T/T_1} $\exp(-\nu_1 t)$. According to the model of [2], the observed power law dependence for I(t), t > 100ns, is the product of a radiative and a nonradiative part. The approximations leading to the radiative part ($\sim t^{-1}$) do not apply for $t \lesssim 1$ns. In that case we expect an exponential decay dominated by ν_1. The nonradiative part should still follow t^{-T/T_1} in the time regime of our experiment because the maximum nonradiative rate is assumed to be $\sim 10^{13}$ sec^{-1}, an internal mode optical phonon frequency [2]. However we could not justify fitting our data to $t^{-T/T_1}\exp(-\nu_1 t)$ for any reasonable and temperature independent choice of the parameters T_1 and ν_1. Invariably the fits degraded as temperature was varied because the t^{-T/T_1} factor varies too rapidly at small t and too slowly at large t compared to our data. Therefore it seems that the model of [2] for nonradiative processes, although phenomenologically correct at long times, should be modified for $t \lesssim 1$ns to be applicable to our case. Perhaps lower frequency phonons, such as the rigid-layer like cluster modes suggested by PHILLIPS [4], should be incorporated. More fundamentally, it seems physically unrealistic to assume (as does this model) that the distributions of radiative and nonradiative rates will be uncorrelated for a carrier localized at a particular luminescence center [4].

In Fig. 3 the data were fit to the equation $\nu(T) = \nu_1 + \bar{\nu}_{nr}(o) \exp(T/T_o)$. Here it is assumed that the observed decay rate has a temperature independent radiative component ν_1 describing the fastest PL processes in a-As_2S_3, and a temperature dependent nonradiative component $\bar{\nu}_{nr}(T)$ which competes with ν_1. Our 3-parameter least squares fit (solid curve in Fig. 3) yields $\nu_1 = (4\pm1) \times 10^8$Hz, $\bar{\nu}_{nr}(o) = (4\pm1) \times 10^8$Hz, and T_o = (53±5)K. The functional form for $\bar{\nu}_{nr}(T)$ is connected with the well known [1] T^2 activated behavior exhibited by the CW quantum efficiency η_{cw} (over 10-230K). Consider the effective quantum efficiency $\eta_1 = \nu_1/\nu(T)$ due to nonradiative competition with ν_1. Our data show that for T>100K, $\eta_1 \propto \exp(-T/T_o)$ similar to η_{cw} except that $T_o \approx 26$K [1] in the CW case. Since η_{cw} is dominated by slow PL while η_1 corresponds to the maximum rate ν_1, the observed similarity in functional dependence on T suggests that similar mechanisms could account for the nonradiative decay of slow and fast PL in a-As_2S_3. We note that at low-T nonradiative competition with ν_1 is appreciable because $\bar{\nu}_{nr}(o) \sim \nu_1$. Therefore, it is likely that tunneling of some type, which does not vanish as T→0K [10], makes a significant contribution to the nonradiative decay of the fastest emission processes.

Recent picosecond photoinduced absorption (PA) studies [8,9] in chalcogenide glasses have shown that the decay of absorption due to photoexcited carriers is quite different from that reported here for picosecond PL. For a-$As_2S_{2.25}Se_{0.75}$ with band-tail excitation (same absorption coefficient as in our experiments) PA decays in ~3ps at 85K increasing to 12ps at 300K [9]. In contrast we observe that PL decays in 410 ps at 85K decreasing to < 10ps at 300K. We believe these differences indicate that the subsets of carriers contributing to PA and PL are separate. During or just after thermalization a fraction $\sim\eta_{CW}(0K)$ = 10-20% of the carriers are trapped at luminescence centers in states with a low absorption cross section; the remainder contribute to PA. We suggest that a portion of the initial rapid decay of PA observed in [9] may derive from carrier capture at PL centers. This would place the onset time for PL between 0.1-1ps.

Because of the disorder inherent in an amorphous solid we expect that after thermalization there will be a broad distribution of electron-hole separations r. We can conceive of two situations - either the electron and hole are spatially separated, or their wavefunctions overlap sufficiently to form a localized exciton. In the separated case emission is thought to proceed via radiative tunneling [1] with the rate $\nu_T \exp(-2\alpha r)$, where α^{-1} is the wavefunction extent. For the localized exciton case it is more appropriate to use the expression $\nu_d = \sqrt{\varepsilon}4/3(e^2/\hbar c)\ \omega^3/c^2|\langle r_d\rangle|^2$ for the emission of a dipole er_d imbedded in a dielectric ($\varepsilon = 5.9$ for a-As_2S_3). We favor the localized exciton picture here for ν_1 because it is plausible that the maximum radiative rate $\nu_1 = 4\times10^8$ Hz corresponds to recombination of photoexcited pairs with the minimum electron-hole separation. Furthermore, the localized exciton picture can provide a consistent short time limit to the slower tunneling recombination involving distant pairs. Using the dipole expression we find $\nu_d = \nu_1$ for r_d = 2.9A. This sets a lower limit on α^{-1} of ~3A, in good agreement with other estimates [1,2,9].

Work partially supported by the Sponsors of the Laser Fusion Feasibility Project of the University of Rochester and by NSF Grant No. PCM-80-18488.

References

1. R.A. Street, Advances in Physics 25, 397 (1976); Solid State Commun. 34, 157 (1980).
2. G. Higashi and M. Kastner, J. Phys. C: Solid State 12, L821 (1979); M. Kastner, J. Phys. C: Solid State 13, 3319 (1980); G. Higashi and M. Kastner, Phys. Rev. B 24, 2295 (1981).
3. M.A. Bosch and J. Shah, Phys. Rev. Lett. 42, 118 (1979); J. Shah, Phys. Rev. B 21, 4751 (1980); K. Murayama and M. Bosch, Journal de Physique 42, C4-343 (1981).
4. J.C. Phillips, J. Non-Cryst. Solids 41, 179 (1980); Phys. Rev. B 24, 1744 (1981).
5. D. Emin, J. Non-Cryst. Solids 35, 969 (1980); M.A. Bosch, R.W. Epworth and D. Emin, J. Non-Cryst. Solids 40, 587 (1980).
6. W. Seka and J. Bunkenburg, J. Appl. Phys. 49, 2277 (1978).
7. W. Knox and G. Mourou, Opt. Commun. 37, 203 (1981).
8. R. Fork, C. Shank, A. Glass, A. Migus, M. Bosch and J. Shah, Phys. Rev. Lett. 43, 394 (1979).
9. D.E. Ackley, J. Tauc and W. Paul, Phys. Rev. Lett. 43, 715 (1979); J. Non. Cryst. Solids 35, 957 (1980).
10. H. Scher and T. Holstein, Phil. Mag. B 44, 343 (1981).

Index of Contributors

MIX
Papier aus verantwortungsvollen Quellen
Paper from responsible sources
FSC® C105338

If you have any concerns about our products,
you can contact us on
ProductSafety@springernature.com

In case Publisher is established outside the EU,
the EU authorized representative is:
Springer Nature Customer Service Center GmbH
Europaplatz 3, 69115 Heidelberg, Germany

Printed by Libri Plureos GmbH
in Hamburg, Germany